Handbook of Petroleum Exploration and Production

Handbook of Petroleum Exploration and Production

3

Series Editor

JOHN CUBITT

ELSEVIER

Amsterdam - Boston - London - New York - Oxford - Paris
San Diego - San Francisco - Singapore - Sydney - Tokyo

WELL TEST ANALYSIS: THE USE OF ADVANCED INTERPRETATION MODELS

DOMINIQUE BOURDET

Consultant, Paris, France

2002

ELSEVIER

Amsterdam - Boston - London - New York - Oxford - Paris
San Diego - San Francisco - Singapore - Sydney - Tokyo

ELSEVIER SCIENCE B.V.
Sara Burgerhartstraat 25
P.O. Box 211, 1000 AE Amsterdam, The Netherlands

First edition 2002

British Library Cataloguing in Publication Data

Bourdet, Dominique
Well test analysis : the use of advanced interpretation models. - (Handbook of petroleum exploration and production ; 3)
1.Oil wells - Testing - Data processing 2.Gas wells - Testing - Data processing
I.Title
622.3'382

ISBN 0444509682

Library of Congress Cataloging in Publication Data

Bourdet, Dominique.
Well test analysis : the use of advanced interpretation models / Dominique Bourdet.-- 1st ed.
p. cm. -- (Handbook of petroleum exploration and production, ISSN 1567-8032 ; 3)
Includes bibliographical references and indexes.
ISBN 0-444-50968-2 (alk. paper)
1. Oil wells--Testing. I. Title. II. Series.

TN871 .B54 2002
622'.3382--dc21

2002026548

ISBN: 0-444-50968-2
ISSN: 1567-8032 (Series)

Cover illustration by Jenny Ryrie

'White Sands; Moon' is by British artist Jenny Ryrie, who specialises in works on paper. The painting is one of a series inspired by the deserts of southern New Mexico, U.S.A.
The artist exhibits internationally and can be contacted at: jenny_ryrie@hotmail.com or tel: +44 (0) 1244 881 078.

∞ The paper used in this publication meets the requirements of ANSI/NISO Z39.48-1992 (Permanence of Paper).
Printed in The Netherlands.

PREFACE

During the last two decades, well test analysis techniques have changed significantly. With the introduction of high accuracy pressure measurements and powerful computers, information that is more accurate and useful is extracted from well tests. The new interpretation methods, using the derivative of the pressure, magnify the characteristic features of the many different types of wells and reservoirs, including groundwater hydrogeology systems. Due to the improved diagnosis of well test data, the number of theoretical solutions available to the interpretation engineer is expanding all the time. Today well test interpretation computer programs offer a wide range of complex well and reservoir configurations for the analysis of pressure transient test responses.

This book covers all aspects of well test analysis for the today's engineer who has access to powerful computers. The most recent advanced interpretation models are presented in detail and their application to field measurements is documented. Practical analyses of well test data is thoroughly discussed. Should the recorded test data deviate from the theory due to operational conditions, guidelines are established for proper analysis. The basic well test analysis technique and the associated theory have been abundantly discussed in the literature and they are simply summarized here. Focus is placed on computerized interpretation of complex systems. With this book, well test interpretation engineers will not only find answers to the different questions encountered in the course of analysis, but also gain a better understanding of the physical process involved, and the meaning and limitations of the results.

The interpretation methodology is briefly presented in the two first Chapters. In Chapter 1, the different types of tests are described, several typical well pressure responses are documented, and the usual well test analysis terminology is introduced. Chapter 2 presents the various interpretation techniques, and defines the limitations of the different methods.

In Chapters 3 to 5, the basic interpretation models are reviewed for well, reservoir and boundary conditions. The analysis of interference tests is described in Chapter 6. The interpretation methods and models are presented from Chapter 1 to 6 for a single-phase liquid flow (oil or water), tests in gas and multiphase reservoirs are discussed in Chapters 7 and 8. In Chapter 9, the analysis of different type of tests, such as drill stem tests, single well vertical interference tests, is explained. Several factors complicating well test analysis are discussed in the conclusion Chapter 10, together with a summary of the test interpretation methodology.

All figures and equations are presented in the usual oil field system of units, which is still currently used in the industry despite numerous attempts to have a metric system accepted. In the Appendix, the equations are presented with the practical metric system.

Several field examples presented in this book have been published in technical articles. The test data and interpretation analysis results are available in the website http://www.elsevier.com/locate/welltest.

A lot of the material presented in this book is based on discussions with many well test interpretation specialists during the last 25 years. In particular, I want to thank Alain Gringarten for the many comments and his encouragement during the preparation of this work, and Piers Johnson for the complete revision of the book.

Two interpretation software packages have been used for the preparation of the Figures: PIE of WTS and SAPHIR of KAPPA Engineering.

Paris, December 2001

Dominique Bourdet has worked over 25 years with Oil and Gas Industry. He is the author of several publications on different aspects of well test analysis. His main research interest is in the analysis of fissured reservoirs. He developed the pressure derivative analysis method that became a standard in the Industry. He has been involved in the research of new well test analysis techniques, in the technical support to field operations, and in the development of well test interpretation software packages. He is an Independent Consultant, specializing in the design, supervision, and interpretation of well tests in unconventional systems. During the last twenty years, he has continuously taught well test interpretation in the Industry and in Universities.

CONTENTS

CHAPTER 1

PRINCIPLES OF TRANSIENT TESTING

In this first Chapter, the terminology used in well testing and interpretation is presented, and different testing procedures are explained. In the second part, characteristic well pressure behavior is illustrated, and the corresponding analysis methods are introduced. It is shown that well test responses follow chronological characteristic behavior at different times, depending upon the well and reservoir configuration. The interpretation techniques are presented in detail in the following Chapter 2.

1.1 INTRODUCTION

1.1.1 Description of a well test

During a well test, a transient pressure response is created by a temporary change in production rate. The well response is usually monitored during a relatively short period of time compared to the life of the reservoir, depending upon the test objectives. For well evaluation, tests are frequently achieved in less than two days. In the case of reservoir limit testing, several months of pressure data may be needed.

In most cases, the flow rate is measured at *surface* while the pressure is recorded *down-hole*. Before opening, the initial pressure p_i is constant and uniform in the reservoir. During the flowing period, the *drawdown* pressure response Δp is defined as follows:

$$\Delta p = p_i - p(t) \tag{1.1}$$

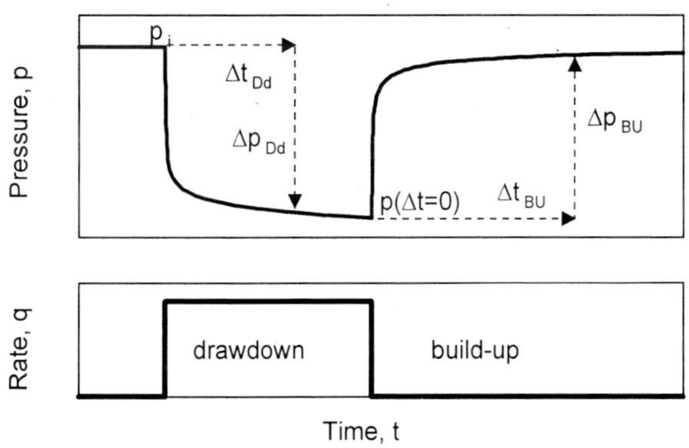

Figure 1.1. Drawdown and build-up test sequence.

When the well is shut-in, the *build-up* pressure change Δp is estimated from the last flowing pressure $p(\Delta t=0)$:

$$\Delta p = p(t) - p(\Delta t = 0) \tag{1.2}$$

The pressure response is analyzed versus the *elapsed time* Δt since the start of the period (time of opening or shut-in).

Well test objectives

Well test analysis provides information on the reservoir and on the well. Geological, geophysical and petrophysical information is used where possible in conjunction with the well test information to build a reservoir model for prediction of the field behavior and fluid recovery for different operating scenarios. The quality of the communication between the well and the reservoir indicates the possibility to improve the well productivity. Usually, the test objectives can be summarized as follows:

Exploration well: On initial wells, well testing is used to confirm the exploration hypothesis and to establish a first production forecast: nature and rate of produced fluids, initial pressure and well and reservoir properties. Tests may be limited to drill stem testing only.

Appraisal well: The previous well and reservoir description can be refined by testing appraisal wells to confirm well productivity, reservoir heterogeneities and boundaries, drive mechanisms etc. Bottom hole fluid samples are taken for PVT laboratory analysis. Longer duration testing (production testing) is usually carried out.

Development well: On producing wells, periodic tests are made to adjust the reservoir description and to evaluate the need for well treatment, such as work-over, perforation strategy or completion design, to maximize the well's production life. Communication between wells (interference testing), monitoring of the average reservoir pressure are some usual objectives of development well testing.

Information obtained from well testing

Well test responses characterize the ability of the fluid to flow through the reservoir and to the well. Tests provide a description of the reservoir in *dynamic conditions*, as opposed to geological and log data. As the investigated reservoir volume is relatively large, the estimated parameters are *average* values. From pressure curve analysis, it is possible to determine the following properties:

Reservoir description:

- Permeability (horizontal k and vertical k_V),
- Reservoir heterogeneities (natural fractures, layering, change of characteristics),
- Boundaries (distance, size and shape),
- Pressures (initial p_i and average $\bar{p}$).

Well description:

- Production potential (productivity index PI and skin factor S),
- Well geometry.

By comparing the result of routine tests, changes of productivity and rate of decrease of the average reservoir pressure can be established.

Test procedure

Drawdown test: the flowing bottom hole pressure is used for analysis. Ideally, the well should be producing at constant rate but in practice, this is difficult to achieve and drawdown pressure data is erratic. The analysis of flowing periods (drawdown) is frequently difficult and inaccurate.

Build-up test: the increase of bottom hole pressure after shut-in is used for analysis. Before the build-up test, the well must have been flowing long enough to reach stabilized rate. During shut-in periods, the flow rate is accurately controlled (zero). It is for this reason build up tests should be performed.

Injection test / fall-off test: when fluid is injected into the reservoir, the bottom hole pressure increases and, after shut-in, it drops during the fall-off period. The properties of the injected fluid are in general different from that of the reservoir fluid, interpretation of injection and fall-off tests requires more attention to detail than for producers.

Interference test and pulse testing: the bottom hole pressure is monitored in a shut-in observation well some distance away from the producer. Interference tests are designed to evaluate communication between wells. With pulse tests, the active well is produced with a series of short flow / shut-in periods and the resulting pressure oscillations in the observation well are analyzed.

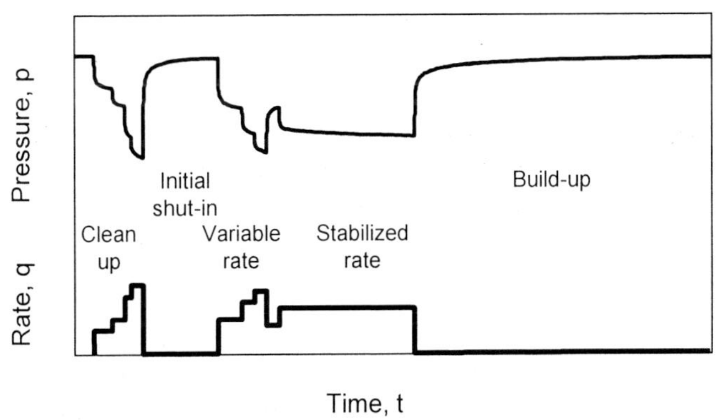

Figure 1.2. Typical test sequence. Oil well.

Gas well test: specific testing methods are used to evaluate the deliverability of gas wells (Absolute Open Flow Potential, AOFP) and the possibility of non-Darcy flow condition (rate dependent skin factor *S'*). The usual procedures are Back Pressure test (Flow after Flow), Isochronal and Modified Isochronal tests.

In Figure 1.2, the typical test sequence of an exploration oil well is presented. Initially, the well is cleaned up by producing at different rates, until the fluid produced at surface corresponds to the reservoir fluid. The well is then shut-in to run the down hole pressure gauges, and reopened for the main flow. The flow rate is controlled by producing through a calibrated orifice on the choke manifold. Several choke diameters are frequently used, until stabilized flowing conditions are reached. After some flow time at a constant rate, the well is shut-in for the final build-up test.

Well completion

Production test: during such tests, the well is completed as a production well with a cased hole and a permanent completion. The well is monitored at surface, from the wellhead.

Drill stem test (DST): the well is completed temporarily with a down-hole shut-in valve. Frequently the well is cased but DST can be made also in open hole. During a DST, the well is closed down-hole. The drill stem testing procedure is used only for relatively short tests. The drill string (drill-pipe) is not used any more, and production tubing is employed.

1.1.2 Methodology

The inverse problem

The objective of well test analysis is to describe an unknown system **S** (well + reservoir) by indirect measurements (**O** the pressure response to **I** a change of rate). Solving **S=O/I** is a typical *inverse problem* (Gringarten et al., 1979).

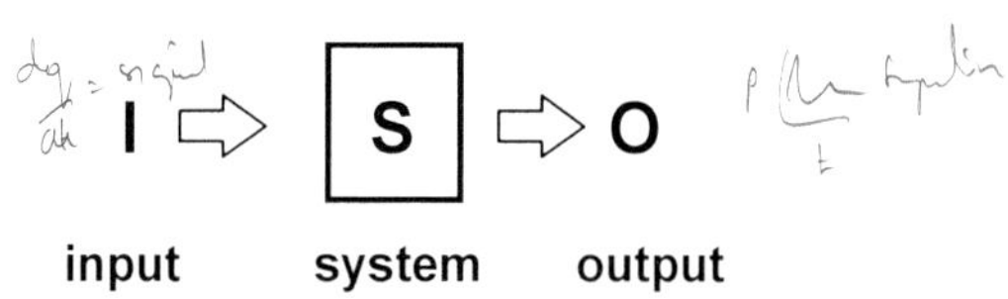

As opposed to the direct problem (**O=IxS**), the solution of the inverse problem is usually *not unique*. It implies an identification process, and the interpretation provides the model(s) whose behavior is identical to the behavior of the actual reservoir.

Interpretation models

The models used in well test interpretation can be described as a *transfer function*; they only define the behavior (homogeneous or heterogeneous, bounded or infinite). Well test interpretation models are often different from the geological or log models, due to the averaging of the reservoir properties. Layered reservoirs for example frequently show a homogeneous behavior during tests.

Interpretation models are made of several components, which are relatively independent, and exhibit different characteristics at different time of the response. Once all components have been identified, the interpretation model is defined. Analytical solutions or numerical models are used to generate pressure responses to the specific production rate history **I** of the well, and the model parameters are adjusted until the model behavior **O** is identical to the behavior of **S**.

In the case of complex reservoir behavior, several models are frequently applicable to describe the test pressure response. The non-uniqueness of the inverse problem solution can be reduced by using additional information, such as geological or geophysical data for example. In some cases, a new test may be specifically designed in order to complete or improve an ambiguous test pressure response.

Input data required for well test analysis

Test data: flow rate and bottom hole pressure as a function of time. The test sequence of events must be detailed, including any operational problems that may affect the well response. Results of analysis are dependent upon the accuracy of the well test data. When the production rate has not been measured during some flow periods, it must be accurately estimated. Errors in rate or pressure data are discussed in Chapter 10.

Well data: wellbore radius r_w, well geometry (such as inclined or horizontal well), depths (formation, gauges).

Reservoir and fluid parameters: formation thickness h (net), porosity ϕ, compressibility of oil c_o, water c_w and formation c_f, water saturation S_w, oil viscosity μ and formation volume factor B. The total system compressibility c_t is expressed as:

$$c_t = c_o(1 - S_w) + c_w S_w + c_f \tag{1.3}$$

The above reservoir and fluid parameters are used for calculation of the results. After a first interpretation, they may always be changed or adjusted if needed to refine the results, for the same theoretical interpretation model.

Additional data can be useful in some cases: production log, gradient surveys, reservoir temperature, bubble point pressure etc. General information obtained from geologist and geophysicists are required to validate the well test interpretation results.

1.2 TYPICAL FLOW REGIMES

1.2.1 Types of flow behavior

The different flow behaviors are usually classified in terms of rate of change of pressure with respect to time.

Steady state

During steady-state flow, the pressure does *not change* with time. This is observed for example when a constant pressure effect, such as resulting from a gas cap or some types of water drive, ensures a pressure maintenance in the producing formation.

$$\frac{\partial p}{\partial t} = 0 \tag{1.4}$$

Pseudo steady state

The pseudo steady state regime characterizes a closed system response. With a constant rate production, the drop of pressure becomes *constant* for each unit of time.

$$\frac{\partial p}{\partial t} = \text{constant} \tag{1.5}$$

Transient state

Transient responses are observed before constant pressure or closed boundary effects are reached. The *pressure variation* with time is a function of the well geometry and the reservoir properties, such as permeability and heterogeneity.

$$\frac{\partial p}{\partial t} = f\,(x, y, z, t) \tag{1.6}$$

Usually, well test interpretation focuses on the transient pressure response. Near wellbore conditions are seen first and later, when the drainage area expands, the pressure response is characteristic of the reservoir properties until boundary effects are seen at late time (then the flow regime changes to pseudo steady or steady state). In the following, several characteristic examples of well behavior are introduced, for illustration of typical well test responses.

1.2.2 Wellbore storage

When a well is opened, the production at surface is initially due to the *expansion* of the fluid stored in the wellbore, and the reservoir contribution is initially negligible. This characteristic flow regime, called the *pure wellbore storage effect*, can last from a few seconds to a few minutes. Then, the reservoir production starts and the sand face rate increases until it becomes the same as the surface rate. When this condition is reached, the wellbore storage has no effect any more on the bottom hole pressure response, the data describes the reservoir behavior and it can be used for transient analysis.

During shut-in periods, the wellbore storage effect is also called *afterflow*: after the well has been shut-in, the reservoir *continues to produce* at the sand face and the fluid stored in the wellbore is recompressed. The same sequence with three different pressure behaviors can be observed: the pure wellbore storage effect, transition when the sand face rate declines, and the end of the wellbore storage effect when the sand face rate becomes negligible and eventually zero.

After any change in the well flowing conditions, there is a time lag between the surface production and the sand face rate. The effect of wellbore storage affects well pressure responses during the first instants of each test period. In the present Chapter, only drawdown responses are illustrated. Multiple-rate sequences and shut-in periods are discussed in Chapter 2.

Wellbore storage coefficient

The wellbore storage coefficient defines the rate of pressure change during the pure wellbore storage regime. For a well full of a single-phase fluid, well bore storage is represented by a compressibility term (van Everdingen and Hurst, 1949):

$$C = -\frac{\Delta V}{\Delta p} = c_o V_w \tag{1.7}$$

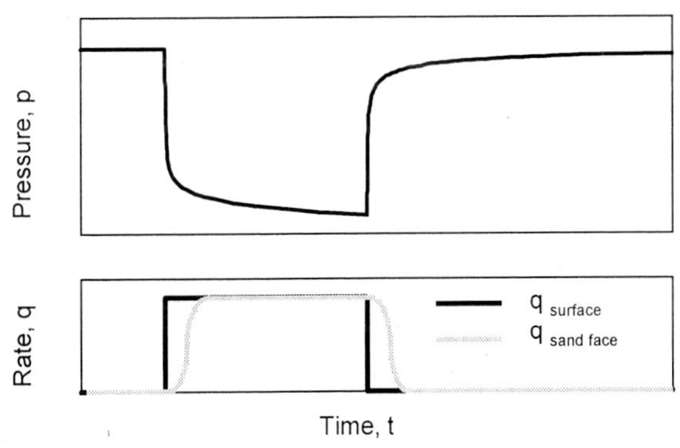

Figure 1.3. Wellbore storage effect. Sand face and surface rates.

where :
c_o : liquid compressibility
V_w : wellbore volume in Bbl

When there is a liquid level (Earlougher, 1977), the level depth is related to the flow rate with $\Delta V = V_u \Delta h$, and the down hole pressure change is expressed $\Delta p = \rho \, \Delta h (g/g_c)$, where

ρ: liquid density (lb/cu ft)
g/g_c : gravitational acceleration (lb_f/lb_m)
V_u : wellbore volume per unit length (Bbl/ft)

$$C = 144 \frac{V_u}{\rho (g/g_c)} \tag{1.8}$$

Specialized analysis

During the pure wellbore regime, the well is acting as a closed volume and, with a constant surface rate condition, the pressure changes linearly with time. The wellbore storage coefficient can be estimated on a plot of the pressure change Δp versus the elapsed time Δt time on a *linear scale* (van Everdingen, and Hurst, 1949). At early time, the response follows a straight line of slope m_{WBS}, intercepting the origin.

$$\Delta p = \frac{qB}{24C} \Delta t \tag{1.9}$$

The *wellbore storage coefficient* C is estimated from the straight-line slope m_{WBS}:

$$C = \frac{qB}{24 m_{WBS}} \tag{1.10}$$

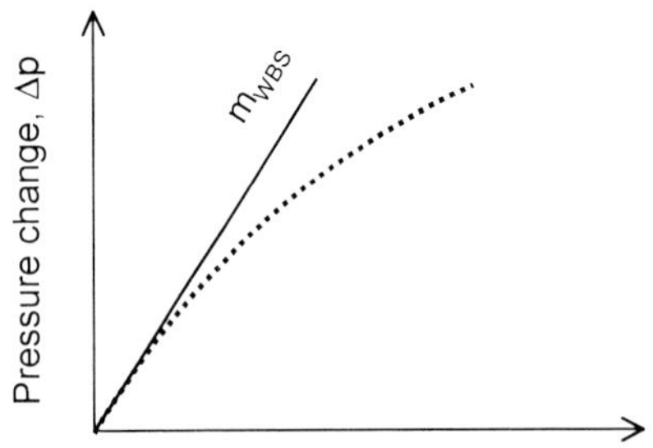

Figure 1.4. Wellbore storage effect. Specialized analysis on a linear scale.

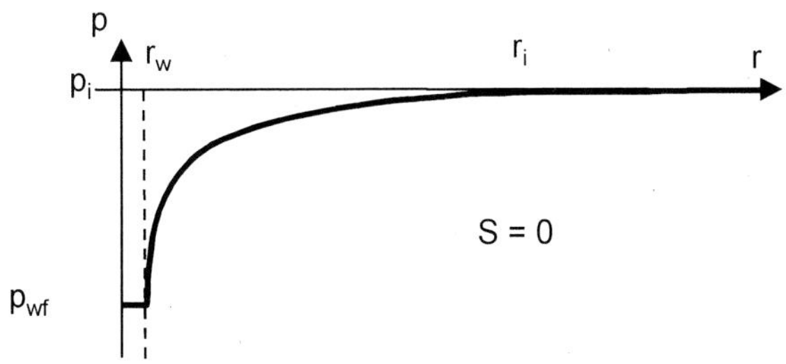

Figure 1.5. Radial flow regime. Pressure profile. Zero skin.

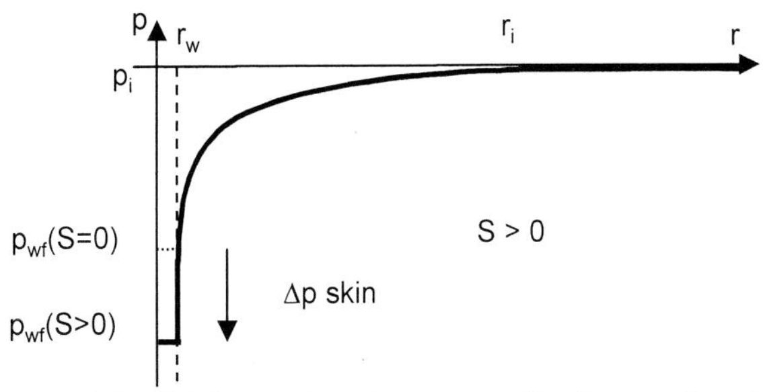

Figure 1.6. Radial flow regime. Pressure profile. Damaged well, positive skin factor.

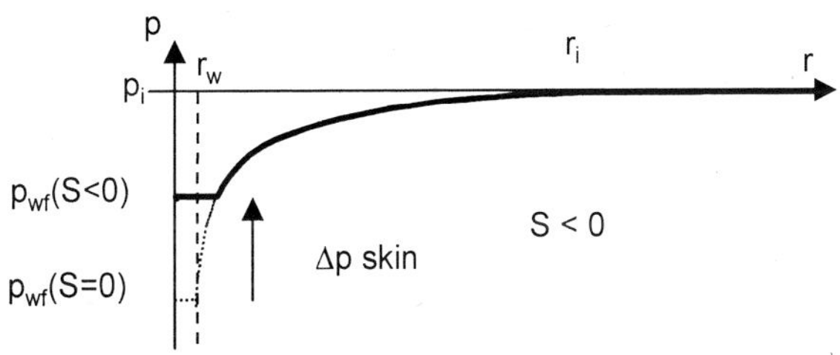

Figure 1.7. Radial flow regime. Pressure profile. Stimulated well, negative skin factor.

1.2.3 Radial flow regime, skin (homogeneous behavior)

When the reservoir production is established, the flow-lines converge towards the well with a radial geometry. In the reservoir, the pressure is a function of the *time* and the *distance* to the well. As the production time increases, the well bottom-hole pressure p_{wf} drops, and the circular drainage area of radius r_i expands in the reservoir. The radius of investigation r_i is discussed later in this Section.

Skin

In the case of a damaged well, a flow restriction is present at the interface between the reservoir and the wellbore, producing an *additional* pressure drop Δp_{skin} when the fluid enters into the well. For a stimulated well, the flowing condition is improved near the well, and the pressure decline is *reduced* in a cylindrical near wellbore reservoir region.

For comparison between wells, the magnitude of the pressure drop near the wellbore has to be normalized. The same Δp_{skin} can describe a low or very high damage, depending on the flow rate and the reservoir permeability. The skin factor S is a *dimensionless parameter* (van Everdingen 1953), and it characterizes the well condition: for a damaged well $S > 0$ and, by extension, $S < 0$ for a stimulated well.

$$S = \frac{kh}{141.2qB\mu} \Delta p_{Skin} \tag{1.11}$$

Typical examples of a *damaged well* ($S > 0$) are poor contact between the well and the reservoir (mud-cake, insufficient perforation density, partial penetration) or invaded zone.

Stimulated well ($S < 0$) behavior is observed when the surface of contact between the well and the reservoir has been increased compared to the basic cylindrical vertical wellbore geometry (fractured well, slanted and horizontal well) or acid stimulated zone.

For a given pressure drop in the vicinity of the wellbore, it is shown from Equation 1.11 that the corresponding skin damage is larger in high permeability reservoirs, or when the fluid viscosity is low. A large skin factor S indicates the possibility of a strong improvement of the well flowing pressure, or a potential significant increase of the flow rate. Dimensionless terms are currently used in well test analysis. They are further discussed in Chapter 2.

In the case of an invaded or stimulated circular zone around the well, the resulting positive or negative skin can be expressed by the difference between the pressure profile corresponding to the original reservoir permeability k, and the actual pressure profile due to the modified reservoir permeability k_S. In the circular zone near the well, the flow is in steady state regime:

$$p_{w,S} - p_{w,S=0} = \frac{141.2qB\mu}{k_S h} \ln \frac{r_S}{r_w} - \frac{141.2qB\mu}{kh} \ln \frac{r_S}{r_w} \tag{1.12}$$

The skin is expressed :

$$S = \left(\frac{k}{k_S} - 1 \right) \ln \frac{r_S}{r_w} \tag{1.13}$$

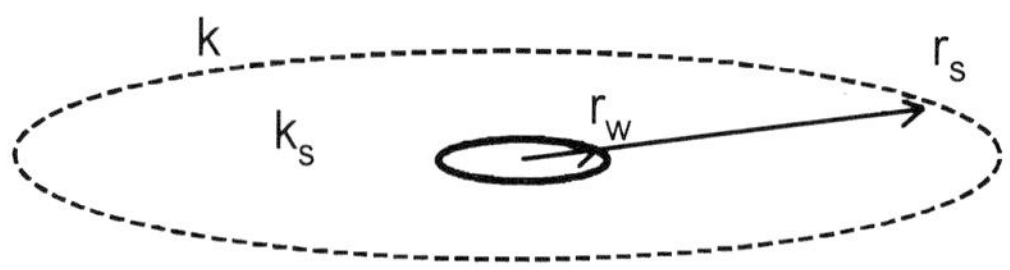

Figure 1.8. Flow through a circular reservoir region.

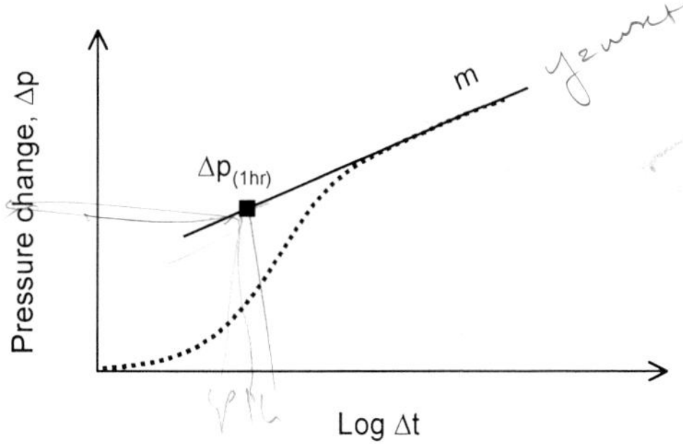

Figure 1.9. Radial flow regime. Specialized analysis on semi-log scale.

The equivalent wellbore radius is defined with no pressure loss ($k_S = \infty$) in the circular zone around the well (Brons and Miller, 1961 a):

$$r_{we} = r_w e^{-S} \tag{1.14}$$

Specialized analysis

During the radial flow regime in reservoirs with homogeneous behavior, the pressure changes with the logarithm of the elapsed time from when the well is opened (Miller et al., 1950). A plot of the bottom hole pressure versus the *logarithm of time* (called MDH plot) follows a straight line when all wellbore storage transitional effect are finished. The slope *m* of semi-log straight line is used to estimate the reservoir *permeability thickness* product *kh*, and the *skin* coefficient *S* is evaluated from the location of the straight line along the *y*-axis.

$$\Delta p = 162.6 \frac{qB\mu}{kh}\left[\log \Delta t + \log \frac{k}{\phi \mu c_t r_w^2} - 3.23 + 0.87S\right] \tag{1.15}$$

Traditionally, the semi-log straight-line location is characterized by the straight-line pressure at 1 hour ($\Delta p_{1\,\mathrm{hr}}$).

$$kh = 162.6 \frac{qB\mu}{m} \tag{1.16}$$

$$S = 1.151\left(\frac{\Delta p_{1\,\mathrm{hr}}}{m} - \log \frac{k}{\phi \mu c_t r_w^2} + 3.23\right) \tag{1.17}$$

Radius of investigation

As illustrated on Figure 1.5, the pressure distribution in the reservoir is a function of the time and the distance to the producing well. It can be expressed with the Exponential Integral function (see Chapter 6):

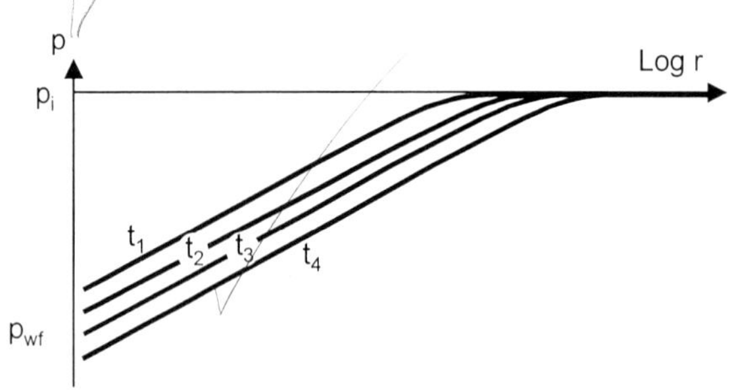

Figure 1.10. Pressure profile versus the logarithm of the distance to the well.

$$\Delta p(\Delta t, r) = -0.5\frac{141.2qB\mu}{kh}\mathrm{Ei}\left(-\frac{\phi\mu c_t r^2}{0.001056k\Delta t}\right) \qquad (1.18)$$

For small x, $\mathrm{Ei}(-x) = -\ln(\gamma x)$: the Exponential Integral can be approximated by a log (with $\gamma = 1.78$, Euler's constant).

$$\Delta p(\Delta t, r) = \frac{162.6\,qB\mu}{kh}\left[\log\frac{0.000264k\Delta t}{\phi\mu c_t r^2} + 0.809\right] \qquad (1.19)$$

(The skin calculation Equation 1.12 is based on Equation 1.19 and, for the semi-log straight line Equation 1.15 the radial distance is set at $r = r_w$).

When plotted versus $\log(r)$, the flowing pressure profile at a given time is a straight line until the distance becomes too large for the logarithm approximation of the Exponential Integral. Beyond this limit, the profile flattens, and tends asymptotically towards the initial pressure as shown on Figure 1.10 (where the pressure profile of Figure 1.5 is presented for different production times versus the logarithm of the distance r).

The radius of investigation r_i tentatively describes the distance that the pressure transient has moved into the formation. Several definitions have been proposed (see discussion in Section 10.3.3), in general r_i is defined with one of the two relationships (Earlougher, 1977; Muskat, 1934 and Lee, 1982):

$$\frac{0.000264k\Delta t}{\phi\mu c_t r_i^2} = \frac{1}{\gamma^2} \text{ or } \frac{1}{4} \qquad (1.20)$$

In dimensionless terms, radius are expressed as

$$r_D = \frac{r}{r_w} \qquad (1.21)$$

With the dimensionless time defined in Equation 2.4, Equation 1.20 is simply:

$$t_D / r_{iD}^2 = \frac{1}{\gamma^2} \text{ or } t_D / r_{iD}^2 = \frac{1}{4}. \tag{1.22}$$

This gives respectively,

$$r_i = 0.029\sqrt{k\Delta t / \phi \mu c_t} \tag{1.23}$$

and

$$r_i = 0.032\sqrt{k\Delta t / \phi \mu c_t} \tag{1.24}$$

Equations 1.23 or 1.24 are based on the assumption of homogeneous isotropic reservoir permeability. Permeability anisotropy is discussed in Section 3.1.5, and radial changes of permeability in Section 4.3.4. It should be noted that the radius of investigation is *independent of the flow rate.*

1.2.4 Fractured well (infinite conductivity fracture): linear flow regime

A common well stimulation method consists of creating a hydraulic vertical fracture from the wellbore to the formation (Russell and Truitt, 1964). The reservoir / well surface of contact is significantly increased, thus producing a negative skin factor. Two main types of fractured well behavior are observed: infinite or finite conductivity fracture. Both are discussed in the Well Model Chapter 3. In the following, fractured well responses are briefly introduced to illustrate two characteristic flow regimes.

The fracture is symmetrical on both sides of the well and it intercepts the complete formation thickness. x_f is the half fracture length. With the *infinite conductivity* fracture model, it is assumed that the fluid flows along the fracture without any pressure drop.

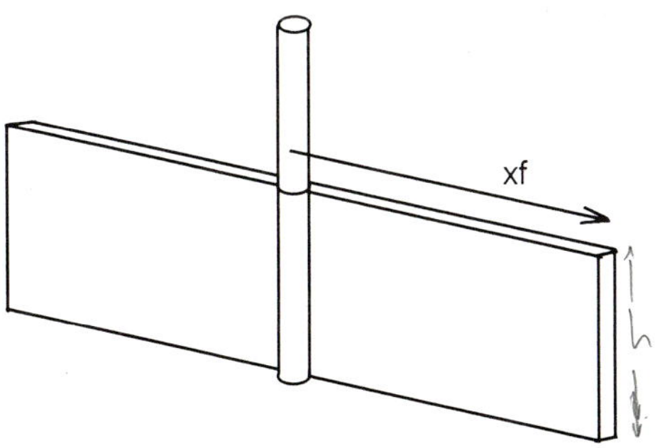

Figure 1.11. Fractured well. Fracture geometry.

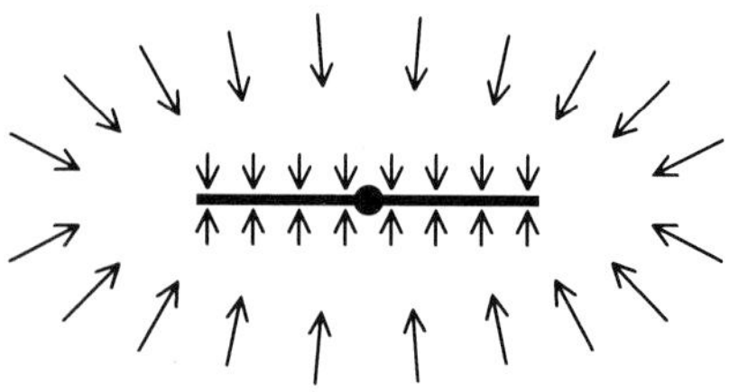

Figure 1.12. Infinite conductivity fracture. Geometry of the flow lines.
Linear and pseudo radial flow regimes.

At early time, the flow-lines are perpendicular to the fracture plane. This is called a *linear flow* regime. Later, the reservoir regions at the two ends of the fracture starts to contribute significantly to the flow, the linear flow regime ends, to change into an *elliptical flow* geometry. Ultimately, the well response shows the characteristic radial flow regime behavior.

During linear flow, the pressure change is proportional to the square root of the elapsed time since the well was opened (Clark, 1968 and Gringarten et al., 1974 a).

$$\Delta p = 4.06 \frac{qB}{hx_f} \sqrt{\frac{\mu}{\phi c_t k}} \sqrt{\Delta t} \qquad (1.25)$$

Specialized analysis

The linear flow regime can be analyzed with a plot of the pressure change Δp versus the *square root* of elapsed time $\sqrt{\Delta t}$: the response follows a straight line of slope m_{LF}, intercepting the origin.

When the reservoir permeability is known from the analysis of the subsequent radial flow regime, the slope m_{LF} of the linear flow straight line is used to estimate the *half fracture length* x_f:

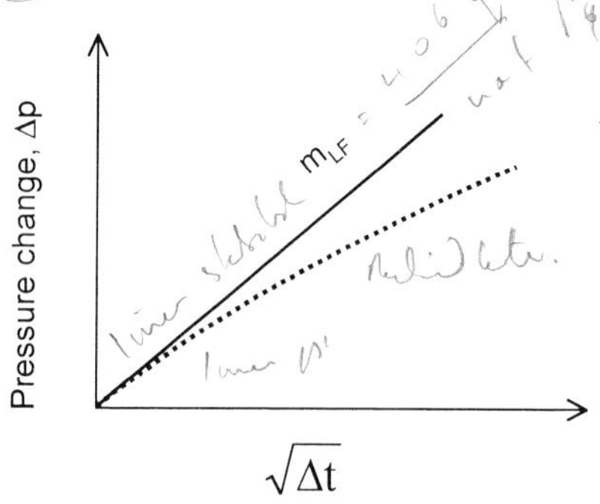

Figure 1.13. Infinite conductivity fracture.
Specialized analysis with the pressure versus the square root of time.

$$x_f = 4.06\sqrt{\frac{\mu}{\phi c_t k}\frac{qB}{hm_{\mathrm{LF}}}} \tag{1.26}$$

1.2.5 Fractured well (finite conductivity fracture): bi-linear flow regime

When the pressure drop in the fracture plane is not negligible, a second linear flow regime is established along the fracture extension. Before the two ends of the fracture are reached, this well configuration produces the so-called *bi-linear flow* regime.

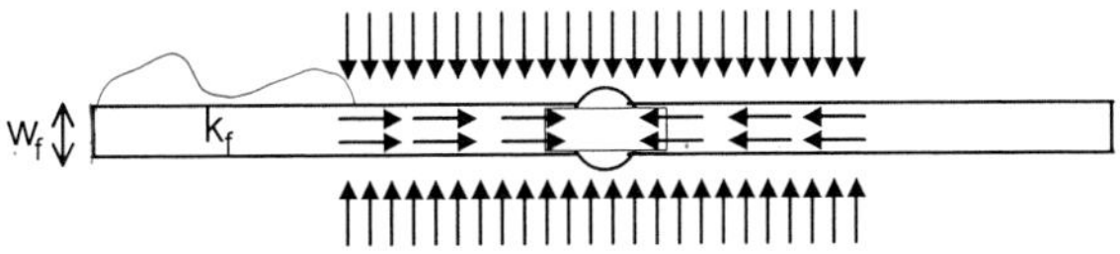

Figure 1.14. Finite conductivity fracture.
Geometry of the flow lines during the bi-linear flow regime.

During bilinear flow, the pressure change is proportional to the fourth root of the elapsed time since the well was opened (Cinco-Ley et al., 1978 a). With w_f the width of the finite conductivity fracture and k_f the permeability in the fracture:

$$\Delta p = 44.11 \frac{qB\mu}{h\sqrt{k_f w_f}\sqrt[4]{\phi\mu c_t k}}\sqrt[4]{\Delta t} \tag{1.27}$$

Specialized analysis

On a plot of the pressure change Δp versus the *fourth root* of elapsed time $\sqrt[4]{\Delta t}$, pressure response follows a straight line of slope m_{BLF}, intercepting the origin, during the bilinear flow regime

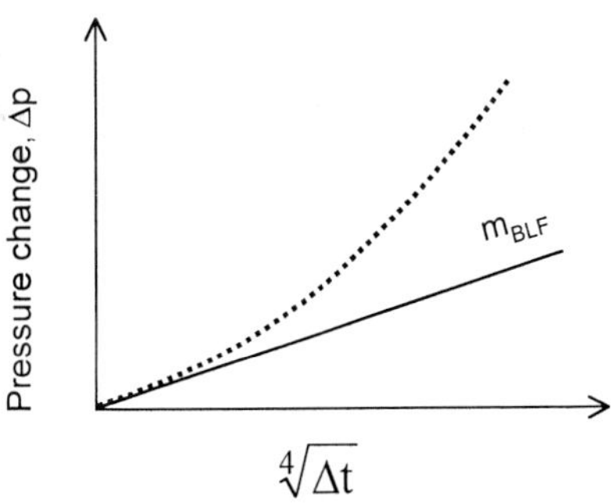

Figure 1.15. Finite conductivity fracture.
Specialized analysis with the pressure versus the fourth root of time.

As for the linear flow analysis, provided the reservoir permeability can be estimated from semi-log analysis of the late time response, the slope m_{BLF} of the bilinear flow straight line is used to estimate the controlling parameter, namely the *fracture conductivity* $k_f w_f$.

$$k_f w_f = 1944.8\sqrt{\frac{1}{\phi\mu c_t k}}\left(\frac{qB\mu}{hm_{BLF}}\right)^2 \tag{1.28}$$

1.2.6 Well in partial penetration: spherical flow regime

With a well in partial penetration, the well is connected to the producing interval on one fraction only of the zone thickness. The reservoir / well surface of contact being reduced, partially penetrating wells are characterized by a positive skin factor as discussed in the Well Model Chapter 3. In the following, this well configuration is introduced to illustrate another example of characteristic flow regime.

The ratio h_w/h of the length of the perforated interval to the formation thickness is called the penetration ratio, k_H and k_V are the horizontal and vertical permeability (Figure 1.16).

In a well in partial penetration, after an initial radial flow regime in front of the perforated interval, the flow lines are established in both the horizontal and vertical directions, until the top and bottom boundaries are reached. A *spherical flow* regime can thus be observed before the flow becomes radial in the complete formation thickness.

During the spherical flow regime, the pressure changes with $1/\sqrt{\Delta t}$ (Brons and Marting, 1961 b, Moran and Finklea, 1962).

$$\Delta p = 70.6\frac{qB\mu}{k_S r_S} - 2452.9\frac{qB\mu\sqrt{\phi\mu c_t}}{k_S^{3/2}\sqrt{\Delta t}} \tag{1.29}$$

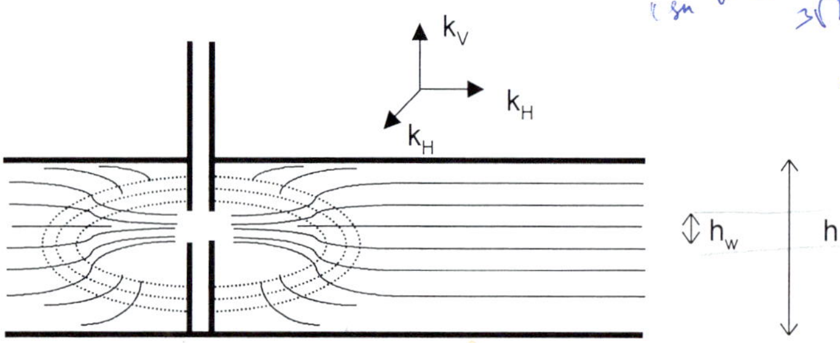

Figure 1.16. Well in partial penetration.
Geometry of the flow lines. Radial, spherical and radial flow regimes.

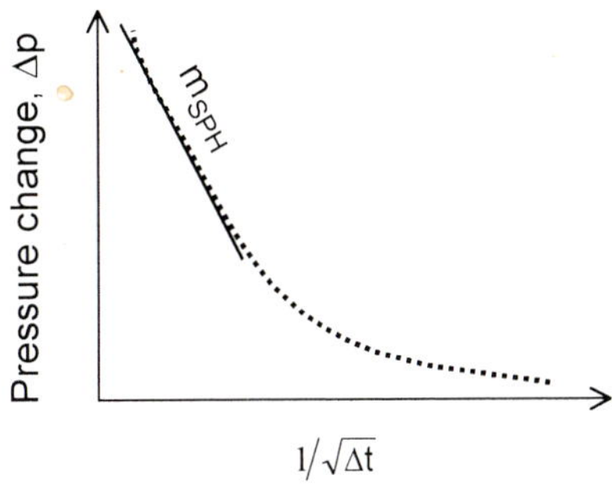

Figure 1.17. Well in partial penetration.
Specialized analysis with the pressure versus 1/ the square root of time.

where k_S is the *spherical permeability* defined as

$$k_S = \sqrt[3]{k_x k_y k_z} = \sqrt[3]{k_H^2 k_V} \tag{1.30}$$

Specialized analysis

On a plot of the pressure versus the *reciprocal of the square root* of time $1/\sqrt{\Delta t}$, a straight line of slope m_{SPH} develops during the spherical flow regime. The spherical permeability k_S can be estimated with:

$$k_S = \left(2452.9 qB\mu \frac{\sqrt{\phi\mu c_t}}{m_{\mathrm{SPH}}} \right)^{2/3} \tag{1.31}$$

Knowing the horizontal permeability from the late time radial flow regime, the vertical to horizontal *permeability anisotropy* is defined as:

$$\frac{k_V}{k_H} = \left(\frac{k_S}{k_H} \right)^3 \tag{1.32}$$

1.2.7 Limited reservoir (one sealing fault)

In the following example, the reservoir is limited in one direction by a vertical plane-sealing boundary at a distance L from the well. During the production, the radius of investigation expands until the sealing boundary is reached. Since no flow-lines support the production on the opposite side of the boundary, the pressure profile for an infinite reservoir described on Figure 1.5 is distorted as shown on Figures 1.19 to 1.21. The sequence of pressure regimes is the following:

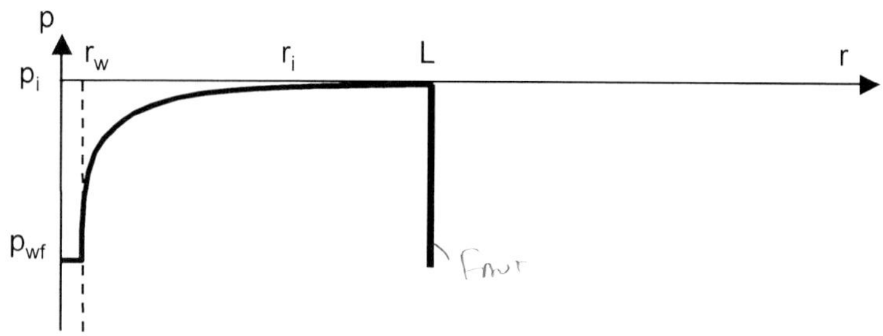

Figure 1.18. One sealing fault. Pressure profile at time t_1.
The fault is not reached, infinite reservoir behavior.

At early time, the radius of investigation is smaller than the fault distance L and the pressure profile in the reservoir corresponds to that of an *infinite system* (Figure 1.18 and time t_1 on Figure 1.22).

Later, the radius of investigation reaches the fault and the pressure profile deviates from the infinite reservoir behavior as shown Figure 1.19. On this diagram, the thin dotted curve corresponds to the theoretical pressure profile in an infinite reservoir. Because of the reservoir limit, the curve beyond the distance L, on the right side of the boundary, is reflected back into the producing area like a mirror effect. In the reservoir region between the well and the sealing boundary, the two pressure drops are combined to produce the actual reservoir pressure profile (the actual pressure drop is the sum of the two responses, namely the infinite reservoir curve and the reflected image curve).

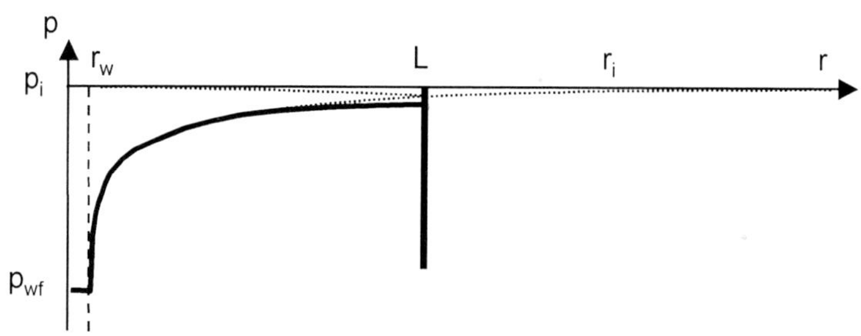

Figure 1.19. One sealing fault. Pressure profile at time t_2.
The fault is reached, but it is not seen at the well. Infinite reservoir behavior.

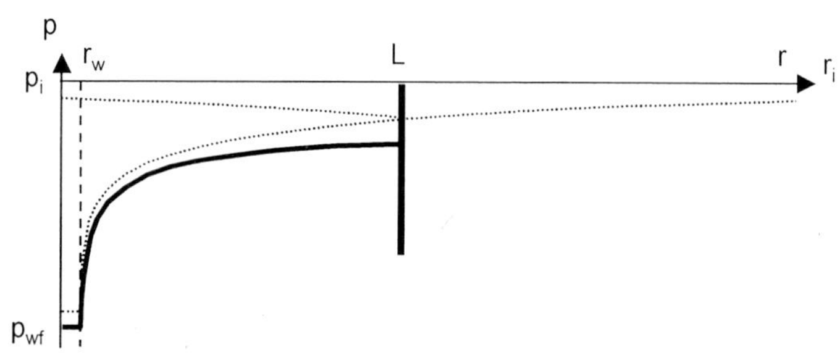

Figure 1.20. One sealing fault. Pressure profile at time t_3.
The fault is reached, and it is seen at the well. Start of boundary effect.

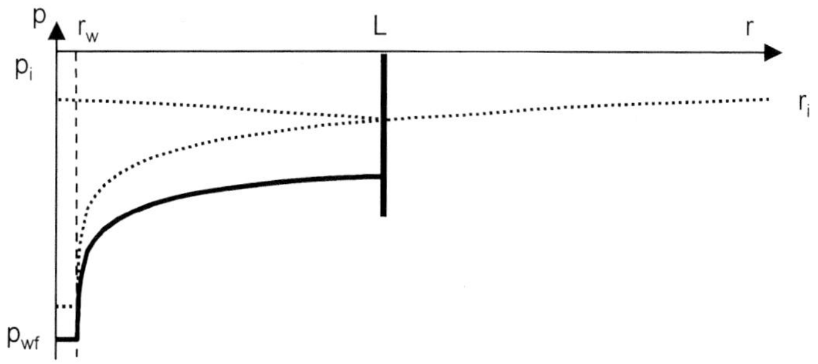

Figure 1.21. One sealing fault. Pressure profile at time t_4.
The fault is reached, and it is seen at the well. Hemi-radial flow.

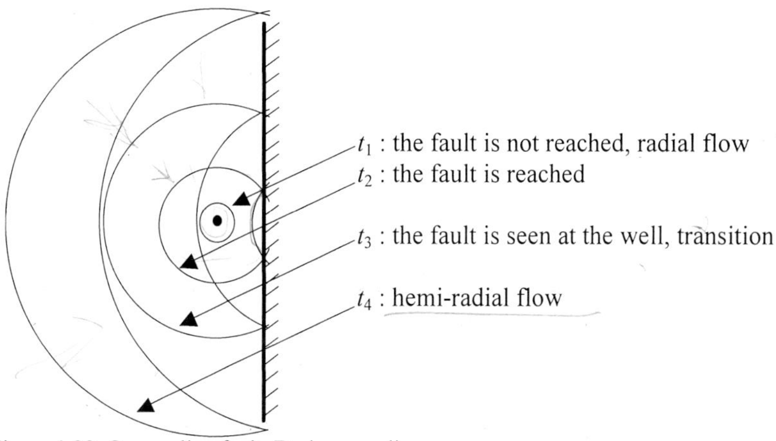

Figure 1.22. One sealing fault. Drainage radius.

This pressure regime corresponds to time t_2 on Figure 1.22: the boundary has been reached and the pressure profile is *distorted in the reservoir*, but the image curve has *not changed* the well flowing pressure. As the flow time increases, the radius of investigation of the theoretical infinite reservoir curve continues to expand, and the image curve *reaches the well* (time t_3 on Figure 1.22 and Figure 1.20). The well bottom hole pressure starts to *deviate* from the infinite reservoir response, and drops faster.

Ultimately, when the well has been flowing long enough, the two profiles tends to merge (after time t_4 on Figure 1.22) and the *hemi-radial flow regime* is reached: the flow lines converge to the well with a half circle geometry.

Specialized analysis

During the hemi-radial flow regime, the pressure changes with the logarithm of the elapsed time but the slope of the semi-log straight line is *double* ($2m$) that of the infinite acting radial flow (van Everdingen and Hurst, 1949, Horner, 1951).

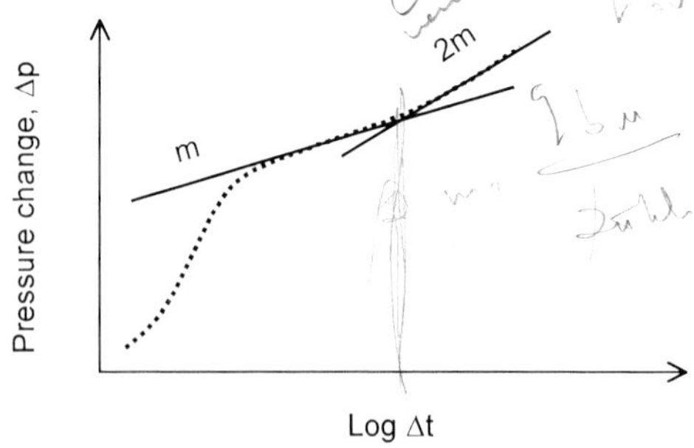

Figure 1.23. One sealing fault. Specialized analysis on semi-log scale.

It should be noted that with a consistent system of units (such as cgs Units), the slope of the semi-log straight line on a Δp versus $\ln(\Delta t)$ plot is expressed $qB\mu/2\pi kh$ during the infinite acting radial flow regime. For hemi-radial flow, the angle is changed to π.

On the semi-log plot, two straight lines are present, with a slope respectively m and $2m$. The time intersect Δt_x between the two lines is used to estimate the *fault distance L* (Gray, 1965).

$$L = 0.01217\sqrt{\frac{k\Delta t_x}{\phi\mu c_t}} \tag{1.33}$$

Interestingly, the radius of investigation at Δt_x is approximately double the fault distance (Equation 1.24): the pressure transient reaches the fault 4 times earlier than the boundary can be observed on the producing well pressure (see discussion Section 10.3.3).

In the next chapters, distances are expressed in dimensionless terms as:

$$L_D = \frac{L}{r_w} \tag{1.34}$$

1.2.8 Closed reservoir: pseudo steady state regime

In closed reservoirs, when all boundaries have been reached, the flow regime changes to *pseudo steady state*: i.e at any point in the reservoir the rate of pressure decline is proportional to time.

As long as the reservoir is infinite acting (time t_1 on the example Figure 1.24 for a circular closed reservoir), the pressure profile expands around the well during the production (in case of radial flow, the well bottom hole pressure drops with the logarithm of time).

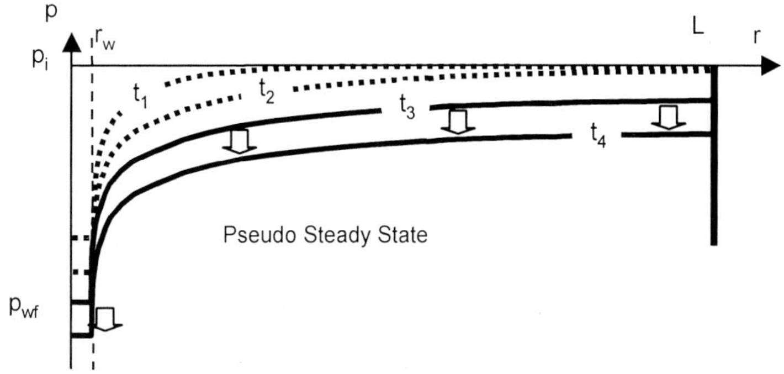

Figure 1.24. Circular closed reservoir. Pressure profiles.
Time t_1: the boundaries are not reached, infinite reservoir behavior: the pressure profile expands.
Time t_2: boundaries reached, end of infinite reservoir behavior.
Times t_3 and t_4: pseudo steady state regime, the pressure profile drops.

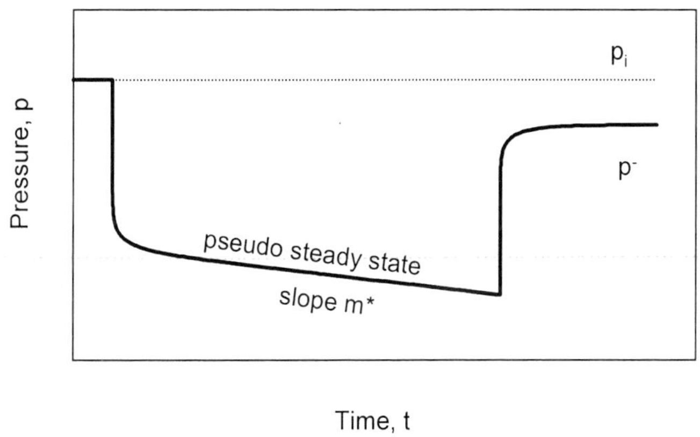

Figure 1.25. Drawdown and build-up pressure responses. Closed system. Linear scale.

When all boundaries have been reached (after time t_2), the *shape* of the pressure profile becomes *constant* with time, and it simply drops as the reservoir is being depleted (times t_3 and t_4). During the pseudo steady state regime, the bottom hole flowing pressure is a linear function of the elapsed time.

During shut-in, the pressure stabilizes in the reservoir and reaches the *average reservoir pressure* $\bar{p}\ (< p_i)$.

Specialized analysis

During drawdown, the Pseudo Steady State regime is analyzed with a plot of the pressure versus elapsed time Δt on a *linear scale* (Jones, 1956). At late time, the straight line of slope m^* is used to estimate the *reservoir pore volume* ϕhA.

$$\Delta p = 0.234 \frac{qB}{\phi c_t hA} \Delta t + 162.6 \frac{qB\mu}{kh} \left[\log \frac{A}{r_w^2} - \log(C_A) + 0.351 + 0.87S \right] \quad (1.35)$$

$$\phi hA = 0.234 \frac{qB}{c_t m^*} \quad (1.36)$$

1.2.9 Constant pressure boundary: steady state regime

The *steady state regime* is observed in wells near a gas cap or an active water drive: the high mobility of the fluid is seen as a constant pressure support. Once the constant pressure boundary is reached, the reservoir pressure profile and the well bottom hole pressure become constant.

1.3 WELL AND RESERVOIR CHARACTERIZATION

1.3.1 Well responses

A limited number of flow line geometries produce a characteristic pressure behavior: radial, linear, bi-linear, spherical etc. For each flow regime, the pressure follows a well-defined time function: $\log \Delta t$, $\sqrt{\Delta t}$, $\sqrt[4]{\Delta t}$, $1/\sqrt{\Delta t}$ etc. A straight line can be drawn on a specialized pressure versus time plot, to access the corresponding well or reservoir parameter.

A complete well response is defined as a *sequence* of flow regimes. By identification of the characteristic pressure behaviors present on the response, the chronology and time limits of the different flow regimes are established, defining the *interpretation model*. Two basic well test interpretation models examples are presented in the following for illustration.

For a fractured well for example, the sequence of regimes is (Gringarten et al., 1974 a):

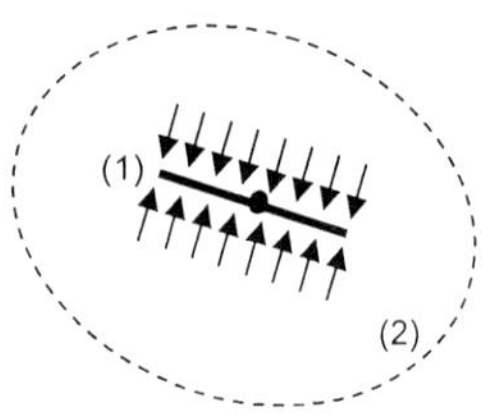

1. Linear flow
2. Radial flow

Figure 1.26. Fractured well example.

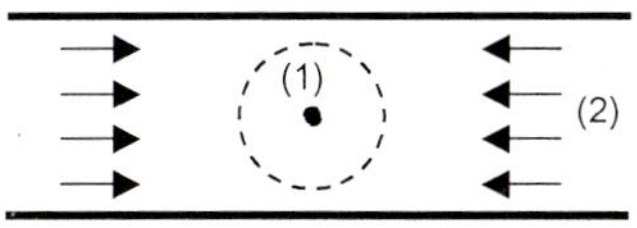

1. Radial flow
2. Linear flow

Figure 1.27. Example of a well in a channel reservoir.

In the case of a well in a channel reservoir, the same regimes are present but the chronology is inverse (Larsen, 1987).

1.3.2 Productivity Index

The Productivity Index is the ratio of the *flow rate* by the *drawdown* pressure drop, expressed from the average reservoir pressure $\bar{p}$.

$$\mathrm{PI}=\frac{q}{\left(\bar{p}-p_{wf}\right)} \tag{1.37}$$

The Ideal Productivity Index defines the productivity if the skin of the well is zero (Matthews and Russell, 1967).

$$\mathrm{PI}_{(S=0)}=\frac{q}{\left(\bar{p}-p_{wf}\right)-\Delta p_{skin}} \tag{1.38}$$

During the infinite acting period $\bar{p} \approx p_i$, the Transient Productivity Index is *decreasing* with time.

$$\mathrm{PI}=\frac{kh}{162.6B\mu\left(\log \Delta t+\log \frac{k}{\phi\mu c_t r_w^2}-3.23+0.87S\right)} \tag{1.39}$$

The Pseudo Steady State Productivity Index is a *constant*:

$$\mathrm{PI}=\frac{kh}{162.6B\mu\left(\log \frac{A}{r_w^2}-\log\left(C_A\right)+0.351+0.87S\right)} \tag{1.40}$$

CHAPTER 2

THE ANALYSIS METHODS

A complete production test is made up of several characteristic flow regimes, initially wellbore storage and near wellbore conditions, to late time boundary effects. Most of the recorded pressure data describes transitional behavior from one regime to the next, and straight lines are difficult to identify on the specialized scale plots described in Chapter 1. The log-log scale is preferred for well test interpretation: all flow regimes can be characterized on a single plot, providing a diagnosis of the complete well behavior and thus defining the appropriate interpretation model(s).

In this Chapter, the curve matching analysis methods are presented. Straight line methods, briefly described in Chapter 1, have been well documented (Earlougher, 1977; Bourdarot, 1998) and are not discussed in detail here. The use of pressure type curves on log-log scales is reviewed and application to multiple-rate and shut-in periods are discussed. Next, the derivative approach is introduced, the characteristic signatures of the different flow regimes are illustrated and the application of the method to practical testing conditions is detailed.

2.1 LOG-LOG SCALE

For a given period of the test, the change in pressure, Δp, is plotted on log-log scales versus the elapsed time Δt, as illustrated on Figure 2.1 (Theis, 1935 and Ramey, 1970). A test period is defined as a period of constant flowing conditions (constant flow rate for a drawdown and shut-in period for a build-up test, see Figure 1.1). The complete set of pressure data between two rate changes is used, from very early time to the latest recorded pressure point. The log-log analysis is a *global approach* as opposed to straight-line methods that make use of only one fraction of the data, corresponding to a specific flow regime.

By comparing the log-log data plot to a set of theoretical curves, the model that best describes the pressure response is defined.

Usually, theoretical curves are expressed in dimensionless terms because the pressure responses become independent of the physical parameters magnitude (such as flow rate, fluid or rock properties). An example of dimensionless term has been discussed in Section 1.2.3 with skin factor S which is much more meaningful than the actual pressure drop near the wellbore Δp_{skin}. As shown in Equation 2.1, the dimensionless pressure p_D and time t_D are linear functions of Δp and Δt, the coefficients A and B being dependent upon different parameters such as the permeability k.

$$p_D = A\Delta p, \quad A = f(kh,...)$$
$$t_D = B\Delta t, \quad B = g(k, C, S,...) \tag{2.1}$$

On log-log scales, the *shape* of the response curve is *characteristic*: the product of one of the variables by a constant term is changed into a displacement on the logarithmic axes. If the flow rate is doubled, for example, the amplitude of the response Δp is doubled also, but the graph of log (Δp) is only shifted by log (2) along the pressure axis.

$$\log p_D = \log A + \log \Delta p$$
$$\log t_D = \log B + \log \Delta t \tag{2.2}$$

The shape of the global log-log data plot is used for the *diagnosis* of the interpretation model(s). It should be noted that the scale expands the response at early time, and compresses the late time data.

2.2 PRESSURE CURVES ANALYSIS

2.2.1 Example of pressure type-curve: "Well with wellbore storage and skin in a homogeneous reservoir"

Log-log analysis technique is illustrated with the basic interpretation model "Well with wellbore storage and skin in a homogeneous reservoir". The corresponding set of dimensionless theoretical curves (called *type curves*), presented by Gringarten et al. in 1979, is illustrated in Figure 2.2.

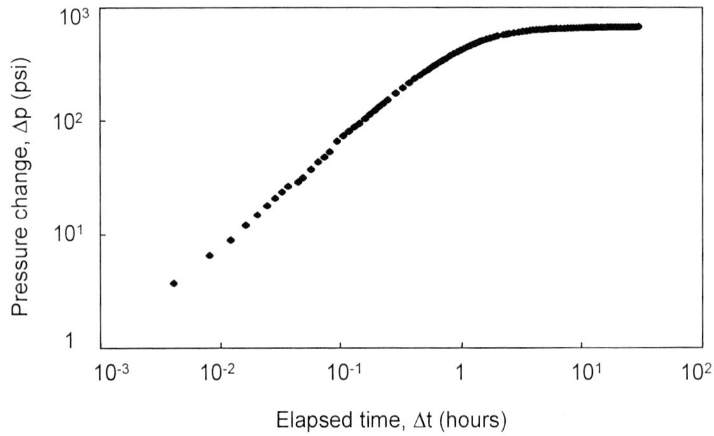

Figure 2.1. Example of log-log data plot.

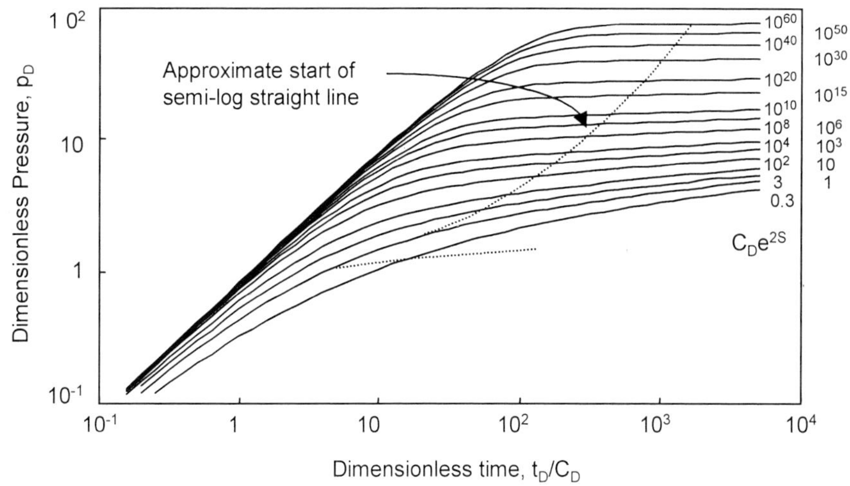

Figure 2.2. Pressure type-curve: Well with wellbore storage and skin, homogeneous reservoir. Log-log scales, p_D versus t_D/C_D. $C_D e^{2S} = 10^{60}$ to 0.3.

Dimensionless terms

The dimensionless pressure and time are defined as:

$$p_D = \frac{kh}{141.2qB\mu}\Delta p \tag{2.3}$$

$$t_D = \frac{0.000264k}{\phi\mu c_t r_w^2}\Delta t \tag{2.4}$$

The dimensionless wellbore storage coefficient is

$$C_D = \frac{0.8936C}{\phi c_t h r_w^2} \tag{2.5}$$

Several type curve presentations have been proposed for this interpretation model. For practical reasons, Gringarten et al. (1979) proposed using a dimensionless time group defined as:

$$\frac{t_D}{C_D} = 0.000295\frac{kh}{\mu}\frac{\Delta t}{C} \tag{2.6}$$

The different curves are labeled with the dimensionless group:

$$C_D e^{2S} = \frac{0.8936C}{\phi c_t h r_w^2} e^{2S} \tag{2.7}$$

The curve label $C_D e^{2S}$ defines the well condition. It ranges from $C_D e^{2S}$ =0.3 for stimulated wells, up to 10^{60} for very damaged wells.

Two characteristic flow regimes can be present in the response of a well with wellbore storage and skin in a homogeneous reservoir:

1. At early time, during the *pure wellbore* regime, the relationship Equation 1.9 can be expressed as:

$$\log\Delta p = \log\frac{qB}{24C} + \log\Delta t \tag{2.8}$$

On log-log scales, the data curve follows a *unit slope* straight line as described by the early time 45° asymptote on the type curve Figure 2.2.

2. When the infinite acting *radial flow* regime is reached, the pressure response follows the semi-log relationship of Equation 1.15 that does not produce a characteristic shape on log-log coordinates. The limit "Approximate start of the semi-log straight line" has been introduced on the type curve Figure 2.2 for the identification of the radial flow regime.

Between the two flow regimes, shown by the initial wellbore storage unit slope straight line and the start of the radial flow regime in Figure 2.2, the response describes a *transitional* behavior when the sand face rate changes, as describes in Section 1.2.2.

Log-log matching procedure

The log-log data plot Δp, Δt of Figure 2.1 is superimposed on the set of dimensionless type-curves p_D, t_D/C_D of Figure 2.2. The early time unit slope straight line is matched on the "wellbore storage" asymptote but the final choice of the $C_D e^{2S}$ curve is frequently not unique. In the example presented in Figure 2.3, all curves above $C_D e^{2S} = 10^8$ provide an acceptable match (the test data, for this illustrative example used through Chapter 2, has been published by Bourdet et al. in 1983 a).

Results of log-log analysis

The *pressure match* defines the displacement between the *y*-axis of the two log-log plots, as the ratio $\mathrm{PM} = p_D/\Delta p$. The *permeability thickness* product can be estimated from Equation 2.3:

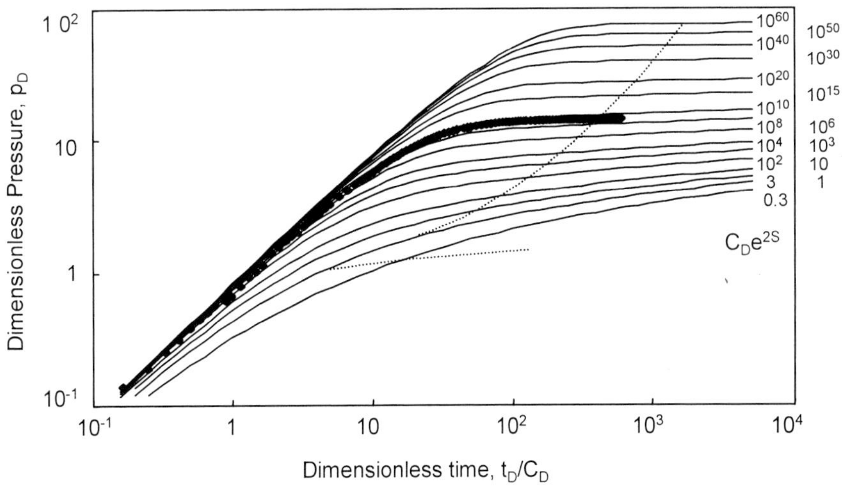

Figure 2.3. Build-up example. Log-log match, p_D versus t_D/C_D.

$$kh = 141.2 q B \mu (\text{PM}) \tag{2.9}$$

The *time match* $\text{TM} = (t_D/C_D)/\Delta t$ gives the *wellbore storage coefficient* with Equation 2.6:

$$C = 0.000295 \frac{kh}{\mu}\left(\frac{1}{\text{TM}}\right) \tag{2.10}$$

The *skin factor* is evaluated from the $C_D e^{2S}$ label of the selected curve (the *curve match*). From Equation 2.5,

$$S = 0.5 \ln \frac{C_D e^{2S}{}_{\text{Match}}}{C_D} \tag{2.11}$$

2.2.2 Shut-in periods

Drawdown periods are in general not suitable for analysis because it is difficult to ascertain a constant flow rate. The response is distorted, especially with the log-log scales that expand the response at early time. Preferably, build-up periods are used where the flow rate is zero, therefore the well is controlled.

Example of a shut-in after a single rate drawdown

Build-up responses do not show the same behavior as the first drawdown in a virgin reservoir at initial pressure. After a flow period of duration t_p, the well shows a pressure drop of $\Delta p(t_p)$. In the case of an infinite reservoir, after shut-in it takes an infinite time to reach the initial pressure during build-up, and to produce a pressure change $\Delta p_{BU}(t{=}\infty)$ of magnitude $\Delta p(t_p)$. As described on Figure 2.4, the shape of pressure build-up curves depends upon the previous rate history.

The diffusivity equation used to generate the well test analysis solutions is linear. It is possible to add several pressure responses, and therefore to describe the well behavior after any rate change.

This is the *superposition principle* (van Everdingen and Hurst, 1949). For a build-up after a single drawdown period at rate q during t_p, the rate history is changed by superposing an *injection* period at rate $-q$ from time t_p, to the flow period from time $t{=}0$ *extended* into the shut-in times $t_p + \Delta t$ (see Figure 2.5).

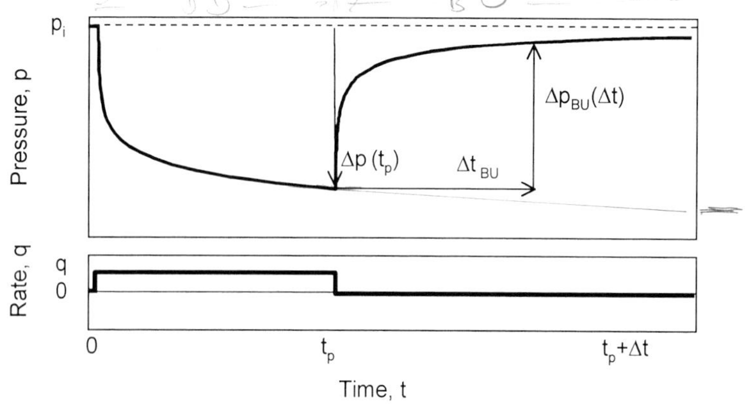

Figure 2.4. Test history : drawdown - shut-in.

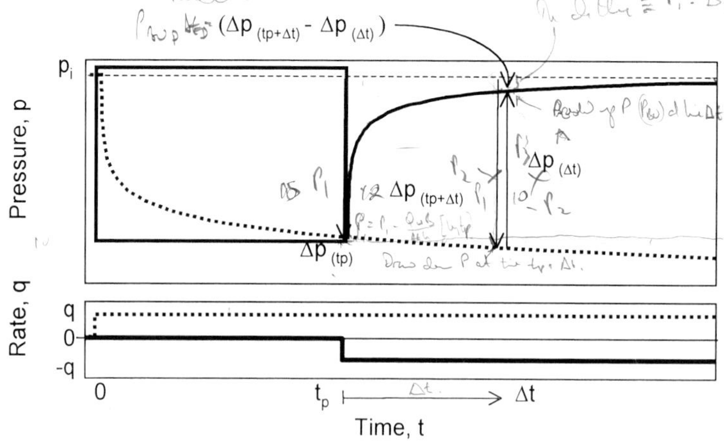

Figure 2.5. Equivalent flow history : extended drawdown (dashed line) + injection (plain line).

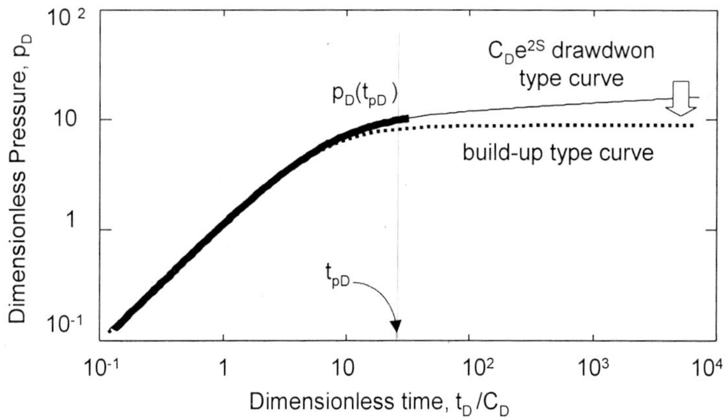

Figure 2.6. Drawdown and build-up type curves. Log-log scales, p_D versus t_D/C_D.

Log-log analysis: build-up type curve

Using the superposition principle, build-up type curves can be generated for any production history. In the case of Figure 2.5 with a single constant rate drawdown of t_p before shut-in, the build-up type curve $p_{BU\,D}(\Delta t)_D$ is simply obtained by subtracting the quantity $\lfloor p_D(t_p + \Delta t)_D - p_D(\Delta t)_D \rfloor$ from the pressure change at the time of shut-in $p_D(t_p)_D$.

$$p_{BU\,D}(\Delta t)_D = p_D(\Delta t)_D - p_D(t_p + \Delta t)_D + p_D(t_p)_D \tag{2.12}$$

Equation 2.12 shows that the build-up type curves differ from the original drawdown curve by $p_D(t_p)_D - p_D(t_p + \Delta t)_D$. On a log-log scales, build-up type curves are below the original drawdown type curve and, when Δt is large compared to the production time t_p, they tend to flatten at $p_D(t_p)_D$. As shown in Figure 2.6, the data is *compressed* on the y-axis, especially when $\Delta t >> t_p$, thus reducing the definition of the diagnostic plot.

Semi-log analysis: superposition time

When the pressure response is in radial flow at times t_p, Δt and $t_p + \Delta t$ (at time of shut-in, during the build-up period and during the extrapolated drawdown), the three p_D terms of Equation 2.12 can be changed into the semi-log approximation. Using the real pressure Equation 1.15 and grouping the three logarithm functions, the resulting build-up Equation 2.13 shows that a semi-log plot of shut-in pressure also display a straight line of slope m when the time is changed into the superposition time (also called *effective* Agarwal time, 1980).

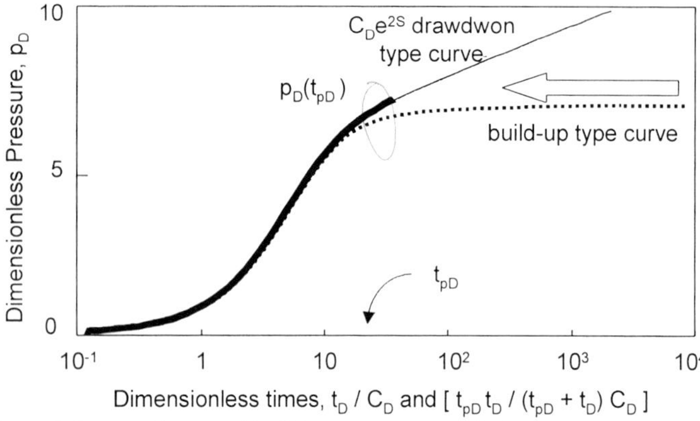

Figure 2.7. Drawdown and build-up type curves of Figure 2.6 on semi-log scales. The thick curve describes the drawdown response and the build-up expressed with the effective time.

$$\Delta p_{BU}(\Delta t) = 162.6\frac{qB\mu}{kh}\left[\log\frac{t_p\Delta t}{t_p + \Delta t} + \log\frac{k}{\phi\mu c_t r_w^2} - 3.23 + 0.87S\right] \tag{2.13}$$

With the superposition time, the build-up correction method *compresses* the time scale.

Horner method

With the Horner method (1951), a simplified superposition time is used: the constant t_p is ignored, and the shut-in pressure is plotted as a function of $\log\left[(t_p + \Delta t)/\Delta t\right]$. On the Horner scale, the shape of the build-up response is symmetrical to that of the superposition plot Figure 2.7, early time data is on the right side of the plot (large Horner time) and, at infinite shut-in time, $(t_p + \Delta t)/\Delta t = 1$.

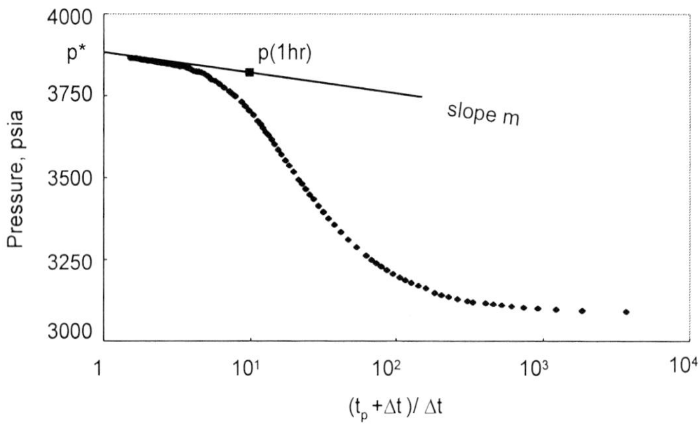

Figure 2.8. Horner plot of build-up example of Figure 2.1.

$$p_{ws} = p_i - 162.6\frac{qB\mu}{kh}\log\frac{t_p + \Delta t}{\Delta t} \qquad (2.14)$$

On a Horner plot of build-up data, the straight line slope m, the pressure at 1 hour on the straight line $\Delta p(\Delta t = 1\text{hr})$, and the *extrapolated straight line* pressure p^* at infinite shut-in time ($\Delta t = \infty$) are estimated. The results of analysis are:

$$kh = 162.6\frac{qB\mu}{m} \qquad (1.16)$$

$$S = 1.151\left(\frac{\Delta p_{1\,\text{hr}}}{m} - \log\frac{k}{\phi\mu c_t r_w^2} + \log\frac{t_p + 1}{t_p} + 3.23\right) \qquad (2.15)$$

In an infinite system, the straight line extrapolates to the initial pressure and $p^* = p_i$.

When the production time is large compared to the shut-in time $t_p >> \Delta t$, the Horner time can be simplified with:

$$\log\frac{t_p + \Delta t}{\Delta t} \approx \log t_p - \log\Delta t \qquad (2.16)$$

The compression of the time scale becomes negligible, the Horner straight-line slope m is independent of the production time and the build-up data can be analyzed on a MDH. semi-log scale such as in Figure 1.9.

Multiple-rate superposition

In the case of a multiple-rate test sequence such as on Figure 2.9, a new flow period is created for all rate changes (defined with the time at start t_i and the rate q_i), and the complete rate history prior to the analyzed period is used. Each previous period is superposed with the same principle as on the basic example of Figure 2.5.

At time Δt of flow period # n, the multi-rate type curve is:

$$p_{\text{MR}\,D}(\Delta t)_D = \sum_{i=1}^{n-1}\frac{q_i - q_{i-1}}{q_{n-1} - q_n}\left[p_D(t_n - t_i)_D - p_D(t_n + \Delta t - t_i)_D\right] + p_D(\Delta t)_D \qquad (2.17)$$

For semi-log analysis, the multiple-rate superposition time is expressed:

$$p_{ws}(\Delta t) = p_i - 162.6\frac{B\mu}{kh}\sum_{i=1}^{n-1}(q_i - q_{i-1})\log(t_n + \Delta t - t_i) + (q_n - q_{n-1})\log(\Delta t) \qquad (2.18)$$

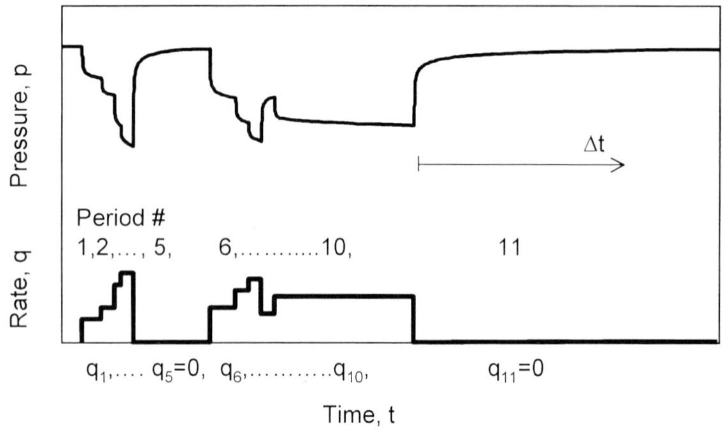

Figure 2.9. Multiple- rate history. Example with 10 periods before shut-in.

Limitations of the time superposition: the sealing fault example

In the example of Figure 2.10, the well is produced 60 hours and shut-in for a pressure build-up. A sealing fault is present near the well and, at 80 hours (20 hours after shut-in), the infinite acting radial flow regime ends to change slowly to the hemi-radial flow geometry.

During the 20 initial hours of the shut-in period (cumulative time 60 to 80 hours), both the extended drawdown and the injection periods are in radial flow regime. The superposition time of Equations 2.13 or 2.14 is applicable, and the Horner method is accurate.

At intermediate shut-in times, from 20 to 80 hours (cumulative time 80 to 140 hours), the extended drawdown follows a semi-log straight line of slope $2m$ while the injection is still in radial flow (slope m). It is not possible to group the different logarithm functions and, theoretically, the semi-log approximation of Equation 2.12 with Equation 2.13 is not correct.

Ultimately, the fault influence is also felt during the injection and the two periods follow the same semi-log straight line of slope $2m$ (shut-in time >> 80 hours, cumulative time >> 140 hours). The semi-log superposition time is again applicable.

In practice, when the flow regime deviates from radial flow in the course of the response, the error introduced by the Horner or multiple-rate time superposition method is negligible on pressure curve analysis results. It is more sensitive when the derivative of the pressure is considered (see Section 2.3.5).

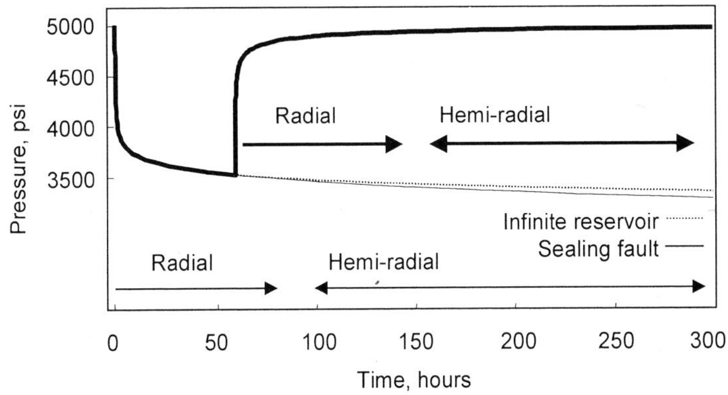

Figure 2.10. History drawdown – build-up. Well near a sealing fault.

Time superposition with other flow regimes

The time superposition is sometimes used with other flow regimes for straight-line analysis. When all test periods follow the same flow behavior, the Horner time can be expressed with the corresponding time function. For fractured wells, Horner time corresponding to linear (Equation 1.25) and bi-linear flow (Equation 1.27) is expressed respectively:

$$(t_p + \Delta t)^{1/2} - (\Delta t)^{1/2} \tag{2.19}$$

$$(t_p + \Delta t)^{1/4} - (\Delta t)^{1/4} \tag{2.20}$$

The Horner time corresponding to spherical flow (Equation 1.29) is sometimes used for the analysis of wireline formation testers pressure data.

$$(\Delta t)^{-1/2} - (t_p + \Delta t)^{-1/2} \tag{2.21}$$

2.2.3 Pressure analysis method

Pressure analysis is made on log-log and specialized plots (Gringarten et al., 1979). The purpose of the specialized analysis is to concentrate on a portion of the data that corresponds to a particular flow behavior. The analysis is carried out by the identification of a straight line on a plot whose scale is specific to the flow regime considered. The *time limits* of the specialized straight lines must have been previously defined by the *log-log diagnosis*.

For the radial flow analysis of a build-up period, the semi-log superposition time is used. The slope m of the Horner / superposition straight line defines the final pressure match of the log-log analysis:

$$\mathrm{PM} = \frac{p_D}{\Delta p} = \frac{1.151}{m} \tag{2.22}$$

Once the pressure match is defined, the $C_D e^{2S}$ curve is known accurately. Results from log-log and specialized analyses must be consistent.

2.3 PRESSURE DERIVATIVE

2.3.1 Definition

With the derivative approach, the time rate of change of pressure during a test period is considered for analysis. In order to emphasize the radial flow regime, the derivative is taken with respect to the logarithm of time (Bourdet et al., 1983 a). By using the *natural logarithm*, the derivative can be expressed as the time derivative, multiplied by the elapsed time Δt since the beginning of the period.

$$\Delta p' = \frac{dp}{d \ln \Delta t} = \Delta t \frac{dp}{dt} \tag{2.23}$$

As pressure analysis, the derivative is plotted on log-log coordinates versus Δt.

2.3.2 Derivative type-curve: "Well with wellbore storage and skin in a homogeneous reservoir"

Radial flow

When the infinite acting radial flow regime is established, the derivative becomes *constant*. This regime does not produce a characteristic log-log shape on the pressure curve, but it can be identified when the derivative of the pressure is considered.

$$\Delta p = 162.6 \frac{qB\mu}{kh} \left[\log \Delta t + \log \frac{k}{\phi \mu c_t r_w^2} - 3.23 + 0.87S \right] \tag{1.15}$$

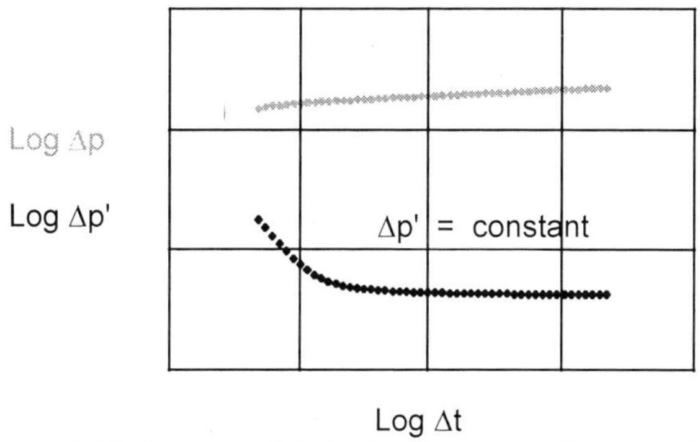

Figure 2.11. Pressure and derivative responses on log-log scales. Radial flow.

$$\Delta p' = 70.6 \frac{qB\mu}{kh} \tag{2.24}$$

In dimensionless terms, the derivative *stabilizes* at 0.5.

$$p'_D = \frac{dp_D}{d\ln(t_D/C_D)} = 0.5 \tag{2.25}$$

Wellbore storage

$$\Delta p = \frac{qB}{24C}\Delta t \tag{1.9}$$

$$\Delta p' = \frac{qB}{24C}\Delta t \tag{2.26}$$

During the pure wellbore storage regime, the pressure change Δp and the pressure derivative $\Delta p'$ are identical. On log-log scales, the pressure and the derivative curves follow a single straight line of *slope equal to unity* (Equation 2.8).

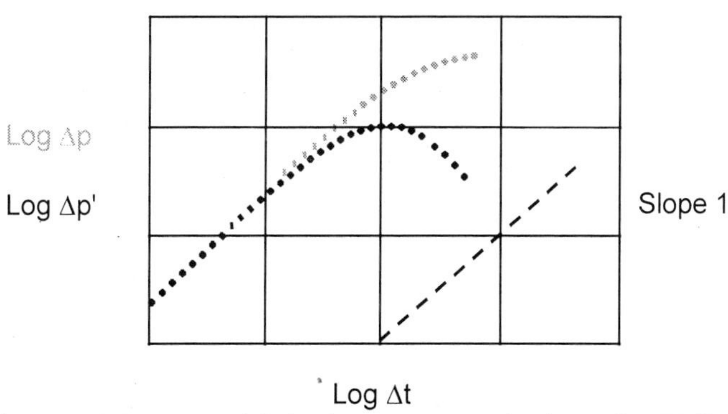

Figure 2.12. Pressure and derivative responses on log-log scales. Wellbore storage

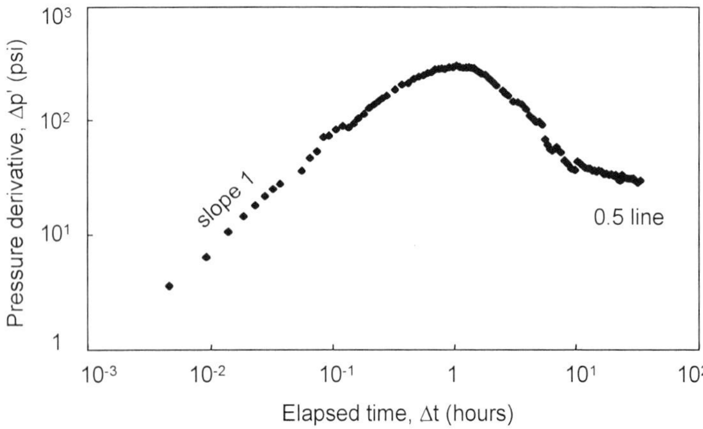

Figure 2.13. Derivative of build-up example Figure 2.1. Log-log scales.

Derivative of Section 2.2 example

In Figure 2.13, the derivative response of Figure 2.1 example shows at early time a unit slope log-log straight line during the pure wellbore storage effect, and later a stabilization when the radial flow regime is reached. At intermediate time between two characteristic flow regimes, the sand face rate is changing as long as the wellbore storage effect is acting (see Section 1.2.2), and the derivative response describes a hump.

Derivative type-curve

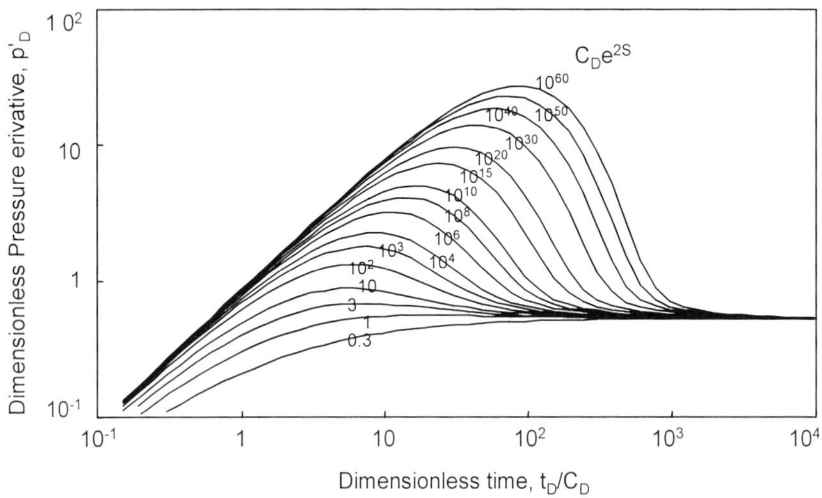

Figure 2.14. "Well with wellbore storage and skin, homogeneous reservoir" Derivative of type-curve Figure 2.2. Log-log scales, p_D versus t_D/C_D. $C_D e^{2S} = 10^{60}$ to 0.3.

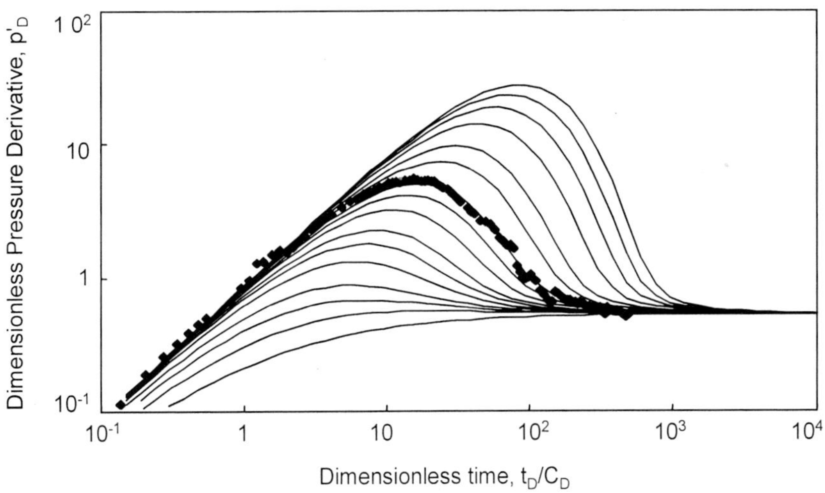

Figure 2.15. Derivative match of example Figure 2.3. Log-log scales, p_D versus t_D/C_D.

In the derivative type-curve of Figure 2.14 for a well with wellbore storage and skin in a homogeneous reservoir (Bourdet et al., 1983 a), the two basic flow regimes are characterized by a unique behavior:

1. At early time, all curves merge on a *unit slope* log-log straight line,
2. During radial flow, the derivative responses *stabilize* at 0.5.

During the transition between the pure wellbore storage effect and the infinite acting radial flow regime, the derivative *hump* can be used to identify the $C_D e^{2S}$ group.

Derivative match

The match point is defined with the unit slope pressure and derivative straight line, and the 0.5 derivative stabilization.

2.3.3 Other characteristic flow regimes

Except for the radial flow regime, during different flow geometries presented in Section 1.2, the pressure changes with the elapsed time power $1/n$:

$$\Delta p = A\left(\Delta t\right)^{1/n} + B \qquad (2.27)$$

With:

- $1/n = 1$ during the pure wellbore storage and the pseudo steady state regimes,
- $1/n = 1/2$ in the case of linear flow,

- $1/n = 1/4$ for bi-linear flow,
- $1/n = -1/2$ when spherical flow is established.

Taking the logarithm derivative (Equation 2.23) of the general Equation 2.27 yields

$$\Delta p' = \frac{dp}{d\ln\Delta t} = \frac{A}{n}(\Delta t)^{1/n} \tag{2.28}$$

On a log-log coordinate system, the relationship Equation 2.28 corresponds to a straight-line slope of $1/n$.

Infinite conductivity fracture (linear flow)

During linear flow, pressure change and the derivative are both proportional to $\Delta t^{1/2}$. On log-log scales, the pressure and derivative follow *two straight lines of slope 1/2* (Alagoa et al., 1985). The level of the derivative half-unit slope line is half that of the pressure.

$$\Delta p = 4.06\frac{qB}{hx_f}\sqrt{\frac{\mu}{\phi c_t k}}\sqrt{\Delta t} \tag{1.25}$$

$$\Delta p' = 2.03\frac{qB}{hx_f}\sqrt{\frac{\mu}{\phi c_t k}}\sqrt{\Delta t} \tag{2.29}$$

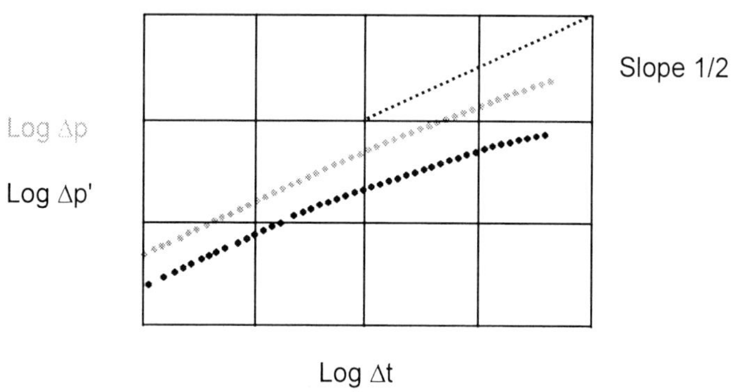

Figure 2.16. Pressure and derivative responses on log-log scales. Infinite conductivity fracture.

Finite conductivity fracture (bi-linear flow)

With the bi-linear flow geometry, pressure and derivative responses are proportional to $\Delta t^{1/4}$. A log-log *straight line of slope 1/4* can be observed on pressure and derivative curves, but the derivative line is four times lower (Wong et al., 1985).

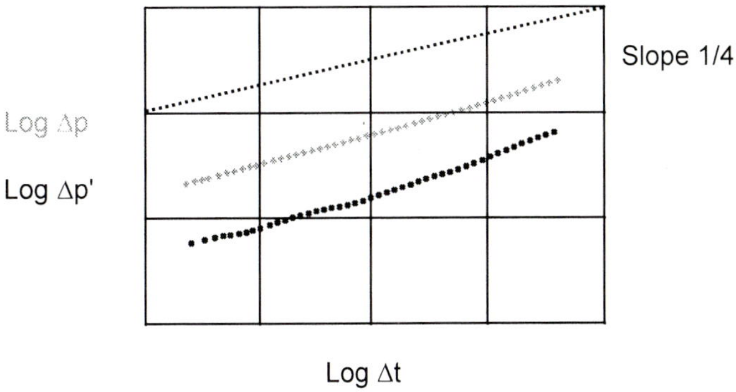

Figure 2.17. Pressure and derivative responses on log-log scales. Finite conductivity fracture.

$$\Delta p = 44.11 \frac{qB\mu}{h\sqrt{k_f w_f}\sqrt[4]{\phi\mu c_t k}}\sqrt[4]{\Delta t} \tag{1.27}$$

$$\Delta p' = 11.03 \frac{qB\mu}{h\sqrt{k_f w_f}\sqrt[4]{\phi\mu c_t k}}\sqrt[4]{\Delta t} \tag{2.30}$$

Well in partial penetration (spherical flow)

During the spherical flow regime, the shape of the log-log pressure curve is not characteristic. The derivative follows a straight line with a *negative half-unit slope*.

$$\Delta p = 70.6 \frac{qB\mu}{k_S r_S} - 2452.9 \frac{qB\mu\sqrt{\phi\mu c_t}}{k_S^{3/2}\sqrt{\Delta t}} \tag{1.29}$$

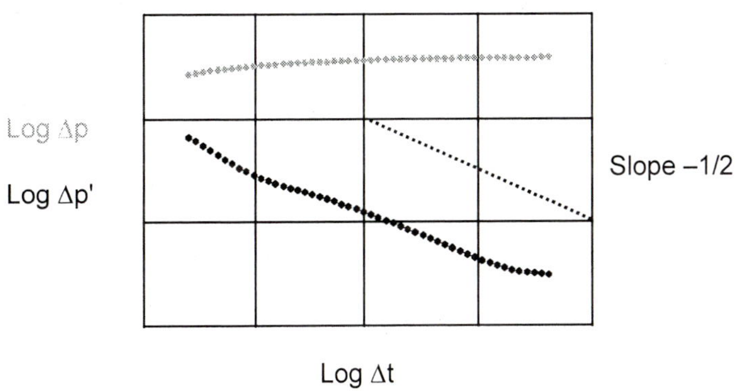

Figure 2.18. Pressure and derivative responses on log-log scales. Well in partial penetration.

$$\Delta p' = 1226.4 \frac{qB\mu\sqrt{\phi\mu c_t}}{k_S^{3/2}\sqrt{\Delta t}} \tag{2.31}$$

Closed system (pseudo steady state)

The late part of the log-log pressure and derivative drawdown curves tend to a *unit-slope straight line* (Clark and Van Golf-Racht, 1985). The derivative exhibits the characteristic straight line before it is seen on the pressure response.

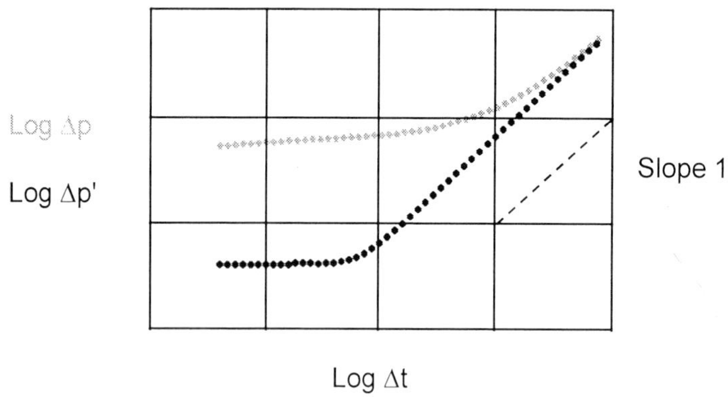

Figure 2.19. Pressure and derivative responses on log-log scales. Closed system (drawdown).

$$\Delta p = 0.234 \frac{qB}{\phi c_t hA} \Delta t + 162.6 \frac{qB\mu}{kh} \left[\log \frac{A}{r_w^2} - \log(C_A) + 0.351 + 0.87S \right] \tag{1.35}$$

$$\Delta p' = 0.234 \frac{qB}{\phi c_t hA} \Delta t \tag{2.32}$$

2.3.4 Build-up analysis

For a shut-in after a single drawdown period (the Horner method is applicable), the derivative is generated with respect to the effective Agarwal time given in the superposition Equation 2.13:

$$\Delta p' = \frac{dp}{d \ln \frac{t_p \Delta t}{t_p + \Delta t}} = \frac{t_p + \Delta t}{t_p} \Delta t \frac{dp}{dt} \tag{2.33}$$

For a complex rate history, the multiple-rate superposition time is used.

In all cases, the derivative is plotted versus the usual *elapsed time* Δt and matched against a drawdown derivative type-curve, such as in Figure 2.14 for example. It should be noted that a log-log build-up derivative curve is dependent upon the rate history introduced in the time superposition calculations, both the elapsed time and the superposition time are used in this plot. The derivative response is *not a raw data plot*. Errors may be introduced in the case of poor data preparation (see discussion next of the data differentiation, and the rate history definition in Section 10.1.1).

Limitations if the time superposition: the sealing fault example

When the response deviates from the infinite acting radial flow regime, taking the derivative with respect to the time superposition does not correct perfectly the build-up effect, and a *distortion* can be introduced on the response. On Figure 2.20, the log-log derivative of the build-up example of Figure 2.10 for a well near a sealing fault is compared to the drawdown curve. During shut-in, the effect of the sealing fault appears delayed compared to the theoretical drawdown derivative response.

With well test analysis software, the *same treatment* is applied on theoretical curves and on data. The test derivative response is matched on a derivative type curve generated by differentiating the theoretical build-up type curve with respect to the same time superposition function as the data, and the data responds in a similar way to the theoretical type curve (the same distortion is introduced on the data and on the model curves).

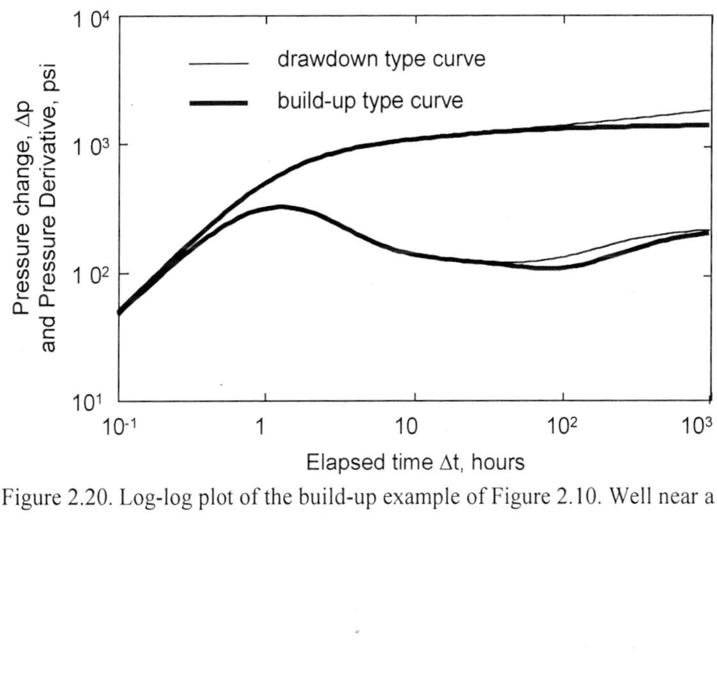

Figure 2.20. Log-log plot of the build-up example of Figure 2.10. Well near a sealing fault.

In some cases, the distortion can produce a temporary decline in the build-up derivative response, and produce a valley shape before the late time response (with heterogeneous reservoirs or boundary effects for example, see Chapters 4 and 5). When recorded test data stops at the time corresponding to the downturn in the derivative, and the upturn of the valley is not seen, interpretation of the late time trend can be difficult. Extrapolation of a small late time downward trend of the derivative response can be hazardous, it can simply correspond to the temporary distortion produced by the build-up derivative calculation or, as discussed next, to the effect produced by smoothing.

2.3.5 Data differentiation

As depicted in Figure 2.21, the data differentiation algorithm uses *three points*: one point before (left = 1) and one after (right = 2) the point i of interest. It estimates the left and right slopes, and attributes their weighted mean to the point i (Bourdet et al., 1983 a). On a p vs. x semi-log plot,

$$\frac{dp}{dx} = \frac{\left(\frac{\Delta p}{\Delta x}\right)_1 \Delta x_2 + \left(\frac{\Delta p}{\Delta x}\right)_2 \Delta x_1}{\Delta x_1 + \Delta x_2} \tag{2.34}$$

It is recommended to start by using consecutive points. If the resulting derivative curve is too noisy, smoothing is applied by increasing the distance Δx between the point i and points 1 and 2. The smoothing is defined as a distance L, expressed on the time axis scales. The points 1 and 2 are the first at distance $\Delta x_{1,2} > L$.

The smoothing coefficient L is increased until the derivative response is smooth enough but no more, *over smoothing* the data introduces *distortions*. Usual values for the smoothing coefficient L depends upon the software used to generate the derivative. Generally, L is less than 0.2 or 0.3.

When a large smoothing is required in order to produce a reasonably well defined derivative curve, it is recommended to examine the data on semi-log superposition scales with an expanded pressure definition on the y-axis, such as on the zoomed data plot inserted in Figure 2.21. Any unusual pressure behavior can be identified and analyzed, for determination of the best smoothing coefficient L. For example, when the pressure data is disturbed only during a small subset of the test period, it is best to ignore this data, and to adapt the derivative smoothing over the remaining good quality pressure data.

At the end of the test period, point i becomes closer to last recorded point than the distance L. Smoothing is not possible any more to the right side, the *end effect* is reached. This effect can introduce distortions at the end of the derivative response. Additional problems linked to the derivative calculation are further discussed in Section 10.1.1.

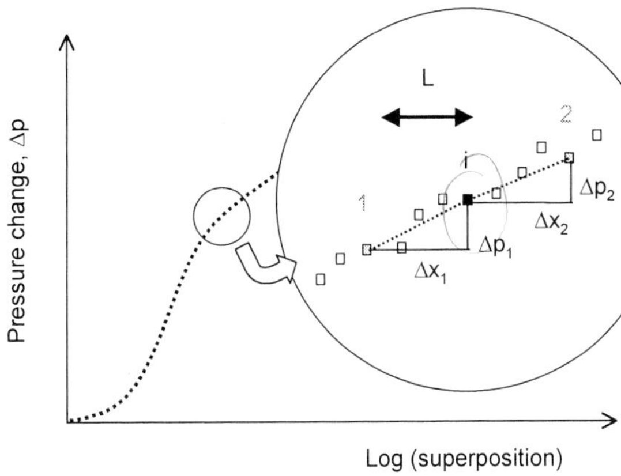

Figure 2.21. Differentiation of a set of pressure data. Semi-log scales.

2.3.6 Derivative responses

As discussed in Section 1.2.1, well responses are made up of several flow regimes, whose chronology defines, from early times to late times, the near wellbore conditions to the reservoir limits in the case of a long test. The derivative response exhibits a characteristic shape during all basic flow geometries. By reading the shape of the derivative response, the *sequence of regimes* can be established. Frequently only transitional behaviors between pure regimes are available on the log-log derivative curve, the diagnosis is not unique and several interpretation models have to be considered.

2.4 THE ANALYSIS SCALES

The log-log analysis is made with a simultaneous plot of the pressure and derivative responses of the interpretation period. Time and pressure matches are defined with the derivative, using respectively the *unit slope straight line* and the *derivative stabilization*. The curve match on pressure and derivative data give access to additional well and reservoir parameters. In the example for a well with wellbore storage and skin in a homogeneous reservoir Figure 2.22 (Bourdet et al., 1983 a), the $C_D e^{2S}$ group is identified by the shape of the derivative hump, and by the pressure curve match.

The double log-log match is confirmed with a match of the pressure type-curve on *semi-log scales* to accurately adjust the skin factor and the initial pressure. A simulation of the complete *test history* is presented on linear scale in order to check the rates any changes in the well behavior or the average pressure.

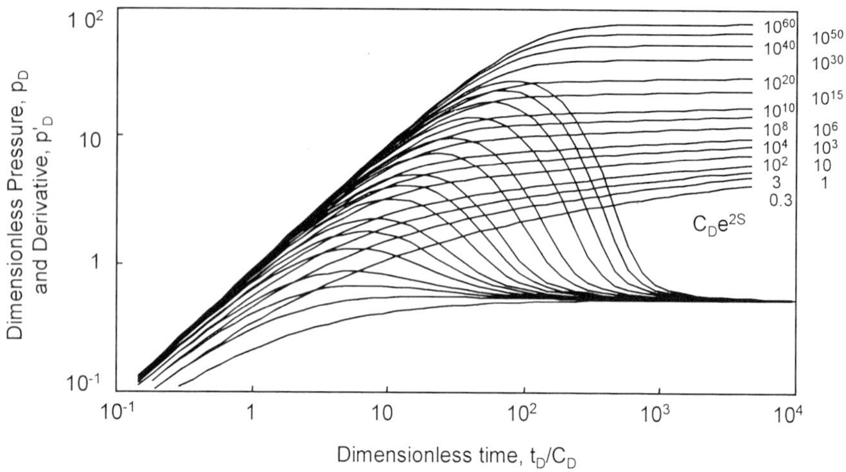

Figure 2.22. Pressure and derivative type-curve for a well with wellbore storage and skin, homogeneous reservoir. Log-log scales, p_D versus t_D/C_D.

The analysis procedure with the type curve shown in Figure 2.22 is discussed further in Section 3.1.3. The complete interpretation methodology is presented in detail in Section 10.2 from the diagnosis to the consistency check. Two examples are used to illustrate that the test history plot is an efficient check of the interpretation model applicability, over a large time range.

CHAPTER 3

WELLBORE CONDITIONS

In this chapter, the effect of the wellbore conditions on pressure responses is described. For simplification, the reservoir is assumed to be infinite and homogeneous.

Different wellbore models have been extensively used in the oil industry following the introduction of type curve analysis. A large catalogue of type curves has been published, using different groups of dimensionless parameters but, for a given wellbore condition, they are in most cases derived from the same mathematical solution. Type curves were designed for manual well test analysis. Today, pressure responses are analyzed with computer programs, and printed type curves are less used.

For each wellbore model, some representative published type curves are briefly described when they exist, and the independent groups of variables are defined. Practical manual analysis is discussed only briefly; type curves are introduced only as a training tool so that model behavior can be understood fully. Examples of the models responses are presented on log-log and specialized scale plots, and the influence of the different parameters is highlighted. Extensions of the models to complex wellbore conditions are discussed, and guide lines are explained for computerized well test analysis.

The interpretation methodology using computer programmes is introduced with the interpretation of a well with wellbore storage and skin in a homogeneous reservoir. A full discussion of computerized well test interpretation is presented at the end of this book, in Section 10.2.

3.1 WELL WITH WELLBORE STORAGE AND SKIN

The interpretation model for a well with wellbore storage and skin in an infinite reservoir with homogeneous behavior is probably the most widely used for transient pressure analysis. A typical example has been presented in previous Chapter (Figures 2.3 and 2.15).

3.1.1 Model description

The well is assumed to be vertical and to penetrate the complete reservoir thickness. Wellbore storage effect and possibly an infinitesimal skin are present.

Characteristic flow regimes

As discussed in former chapter, two characteristic regimes can be observed with this model:
1. *Wellbore storage effect*, with Δp proportional to Δt (Figure 1.4), and a unit slope log-log straight line on pressure and derivative curves (Figure 2.12). The wellbore storage coefficient C can be estimated from the corresponding pressure data.
2. *Radial flow* with Δp proportional to $\log(\Delta t)$ (Figure 1.9), and a constant derivative response (Figure 2.11). Radial flow analysis yields the permeability-thickness product kh and skin factor S.

Analytical solutions

The analytical solution is obtained by introducing the wellbore storage effect and an infinitesimal positive skin on the basic solution to the diffusivity equation for a constant rate drawdown in a finite radius well (van Everdingen and Hurst, 1949). The solutions are generated in the Laplace space and inverted with a numerical algorithm (Stehfest, 1970). Negative skins are introduced on the solution at S=0 with the equivalent effective wellbore radius of Equation 1.14.

3.1.2 Review of pressure and derivative type curves for a well with wellbore storage and skin in a homogeneous reservoir

The first set of wellbore storage and skin pressure type curves were introduced by Agarwal et al. in 1970. The dimensionless pressure p_D of Equation 2.3 is plotted versus the dimensionless time t_D of Equation 2.4. Several families of curves are presented as a function of two dimensionless parameters: the dimensionless wellbore storage coefficient C_D of Equation 2.5, and the skin S of Equation 1.11.

Matching with this type curve is difficult since several combinations of C_D and S can produce a similar shape. Different presentations have been proposed (McKinley, 1971; Earlougher and Kerch, 1974) but the most widely used is the Gringarten et al. type curve of Figure 2.2 (1979) with the dimensionless pressure p_D is presented versus the dimensionless time group t_D / C_D of Equation 2.6. The shape of the dimensionless pressure curves being a function of the $C_D e^{2S}$ group of Equation 2.7, this *independent variable* is used to define the model responses.

It is interesting to note that, assuming a strictly radial geometry around a vertical well, 0.5 is the theoretical lower limit for the dimensionless $C_D e^{2S}$ group (Gringarten et al., 1979). Considering that an acid stimulation treatment can be described as an infinite permeability circular zone of radius r_S around the well (see Figure 1.8 with $k_S = \infty$), the stimulated zone participates to the storage effect, and the apparent wellbore storage coefficient is changed from C to C_S (in bbl/psi):

$$C_S = C + \frac{\pi}{5.6146}\left(r_S^2 - r_w^2\right)\phi c_t h \tag{3.1}$$

Combining Equations 1.14 and 2.7 yields

$$\left(C_D e^{2S}\right)_S = C_D e^{2S} + \frac{1 - e^{2S}}{2} \tag{3.2}$$

whose minimum is 0.5. Strictly speaking, the acid penetrates the formation by opening channels, and the circular stimulated model with infinite permeability is only a rough description of the near wellbore condition for an acid stimulated well (see discussion of the natural negative skin in fissured reservoir, Chapter 4). The 0.5 limit is approximate.

On the Gringarten et al type curve of Figure 3.1, the different $C_D e^{2S}$ curves are classified in terms of well conditions. The curves above $C_D e^{2S}$ =1000 defines a *damaged* well, between 1000 and 5 the curves correspond to approximately a *zero skin* condition, from 5 to 0.5 to an *acidized* well and, below 0.5 to a *fractured* well (on this set of type curves, the wellbore storage and skin solution has been replaced by the solution for an infinite conductivity fracture with wellbore storage presented next, in Section 3.2). The different limits are approximate but they indicate that the shape of the log-log pressure response describe qualitatively the type of wellbore condition.

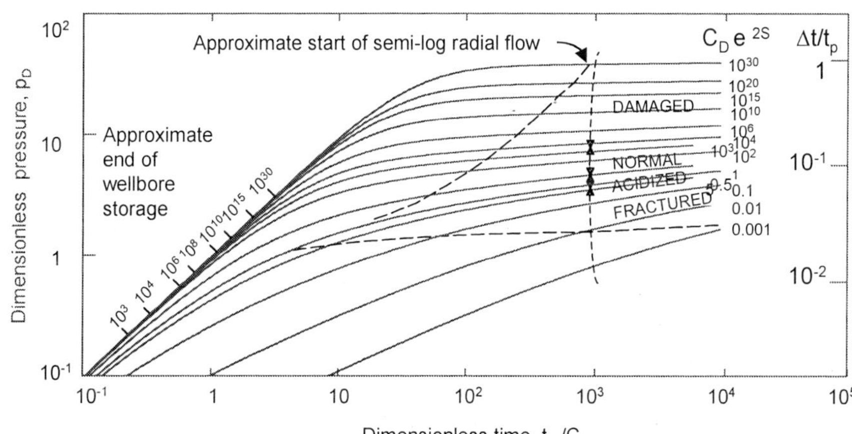

Figure 3.1. Wellbore storage and skin pressure type curve. Log-log scales, p_D versus t_D/C_D. Graph courtesy A.C. Gringarten.

Subsequently, the derivative of the pressure was introduced with the type curves shown in Figure 2.22 of Bourdet et al. (1983 a). The dimensionless parameters are the same as on Figure 3.1, and both the pressure and the derivative responses are superimposed on the same log-log scales. Only the radial flow model to a vertical well is presented on this set of type curves, fractured wells are not accounted for.

As discussed in Section 2.3.2, the derivative curves merge at early and late times (first on the unit slope straight line during wellbore storage effect and on the 0.5 horizontal line during radial flow). At intermediate times, the shape of the derivative hump defines the $C_D e^{2S}$ parameter.

Several variations of the wellbore storage and skin pressure and derivative type curves have been proposed. In addition to the usual pressure and derivative, the pressure derivative of the second order, and the pressure integral have also been considered. The type curves are presented with various ratios of the pressure function, its derivatives and integral (Duong, 1987; Onur and Reynolds, 1988; Blasingame et al., 1989; Blasingame et al., 1990). Bourgeois and Horne (1991) proposed dimensionless wellbore storage and skin type curves in the Laplace space, and not in the usual time space, for the identification of the interpretation model and the estimation of well and reservoir parameters.

3.1.3 Matching procedure on pressure and derivative responses

The matching procedure with the wellbore storage and skin model has been described in Chapter 2. In the following, two characteristic model responses are used for illustration of the model behavior.

Figure 3.2 presents on log-log scales the pressure and derivative responses generated with respectively $C_D e^{2S} = 10^{30}$ and 0.5. On Figure 3.3, the same pressure curves are plotted on semi-log scales. $C_D e^{2S} = 10^{30}$ corresponds to a severely damaged well whereas 0.5 can describe the response of a well with negative skin.

On the log-log plot Figure 3.2, the pressure and derivative curves for the high $C_D e^{2S}$ value follow the unit slope wellbore storage straight line at early times. Whereas, when $C_D e^{2S} = 0.5$, the pure wellbore storage regime finishes earlier and the curves starts its transition between pure wellbore storage and a radial flow regime. When the well is damaged ($C_D e^{2S} = 10^{30}$), this intermediate time flow regime is described by a very *characteristic hump* on the derivative response but not when $S \leq 0$ (there is *no maximum* if the well is not damaged or stimulated).

During the radial flow regime, the *vertical distance* between the derivative and the pressure curves gives an indication of the wellbore damage. With no skin, the pressure curve is one log cycle or less above the derivative stabilization. For a highly damaged well, the distance between the two curves can be more than two log cycles as illustrated by the example $C_D e^{2S} = 10^{30}$.

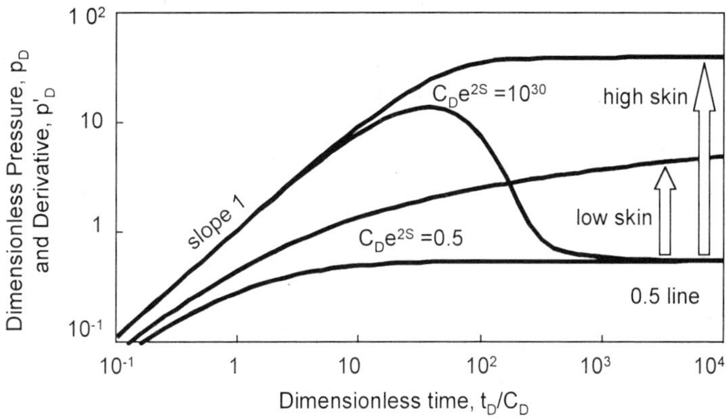

Figure 3.2. Responses for a well with wellbore storage and skin in an infinite homogeneous reservoir. Log-log scales, p_D versus t_D/C_D. $C_D e^{2S} = 10^{30}$ and 0.5.

Comparing the two responses of Figure 3.2 shows that the start time of the radial flow regime, defined by the derivative stabilization at 0.5, is a function of both the wellbore storage coefficient C and the skin factor S. With damaged wells, the effect of wellbore storage lasts longer and the start of the semi-log straight line is delayed. The *lower* the skin factor, the *faster* the reservoir response is reached.

When matching test data on a computer software package, real variables are used in place of dimensionless parameters. Both the test data and the theoretical model are presented with the pressure and derivative responses Δp, $\Delta p'$ versus the elapsed time Δt. The model parameters are adjusted until a good fit is obtained.

1. As discussed in Section 2.2, the permeability thickness product kh defines the pressure match $\mathrm{PM} = p_D/\Delta p$. From Equation 2.3, *increasing* the kh product in the model parameters displaces the theoretical curves *downwards* on the log-log plot, but the shape of the curve is not changed.
2. Once the derivative stabilization is correctly matched, the skin factor S can be evaluated. The pressure curve is preferably used for skin adjustment rather than the derivative transition between the pure wellbore storage and the radial flow regime (the derivative response is frequently noisy during this intermediate time period. Furthermore, it can be distorted by changing wellbore storage effect, see Section 10.1.2).
3. The last parameter to be determined is the wellbore storage coefficient C from the early time response. *Increasing* C displaces the unit slope line (and the derivative hump) towards *later times* on the computer generated curves.

When several periods are available for analysis, by *normalizing* the pressure axis by the flow rate (with $\Delta p/q$ and $\Delta p'/q$), the different periods can be compared on the same log-log plot. This is a very efficient diagnosis tool, especially with noisy data. As the derivative becomes independent of the skin parameter after the end of wellbore storage effect, the derivative curves must overlay after the early time response and, during the radial flow regime, all responses merge on the same derivative stabilization. If not, the flow rate data must be checked (see Section 10.1.1). Once the derivative responses are consistent, the difference between the pressure curves reveals any change in the well condition such as wellbore storage, skin damage or rate dependent skin (discussed in Chapter 7, Figure 7.11 for example).

If the test stops before radial flow has been reached and the derivative stabilization is not available, an approximate analysis can be attempted on the pressure and derivative transition responses. In general the solution is not uniquely defined but, when the derivative is not severely distorted by noise or changing wellbore storage effects, upper and lower limits can be estimated for the permeability thickness product *kh* and the skin factor *S*.

3.1.4 Associated specialized plot straight line and interpretation procedure

Wellbore storage and radial flow analyses have been presented in Chapters 1 and 2 for drawdown and multiple rate test sequences. Today, straight-line methods are not considered to provide accurate results as curve fitting on pressure and derivative responses. Straight-line analysis can be attempted for a quick estimate of the parameters, or for consistency check of the interpretation results. Before specialized analyses, the time limits of the various regimes must have been identified by matching the pressure and derivative responses on the type curves of Figure 2.22.

More importantly, specialized scales are used with computerized interpretation to refine the log-log analysis results. Once the log-log match is accurately defined, the model parameters are further adjusted on a semi-log plot such as Figure 3.3. This scale is not appropriate for *C* and *kh* fine-tuning, but it is more sensitive to small skin errors than the log-log system of coordinates.

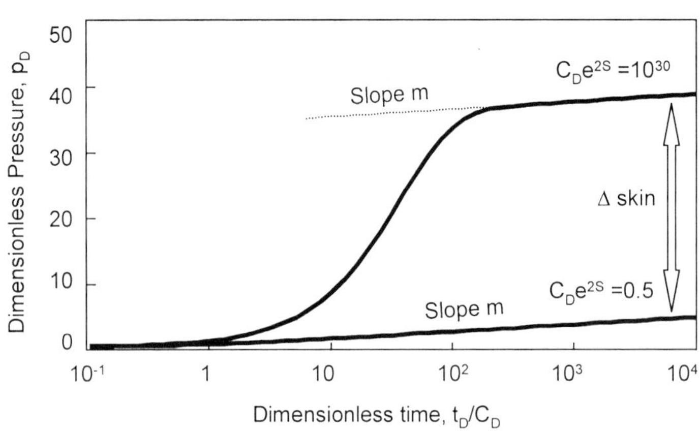

Figure 3.3. Semi-log plot of Figure 3.2.

The next step is to generate a test simulation plot on linear scales. The interpretation model is applied to the complete sequence of flow and shut-in periods, to produce the theoretical response during the well rate history. With the test simulation, the theoretical model is compared to the measured pressure data on a longer time interval than during the analysis of a single period. The test simulation is used to determine the initial pressure at start of the test, and will illustrate clearly any change of well condition or flow rate errors. By extrapolating the model beyond the initial diagnosis period, possible inconsistencies in the model can be identified (see Section 10.2.3).

3.1.5 Skin discussion

Tests are sometimes interpreted with the model for a well with wellbore storage and skin in a homogeneous reservoir when the well or reservoir configuration is known to be different. Wells in partial penetration or slanted wells for example frequently show a typical wellbore storage and skin response, and the well geometry only affects the skin calculated from the match. In the first case, the skin can reach values higher than 30 (see discussion Section 3.4.3), whereas deviated wells can show a slightly negative skin (Section 3.5.2).

When the properties of the near wellbore reservoir region have been changed after drilling or stimulation, an apparent positive or negative skin can be observed as a result of the radial composite reservoir condition (Section 4.3.2). In his discussion of the two usual approximations for the skin effect, namely the infinitesimal skin thickness and the equivalent wellbore radius concept, Daviau (1986) concludes that the homogeneous reservoir model is applicable to most skin configurations, unless the wellbore storage coefficient is extremely low.

As discussed in Section 4.1.5, wells in naturally fractured reservoirs also can show a homogeneous behavior, associated possibly to a natural negative skin and to a very large wellbore storage coefficient. Reservoirs with horizontal permeability anisotropy as well, generate a small natural negative skin on test responses. The different possible skin components are summarized at the end of this Chapter, in Table 3.7.

Permeability anisotropy

In the case of a reservoir with horizontal permeability anisotropy, the pressure response of a producing well can be described by an equivalent isotropic reservoir model of average radial permeability (Earlougher, 1977). With the maximum k_{max} and the minimum k_{min} permeability oriented 90° apart, the average permeability is:

$$\bar{k} = \sqrt{k_{max} k_{min}} \tag{3.3}$$

An equivalent transformed isotropic system can be used to describe the pressure behavior of the reservoir by changing the dimensions in the two main directions of permeability. The transformation of variables is, respectively for the maximum and the minimum permeability directions:

$$x' = x\sqrt{\frac{\bar{k}}{k_{\max}}} = x\sqrt[4]{\frac{k_{\min}}{k_{\max}}} \tag{3.4}$$

$$y' = y\sqrt{\frac{\bar{k}}{k_{\min}}} = y\sqrt[4]{\frac{k_{\max}}{k_{\min}}} \tag{3.5}$$

In the equivalent isotropic system, the wellbore is changed into an ellipse whose major axis, in the low permeability direction, is $r_w\sqrt[4]{k_{\max}/k_{\min}}$. In the high permeability direction, the minor axis is $r_w\sqrt[4]{k_{\min}/k_{\max}}$.

The area of the well is the same in the original and transformed systems, but the perimeter is increased. The elliptical well behaves like a cylindrical hole whose equivalent radius is the average of the major and minor axes (Brigham, 1990):

$$r_{we} = \frac{1}{2} r_w \left[\sqrt[4]{k_{\min}/k_{\max}} + \sqrt[4]{k_{\max}/k_{\min}} \right] \tag{3.6}$$

Since the analysis results are calculated with reference to the actual wellbore radius r_w, the reservoir anisotropy produces an apparent negative skin component:

$$\begin{aligned} S_{ani} &= -\ln \frac{\sqrt[4]{k_{\min}/k_{\max}} + \sqrt[4]{k_{\max}/k_{\min}}}{2} \\ &= -\ln \frac{\sqrt{k_{\min}} + \sqrt{k_{\max}}}{2\sqrt{\bar{k}}} \end{aligned} \tag{3.7}$$

With typical permeability anisotropy values in the horizontal plane, the negative geometrical skin effect is low. For horizontal wells (see Section 3.6), the effect of permeability anisotropy between the vertical and horizontal directions can be much larger, and apparent negative skins of S_{ani} =-1 may be observed.

Table 3.1. Anisotropy skin S_{ani}

$k_{\max} / k_{\min}$	10	100	1000
S_{ani}	-0.157	-0.55	-1.06

3.2 INFINITE CONDUCTIVITY OR UNIFORM FLUX VERTICAL FRACTURE

The hydraulic fracturing technique has been used from the 1950's to improve the productivity of damaged wells, or wells producing from low-permeability reservoirs. By injecting fluid into the formation, a vertical plane fracture is created and filled with propping agents to prevent closure.

Fractured well models for high and low conductivity, and the corresponding specialized analysis, have been presented in Sections 1.2.4 and 1.2.5. Log-log pressure and derivative responses are illustrated in Figures 2.16 and 2.17 (Section 2.3.3). The high conductivity fracture model is discussed in this Section, fractures with a low conductivity are considered in Section 3.3.

3.2.1 Model description

The well intercepts a symmetrical vertical plane fracture of half-length x_f (Figure 1.11). The well and the fracture penetrate totally the reservoir thickness and there is *no pressure loss* along the fracture plane. Wellbore storage effects can be present in the well, and the fracture can be affected by a skin damage.

Characteristic flow regimes

Two characteristic regimes can be observed after the wellbore storage early time effect, as illustrated on Figure 1.12:

1. *Linear flow*, with Δp proportional to $\Delta t^{1/2}$ and a half unit slope straight line on pressure and derivative log-log curves (Figure 2.16). The linear flow regime defines the $k(x_f)^2$ product, and therefore the fracture half-length x_f.
2. *Pseudo-radial flow* regime when the flow lines converge from all reservoir directions. During the pseudo-radial flow regime, the pressure follows a semi-log straight-line behavior, as during the usual radial flow regime towards a cylindrical vertical well. The fracture influence is then described by a geometrical negative skin and the pseudo-radial flow analysis provides the permeability thickness product kh and S_G.

Analytical solutions

The analytical solutions for fractured wells have been developed by Gringarten et al. (1974 a, 1975 a) for the uniform flux and the infinite conductivity fractures. With the uniform flux solution, the flow per unit of fracture surface is assumed constant along the fracture length, while the infinite conductivity model is based on the assumption that the pressure is uniform in the fracture. The solutions are obtained by dividing the fracture length into M segments, and using the Green's function and product solution method

(Gringarten and Ramey, 1974 b). When uniform flux is established along the fracture extension, each segment produces at the same rate. The uniform flux solution was designed as a first approximation of the behavior of a fractured well, and is the exact solution only at early times (when the flux in an infinite conductivity fracture is uniform).

Gringarten et al. (1974 a) obtained the infinite conductivity model by measuring the pressure drop at the point $x_D = x/x_f = 0.732$ in the uniform flux fracture. The authors showed by numerical simulation that, in infinite conductivity fractures, the flux distribution changes after the early time response, and reaches a stabilized profile along the fracture length. During the pseudo-radial flow regime, the two ends of the fracture are the most productive segments, as it will be shown also for horizontal wells (see Section 3.6). The shape flux distribution for the different fractured well models is discussed further in Section 3.3.6.

3.2.2 Review of pressure and derivative type curves for a well with infinite-conductivity fracture

In the mid 1970's, the catalog of available pressure type curves was extended to the fractured well condition. The curves offer the possibility of identifying the nature of the wellbore stimulation, and to quantitatively analyze the effect of fracturing on well responses. In addition to the permeability, the analysis provides information concerning the fracture characteristics, and possibly the distance to boundaries. Different type curve presentations have been proposed, they are briefly summarized as follows.

Gringarten (1978) presented a pressure type curve for a well with an infinite-conductivity fracture at the center of a closed rectangle, shown in Figure 3.4. The dimensionless pressure p_D of Equation 2.3 is expressed as a function of the dimensionless time t_{Df}, defined with respect to the half fracture length x_f:

$$t_{Df} = \frac{0.000264k}{\phi \mu c_t x_f^2} \Delta t \tag{3.8}$$

No skin damage and no wellbore storage effects are assumed. The permeability thickness product is obtained from the pressure match (Equation 2.9), the fracture half-length x_f from the time match:

$$x_f = \sqrt{\frac{0.000264k}{\phi \mu c_t} \frac{1}{\text{TM}}} \tag{3.9}$$

The type curves of Figure 3.4 can be used for the analysis of fractured well limit testing ("A" is the area of the rectangular reservoir, x_e the half-length of the reservoir in the direction parallel to the fracture, and y_e the half-width of the reservoir). In addition, several build-up type curves are presented for the infinite reservoir case.

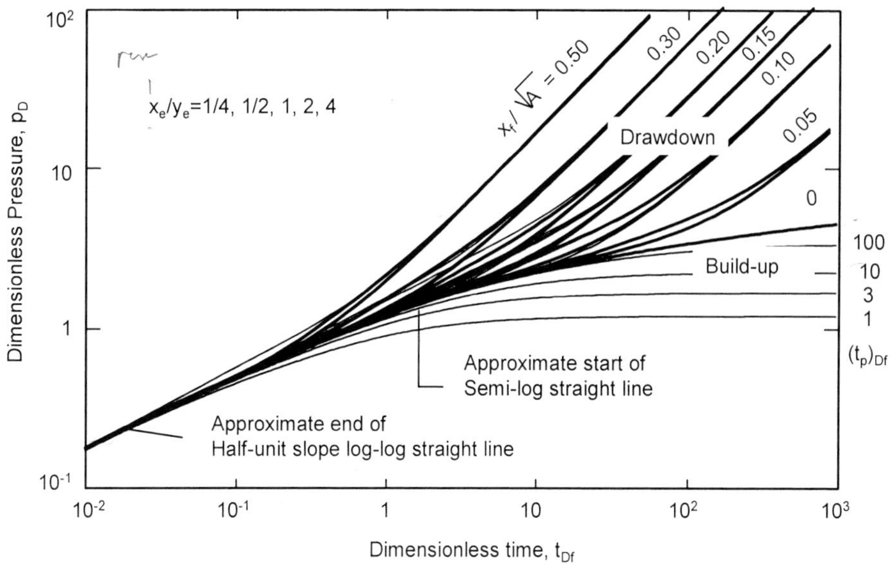

Figure 3.4. Pressure type curve for an infinite conductivity vertical fracture at the center of a closed rectangle. Log-log scales, p_D versus t_{Df}. Graph courtesy A.C. Gringarten.

When wellbore storage dominates the early times response, the type curve of Figure 3.1 with p_D versus the dimensionless time group t_D / C_D, can be used. On this type curve, values of $C_D e^{2S}$ smaller than 0.5 are generated with the infinite conductivity fracture model and wellbore storage effect.

A match on a low $C_D e^{2S}$ curve yields the permeability thickness product kh from the pressure match, the wellbore storage coefficient C from the time match and the geometrical skin S_{HKF} from the curve match. From equation 3.12, the geometrical skin can be used to estimate the fracture half-length x_f.

The pressure derivative behavior for uniform flux and infinite conductivity fractures has been investigated in 1988 by Tiab and Puthigai by differentiating the pressure with respect to time. In the following text, the logarithmic derivative presented in Chapter 2 is considered, with the Alagoa et al. type curves (1985).

These type curves combine the pressure and derivative presentations. The effect of wellbore storage is considered, the various curves are expressed with p_D versus t_{Df}, and they are labeled in terms of C_{Df}.

$$C_{Df} = \frac{0.8936C}{\phi c_t h x_f^2} \tag{3.10}$$

The matching procedure is similar to p_D versus t_{Df} the Gringarten's type curves, and the wellbore storage coefficient C can be estimated from C_{Df} and x_f.

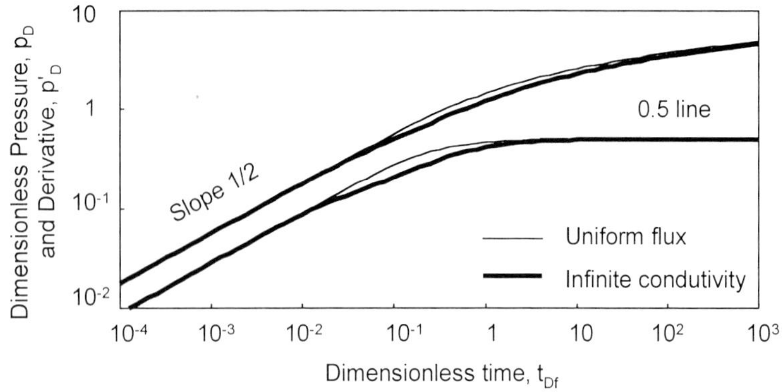

Figure 3.5. Responses for a well intercepting a high conductivity fracture. Log-log scales, p_D versus t_{Df}. No wellbore storage effect, $C_D = 0$. Infinite conductivity and uniform flux models.

The type curves for the infinite conductivity model and for the uniform flux fracture solution exhibits the same characteristic shapes during the three typical behaviors and, in fact, present very similar log-log pressure response curves. Only the derivative curves show some difference during the transition from linear to radial flow as discussed next.

3.2.3 Matching procedure on pressure and derivative responses

Infinite conductivity and uniform flux models

In practice, wellbore storage is short lived in fractured wells, and frequently is not observed on the recorded data: the response starts to follows the half unit slope pressure and derivative straight lines from early time as illustrated on the examples Figure 3.5 where the curves are generated with $C_D = 0$. The two high conductivity fracture models are slightly different at intermediate times, between linear flow and radial flow. With the uniform flux model, the transition from the half unit slope straight line to the 0.5 line is shorter, and the angle between the two regimes is more pronounced. The pressure curve is slightly higher.

When matching test data against a high conductivity fracture model such as on Figure 3.5, the *derivative stabilization* during the pseudo radial flow regime is used to determine the pressure match (giving the permeability thickness product kh). The location of the *half unit slope* pressure and derivative straight lines provides the half fracture length x_f with Equation 3.9 (the longer the fracture, the later the start of the pseudo radial flow regime).

When the log-log match is adjusted with a well test interpretation software package, the pressure and derivative data curves are presented versus the elapsed time Δt, and the theoretical model is graphed on the same scale. *Increasing* the fracture half length x_f of the model displaces its half unit slope straight lines toward *later times*, to the right of the plot (Equations 1.25 and 2.29).

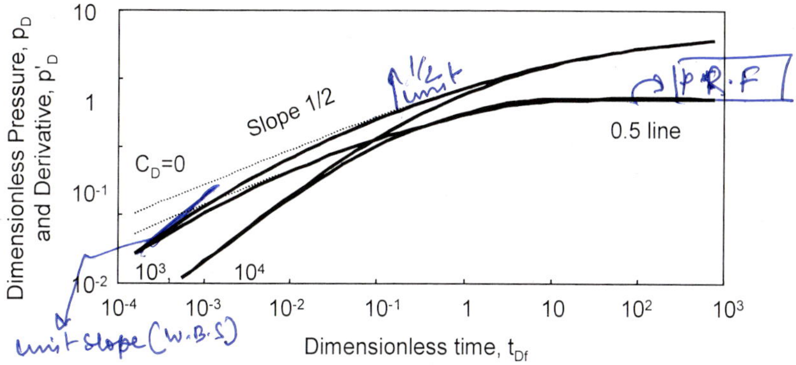

Figure 3.6. Responses for a fractured well with wellbore storage. Infinite conductivity fracture. Log-log scales, p_D versus t_{Df}. $C_D = 0$, 10^3, 10^4.

Effect of wellbore storage

On Figure 3.6, wellbore storage effect is introduced on infinite conductivity fracture responses (C_D of Equation 2.5 is used). For C_D values of 10^3 or above, the wellbore storage effect is indicated by a deviation below the half unit slope lines, before linear flow becomes evident. In case of high C_D, the wellbore storage effect masks the half unit slope pressure and derivative straight lines, the choice between a high or a low conductivity model (Section 3.3) is difficult, and x_f is not uniquely defined from early time data analysis.

Damaged fracture

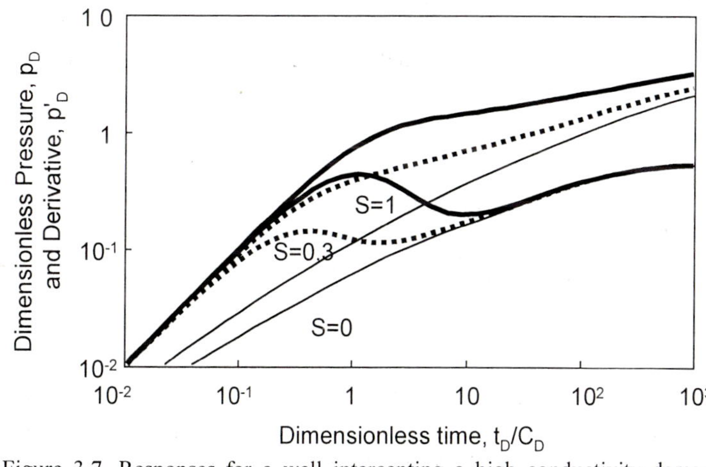

Figure 3.7. Responses for a well intercepting a high conductivity damaged fracture. Log-log scales, p_D versus t_D/C_D.

Two types of damaged fracture have been considered (Cinco-Ley and Samaniego-V, 1981). Either an infinitesimal skin is located *around the fracture* (when a zone of reduced permeability has been created around the fracture by fracturing fluid loss), or the damaged region is located *within the fracture* near the wellbore (this configuration is called choked fracture).

With damaged fractures, the duration of the wellbore storage effect is extended and the response follows a unit slope straight line at early time, as illustrated in Figure 3.7. Later, the derivative describes a hump until the sand face rate is fully established. Then, the reservoir response shows the linear, followed by the pseudo radial flow, characteristic derivative behaviors.

3.2.4 Associated specialized plot straight lines

The limits of the various regimes are identified from the straight-line portions of pressure and derivative curves of Figures 3.5 to 3.7. The first straight line, with unit slope, is usually not seen on non-damaged fractured well data: wellbore storage analysis is in most cases not justified.

Linear flow analysis

Linear flow specialized analysis (Section 1.2.4) is carried out on the data points matching the pressure and derivative type curves on the two half unit slope straight lines: a plot of the same points with Δp versus the square root of Δt exhibits a straight line passing through the *origin* (Figure 1.13). The slope m_{LF} is used to provide an estimate of the fracture half-length x_f. If k is known from radial flow analysis,

$$x_f = 4.06\sqrt{\frac{\mu}{\phi c_t k}\,\frac{qB}{h m_{LF}}} \tag{1.26}$$

When the first points of the well response are affected by the wellbore storage, the data plot does not follow linear flow behavior at very early time: it reaches the straight line only when all well bore storage has become negligible. On the specialized square root of time scale, both the data plot and the straight line of slope m_{LF} pass through the origin, but the data plot curves below the straight line until wellbore storage has died out.

In the case of a shut-in period when the previous drawdown was stopped during the linear flow regime, the Horner time for linear flow of Equation 2.19 can be used.

Pseudo-radial flow analysis

After linear flow, the response exhibits a pseudo radial flow behavior that displays the usual semi-log straight line behavior. Strictly speaking, flow lines to the fracture are not radial, and this regime is affected by the near wellbore condition as illustrated on Figure 3.8. Cinco (1982) reports the drainage becomes approximately circular when the radius is more than 3 times the fracture length.

In a vertical cylindrical well, the flow lines converge with a radial geometry until the fluid enters into the wellbore. With fractured well, the flow is distributed along the fracture length, and the density of flow lines is therefore reduced in the near wellbore region. By breaking the formation, the communication between the reservoir and the well is improved and less pressure loss occurs in this reservoir region: fractured wells exhibit a *negative geometrical skin* S_G during radial flow.

This geometrical skin S_G is related to the fracture half-length x_f (Gringarten et al., 1975 a). For the uniform flux solution, the geometrical skin S_{UFF} is:

$$x_f = 2.718 r_w e^{-S_{\mathrm{UFF}}} \tag{3.11}$$

For the infinite conductivity solution, S_{HKF} is expressed:

$$x_f = 2 r_w e^{-S_{\mathrm{HKF}}} \tag{3.12}$$

$x_f \Rightarrow 2 r_w e^{-S_{HKF}}$

Values of skin for fractured wells can be as low as –6 or –7.

In can be noted that, for an infinite conductivity fracture, the effective well radius r_{we} of Equation 1.14 is exactly one-fourth the total fracture length, the $C_D e^{2S}$ parameter is related to C_{Df} of Equation 3.10 with:

$$C_D e^{2S} = C_{De} = 4 C_{Df} \tag{3.13}$$

In Equation 3.13, C_{De} is the dimensionless wellbore storage calculated by using the effective wellbore radius instead of r_w in Equation 2.5.

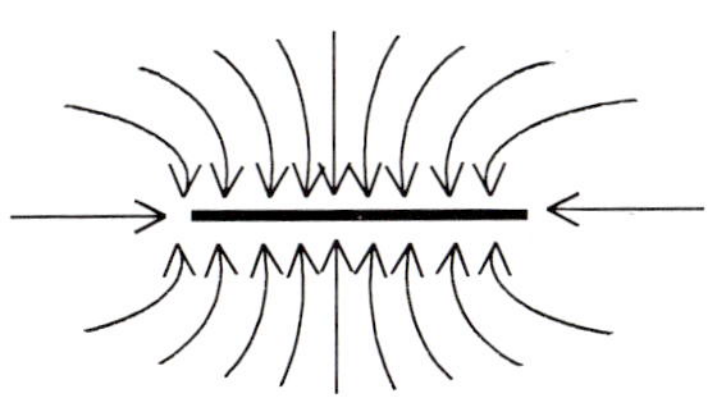

Figure 3.8. Flow line geometry near a fractured well.

3.3 FINITE CONDUCTIVITY VERTICAL FRACTURE

When the *pressure gradient* along the fracture length is not negligible, the low conductivity fracture model has to be used for the analysis of hydraulically fractured wells. This may happen for example when the permeability of the fracture is not very high compared to the permeability of the formation, especially when the fracture is long.

3.3.1 Model description

With the finite conductivity fracture model (Cinco-Ley et al., 1978 a), linear flow is produced within the fracture, in addition to the linear flow regime from the pay zone into the fracture plane. The fracture geometry is defined on Figure 1.14: the well intercepts a symmetrical vertical plane fracture of half length x_f, w_f is the width, k_f is the fracture permeability and $k_f w_f$ is the fracture conductivity.

Characteristic flow regimes

Three characteristic regimes (Cinco-Ley. and Samaniego-V., 1978 b) can be observed after the wellbore storage effect:

1. At early times, as long as the fracture tips have not been reached, the combination of fracture linear flow and reservoir linear flow produce the so-called *bi-linear flow* regime. The pressure change is then proportional to the fourth root of the elapsed time $\sqrt[4]{\Delta t}$ and, on the log-log plot, both the pressure and derivative responses follow a quarter unit slope straight line (Figure 2.17). When present, the bi-linear flow regime gives access to the fracture conductivity $k_f w_f$ (the wellbore pressure is *independent* of the fracture half-length x_f during bi-linear flow).
2. Later, the pressure behavior becomes equivalent to that of an infinite conductivity fractured well. A *linear flow* regime can be observed, characterized by the usual pressure and derivative half unit slope log-log straight lines. The fracture half-length x_f can be estimated.
3. *Pseudo-radial flow* regime, with the derivative stabilization is observed next, to give the permeability thickness product kh and the geometrical skin S_G.

Analytical solutions

Cinco-Ley et al. (1978 a) used a semi-analytical approach to derive low conductivity fractured well responses. They broke up the fracture into several uniform flux segments, solve a set of equations to generate the flux profile along the complete fracture, and finally estimate the wellbore pressure.

Several extensions of the basic finite conductivity fractured well model have been considered: wellbore storage effect, skin damage on the fracture walls or choked fracture (Cinco-Ley and Samaniego-V, 1981), variable conductivity and length (Bennet et al., 1983), etc.

3.3.2 Review of pressure and derivative type curves for a well with finite-conductivity fracture

Cinco-Ley et al. (1978 a) presented a set of type curves with the dimensionless pressure p_D expressed as a function of the dimensionless time t_{Df} of equation 3.8 for different values of the dimensionless fracture conductivity $k_{fD}w_{fD}$, expressed as:

$$k_{fD}w_{fD} = \frac{k_f w_f}{k x_f} \tag{3.14}$$

Equation 3.14 shows that $k_{fD}w_{fD}$ is directly proportional to the fracture conductivity and inversely proportional to the reservoir permeability and fracture half-length. It can be concluded that the effect of low conductivity in the fracture is magnified when the reservoir permeability is high, or the fracture long (large $k\,x_f$ product). The authors indicate that the infinite conductivity assumption is valid when the dimensionless fracture conductivity $k_{fD}w_{fD}$ is greater than 300. For lower values, the wellbore pressure is affected by the fracture conductivity and, as a result, the geometrical skin S_{LKF} is less negative than for an infinite conductivity fracture.

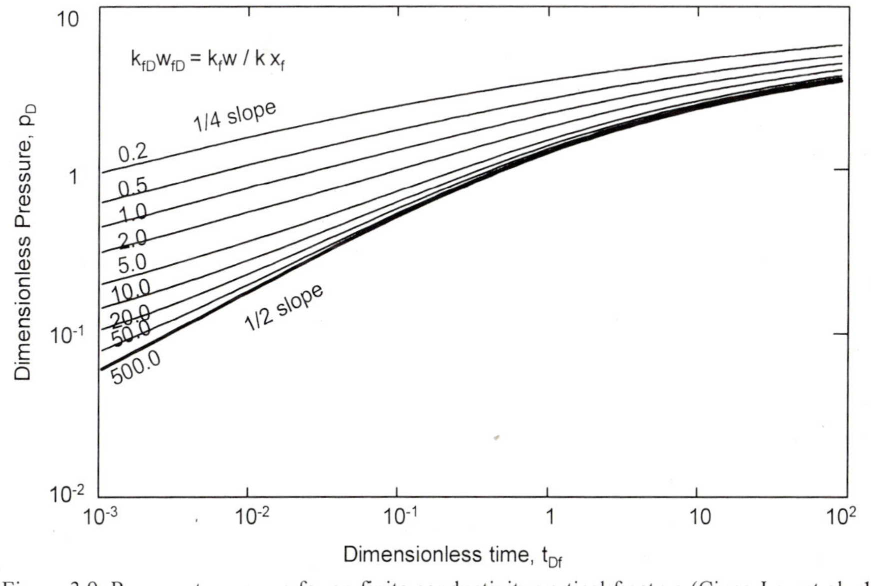

Figure 3.9. Pressure type curve for an finite conductivity vertical fracture (Cinco-Ley et al., 1978 a). Log-log scales, p_D versus t_{Df}. Graph courtesy A.C. Gringarten.

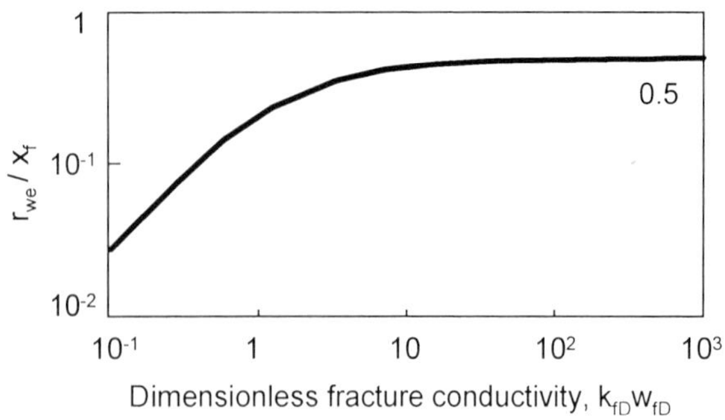

Figure 3.10. Effective wellbore radius for a well with a finite conductivity fracture. Log-log scales.

Other type curves presentation have been proposed, including wellbore storage and fracture skin effects. The derivative of finite conductivity fractured wells with wellbore storage has been considered by Wong et al. (1985).

3.3.3 Skin discussion

For a finite conductivity fracture, the skin is defined by two terms (Cinco-Ley and Samaniego-V, 1981): the *geometrical skin* S_{HKF} assuming an infinite conductivity fracture (Equation 3.12), and a *correction parameter* G to account for the pressure losses resulting from the low fracture conductivity.

$$S_{LKF} = G\left(\frac{k_f w_f}{k x_f}\right) + \ln\frac{2r_w}{x_f} \tag{3.15}$$

On Figure 3.10, the ratio of the effective well radius of Equation 1.14 ($r_{we} = r_w e^{-S_{LKF}}$) to the fracture half-length x_f, is presented on log-log scales as a function of the dimensionless fracture conductivity $k_f w_f / k x_f$. When $k_f w_f / k x_f$ is greater than 300, the fracture behaves as an infinite fracture, $S_{LKF} = S_{HKF}$ and $r_{we}/x_f = 0.5$. For lower conductivities, the skin is less negative and r_{we}/x_f is smaller.

3.3.4 Matching procedure on pressure and derivative responses

On Figure 3.11, an example of a low conductivity fracture response is presented with p_D versus t_D/C_D. The wellbore storage effect is not visible on this example, and three subsequent flow regimes can be identified:

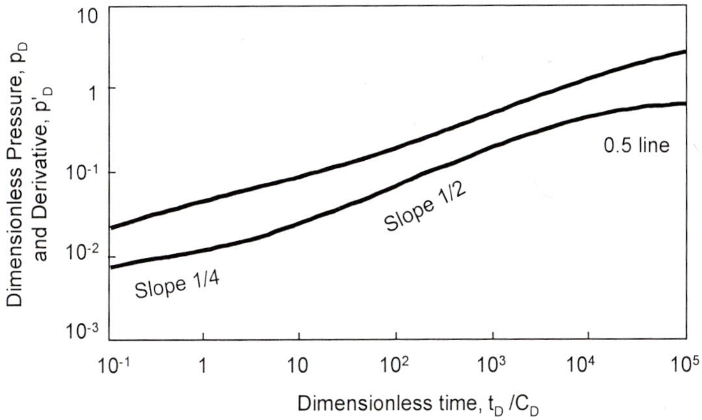

Figure 3.11. Response for a well intercepting a finite conductivity fracture. Log-log scales, p_D versus t_D/C_D. No fracture skin, $k_{fD}w_{fD} = 25$.

1. At early times, during bilinear flow, pressure and derivative curves follow two *parallel straight lines of slope* 1/4.
2. During the linear flow regime, two other *parallel straight lines of slope* 1/2 are evident.
3. When radial flow is reached, the *derivative stabilizes on* 0.5 (in dimensionless scales).

The distance between the two quarter-unit slope straight lines is log (4), and the distance between two half-unit slope lines is log (2).

For *large* fracture conductivity $k_{fD}w_{fD}$, the bilinear flow regime is short lived and the 1/4 slope pressure and derivative straight lines are moved *downwards*. The behavior tends to a high conductivity fracture response.

Conversely, when the dimensionless fracture conductivity is low (curve $k_{fD}w_{fD}$ =1 on Figure 3.12), the linear flow regime is not present and the response changes directly from bi-linear flow to the pseudo radial flow regime, through a transition that never describes the half unit slope line. In such configuration, the pressure loss in the fracture is large, and two segments of the fracture near the *tips* are not participating to the flow.

With real test data (Δp, $\Delta p'$ vs. Δt), when all flow regimes are clearly defined, the match against a low conductivity fracture model such as on Figure 3-11 provides the *kh* product from the pressure match, the fracture half-length x_f and the fracture conductivity k_fw_f from the location of the half unit and quarter unit slope derivative straight lines respectively.

The example response of Figure 3.11 is displayed over 6 time log-cycles (t_D/C_D from 10^{-1} to 10^5). Frequently, log-log plots of actual build-up data describe the response during a smaller time range, and the match is performed only on a fraction of the complete model response. Matching is then difficult and the solution non-unique:

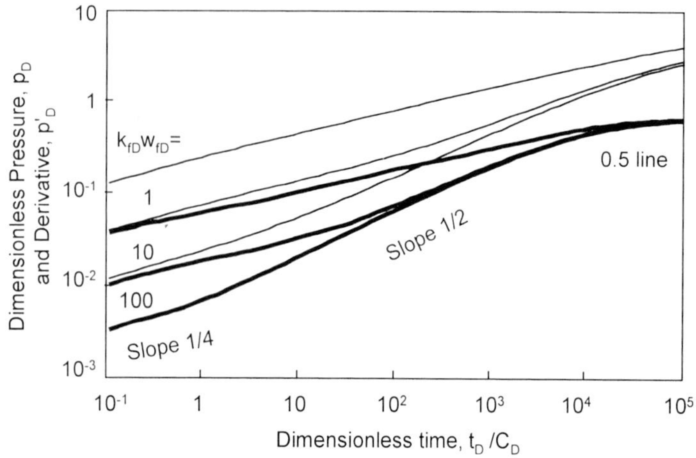

Figure 3.12. Response for a well intercepting a finite conductivity fracture. Log-log scales, p_D versus t_D/C_D. No fracture skin, $k_{fD}w_{fD}$ = 1, 10 and 100.

- With some fractured well responses, it may take several months to reach the start of the pseudo radial flow regime but in practice the data do not go beyond the bilinear or linear flow period.
- Depending upon the wellbore storage and fracture parameters, the different regimes can overlap, and some of them are not shown clearly. When the response is similar to the infinite fracture example with wellbore storage of Figures 3.6 or 3.7, the choice of the type of fracture response, and the resulting fracture parameters, are not uniquely defined.

When a long fracture is planned, it is recommended to test the well before fracturing, in order to obtain an estimate of the permeability thickness product. After fracturing, this parameter may not be defined by transient pressure analysis.

3.3.5 Associated specialized plot straight lines

The three regimes described on the example response of Figure 3.11 can be analyzed by straight-line methods. The time limits of the specialized analysis straight lines are defined by the type curve match, using in particular the improved definition of the derivative presentation. At early time, a fourth regime, corresponding to wellbore storage, can possibly be present.

The three or four typical regimes cannot be expected to be present in a single response, only one or two are in general developed enough for specialized analysis. In some cases, long responses only present transitional behaviors from one regime to the next, and no straight-line analysis can be performed.

Bilinear flow analysis

The quarter unit slope log-log straight line corresponds to a plot of Δp versus the fourth root of the elapsed time Δt (Section 1.2.5): the bilinear flow regime is characterized by a straight line passing through the *origin* (Figure 1.15). When the formation permeability is known, the slope m_{BLF} gives the fracture permeability-width product from Equation 1.28:

$$k_f w_f = 1944.8\sqrt{\frac{1}{\phi\mu c_t k}}\left(\frac{qB\mu}{hm_{\mathrm{BLF}}}\right)^2 \qquad (1.28)$$

Linear and Pseudo-radial flow analyses

Specialized analyses of linear and pseudo-radial flow are performed as described in the previous Section 3.2, for a high conductivity fracture. On a Δp, $\sqrt{\Delta t}$ scale, the straight line passing through the origin is used to estimate kx_f^2 (Equation 1.26). When radial flow has been reached after the initial fracture flow regimes, the semi-log straight line can be used to estimate the permeability thickness product kh and the n*egative geometrical skin* S_{LKF} of Equation 3.15.

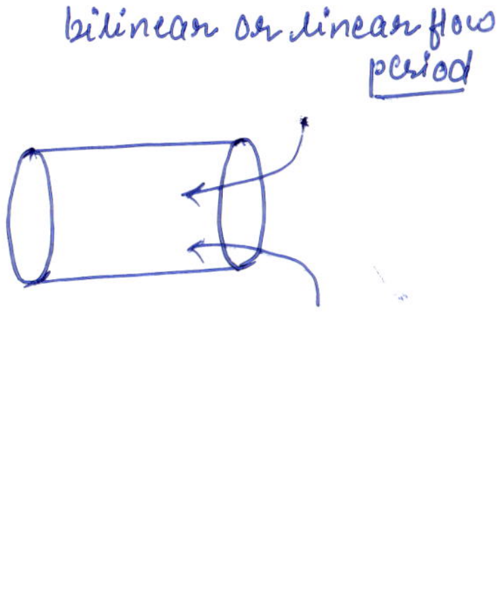

3.3.6 Flux distribution along the fracture

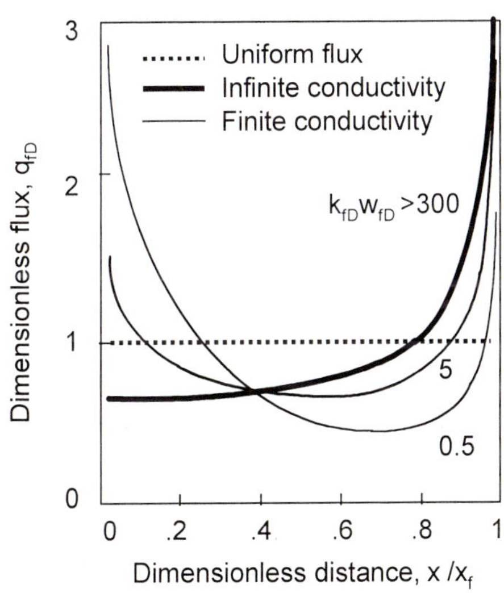

Figure 3.13. Stabilized flux distribution. Uniform flux, Infinite conductivity ($k_{fD}w_{fD}$ > 300) and Finite conductivity fracture ($k_{fD}w_{fD}$ = 0.5 and 5) models.

When the pseudo radial flow regime is reached, the distribution of the flux entering the fracture stabilizes. Figure 3.13 (Cinco-Ley and Samaniego-V, 1981) presents the dimensionless flux distribution for the uniform flux, infinite conductivity and finite conductivity fracture models. The flux per unit of fracture length being $q(x, \Delta t)$,

$$q_{fD} = 2\frac{q(x,\Delta t)x_f}{q_w} \tag{3.16}$$

With the uniform flux model, $q_{fD} = 1$. In case of infinite conductivity fracture ($k_{fD}w_{fD}$ >300), the fluid enters the fracture mostly in the region near the tips. When the fracture conductivity is decreased, the fracture section near the wellbore becomes more productive.

3.3.7 Field example

Figure 3.14 presents the pressure and derivative responses of in a well intercepting a low conductivity fracture. During this two day build-up test, only the bilinear flow regime is evident with a long quarter unit slope straight-line. At very early time, the data is apparently affected by the end of wellbore storage but, on the late time data, no transition towards the linear or the radial flow regime can be identified.

The test is too short for estimating the reservoir permeability and only the group $\sqrt{k}\,(k_f w_f)$ can be accessed from the analysis of the bilinear flow regime (with Equation 1.28).

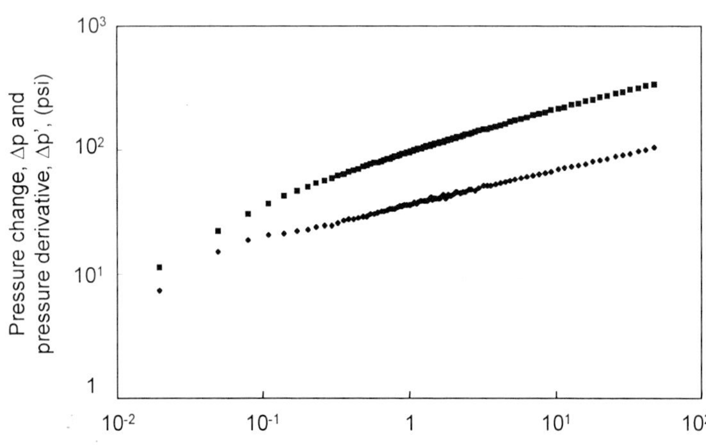

Figure 3.14. Build-up test in a well intercepting a low conductivity fracture. Log-log scales.

3.4 WELL IN PARTIAL PENETRATION

In the case of limited entry or partial penetration, the well communicates with only a fraction of the producing zone thickness. This could be due to plugged perforations for example, but partial penetration is also sometimes decided on for production efficiency. This happens when a gas cap or a bottom water drive is active at the upper or lower boundary, the well being perforated as far away as possible from the constant pressure surface. A discontinuity in the upper or lower sealing boundaries, which create communications between the open interval and another zone, also produces partial penetration behavior. This latter configuration can be analyzed with the two-permeability solution (Chapter 4).

Partial penetration corresponds to a reduction of the surface of contact between the well and the reservoir, as opposed to the other wellbore conditions discussed in this chapter, such as fractured wells, slanted and horizontal wells. They all increase this contact, and are characterized by a negative geometrical skin. Partial penetration effects produce a positive geometrical skin, resulting from the distortion of the flow lines when converging towards the perforated interval (see Figure 1.16).

3.4.1 Model description

A schematic of a well with limited entry is shown in Figure 3.15. The interval open to flow has a thickness h_w, which is a fraction of the reservoir thickness h. The center of the open interval is at a distance z_w from the lower reservoir boundary. The permeability is k_H in the horizontal direction and k_V in the vertical direction.

Characteristic flow regimes

When the lower and upper boundaries are impermeable such as on the flow diagram of Figure 1.16, three characteristic regimes can be observed after the wellbore storage early time effect:

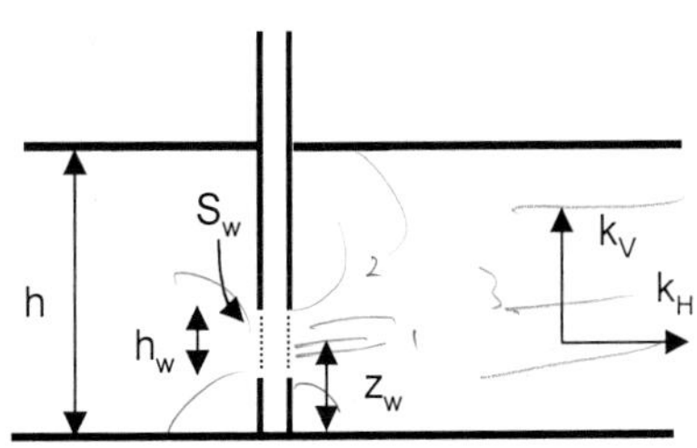

Figure 3.15. Geometry of a partially penetrating well.

1. *Radial flow* over the open interval h_w, with Δp proportional to $\log(\Delta t)$ and a first derivative plateau. Analysis of the initial radial flow regime yields the permeability-thickness product for the open interval $k_H\, h_w$, and the infinitesimal skin of the well, S_w.
2. *Spherical flow* with Δp proportional to $\Delta t^{-1/2}$ and a negative half unit slope straight line on the derivative log-log curve (Figure 2.18). The spherical flow regime lasts until the lower and upper boundaries are reached. Analysis yields the permeability anisotropy k_V / k_H (Section 1.3.6).
3. *Radial flow* over the entire reservoir thickness with Δp proportional to $\log(\Delta t)$ and a second derivative stabilization. The reservoir permeability-thickness product $k_H\, h$, and the total skin S_T can be estimated from the second radial flow regime

If the top or bottom boundary is a constant pressure interface, the pressure stabilizes and the derivative drops after the spherical flow regime.

Analytical solutions

The partial penetration solution for uniform flux and infinite conductivity wellbore was presented by Gringarten and Ramey (1975 b), using the same approach as for the fractured well models (see Section 3.2). When there is no pressure loss in the wellbore, the infinite conductivity well is obtained from the uniform flux partial penetration solution by computing the wellbore pressure at an effective point located at 0.732 of the half-length of the producing interval.

3.4.2 Model responses, sealing upper and lower limits

Due to the lack of distinguishable features on pressure responses, no type curves for wells in partial penetration are currently used, although some have been presented in the literature (Kuchuk and Kirwan, 1987). The derivative is firstly used for the analysis of limited entry wells, and a computer program is required for generating and matching the model.

Influence of k_V / k_H

Typical responses of partial penetration in a reservoir with sealing upper and lower limits are presented in Figures 3.16 with the usual p_D versus t_D/C_D dimensionless variables. Only 20% of the thickness is communicating with the well and the producing segment has no skin and is centered in the formation. Three permeability anisotropy are considered with $k_V / k_H = 10^{-1}$, 10^{-2} and 10^{-3}.

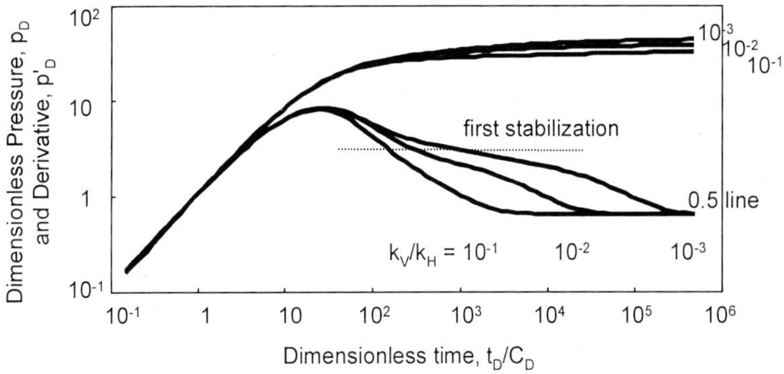

Figure 3.16. Responses for a well in partial penetration with wellbore storage. Log-log scales, p_D versus t_D/C_D. $h_w/h = 1/5$ in center of the interval, $C_D = 33$, S_w=0, $k_V/k_H = 0.1$, 0.01 and 0.001.

During the final radial flow regime for the complete interval thickness, the derivative stabilizes at 0.5 whereas it stabilizes at 0.5 h/h_w during the first radial flow regime in front of the perforated segment. The negative half-unit slope straight-line transition between the two derivative plateaus defines the quality of the vertical communication in the reservoir. With *low* k_V/k_H, vertical flow is limited and the spherical flow regime is seen *late*. On the example $k_V/k_H = 10^{-3}$, the first stabilization is well developed before the subsequent negative half-unit slope line. More frequently, partially penetrating well responses are similar to curve $k_V/k_H = 10^{-1}$, where the first stabilization does not exist and spherical flow is reached as soon as the wellbore storage effect ends.

With low vertical permeability, a significant pressure drop is created when the flow lines converge towards the producing segment, generating a large positive geometrical skin S_{PP}. The total skin S_T, measured during the final radial flow regime, combines the well infinitesimal skin S_w, and the additional completion skin S_{PP} with :

$$S_T = \frac{h}{h_w} S_w + S_{PP} \tag{3.17}$$

On Figures 3-16, the largest geometrical skin S_{PP} corresponds to $k_V/k_H = 10^{-3}$: this pressure curve is above that for $k_V/k_H = 10^{-1}$ during the final radial flow. The different skin factors for limited entry wells are further discussed in Section 3.4.3.

Influence of z_w/h

If the perforated segment is not centered in the producing formation, the spherical flow regime ends when the closest upper or lower boundary is reached. A hemi-spherical flow geometry is then developed, until the second boundary is seen.

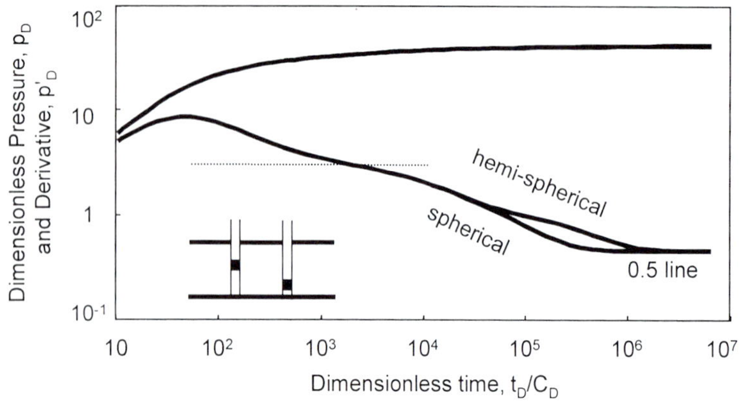

Figure 3.17. Responses for a well in partial penetration with wellbore storage. Log-log scales, p_D versus t_D/C_D. $h_w/h = 1/10$, $C_D = 6$, $S_w=0$, $k_V/k_H = 0.005$, $z_w/h = 0.5$ and 0.2.

As shown on Figure 3.17, the derivative also follows a negative half-unit slope straight line during the hemi-spherical flow regime, but displaced above the first corresponding to spherical flow. A similar behavior has been discussed for radial flow in Chapter 2 with the sealing fault example in Figure 2.20. In the case of spherical geometry, the hemi-spherical flow regime shows the characteristic behavior of spherical flow, but the apparent permeability is half the true k_S of Equation 1.29.

The example response of Figure 3.17 with a centered perforated segment ($z_w/h = 0.5$) is used in the Section 9.4.2, for the vertical interference test discussion.

3.4.3 Skin discussion

Several methods have been proposed to estimate the geometrical skin S_{PP} from the system parameters. Gringarten and Ramey (1974 b) and Streltsova (1979) use infinite series, Brons and Marting (1961 a) presented several graphs, and Odeh proposed an empirical equation (1980).

In 1987, Papatzacos derived a formula, using the penetration ratio h_w/h, the dimensionless reservoir thickness-anisotropy group h_D, and the distance z_w from the center of the open interval to the lower or upper boundary:

$$S_{pp} = \left(\frac{h}{h_w} - 1\right)\ln\left(\frac{\pi h_D}{2}\right) + \frac{h}{h_w}\ln\left[\frac{\frac{h_w}{h}}{2 + \frac{h_w}{h}}\sqrt{\frac{(z_w + h_w/4)(h - z_w + h_w/4)}{(z_w - h_w/4)(h - z_w - h_w/4)}}\right] \tag{3.18}$$

where

$$h_D = \frac{h}{r_w}\sqrt{\frac{k_H}{k_V}} \tag{3.19}$$

In 1991, Vrbik proposed another approximation:

$$S_{\text{pp}} = \left(\frac{h}{h_w} - 1\right)\left(1.2704 - \ln h_D\right) - \left(\frac{h}{h_w}\right)^2 \left[f(0) - f\left(\frac{h_w}{h}\right) + f\left(2\frac{z_w}{h}\right) - 0.5 f\left(\frac{2z_w + h_w}{h}\right) - 0.5 f\left(\frac{2z_w - h_w}{h}\right) \right] \tag{3.20}$$

where

$$f(y) = y \ln y + (2 - y)\ln(2 - y) + \frac{1}{\pi h_D} \ln\left[\sin^2(\pi y/2) + 0.1053/h_D^2\right] \tag{3.21}$$

Equations 3.18 and 3.20 provide a good approximation of S_{PP} for typical partial penetration well configurations (Papatzacos indicates that h_w/h and $(h_w/r_w)\sqrt{k_H/k_V}$ should not be very small). Ding and Reynolds (1994) extended the equations 3.18 and 3.20 to a multi-layer case, and compared the resulting skin approximations to the skin estimated with a finite-difference simulator. They confirmed the limits of validity of the approximations, but they concluded that Equation 3.18 seems to give accurate results for a wider range of parameters.

In the following tables, the geometrical skin S_{PP} is estimated for a reservoir of thickness h =1000r_w. Two well locations are envisaged and different penetration ratios and permeability anisotropies are considered. The tables present firstly results from Equation 3.18 and secondly results from Equation 3.20.

Table 3.2. Geometrical skin S_{PP} for a centered partial penetration well (z_w/h=0.5).

k_V / k_H	1	10^{-1}	10^{-2}	10^{-3}
h_w/h =0.1	36.8-35.3	47.1-45.4	57.5-55.7	67.9-66.0
h_w/h =0.25	14.3-13.6	17.7-17.0	21.2-20.4	24.7-23.9
h_w/h =0.5	5.2-4.7	6.3-5.9	7.5-7.0	8.6-8.2

Table 3.3. Geometrical skin S_{PP} for an off-centered partial penetration well (the perforated interval is on top or at bottom).

k_V / k_H	1	10^{-1}	10^{-2}	10^{-3}
h_w/h =0.1	41.5-41.3	51.9-51.5	62.3-61.8	72.6-72.2
h_w/h =0.25	15.8-15.6	19.2-19.0	22.7-22.4	26.1-25.9
h_w/h =0.5	5.6-5.4	6.7-6.6	7.9-7.7	9.0-8.9

The two tables show that, even with a small penetration ratio, the geometrical skin S_{PP} is seldom larger than 30 or 50. If the producing segment is in addition damaged, the product $(h/h_w)S_w$ of Equation. 3.17 can be very large, and the total skin S_T can reach values of several hundreds. On limited entry wells, wellbore damage is *amplified*. Removing a wellbore damage S_w, by acid stimulation, can significantly increase the productivity of a limited entry well.

For a fully penetrating damaged well, S could be as high as 30, but not significantly above. If a well shows a skin value higher than 30, partial penetration should be envisaged even if the derivative response does not display the characteristic behavior of a limited entry well.

3.4.4 Matching procedure on pressure and derivative responses

When matching test data from a partially penetrating well, reservoir and well parameters can be estimated provided the three characteristic flow regimes are clearly defined. The pressure match, adjusted with the final derivative stabilization during the radial flow regime over the complete thickness, is used to determine the permeability thickness product $k_H\,h$ and the total skin S_T. The wellbore skin S_w and the penetration ratio h_w/h are estimated from the first radial flow (derivative plateau at 0.5 h/h_w in dimensionless terms). Once the $k_H\,h$ product is fixed, *reducing* h_w/h moves *upwards* the first derivative plateau $\Delta p_{\text{1st stab}}$:

$$\frac{h_w}{h} = \frac{\Delta p_{\text{2nd stab.}}}{\Delta p_{\text{1st stab.}}} \tag{3.22}$$

The permeability anisotropy k_V/k_H and location of the open interval are estimated from the spherical flow match on the -1/2 slope straight-line. *Decreasing* k_V/k_H displaces the negative half unit slope straight lines toward *late times*, on the right of the plot.

In practice, partial penetration responses rarely exhibit the three individual flow regimes.

- The first radial flow is often masked by wellbore storage effect. In such cases, h_w/h and S_w. are not uniquely defined.
- The transition does not always follow a pure spherical flow behavior. It can for example start in spherical geometry, and change to a hemi-spherical regime when the well is not centered in the zone thickness and one of the sealing horizontal limits is reached before the other. The definition of z_w/h from the spherical flow transition between the two derivative plateaus is in many cases approximate.

Discussion of the match results

When the relationship Equation 3.22 is used to estimate the penetration ratio h_w/h, a constant permeability is assumed along the reservoir thickness. If the permeability varies with the depth, the apparent perforated interval length h_w estimated from the partial penetration match is effectively longer or shorter than the real producing segment when the well is perforated in front of a high (low) permeability interval. A layered reservoir model can be used to refine the analysis (Section 4.2.3).

Similarly, the permeability anisotropy k_V/k_H estimated from the derivative match is the ratio of the average vertical permeability to the average horizontal permeability. In the vertical direction, permeability is averaged with a harmonic mean (see discussion of Equation 3.56, Section 3.6.11). In case of low vertical permeability streaks in the formation, the vertical permeability resulting from the match can be significantly lower than the average vertical permeability estimated from core analysis. Values of k_V/k_H as low as 10^{-3} are not exceptional in transient test analysis results.

3.4.5 Associated specialized plot straight lines

Associated specialized plot straight lines

After the usual wellbore storage regime, each of the three characteristic regimes of a limited entry well can be analyzed by straight-line methods, provided the derivative data plot indicates that they are well defined and not dominated by transitional behaviors from one regime to next. The following Figures 3.18 and 3.19 are the radial flow and spherical flow specialized plots of the three examples Figure 3.16.

Radial flow analysis

The two radial flow regimes can be analyzed on the usual semi-log scales. The first radial flow relative to the open interval is frequently poorly defined as shown by the examples in Figure 3.18 and semi-log analysis is mostly made on late time data, during the radial flow regime over the complete reservoir thickness.

The analysis of the first semi-log straight line provides an estimate of the permeability thickness product $k_H h_w$ and the wellbore skin coefficient S_w. For shut-in periods, this straight line extrapolates to a value higher than the initial pressure. The true extrapolated build-up pressure p^* is estimated from the second semi-log straight line, which also defines the permeability-thickness product $k_H h$ for the complete reservoir thickness and the total skin S_T of Equation 3.17.

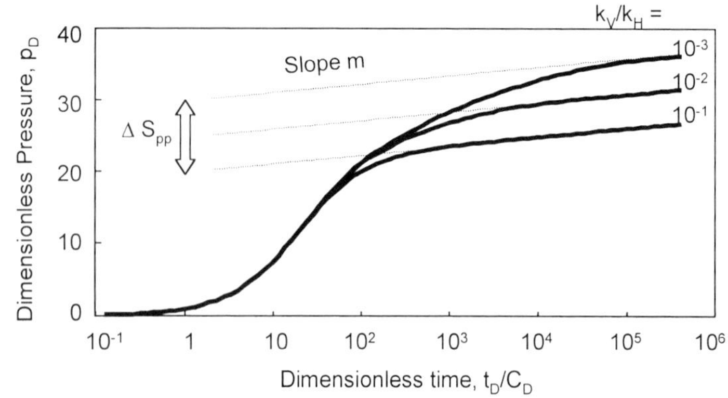

Figure 3.18. Semi-log plot of Figure 3.16 examples. Influence of k_V/k_H on S_{PP} (S_w=0).

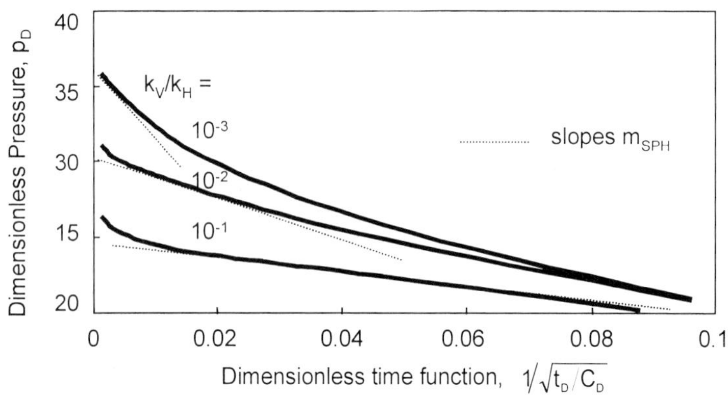

Figure 3.19. Spherical flow analysis of Figure 3.16 examples. $1/\sqrt{\Delta t}$ plot.

Spherical flow analysis

During spherical flow, the pressure is proportional to the inverse of the square root of time. In some cases, the $1/\sqrt{\Delta t}$ time scale of the specialized plot significantly compresses the data, and the spherical flow straight line can be difficult to identify, as illustrated in Figure 3.19. With the curve $k_V/k_H = 10^3$ for example, the derivative curve shows that the spherical flow regime is established between t_D/C_D =2*10^4 and 2*10^5. The limits of the straight line are therefore $1/\sqrt{t_D/C_D}$ =0.002 and 0.007. With the curve $k_V/k_H = 10^1$, the straight line, between $1/\sqrt{t_D/C_D}$ =0.02 and 0.07, is easier to identify.

When the open interval is in the middle of the formation, the slope m_{SPH} of the spherical flow straight line gives the permeability anisotropy from Equations 1.31 and 1.32. If the open interval is close to the top or bottom-sealing boundary, flow is hemi-spherical and the slope m_{SPH} must be divided by two in Equation 1.31. The intercept of the spherical flow straight line with the pressure axis is not used.

3.4.6 Influence of the number of open segments

When the open interval is distributed in several segments, the ability of vertical flow is improved compared to the single segment partially penetrating the well with the same h_w. On the examples Figure 3.20 with 1, 2 and 4 segments, the negative half unit slope straight line is displaced towards *early time* when the number of segments is *increased.* The geometrical skin of the single segment curve is S_{PP} =17.9 (18.5 and 17.7 with Equations 3.18 and 3.20), but it is respectively 15.9 and 13.9 with 2 and 4 segments.

3.4.7 Constant pressure upper or lower limit

In the example Figure 3.21, the bottom boundary, corresponding to a water / oil contact, behaves like a constant pressure surface. No final radial flow regime develops after the spherical flow regime, the pressure stabilizes and the derivative drops (Abbaszadeh and Hegeman, 1988).

The effect of a gas cap or a bottom water drive on well responses is further discussed in the boundaries Chapter, Section 5.10.3.

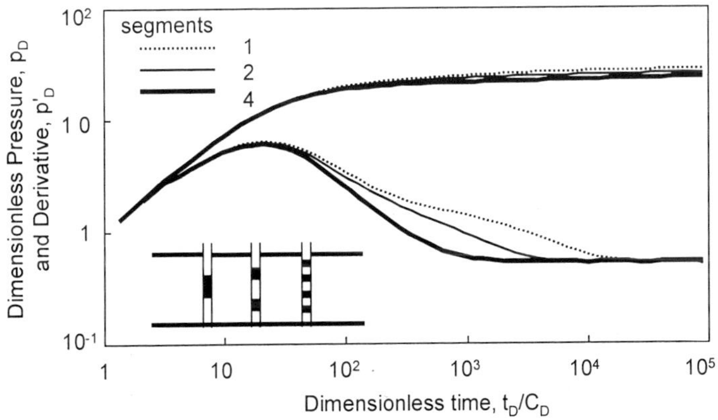

Figure 3.20. Responses for a well in partial penetration with wellbore storage. Log-log scales, p_D versus t_D/C_D. h_w/h = 1/4, one, two or four segments. C_D = 100, S_w=0, k_V/k_H = 0.01, one segment centered, two or four segments uniformly distributed in the interval.

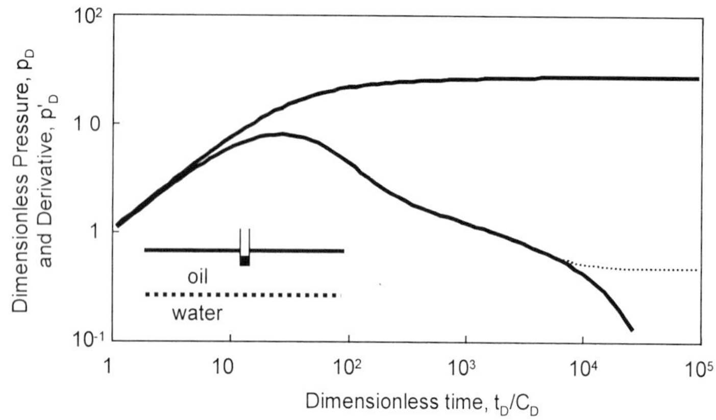

Figure 3.21. Responses for a well in partial penetration with a bottom constant pressure boundary. Log-log scales, p_D versus t_D/C_D. $h_w/h = 1/5$, $C_D = 1000$, $S_w=0$, $k_V/k_H = 0.005$, one segment on top. The dotted derivative curve describes the response with sealing upper and lower boundaries.

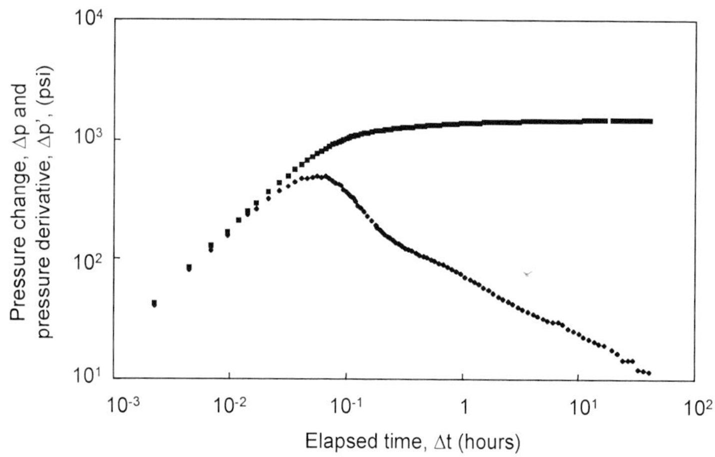

Figure 3.22. Build-up test in a partially penetrating well. Log-log scales.

3.4.8 Field examples

In the following section, two partial penetration build-up examples are briefly discussed. For the example shown in Figure 3.22, the well is shut-in at surface. The response describes a long wellbore storage unit slope straight-line, followed by the characteristic derivative hump, and by a negative half unit slope straight line. No derivative stabilization is seen, neither of the two radial flow regimes is present on the well response.

The spherical flow regime starts after 15 minutes of shut-in, and is not finished at the end of the two days build-up test. During the long spherical flow regime, the derivative drops from 10^2 to 10 psi. Clearly, the penetration ratio h_w/h must be less than 10^{-1}. The horizontal and vertical permeability cannot be estimated from this test data, only the spherical permeability of Equations 1.29 and 1.30 can be evaluated.

For the 10 hours build-up example of Figure 3.23 the well was shut-in down hole. The wellbore storage effect ends after one minute, a long declining derivative transition follows immediately, and the final stabilization for radial flow is reached at the end of the test response. The shape of the derivative curve suggests a partial penetration behavior but, it does not follows the negative half unit slope straight line. The response is in transition between the two theoretical derivative stabilizations (the first is masked by the wellbore storage hump). The penetration ratio is large ($h_w/h \approx 0.3$).

With this test, the vertical and horizontal permeability are defined, only the perforated interval thickness h_w and the wellbore skin S_w are approximate.

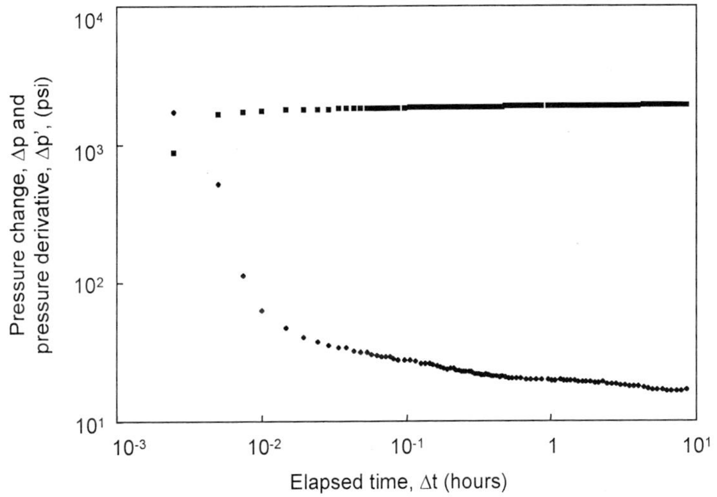

Figure 3.23. Build-up test in a partially penetrating well. Log-log scales.

3.5 SLANTED WELL

With directional drilling, many wells are not vertical when the formation is reached. In the following section, we discuss the influence of the angle of slanted wells on pressure behavior. We assume that the reservoir is homogeneous, the slanted well penetrates and is perforated over the full formation thickness. The well deviation is defined by the angle θ with respect to the vertical plane.

3.5.1 Behavior

Two radial flow regimes can be expected on slanted well responses: the first, at early times, is in the plane normal to the slanted well. The second, after a transition period, corresponds to horizontal radial flow in the producing zone. In practice, except when the angle θ is very large, the first regime is masked by wellbore storage effects and the well behavior is similar to the response of a vertical well. In case of a fully penetrating slanted well, the surface area exposed to flow is increased compared to a vertical well, thus the slant produces an apparent geometrical negative skin. For very large angles, the response tends towards a horizontal well response (Abbaszadeh and Hegeman, 1988).

3.5.2 Skin discussion

Several components are contributing to the total skin factor S_T estimated during the horizontal radial flow regime: the wellbore skin S_w, the negative skin due to anisotropy between vertical and horizontal permeability S_{ani} (see discussion of permeability anisotropy Section 3.1.5), and the geometrical skin effect S_θ.

Cinco-Ley et al. (1975) give an approximated equation for the geometrical skin S_θ:

$$S_\theta = -\left(\frac{\theta_w'}{41}\right)^{2.06} - \left(\frac{\theta_w'}{56}\right)^{1.865} \log \frac{h}{100\, r_w}\sqrt{\frac{k_H}{k_V}} \quad (3.23)$$

where θ_w' is an equivalent angle, for the transformed isotropic system :

$$\theta_w' = \tan^{-1}\left(\sqrt{\frac{k_V}{k_H}}\tan\theta\right) \quad (3.24)$$

Equation 3.23 shows that the larger the thickness, the more negative is the geometrical skin. The authors report that the approximation is valid when $0° \le \theta_w' \le 75°$.

After transformation of the vertical distances to correct the permeability anisotropy, Abbaszadeh and Hegeman (1988) express the skin S_{ani} describing the elliptical wellbore as:

$$S_{ani} = -\ln \frac{1 + 1/\sqrt{\cos^2\theta_w + (k_V/k_H)\sin^2\theta_w}}{2} \quad (3.25)$$

Pucknell and Clifford (1991) define the total skin factor S_T :

$$S_T = \frac{\cos\theta_w}{\sqrt{\cos^2\theta_w + (k_V/k_H)\sin^2\theta_w}}\left(S_w + S_{ani}\right) + S_\theta \tag{3.26}$$

For usual angles, the skin S_θ is not less than -2 or -3. For very large angles, the response tends towards the horizontal well response, and S_θ can be lower. When the vertical permeability k_V is low compared to k_H, θ_w' is small and the geometrical skin S_θ becomes negligible. In such cases, the effect of the anisotropy is more pronounced and S_{ani} can be more negative than S_θ.

In the following tables, geometrical skin S_θ and S_{ani} are estimated in a reservoir of thickness $h = 1000\ r_w$.

Table 3.4. Geometrical skin S_θ

k_V / k_H	1	10^{-1}	10^{-2}	10^{-3}
θ=30°	-0.8	-0.1	0	0
θ=60°	-3.3	-0.9	-0.1	0

Table 3.5. Anisotropy skin S_{ani}

k_V / k_H	1	10^{-1}	10^{-2}	10^{-3}
θ=30°	0	0	-0.1	-0.1
θ=60°	0	-0.3	-0.4	-0.4

3.5.3 Associated specialized plot straight lines

In theory, the two radial flow regimes can be analyzed using semi-log straight line techniques. The first defines the average permeability in the plane normal to the well, multiplied by the well penetration length. In practice, only the second regime, corresponding to horizontal flow from the producing interval, is seen. Semi-log analysis yields the permeability thickness product $k_H\ h$ of the producing zone and the total skin factor S_T.

3.6 HORIZONTAL WELL

Advances in drilling and completion technologies have placed horizontal wells among the techniques used to improve production performance. For example in the case of gas cap or bottom water drive, horizontal wells prevent coning without introducing the flow restriction seen in partial penetration wells. Horizontal drilling is also efficient to increase the well surface area for fluid withdrawal, thus improving the productivity.

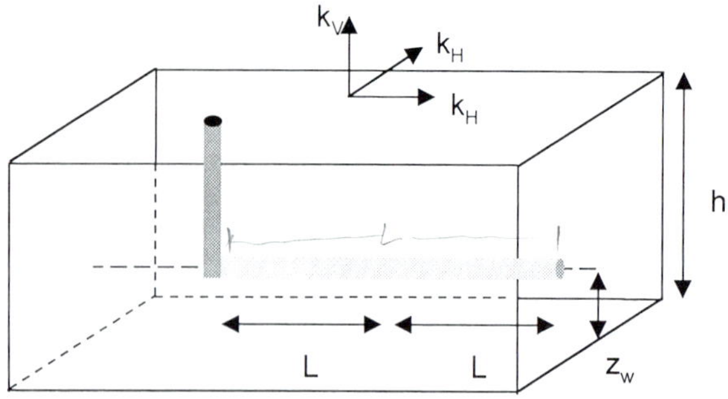

Figure 3.24. Horizontal well geometry.

3.6.1 Model description

In this section, we consider first the pressure behavior of horizontal wells in homogeneous reservoirs with sealing upper and lower boundaries. As shown in Figure 3.24 the well is strictly horizontal, the penetration half-length is L and z_w defines the distance between the drain hole and the bottom-sealing boundary. The vertical part of the well is not perforated, there is no flow towards the end of the well and the well conductivity is infinite. k_H and k_V are the horizontal and the vertical permeability.

Characteristic flow regimes

In an infinite system, the geometry of the flow lines towards a horizontal well produces a sequence of three typical regimes, as depicted in Figure 3.25. On the corresponding pressure and derivative response illustrated in Figure 3.26, three characteristic behaviors are displayed after the wellbore storage unit slope straight line:

1. The first regime is *radial flow in the vertical plane*. On a log-log derivative plot, the wellbore storage hump is followed by a first stabilization. During this radial flow regime, the permeability-thickness product $2\sqrt{k_V k_H}\,L$ is defined with the average permeability in the vertical plane, and the well effective length $2L$.

2. When the sealing upper and lower limits are reached, a *linear flow* behavior is established. The derivative follows a half-unit slope log-log straight line.

3. Later, the flow lines converge from all reservoir directions towards the well, producing a *horizontal radial flow* regime. The derivative stabilization corresponds to the infinite acting radial flow in the reservoir, the permeability-thickness product is $k_H h$.

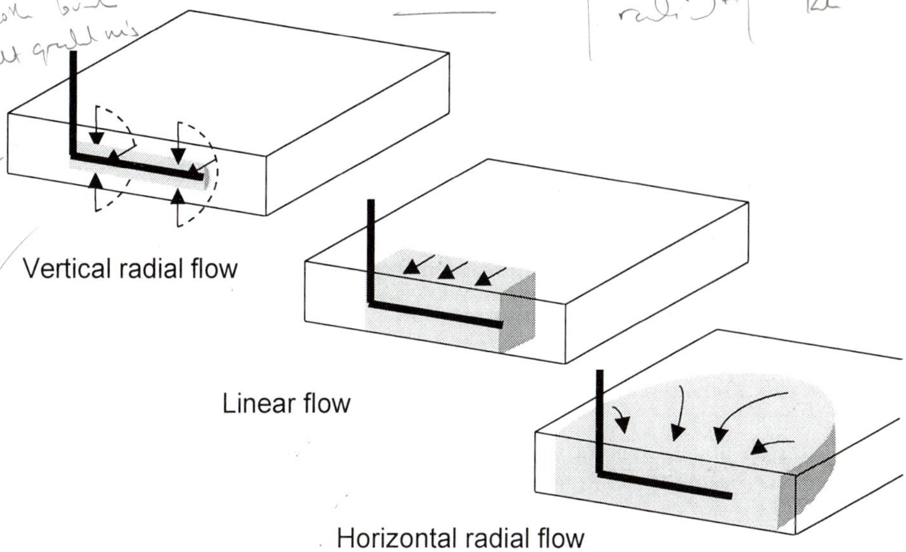

Figure 3.25. Flow geometry to an horizontal well.

Extensions of the model

In practice, the well geometry is not as simple as in the ideal configuration described on Figure 3.24. Most horizontal drain holes are not straight and parallel to the upper and lower boundaries, but show several oscillations over the formation thickness. Frequently, the skin is not uniform along the drain hole and in many cases the well does not produce on the complete length but in one or several segments. When the pressure gradient in the wellbore become large, the infinite conductivity hypothesis is not applicable and the horizontal well shows a finite conductivity behavior.

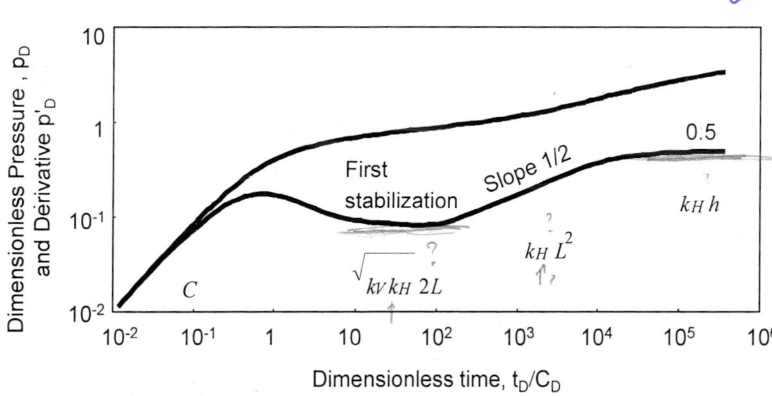

Figure 3.26. Horizontal well with wellbore storage and skin, homogeneous reservoir. Log-log scales, p_D versus t_D/C_D. C_D=1000, S_w =0, L =1000ft, h =100ft, r_w =0.25ft, z_w/h =0.5, k_V/k_H =0.1.

The basic horizontal well model is presented in details Sections 3.6.1 to 3.6.7. Variations from the ideal horizontal well geometry are discussed in Section 3.6.9, fractured and multilateral horizontal well responses are described in Sections 3.6.10 and 3.6.12. In Section 3.6.11, the influence of changes of reservoir properties in the horizontal or vertical directions, or change of fluid properties in the formation, are briefly reviewed. It is shown that when the basic horizontal well model depicted in Figure 3.24 is used to describe complex well or reservoir configurations, the effective well length and the average vertical permeability k_V resulting from analysis can be significantly in error. With complex wellbore conditions, k_V is frequently under-estimated whereas it can be over-estimated in layered systems with semi-permeable interbeds.

Analytical solutions

The first analytical solutions for uniform flux and infinite conductivity horizontal well responses have been derived in the mid 80's: Daviau et al. (1985), Clonts and Ramey (1986) and Rosa and Carvalho (1989) have used source and Green's functions whereas Goode and Thambynayagam (1987) and Kuchuk et al. (1991 a) obtained a solution by application of Laplace and Fourier transforms. With the infinite conductivity horizontal well model, the pressure is assumed constant along the wellbore. This is obtained by measuring the pressure of a uniform flux horizontal drain at an equivalent point in the well (Daviau, Clonts, Rosa), or by averaging the pressure along the length of the well (Goode, Kuchuk). The effect of pressure drop within the horizontal section, and the validity of the infinite conductivity assumption are discussed in Section 3.6.9.

Horizontal well solutions are approximate. They are generated using the line-source solution, which is valid only when $t_D / r_{we}^2 > 25$. For large negative skin, this condition is not satisfied at early time. Furthermore, when the anisotropy between vertical and horizontal permeability is large, small discrepancies can be observed between different horizontal well solutions. With the uniform flux distribution, the pressure is not uniform around the wellbore circumference, and the choice of the reference point on the wellbore can influence the result slightly.

Dimensionless variables

For a horizontal well with wellbore storage and skin, the dimensionless variables are defined with respect to the total formation thickness. Equation 2.3 gives the dimensionless pressure.

In the case of permeability anisotropy between vertical and horizontal directions, an equivalent isotropic solution is used by introducing the anisotropy term k_V / k_H in the definition of the dimensionless vertical distances (see discussion of horizontal permeability anisotropy Section 3.1.5): when the vertical permeability k_V is low, the apparent vertical distances are increased. The apparent open interval thickness h_a and position of the horizontal drain hole with respect to the lower boundary of the zone z_{wa} are defined respectively :

$$h_a = h\sqrt{\frac{k_H}{k_V}} \tag{3.27}$$

(Apparent open interval thickness)

$$z_{wa} = z_w\sqrt{\frac{k_H}{k_V}} \tag{3.28}$$

The circular section of the horizontal well is changed into an ellipse and the horizontal well behaves like a cylinder with the apparent larger equivalent radius r_{we} of Equation 3.6. With large anisotropy k_V/k_H, r_{we} can be 2 or 3 times larger than the actual wellbore radius and the resulting anisotropy skin S_{ani} clearly negative (see Table 3.1).

Several skin coefficients are defined for horizontal wells: the mechanical infinitesimal skin S_w, the anisotropy skin S_{ani}, the apparent skin during the vertical radial flow regime S_{TV}, the geometrical skin S_G and the total skin during the horizontal radial flow S_{TH}. The definitions of all skins are presented in detail in the subsequent sections.

In the definition of the dimensionless terms, several well parameters can be used for the reference length, considering the wellbore radius r_w or, by analogy with wells intercepting a fracture of half-length x_f (see Section 3.2, Equation 3.8), with the well half-length L. For the dimensionless time for example, t_D can be expressed by Equation 2.4 or by:

$$t_{DL} = \frac{0.000264k}{\phi\mu c_t L^2}\Delta t \tag{3.29}$$

No group of independent variables has been identified to provide a universal description of horizontal well responses, as it has been possible with most well models. Many authors use the ratio h_D of the apparent thickness h_a of Equation 3.27, by the well half length L, as a leading parameter of horizontal well behavior (similar to Equation 3.19):

$$h_D = \frac{h_a}{L} = \frac{h}{L}\sqrt{\frac{k_H}{k_V}} \tag{3.30}$$

In the following examples, the wellbore radius r_w is used in the dimensionless parameters definition. The dimensionless wellbore storage coefficient and the dimensionless time group t_D/C_D are given respectively in Equation 2.5 and 2.6. All examples presented below are generated with h = 100 ft and r_w = 0.25 ft and the dimensionless pressure p_D is presented versus the dimensionless time group t_D/C_D.

The question of the reference in the definition of the dimensionless terms is further discussed in subsequent sections for the different skin parameters estimated on horizontal well responses.

3.6.2 Equations for the characteristic regimes

In the following sections, the different limiting forms of the Kuchuk et al. (1991 a) solution are presented, and the different skin coefficients defined from horizontal well responses are described.

Radial flow in the vertical plane

During the vertical radial flow regime, the equation of the semi-log straight line is expressed (Kuchuk, 1995):

$$\Delta p = \frac{162.6 q B \mu}{2\sqrt{k_V k_H}\, L}\left[\log \frac{\sqrt{k_V k_H}\,\Delta t}{\phi \mu c_t r_w^{\acute{e}}} - 3.23 + 0.87 S_w - 2\log \frac{1}{2}\left(\sqrt[4]{\frac{k_V}{k_H}} + \sqrt[4]{\frac{k_H}{k_V}}\right)\right] \tag{3.31}$$

The second logarithm of Equation 3.32 corresponds to the negative anisotropy skin S_{ani} resulting from the equivalent wellbore radius r_{we} of Equation 3.6. The total skin factor S_{TV} measured from the early time radial flow analysis combines the wellbore mechanical skin factor S_w and S_{ani}.

$$S_{\text{TV}} = S_w + S_{ani} = S_w - \ln \frac{\sqrt[4]{k_V / k_H} + \sqrt[4]{k_H / k_V}}{2} \tag{3.32}$$

In the following text, it is assumed that the wellbore mechanical skin factor S_w is uniform along the well length. The influence of non-uniform damage is discussed in Section 3.6.9.

Linear flow regime

During the linear flow regime, the pressure changes as the square root of the elapsed time:

$$\Delta p = \frac{8.128 q B}{2Lh}\sqrt{\frac{\mu \Delta t}{\phi c_t k_H}} + \frac{141.2 q B \mu}{2\sqrt{k_V k_H}\, L} S_w + \frac{141.2 q B \mu}{k_H h} S_z \tag{3.33}$$

The first term of Equation 3.33 is similar to Equation 1.25 for a well intercepting a fully penetrating vertical fracture. With a horizontal well, the flow lines have to converge towards the well located at z_w in the formation thickness. This partial penetration effect produces a pressure drop, expressed with the skin S_z. During the linear flow regime, the two skin effects S_w and S_z are additive.

Sw and Sz are additive
→ due to partial penetration

$$S_z = -1.151\sqrt{\frac{k_H}{k_V}}\frac{h}{L}\log\left[\frac{\pi r_w}{h}\left(1+\sqrt{\frac{k_V}{k_H}}\right)\sin\left(\frac{\pi z_w}{h}\right)\right] \tag{3.34}$$

Equation 3.34 is approximate and only valid when the length of the well is long compared to the apparent thickness (Equation 3.30, $h_D \leq 2.5$).

Pseudo-radial flow from the reservoir

Using the well half-length L as the reference for semi-log analysis of horizontal radial flow, Kuchuk et al. define:

$$\Delta p = 162.6\frac{qB\mu}{k_H h}\left[\log\frac{k_H \Delta t}{\phi\mu c_t L^2} - 2.53\right] + \frac{141.2qB\mu}{2\sqrt{k_V k_H}\,L}S_w + \frac{141.2qB\mu}{k_H h}S_{zT} \tag{3.35}$$

where S_{zT} is :

$$S_{zT} = S_z - 0.5\frac{k_H}{k_V}\frac{h^2}{L^2}\left(\frac{1}{3} - \frac{z_w}{h} + \frac{z_w^2}{h^2}\right) \tag{3.36}$$

In practice, the efficiency of horizontal wells is frequently described by the total skin S_{TH} defined with reference to a fully penetrating vertical well of radius r_w. With the usual radial flow relationship,

$$\Delta p = 162.6\frac{qB\mu}{k_H h}\left[\log\frac{k_H \Delta t}{\phi\mu c_t r_w^2} - 3.23 + 0.87S_{TH}\right] \tag{3.37}$$

the total skin factor S_{TH} combines the wellbore mechanical skin factor S_w and the geometrical skin S_G. Comparing Equations 3.35 and 3.37,

$$\begin{aligned} S_{TH} &= \frac{h}{2L}\sqrt{\frac{k_H}{k_V}}S_w + S_G \\ &= \frac{h}{2L}\sqrt{\frac{k_H}{k_V}}S_w + S_{zT} + 1.151\left(0.70 + 2\log\frac{r_w}{L}\right) \end{aligned} \tag{3.38}$$

the horizontal well geometrical skin S_G is expressed as :

$$S_G = 0.81 - \ln\frac{L}{r_w} + S_{zT}$$

$$= 0.81 - 1.151\left\{2\log\frac{L}{r_w} + \sqrt{\frac{k_H}{k_V}}\frac{h}{L}\log\left[\frac{\pi r_w}{h}\left(1+\sqrt{\frac{k_V}{k_H}}\right)\sin\left(\frac{\pi z_w}{h}\right)\right]\right\}$$
$$- 0.5\frac{k_H}{k_V}\frac{h^2}{L^2}\left(\frac{1}{3}-\frac{z_w}{h}+\frac{z_w^2}{h^2}\right) \quad (3.39)$$

In Equation 3.39, the term $[0.81 - \ln(L/r_w)]$ is very close to the pseudo-skin of a fractured well (Equation 3.12) and S_{zT} (Equation 3.36) describes the pressure drop due to the convergence of the flow lines before reaching the well. This term disappears when h_D of Equation 3.30 is very small, for example in the case of a long well and high vertical permeability k_V. The geometrical skin S_G of horizontal wells is further discussed in Section 3.6.4.

3.6.3 Derivative behavior

Description

Due to the complex behavior of pressure and derivative responses, no type curves are available for horizontal wells. The derivative log-log curve is used for the identification of the characteristic flow regimes, but the analysis is made by generating pressure and derivative responses with a computer or, when applicable, by using specialized plot straight lines.

In the example of a horizontal well response of Figure 3.26, the last derivative stabilization (on the 0.5 line) corresponds to *pseudo radial flow* in the producing zone whereas the first stabilization describes the initial *radial flow in the vertical plane*. The average permeability in the vertical plane is defined as the geometric mean of k_V and k_H and the permeability thickness product is $2\sqrt{k_V k_H}\,L$. In dimensionless terms, the level of the first stabilization is expressed with the dimensionless apparent thickness h_D:

$$(\Delta p_{\text{1st stab.}})_D = 0.25 h_D = 0.25\frac{h}{L}\sqrt{\frac{k_H}{k_V}} \quad (3.40)$$

When both the upper and the lower boundary have been reached, there is no vertical contribution to the flow any more, and expansion of the drainage volume becomes strictly horizontal. If the length of the well is significantly larger than the reservoir thickness, most of the production is due to *linear flow* in front of the horizontal drain, and the flow contribution from the two ends of the well are negligible. During this intermediate time linear flow regime, the derivative follows a half-unit slope straight line.

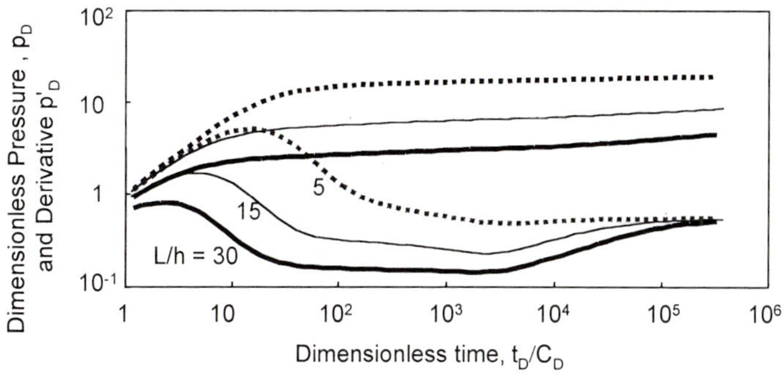

Figure 3.27. Horizontal well. Influence of L. Log-log scales, p_D versus t_D/C_D. C_D =1000, S_w =5, k_V/k_H =0.004, r_w =0.25ft, z_w/h =0.5, L =3000, 1500 and 500ft.

Once the linear flow regime has started, horizontal wells behave like wells intercepting an infinite conductivity vertical fracture of half-length $x_f = L$ (Section 3.2). The presence of an initial vertical radial flow regime before linear flow is simply seen as a skin on the equivalent fracture model (Equations 3.34 during linear flow and 3.39 during pseudo-radial flow).

Influence of L and k_V/k_H

With the t_D/C_D time scale, the location of the half unit slope straight line indicates the effective well half-length L. When L is doubled, the line is *displaced* by a factor of 4 along the time scale and, as the first derivative stabilization is an inverse function of L, $\Delta p_{1\text{st stab.}}$ is *twice as low* (Figure 3.27).

In the examples of Figures 3.28 and 3.29, three well lengths are considered but the permeability anisotropy k_V/k_H is adjusted in order to keep the same derivative stabilization during the vertical radial flow regime. With Figure 3.28, the vertical permeability k_V is relatively large and $(\Delta p_{1\text{st stab.}})_D = 0.223$ is below the radial flow 0.5 line. In such cases, the horizontal drain produces a *negative geometrical skin* (See discussion of the geometrical skin Sections 3.6.4 and 3.7).

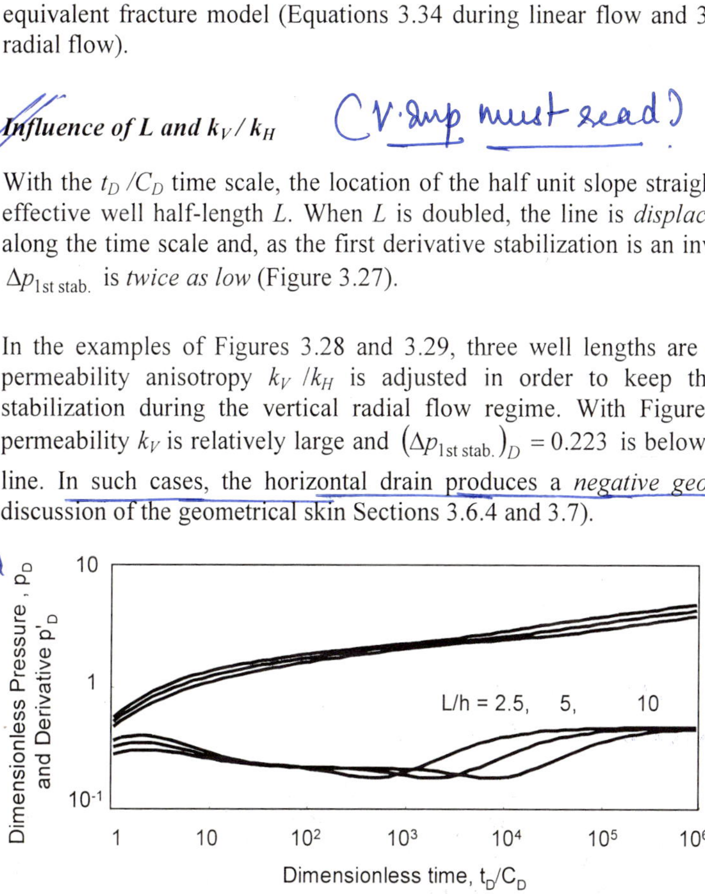

Figure 3.28. Horizontal well. Influence of L, $(\Delta p_{1\text{st stab}})_D$= 0.223. C_D =100, S_w =0, k_V/k_H =0.2, L =250ft; k_V/k_H =0.05, L =500ft; k_V/k_H =0.0125, L =1000ft; h =100ft, r_w =0.25ft, z_w/h =0.5.

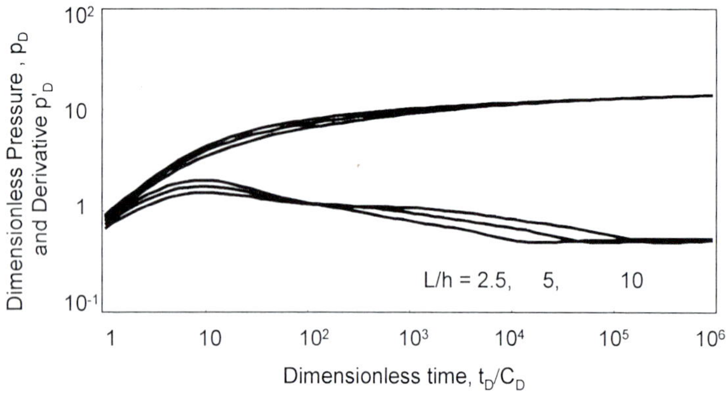

Figure 3.29. Horizontal well. Influence of L, $(\Delta p_{1st\ stab})_D = 1$. $C_D = 100$, $S_w = 0$, $k_V/k_H = 0.01$, $L = 250$ft; $k_V/k_H = 0.0025$, $L = 500$ft; $k_V/k_H = 0.000625$, v=1000ft; $h = 100$ft, $r_w = 0.25$ft, $z_w/h = 0.5$.

Conversely, when k_V is low, the first derivative stabilization is above 0.5 ($(\Delta p_{1st\ stab.})_D = 1$ on the examples Figure 3.29), and the horizontal well behavior tends to be equivalent to a well in partial penetration (Section 3.4). In the case of low vertical permeability, short horizontal wells exhibit a *positive geometrical skin*, and therefore an overall damaged well behavior. This is an important point and demonstrates that not all horizontal wells will increase productivity.

Influence of z_w/h

The first vertical radial flow lasts until one of the upper or the lower boundary is reached. If the horizontal well is not centered in the zone thickness ($z_w/h \neq 0.5$), *a hemi-radial flow* regime can develop when only the closest limit is seen. As long as the second sealing boundary is not reached, the shape of the derivative curve is similar to that of a vertical well near a sealing fault (Section 5.1). The second derivative stabilization is at a level twice the first (of Equation 3.40), as illustrated on the examples of Figure 3.30.

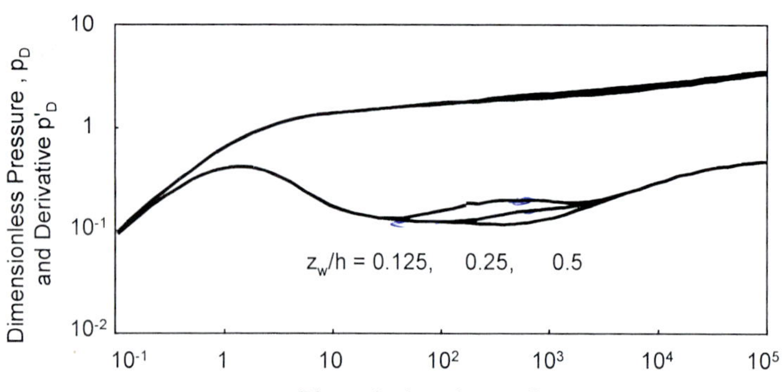

Figure 3.30. Horizontal well. Influence of z_w. Log-log scales, p_D versus t_D/C_D. $C_D = 1000$, $S_w = 2$, $L = 1500$ft, $k_V/k_H = 0.02$, $h = 100$ft, $r_w = 0.25$ft, $z_w/h = 0.5$, 0.25, 0.125.

The influence of z_w/h on horizontal well responses is further discussed next and in Section 3.6.6, with a semi-log plot of Figure 3.30 examples.

3.6.4 Skin of horizontal wells

Since several distinct flow regimes are observed during horizontal well responses, several skin parameters can be defined to describe the different flow geometries, even though the infinitesimal skin damage S_w is constant at the wellbore. Furthermore, since the skin factors are a dimensionless pressure drop, several references can be used to normalize the different Δp_{skin}. In the following, we summarize the influence of the well and reservoir parameters on the three skins usually estimated from analysis. We show that, in the presentation of the analysis results, the reference used to express the skin parameters must be clearly defined.

Mechanical skin S_w

As an extension of the total horizontal radial flow skin concept S_{TH} used in Equation 3.37, the infinitesimal wellbore skin S_w is sometimes also defined with reference to a vertical well of radius r_w and a permeability k_H. The resulting skin parameter S'_w does not define the completion quality as does S_w of Equations 3.31 and 3.32.

$$S'_w = 0.5 h_D S_w = \frac{h}{2L}\sqrt{\frac{k_H}{k_V}} S_w \qquad (3.41)$$

Geometrical skin S_G

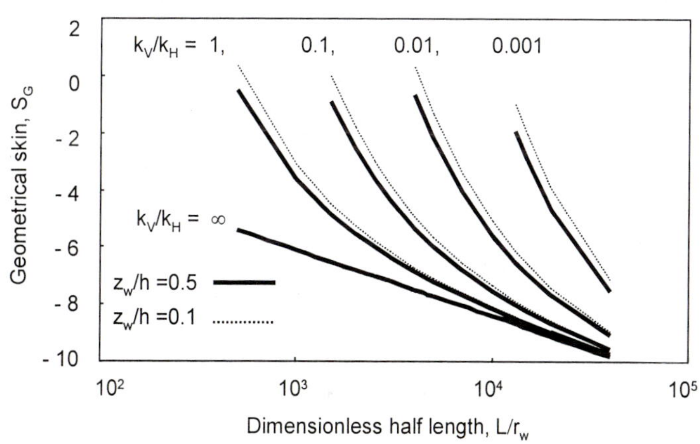

Figure 3.31. Semi-log plot of the geometrical skin S_G versus L/r_w. Influence of k_V/k_H. h/r_w =1000, z_w/h =0.5 and 0.1.

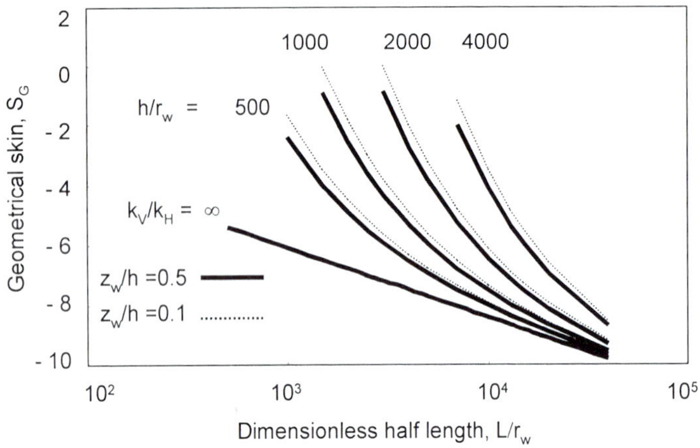

Figure 3.32. Semi-log plot of the geometrical skin S_G versus L/r_w. Influence of h/r_w. k_V/k_H =0.1, z_w/h =0.5 and 0.1.

On Figures 3.31 and 3.32, the geometrical skin S_G of Equation 3.39 is graphed versus $\log(L/r_w)$. With Figure 3.31, several hypothesis of permeability anisotropy k_V/k_H are considered, assuming a constant formation thickness h/r_w. When the vertical permeability is very large ($k_V \to \infty$), the *partial penetration* term S_{zT} cancels out, and the negative geometrical skin S_G is a linear function of $\log(L/r_w)$. When a vertical pressure drop is introduced as on the examples $k_V/k_H \leq 1$, the geometrical skin is less negative, and the curves reach the infinite vertical permeability behavior only when the drain hole is very long.

For a given permeability anisotropy k_V/k_H, increasing the formation thickness h/r_w also produces more partial penetration skin effect as shown on Figure 3.32. Again, when the horizontal well becomes very long, the adverse effect of the vertical pressure drops on the geometrical skin S_G is reduced. Ozkan and Raghavan (1989) indicate that the late time response of horizontal wells tends to be equivalent to that of vertically fractured ones when $h_D \leq 0.25$. V.Imp

The dotted curves on Figures 3.31 and 3.32 show the geometrical skin when the well is not centered in the formation thickness. With z_w/h =0.1, a small additional pressure drop is introduced on the response, and S_G is slightly *less negative* (see discussion of Figure 3.33 in Section 3.6.6).

Total skin S_{TH}

As shown on Equation 3.38, the total skin S_{TH} estimated on horizontal well responses combines the geometrical skin S_G of Equation 3.39 and the mechanical infinitesimal skin S_w normalized by h_D (to give S'_w of Equation 3.41). For long horizontal drain holes, h_D is in general smaller than unity and the effect of a wellbore damage is *reduced.* The opposite effect is observed on partially penetrating wells, where a mechanical skin damage S_w is amplified in the total skin (Equation 3.17). V. Imp

3.6.5 Matching procedure on pressure and derivative responses

Frequently, horizontal well responses do not exhibit the three individual flow regimes. Horizontal wells involve large wellbore volume, therefore a large wellbore storage coefficient and the wellbore storage effect lasts in general longer than in vertical wells. For this reason, the first radial flow may be difficult to identify. The last derivative stabilization is not always present within a normal test duration: the linear flow transition, before pseudo-radial flow regime, can last several log cycles on the time scale. The log-log diagnostic indicates the different flow regimes present on the response, and which parameters, or groups of parameters, can be estimated and which are not defined. Manual log-log analysis is not appropriate with horizontal wells, the match is performed on a computer-generated response.

When the complete sequence of flow regimes is identified on the derivative response, the early time *unit slope* straight line and the *final stabilization* are used to define the time and pressure matches, yielding the permeability-thickness product $k_H h$ from Equation 2.9 and the wellbore storage coefficient C from Equation 2.10. The intermediate time linear flow regime is used to estimate the effective well half-length L, by adjusting the match of the generated curve on the *half unit* slope straight line. k_H and L being defined, the *first derivative stabilization* determines the permeability anisotropy k_V / k_H. The match of the pressure curve during the initial vertical radial flow regime gives the mechanical skin S_w (or S_{TV}). The geometrical skin S_G, and therefore the total skin S_{TH} are defined from the estimated well and reservoir parameters (Equations 3.39). When the analysis is consistent, the theoretical pressure curve matches the data during the complete response.

Frequently, some segments of the well do not produce and the effective length $2L$ resulting from analysis is smaller than the drilled length. In Section 3.6.9, it is shown in the discussion of Figures 3.37 and 3.38 that, when several sections opened to the flow are distributed along the complete drain hole, a good match is frequently obtained by assuming the total drilled length. Then, the estimated vertical permeability k_V can be greatly *under estimated.*

When the vertical radial flow regime is masked by wellbore storage, the permeability anisotropy k_V / k_H cannot be assessed. The late time data give the total skin S_{TH} but, since the geometrical skin S_G is not defined, S_w is not reliable. Different hypothesis of k_V / k_H can change S_w from negative to positive values.

If the test data ends before the final derivative stabilization is reached, the horizontal permeability k_H and the total skin S_{TH} are not fixed, but the half unit slope straight line gives $k_H \; L^2$ (see Equation 3.33). In such case, the vertical permeability k_V can be estimated from the vertical radial flow derivative stabilization, if present. Again, the permeability anisotropy k_V / k_H and the mechanical skin S_w are not accurately defined, but the error on S_w is in general small.

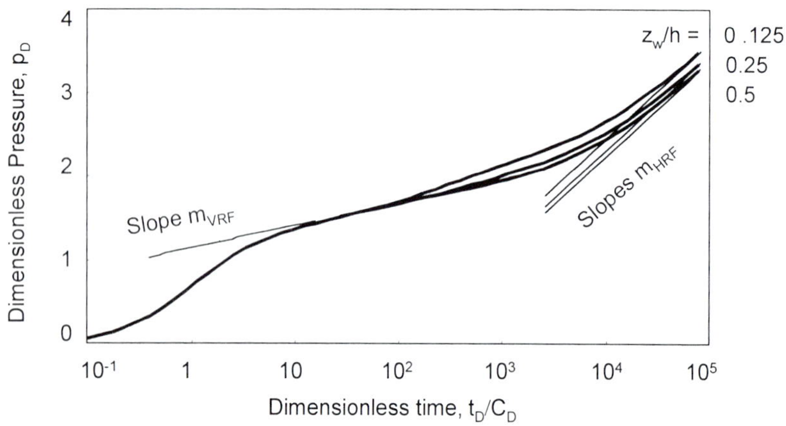

Figure 3.33. Semi-log plot of Figure 3.30 examples.

3.6.6 Associated specialized plot straight lines

Four specialized analyses are possible, depending upon the type and the duration of the regimes defined by the derivative log-log plot. The wellbore storage analysis is the same as for vertical wells (Section 1.3.2). In the following section, straight-line analysis methods are presented for the vertical radial flow, linear flow and the horizontal pseudo radial flow regimes.

Figure 3.33 is a semi-log plot of the Figure 3.30 examples for three different well locations z_w/h. When the well is centered (z_w/h =0.5), the response exhibits two straight lines on semi-log scale and, as the permeability thickness product during the initial vertical radial flow is larger than $k_H h$, the first slope m_{VRF} is lower than the final straight line slope m_{HRF}. When the well is off-centered, an intermediate time straight line of slope 2 m_{VRF} can be observed during the hemi-radial flow in the vertical plane (curve z_w/h =0.125). In such case, the final semi-log straight line is displaced upwards, because of the influence of z_w/h on the geometrical skin S_G of Equation 3.39. A similar effect on late time semi-log straight lines can be observed in reservoirs with multiple boundaries (Figure 5.13 of Chapter 5 for example).

Frequently, after wellbore storage, horizontal well responses only show transitional behaviors between the characteristic flow regimes, and no specialized analysis is possible. Furthermore, with build-up data, the Horner or multiple-rate superposition methods used on the specialized plots can distort the characteristic straight lines, as a result of the changes of flow behavior during the response (see Section 2.3.4). Except for the final horizontal radial flow regime, the straight-line methods presented in the following are seldom used.

Radial flow in the vertical plane

During the first radial flow regime in the vertical plane, the equation of the semi-log straight line is expressed in Equation 3.31. The slope m_{VRF} gives the product of average permeability in the vertical plane $\sqrt{k_H k_V}$, multiplied by the perforated half-length L:

$$\sqrt{k_V k_H}\,L = \frac{81.3 qB\mu}{m_{\mathrm{VRF}}} \tag{3.42}$$

When L and the permeability anisotropy k_V/k_H are known, the skin S_{TV} measured from the first semi-log straight line is used to estimate the infinitesimal skin S_w. From Equation 3.31,

$$S_w = 1.151\left[\frac{p(1\mathrm{hr}) - p(\Delta t = 0)}{m_{\mathrm{VRF}}} - \log\frac{\sqrt{k_V k_H}}{\phi\mu c_t r_w^2} + 2\log\frac{1}{2}\left(\sqrt[4]{\frac{k_V}{k_H}} + \sqrt[4]{\frac{k_H}{k_V}}\right) + 3.23\right] \tag{3.43}$$

Provided the $\sqrt{k_H k_V}\,L$ product is correctly estimated from m_{VRF}, the dependence of S_w on the anisotropy k_V/k_H and on the effective well half-length L are logarithmic. The calculation of the infinitesimal skin with Equation 3.43 is not very sensitive to an error on k_V/k_H or L (in Section 3.1.5, it is shown that S_{ani} is in general between 0 and -1).

When the nearest upper or lower sealing boundary is reached, the flow regime changes to hemi-radial flow and the response deviates from the semi-log slope m_{VRF} to follow a semi-log straight line of slope $2m_{\mathrm{VRF}}$. The time of intercept between the m_{VRF} and $2m_{\mathrm{VRF}}$ straight lines can be used to estimate the vertical permeability k_V with a relationship similar to Equation 1.33 for a sealing fault (see section 5.1.3). Kuchuk et al. (1991 a) propose to use the time Δt_{end} of end of the initial vertical radial flow (i.e. when the derivative deviates from the first stabilization, and not the mid point of the derivative transition as in section 5.1.1) with :

$$k_V = \frac{\phi\mu c_t}{0.000264\pi\Delta t_{end}}\min\left\{z_w^2, (h - z_w)^2\right\} \tag{3.44}$$

For a build-up analysis, the first straight line extrapolated pressure is not used, p^* is estimated from the horizontal radial flow regime (Section 3.6.7).

Linear flow regime

This flow regime results of the influence of the two sealing upper and lower limits. As already mentioned, the horizontal well behaves like an infinite conductivity fractured well, but the linear flow regime can also be described as a boundary effect. In fact, by rotating the horizontal well through 90°, the configuration is similar to a vertical well between two parallel sealing faults (Chapter 5.2). As opposed to a fractured well and channel responses, described by Figures 3.5 and 5.4 for example, none of the curves presented in Figures 3.26 to 3.30 present a long derivative half unit slope straight line. On horizontal well responses, the vertical radial and hemi-radial flow regimes dominate the early time data. Later, the transition between linear flow and the final pseudo radial flow regime is long, a flow contribution from the reservoir region at both ends of the well is felt a long time before the start of the final radial flow regime, and the pure linear flow regime is short lived. In order to see this characteristic regime, the distance between the two derivative stabilizations must be large. From Equation 3.40, it can be seen that the well length $2L$ must be very long compared to the apparent thickness h_a of Equation 3.27 (small h_D, Kuchuk et al., 1990).

When the half unit slope derivative straight line is clearly established, the corresponding pressure points are analyzed on a plot of the pressure versus the square root of the elapsed time, as for a fractured well or a channel reservoir (see Sections 3.2 and 5.2). From Equation 3.33, the slope m_{LF} of the straight line gives $k_H L^2$:

$$k_H L^2 = 16.52\left(\frac{qB}{m_{\mathrm{LF}}\, h}\right)^2 \frac{\mu}{\phi c_t} \tag{3.45}$$

The intercept $p(0\mathrm{hr})$ of the linear flow straight line at time 0 can theoretically be used to estimate the infinitesimal skin S_w (Kuchuk et al., 1990):

$$S_w = \frac{2\sqrt{k_V k_H}\, L}{141.2 qB\mu}\left[p(0hr) - p(\Delta t = 0)\right] + 2.303\log\left[\frac{\pi r_w}{h}\left(1+\sqrt{\frac{k_V}{k_H}}\right)\sin\left(\frac{\pi z_w}{h}\right)\right] \tag{3.46}$$

Alternatively, when S_w is known from previous vertical radial flow regime, z_w/h can be estimated from Equation 3.46 in the same way as, for channel reservoirs, the intercept $p(0\mathrm{hr})$ defines the well location between the faults (see Section 5.2.5). It can be noted that the linear flow partial penetration skin effect S_z of Equation 3.34 has the same form as the geometrical skin of channel reservoir (Equation 5.8), discussed in Section 5.2.5.

Pseudo-radial flow from the reservoir

The analysis of the pseudo-radial flow regime is identical to the semi-log analysis of a vertical well response (Equation 3.37). The straight line slope m_{HRF} gives the horizontal permeability thickness product $k_H\, h$, the straight line intercept at 1 hour is used to estimate the total skin coefficient S_{TH} and, for a build-up periods, the extrapolation to infinite shut-in time gives p^*.

$$k_H h = \frac{162.6 qB\mu}{m_{\mathrm{HRF}}} \tag{3.47}$$

$$S_{\mathrm{TH}} = 1.151\left[\frac{p(1\mathrm{hr}) - p(\Delta t = 0)}{m_{\mathrm{HRF}}} - \log\frac{k_H}{\phi\mu c_t r_w^2} + 3.23\right] \tag{3.48}$$

Either the mechanical skin S_w or the geometrical skin S_G can be estimated from Equations 3.38 and 3.39.

3.6.7 Build-up analysis

On horizontal well responses, the flow geometry changes from early time to late time and three different characteristic regimes can be observed, as illustrated on previous derivative examples. For shut-in periods, the Horner and time superposition methods used for straight line and derivative analysis are based on the assumption that all superposed periods follow the same flow regime (see Section 2.2.2 and 2.3.4). In the case of complex responses, it is likely that the extrapolated periods follow different behaviors, and the multiple-rate superposition method is theoretically invalid.

The resulting build-up derivative can be distorted (see discussion Figure 2.20 for example) but, since the log-log match of horizontal well responses is made on a computer generated multiple-rate pressure and derivative curves, the use of superposition time does not introduce error in the results.

With straight-line methods, it is found in practice that unless the production time is very short and the well has been closed during the vertical radial flow regime, the superposition methods are applicable for all flow regimes.

The semi-log superposition function can be used for radial flow analysis. As the producing time t_p is generally significantly greater than Δt during the early time vertical radial flow regime, the Horner time can be simplified with $\log(t_p + \Delta t/\Delta t) \approx \log t_p - \log \Delta t$ (Equation 2.16), and the result becomes independent of the production history. On a Horner plot of horizontal well response, the first straight line gives the correct $\sqrt{k_H k_V}\, L$ product with Equation 3.42. The first straight line extrapolated pressure is not used, the pressure at infinite shut-in time p^* is estimated from the second straight line during the horizontal radial flow regime, if present.

When the linear flow regime is clearly established, build-up responses can be analyzed with the Horner or multiple-rate superposition time corresponding to this flow regime (Equation 2.19). If the previous drawdown had reached the horizontal pseudo radial flow at time of shut-in, $t_p >> \Delta t$ then the method remains applicable.

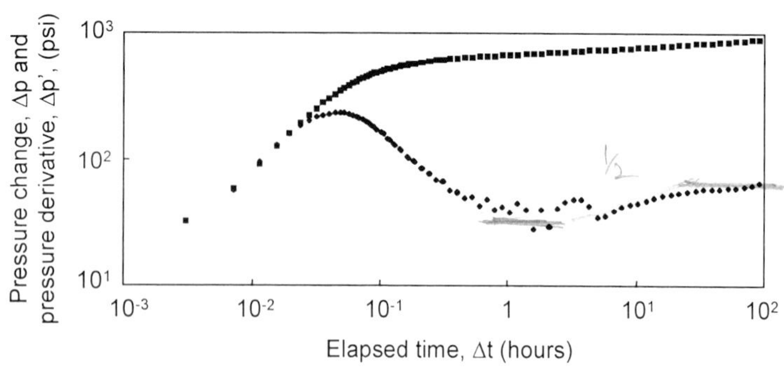

Figure 3.34. Build-up test in a horizontal well. Log-log scales.

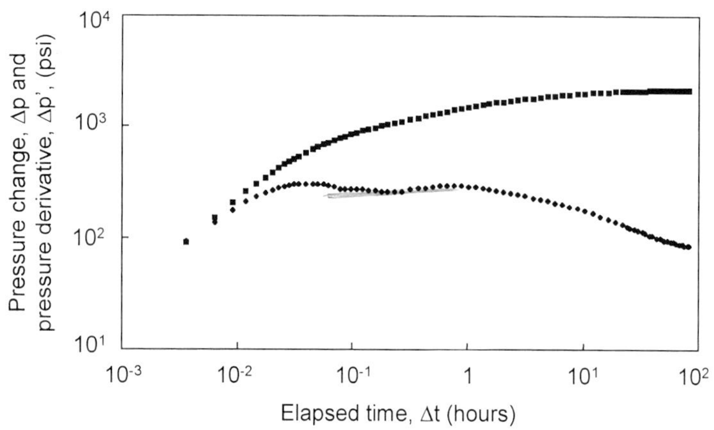

Figure 3.35. Build-up test in a horizontal well. Log-log scales.

3.6.8 Field examples

In Figures 3.34 and 3.35, two examples of horizontal well build-up tests are presented. For the example in Figure 3.34, the response describes the wellbore storage unit slope straight-line, followed successively by the characteristic derivative hump, a first derivative stabilization during the vertical radial flow, an increase of derivative near 10 hours, and the final derivative stabilization during the horizontal radial flow. This well shows a usual horizontal well behavior similar to the responses in Figure 3.26, *all reservoir and well parameters* can be estimated. The geometrical skin of this horizontal well is *negative*.

A completely different response is obtained on the 100 hours build-up example of Figure 3.35. After a short wellbore storage effect, the derivative stabilizes during the first hour, and later it declines slowly until the end of the build-up test. No final derivative stabilization is seen; the horizontal radial flow is not reached. The overall behavior is similar to the low k_V examples of Figure 3.29: the geometrical skin is *positive*. Straight-line analysis of this horizontal well response is only applicable during the *vertical radial flow* regime, to provide $\sqrt{k_H k_V}\,L$ and S_w. When the data is matched against a computer-generated model, a relatively unique analysis is obtained. The mechanical skin S_w is negative (no derivative hump is seen before the stabilization).

3.6.9 Discussion of the horizontal well model

In the following, several variations of the basic horizontal well model are considered. With finite conductivity wells, or when the skin is non-uniform along the well length, and with partially open horizontal wells, a pressure gradient is introduced in the reservoir along the well length. These wellbore conditions can distort the pressure response, especially at early time, and produce an under estimated $\sqrt{k_V k_H}\,L$ product when they are ignored. In case of non-rectilinear wells, the response is affected at intermediate times, with little effect one the estimated parameters.

Finite conductivity horizontal wells

In the previous discussion, the horizontal drain is assumed to be of infinite conductivity. Frequently, highly productive horizontal wells are completed with small diameters and the pressure gradients along the well length cannot be neglected, particularly when the flow becomes turbulent. Several authors have considered the effect of pressure drop in the wellbore on horizontal well responses (Dikken, 1990; Ozkan et al., 1995; Ozkan and Raghavan, 1997).

Using the same approach as Cinco et al. (1978 a) for finite conductivity fractured wells, Ozkan et al. express the pressure drop with an equivalent wellbore permeability in the case of laminar flow. The conductivity of the horizontal well is defined as an inverse function of the well length $2L$. They describe the flux distribution along the wellbore as follows, for high and low conductivity wells:

- When the pressure gradients in the wellbore are negligible compared to the pressure gradients in the reservoir, the well shows a high conductivity behavior. At early time, the flux distribution is uniform along the wellbore. When the flow tends towards the horizontal radial flow regime, the *two ends* of the horizontal drain are the most productive sections, and the flux profile along the well length is described by a U-shaped symmetric distribution, similar to the flux towards a well with an infinite conductivity fracture (Figure 3.13).
- In the case of a low drawdown (such as when the reservoir permeability is high, the thickness small and the horizontal section long), when the wellbore radius is not large enough, the pressure drop in the wellbore can be comparable to the pressure drop in the reservoir. The well behavior deviates from the infinite conductivity response. Due to the pressure gradients in the low conductivity well, most of the fluid enters near the *heel of the well*, resulting in a distortion of the flux profile from the uniform or U-shaped distribution, into an asymmetric shape.

As for the finite conductivity fracture model of Cinco et al., the effect of a finite conductivity horizontal well is more pronounced at early times. The presence of high pressure gradients in the wellbore can distort the pressure response during the vertical radial flow and linear flow regimes, since the flow in the reservoir becomes *three dimensional* (with a component parallel to the well axis). For low conductivity horizontal wells, the derivative is above the vertical radial flow stabilization of Equation 3.40. The effect of wellbore friction is the highest in non-damaged horizontal wells, and it tends to be reduced when the mechanical skin factor S_w is large (Ozkan and Raghavan, 1997).

By neglecting wellbore hydraulics, the product $2\sqrt{k_V k_H}\,L$ can be underestimated by a factor of 3 or more, but the permeability-thickness product $k_H\,h$ should be accurately defined. In the analysis results, both the vertical permeability k_V and the effective well half-length L are too low, whereas the estimated mechanical skin factor S_w is too large.

During the horizontal radial flow regime, the authors explain that the wellbore pressure gradients simply introduce an additional pressure drop and the response of a low conductivity horizontal well becomes similar to that of a *damaged infinite conductivity* horizontal well (with a less negative total skin S_{TH}).

Bourgeois et al. (1996 a) propose to approximate the effect of wellbore friction on the total skin S_{TH} by a rate dependent skin effect similar to the non-Darcy skin of gas wells (Section 7.2.4). The total skin of Equation 3.38 is then changed into :

$$S_{\mathrm{TH}} = \frac{h}{2L}\sqrt{\frac{k_H}{k_V}}\left(S_w + Dq\right) + S_G \qquad (3.49)$$

where Dq describes the friction skin during the horizontal radial flow regime.

Non-uniform mechanical skin

Ozkan and Raghavan (1997) investigated the influence of a non-uniform mechanical skin on infinite conductivity horizontal well responses. They concluded that, in early time response, a change of skin damage along the well length tends to move the derivative above the vertical radial flow stabilization of Equation 3.40. During the horizontal radial flow regime, the derivative stabilization can be used to estimate the $k_H h$ product but the well productivity (or the total skin S_{TH}) is slightly influenced by the skin factor distribution. S_{TH} is more negative when the two ends of the horizontal drain are not damaged, and the mechanical skin is mostly located in the central section of the well. No damage at the *heel and toe* of the well improves the productivity because of the U-shaped flux profile discussed earlier for high conductivity horizontal drains. As described next, a similar conclusion is obtained with partially completed horizontal wells.

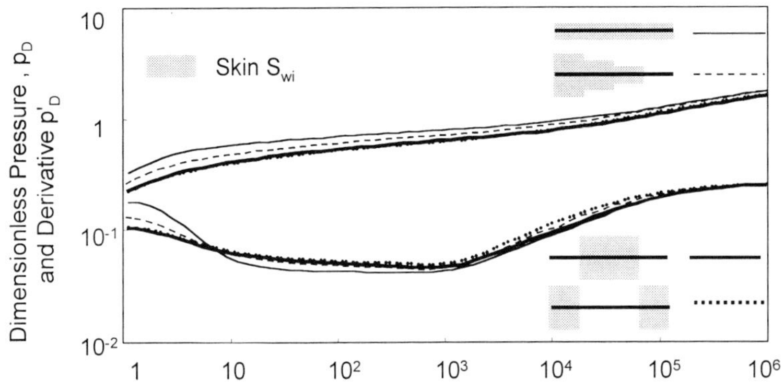

Figure 3.36. Horizontal well. Influence of non-uniform skin. Log-log scales, p_D versus t_D/C_D. C_D =100, L =1000ft, h =100ft, r_w =0.25ft, z_w / h =0.5, k_V / k_H =0.1. Four segments L_{wi}= L /4 with S_{wi} =(4 - 4 - 4 - 4), (8 - 5.3 - 2.6 - 0), (0 - 8 - 8 - 0), (8 - 0 - 0 - 8).

On Figure 3.36, three examples of non-uniform skin distributions are compared to the response of a well with a constant mechanical skin factor S_w = 4. The well length is divided into four equal segments and each segment is affected by a skin factor S_{wi} such as the arithmetic mean of S_{wi} is constant at 4. In one case, the skin is linearly decreasing from one end to the other and, in the two other cases, the damage is either located on the two external segments or on the central sections. The examples Figure 3.36 confirms that, when the ends are not damaged, the total skin of the well S_{TH} is slightly more negative than on the three other responses ($S_{TH} \doteq$ -6.4 instead of -6.2). The authors conclude that stimulation treatments of horizontal wells should preferably concentrate on the *heel and the toe.*

Partially open horizontal wells

Frequently, some sections of the horizontal drain are not contributing to the flow and the effective well half-length L estimated by analysis is smaller than the length of the drilled well. It is shown in the following that the pressure behavior of partially open horizontal wells depends not only upon the effective well half-length L, but also upon the number and the distribution of the open sections along the well-drilled length (Goode and Wilkinson, 1991; Kamal et al., 1993; Yildiz and Ozkan, 1994).

On Figure 3.37, three different repartition examples of the productive segments are compared. For all completion scenarios, the same effective well half-length is assumed with L_{eff}=1/4L of the total drilled length (the response corresponding to the fully open horizontal well is shown with the thin dotted curves). When only one section is producing, the response corresponds to a horizontal well with half-length L_{eff} (thin solid curves).

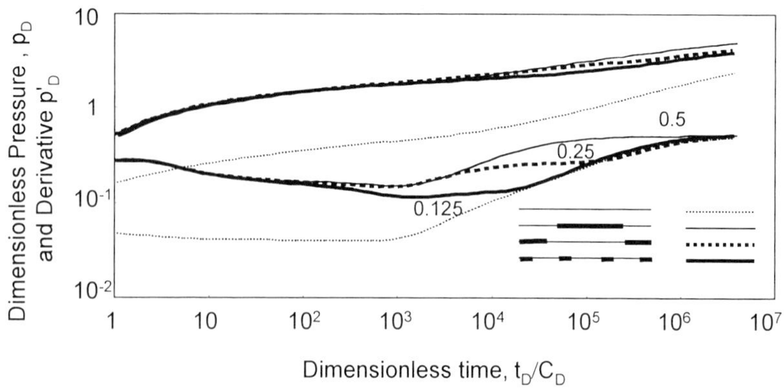

Figure 3.37. Partially open horizontal well. Influence of the number of open segments. C_D =100, 1, 2, 4 segments with S_{wi} =0, ΣL_{eff}= L /4, L =2000ft, h =100ft, r_w =0.25ft, z_w / h =0.5, k_V/k_H =0.1.

Whatever is the repartition of the open sections, only the total length of the producing intervals influences the response during the initial vertical radial flow. At early time, the pressure and derivative curves generated for several producing intervals shows the same behavior as the single producing interval with similar L_{eff}. Later, when the distances between the open intervals are large, each segment acts as a horizontal well, and a horizontal radial flow geometry develops around the different producing sections. Kamal et al. (1993) showed that, during this intermediate time radial flow regime, the derivative stabilizes at 0.5 divided by the number of open segments. When only the heel and toe of the well are producing (thick dashed pressure and derivative curves), the derivative stabilizes at 0.25 and, when four segments are open to flow, it stabilizes at 0.125 (thick solid curves).

Once the interference effect of neighboring segments is felt, the intermediate radial flow regime changes into linear flow and the derivative response reaches that of a single horizontal drain hole whose length corresponds to the *distance between the two ends* of the external open segments. During the final horizontal radial flow, the total skin S_{TH} is slightly more negative when the open section is more *distributed*: with 4 segments, S_{TH}= -6.7 on Figure 3.37 whereas S_{TH} = -6.3 in case of two segments and S_{TH} = -5.4 with only one segment.

When analyzing the example with four segments of Figure 3.37, the horizontal permeability is defined from the final derivative stabilization. The half unit slope derivative straight line gives access to maximum external distance of the open segments, which is 4 times the effective well length in this example. By assuming that 100% of the well length is producing with a single horizontal drain model, Kamal et al. (1993) noted that the vertical permeability value resulting from the vertical radial flow analysis of the first derivative stabilization is *under estimated* (by a factor of 16 in the example).

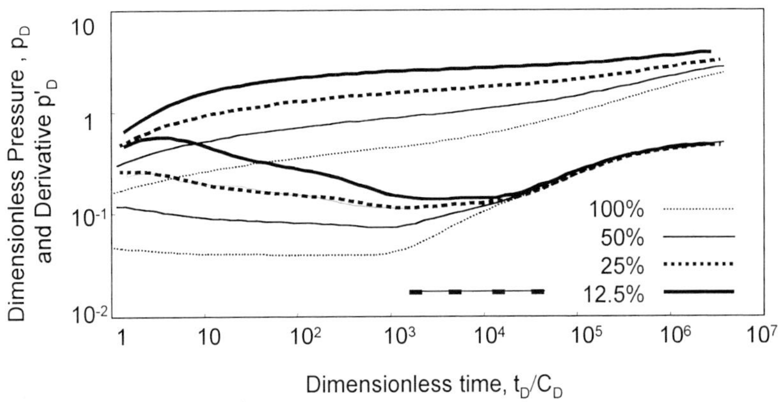

Figure 3.38. Partially open horizontal well. Influence of the penetration ratio. Log-log scales, p_D versus t_D/C_D. C_D =100, 4 segments with S_{wi} =0, ΣL_{eff}= L /8, L /4, L /2 and L, L =2000ft, h =100ft, r_w =0.25ft, z_w /h =0.5, k_V /k_H =0.1.

Figure 3.38 shows the influence of the penetration ratio for a horizontal well with four uniformly distributed segments of equal length. The ratio of the total length of the open segments to the length of the drilled well is respectively 12.5, 25, 50 and 100%. As already observed on Figure 3.37, all derivative curves merge at late time, during linear and pseudo radial flow, on the fully penetrating horizontal well response. Before, the derivative is displaced upwards. In case of low penetration ratio such as on the example 12.5%, the flow is *three-dimensional* at early time (Yildiz and Ozkan, 1994) with a decreasing derivative trend. Assuming no mechanical skin damage, the total skin S_{TH} of the fully penetrating horizontal well of Figure 3.38 is S_{TH} =-7.9. With a penetration ratio of 50, 25 and 12.5%, S_{TH} is still very negative with respectively –7.4, -6.6 and –5.1.

Yildiz and Ozkan (1994) presented a general selectively completed infinite conductivity horizontal well model. They observed that the rate profile and the pressure response are affected at early time by a non-uniform skin distribution between the productive segments and use of vertical radial flow analysis is not possible. They concluded that it is not possible to estimate length and distribution of the open interval from use of transient analysis.

Non-rectilinear horizontal wells

Horizontal wells are in general not parallel to the top and bottom sealing interfaces. In Figure 3.39, two examples of non-rectilinear horizontal well responses are compared to the straight horizontal drain hole model. Two symmetric geometries are considered: half of the well length is either centered in the formation thickness (z_w =0.5h) or close to upper or lower sealing boundary (z_w =0.05h). The other half, distributed in two equal segments at the heel and toe, is close to a boundary in the first case (z_w =0.05h), and centered in the other. The linear horizontal well, shown with a thin pressure and derivative curve, is located at the average distance with z_w =0.275h.

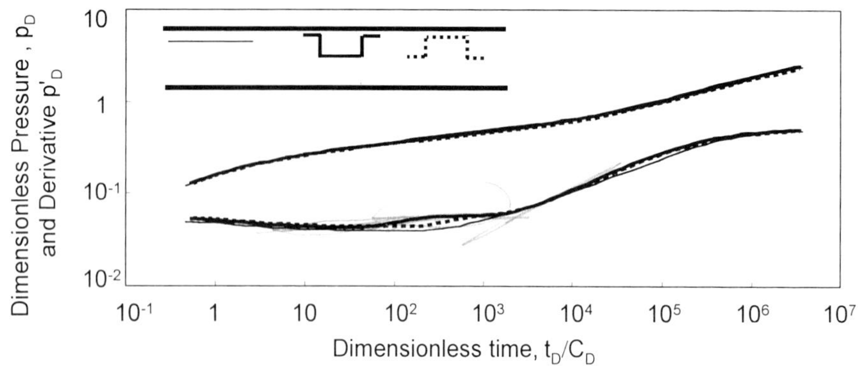

Figure 3.39. Non-rectilinear horizontal well. Log-log scales, p_D versus t_D/C_D. C_D =100, L =2000ft (500+1000+500), S_{wi} =0, h =100ft, r_w =0.25ft, k_V / k_H =0.1, $(z_w / h)_i$=0.5 or 0.95 (average 0.725).

When some segments of the horizontal well are closer to the upper or lower boundary, the vertical radial flow regime ends earlier than when the well is a single linear segment at the average vertical distance. The *transition* from the first derivative stabilization to the half unit slope straight line is slightly distorted but, as shown Figure 3.39, the three derivative responses are very similar. The pressure curves are not distinguishable (the total skin S_{TH} being slightly less negative when the two ends of the well are off-centered).

3.6.10 Fractured horizontal wells

When fracturing horizontal wells, the fracture direction with respect to the wellbore depends upon the orientation of the well compared to the least principal stress. If the well is drilled in the direction of the least stress, several vertical fractures transverse to the well may be created along the well length. When the well is perpendicular to the least stress, the fractures are parallel to the well.

Soliman et al. (1990) presented an approximate analytical solution for horizontal wells in the direction of the least stress, with circular finite conductivity transverse fractures. Larsen and Hegre (1991) investigated both circular transverse, and rectangular longitudinal, finite conductivity fractures. They assume the horizontal wellbore is not perforated outside the fractured segments.

With a transverse fracture, the flow at early time is linear from the formation to the fracture, and radial inside the fracture to the wellbore. Larsen and Hegre (1994 a) note that this *radial-linear flow* geometry is similar to that of transient double porosity reservoirs, slab matrix blocks with a semi-log straight line of slope half that of the radial flow in the fissure system (Section 4.1.3). With transverse fractures, the radial-linear flow regime is characterized by a semi-log straight line of slope m_{RLF} half that of the pure radial flow in the fracture. Therefore, the slope is only a function of the fracture conductivity $k_f w_f$:

$$m_{\text{RLF}} = 81.3 \frac{qB\mu}{k_f w_f} \tag{3.50}$$

With a longitudinal fracture, a *bilinear flow* regime develops at early time, as for a vertical well intercepting a finite conductivity fracture. On a pressure versus $\sqrt[4]{\Delta t}$ plot, the slope m_{BLF}, similar to Equation 1.27, is a function of the fracture half-length x_f along the horizontal well direction. When the reservoir permeability k_H is known, m_{BLF} also gives access to the fracture conductivity $k_f w_f$:

$$m_{\text{BLF}} = 44.11 \frac{qB\mu}{x_f \sqrt{k_f w_f} \sqrt[4]{\phi \mu c_t k_H}} \tag{3.51}$$

In the case of a single fracture, the radial-linear or bilinear flow regime is followed by the *formation linear flow*, and finally the pseudo-radial flow towards the horizontal well. During the linear flow regime, the slope m_{LF} of the pressure versus $\sqrt{\Delta t}$ straight line can be used to estimate the fracture extension if the formation permeability is known. For a transverse circular fracture of radius r_f, the authors express m_{LF} as:

$$m_{\text{LF}} = 5.17 \frac{qB}{h r_f} \sqrt{\frac{\mu}{\phi c_t k_H}} \tag{3.52}$$

For a rectangular fracture of horizontal extension $2x_f$, a relationship similar to Equation 3.45 is obtained:

$$m_{\text{LF}} = 4.06 \frac{qB}{h x_f} \sqrt{\frac{\mu}{\phi c_t k_H}} \tag{3.53}$$

On a log-log derivative plot, the sequence of characteristic straight lines is, after wellbore storage,

1. *first stabilization* in case of transverse fracture (radial-linear flow) or *quarter unit slop*e with longitudinal fracture (bilinear flow),
2. *half unit slope* during formation linear flow
3. *final stabilization* during formation pseudo radial flow.

The fracture conductivity determines the location of the first derivative straight line (stabilization or 1/4 slope). For high conductivity fractures, the derivative response is low during the radial-linear or bilinear flow regimes, the corresponding early time straight line is moved down on the log-log scale, and the formation linear flow develops early. It is shown in Section 3.6.3 that for non-fractured horizontal wells, the linear flow 1/2 slope defines the effective well length. In the case of fractured horizontal wells, it gives the *horizontal extension* of the fracture. With long fractures, the 1/2 slope derivative straight line is displaced towards late times.

For multi-fractured horizontal wells, the different fractures produce independently until interference effects between neighboring fractures are felt. Then, a compound linear flow develops before the final pseudo radial flow regime.

At early time, if the independent fractures have similar characteristics, the response is directly proportional to the number of fractures and can be analyzed with a single fracture model by dividing the flow rate by the number of fractures (Larsen and Hegre, 1994 a; Raghavan et al., 1997). Radial-linear (transverse fractures) or bilinear flow regimes (longitudinal fractures) can be analyzed on such multi-fractured horizontal well responses. Later, linear flow and pseudo radial flow around the different fracture segments (when the distance between the fractures is large) can also be identified. Once the interference between the fractures is felt, the response deviates like in the case of partially open horizontal wells presented in Section 3.6.9. The end of the compound linear flow regime, and start of the final pseudo radial flow, is independent of the number of fractures but depends only on the distance between the outermost fractures.

3.6.11 Horizontal wells in reservoirs with changes of permeability

In the following, it is shown that two types of reservoir heterogeneities affect the analysis results of horizontal well responses, even though the overall well behavior is apparently homogeneous. The influence of horizontal permeability anisotropy is first discussed. In layered reservoir, changes of permeability in the vertical direction can reduce the ability of vertical flow during the early time response.

Horizontal permeability anisotropy

With horizontal wells, it takes frequently a long time before the final horizontal radial flow regime is established. In the case of horizontal permeability anisotropy, the well response is sensitive to the well orientation (Goode and Thambynayagam, 1987; Kamal et al., 1993).

With the three directions of permeability defined on Figure 3.40, the characteristic regimes of an horizontal well response are controlled by a different permeability:

1. At early time, the average permeability during the vertical radial flow is $\sqrt{k_z k_y}$.
2. During the linear flow regime, only the permeability k_y normal the well orientation is acting.
3. The final horizontal radial flow regime defines the average horizontal permeability $\overline{k}_H = \sqrt{k_x k_y}$ of Equation 3.3.

When the isotropic horizontal permeability model is used for analysis, the vertical permeability k_z is unchanged but the apparent effective half-length is:

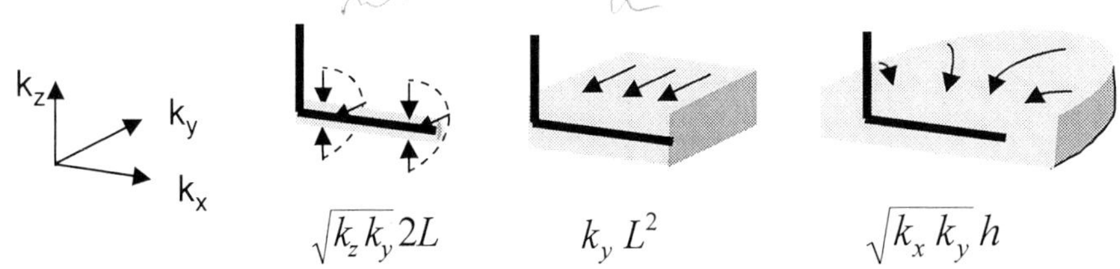

Figure 3.40. Horizontal permeability anisotropy.
Effective permeability during the three characteristic flow regimes towards a horizontal well.

$$L_a = \sqrt[4]{k_y / k_x}\, L \tag{3.54}$$

Equation 3.54 shows that, if the horizontal well is in the *minimum* (maximum) permeability direction, apparent effective length *increased* (decreased).

Horizontal wells in vertically heterogeneous reservoirs

Even though the homogeneous reservoir model is currently used for many well test analysis, most reservoirs are stratified and permeability varies with depth. In most cases, variations of horizontal permeability with depth do not alter significantly the horizontal radial flow regime (see Section 4.2) but, as horizontal wells responses are also sensitive to vertical flow, the changes of vertical permeability over the producing thickness affect the response.

In the following, the horizontal well model of Kuchuk and Habashy (1996) for a multi-layer reservoir with crossflow is used to evaluate the effect of vertical changes of k_V. It is shown that when the heterogeneity between the different layers is moderate, the homogeneous reservoir model can be used to provide average permeability in both horizontal and vertical directions. Conversely, when horizontal wells are completed in formations with several interbeds of reduced permeability between the main layers, the single homogeneous layer model considered in the previous sections is not appropriate for accurate analysis (Suzuki and Nanba, 1991). Finally, as horizontal drilling is a common practice in reservoirs with a gas cap or lower water drive to prevent coning or cresting, the effect of a constant pressure upper or lower boundary is discussed.

On the example Figure 3.41, the reservoir is described as a three-layer system. The horizontal well is centered in layer 2, layers properties are defined in Table 3.6.

Table 3.6. Layered system of Figure 3.41

Layer	h_i	k_{Hi}	k_{Vi}	$(k_V / k_H)_i$
1	30	15	1.2	0.08
2	30	10	0.5	0.05
3	40	8	0.24	0.03

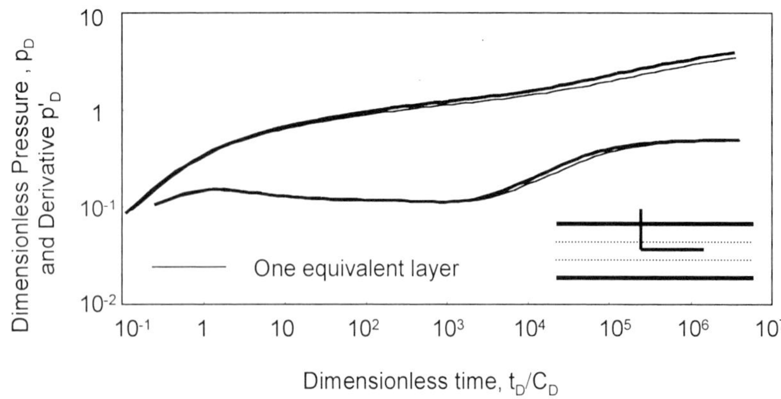

Figure 3.41. Horizontal well in a three layers reservoir with crossflow. Log-log scales, p_D versus t_D/C_D. C_D =100, L =1000ft, S_w =0, h =100ft (30+30+40), r_w =0.25ft, z_w/h =0.55, k_{H1}/k_{H2}=1.5, k_{H3}/k_{H2}=0.8, $(k_V/k_H)_1$=0.08, $(k_V/k_H)_2$=0.05, $(k_V/k_H)_3$=0.03.

The thin curves of Figure 3.41 describe the response of the same horizontal well in the equivalent homogeneous layer. The two model responses appear very similar, the use of the *homogeneous layer* approximation is acceptable. For a n layer system, the average horizontal permeability is defined (Section 4.2.5) as:

$$\bar{k}_H = \sum_1^n k_i h_i \Big/ \sum_1^n h_i \tag{3.55}$$

For the vertical flow, the changes of permeability are acting in series. The resulting average vertical permeability estimated during the vertical radial flow is defined with the *average vertical* permeability above, and below the horizontal drain. If the well is centered in layer j :

$$\bar{k}_V = 0.5\left(\frac{\sum_1^{j-1} h_i + h_j/2}{\sum_1^{j-1} h_i/k_i + h_j/2k_j} + \frac{\sum_{j+1}^{n} h_i + h_j/2}{\sum_{j+1}^{n} h_i/k_i + h_j/2k_j} \right) \tag{3.56}$$

Equations 3.55 and 3.56 are applicable to the example Figure 3.41 with n=3 and j=2: $\bar{k}_H = 10.7$ and $\bar{k}_V = 0.5(0.82 + 0.28) = 0.55$.

On Figure 3.42, a low permeability zone is inserted in the producing interval: the horizontal well is located in layer 3, below the semi-permeable wall (layer 2). The response shows first the vertical radial flow regime around the wellbore in layer 3 and, when both the bottom boundary and the low permeability interbed are reached, it tends to deviates into a linear flow regime as if layer 3 was isolated (the thin dashed curves describe the response of the horizontal well if layer 2 is sealing). Later, a crossflow is established through the semi-permeable wall and layer 1 participates to the production.

The derivative deviates below the half-unit slope straight line in a transition, and finally reaches the stabilization when the flow becomes horizontal pseudo-radial.

A match of the stratified reservoir response with an equivalent homogeneous model is presented with the thin curves. The average horizontal permeability is defined by Equation 3.55. Due to the early deviation above the first derivative stabilization when the semi-permeable wall is reached, the effective well length used for this match is only 55% of the true length. L being *under estimated*, the vertical permeability resulting from the vertical radial flow stabilization is *too large* (3.8 times the vertical permeability in layers 1 and 3, and 6.7 times the average vertical permeability of Equation 3.56).

The presence of interbeds with very low k_V, in a otherwise homogeneous reservoir, affects the shape of horizontal well response curves and consequently the productivity. On the stratified reservoir example Figure 3.42, the total skin S_{TH} of the horizontal well is S_{TH} =-6.48. In case of a non-rectilinear well with a segment of L /2 in layers 1 and 3, the total skin would be lower at S_{TH} =-6.53 and, without layer 2 (homogeneous reservoir) it is S_{TH} =-6.78.

Kuchuk and Habashy (1996) use the layered reservoir model to describe the influence of a gas cap or bottom water drive on horizontal well responses. Since in the model boundaries between layers are horizontal planes, they assume that the interface between the fluids is not moving or distorted by cresting during the production. In the example of Figure 3.43, the horizontal well is located at the bottom of a layer overlaid by a gas cap. The sequence of regimes is vertical radial flow and hemi-radial flow until the gas interface is reached. Later, due to the large mobility and compressibility of the top gas region, the pressure tends to stabilize and shows the influence of a constant pressure boundary similar to the partial penetration example of Figure 3.21. If the thickness of the gas cap is not large enough, the response deviates from the constant pressure upper boundary behavior, and finally stabilizes to describe the total mobility of the oil and the gas zones.

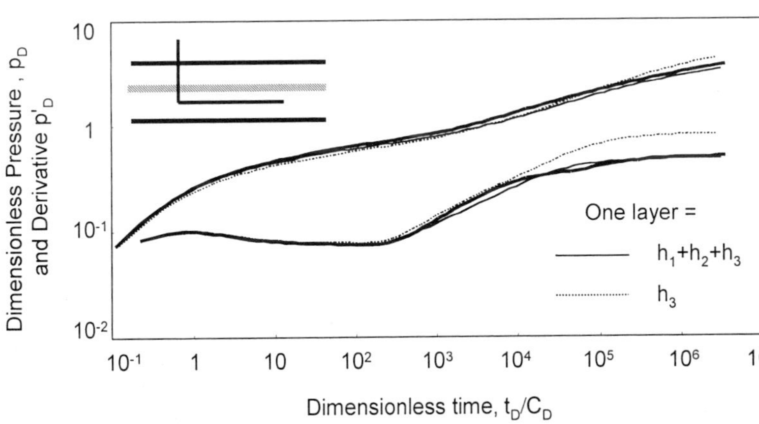

Figure 3.42. Horizontal well close to a low permeability interbed. Log-log scales. C_D =100, L =1000ft, S_w =0, h =100ft (45+5+50), r_w =0.25ft, z_w /h =0.25, k_{H1}=k_{H3}=100 k_{H2}, $(k_V/k_H)_i$=0.1.

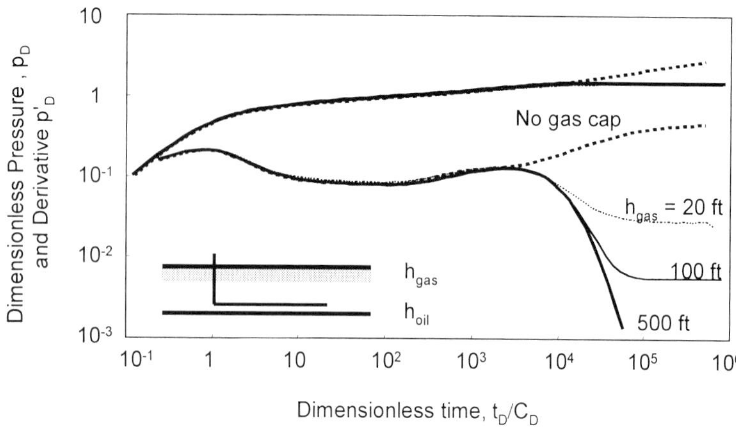

Figure 3.43. Horizontal well in a reservoir with gas cap. C_D =100, L =1000ft, S_w =2, h_{oil}=100ft, r_w =0.25ft, z_w / h_{oil}=0.2, $(k_V / k_H)_i$=0.1, $h_{\text{gas}}/h_{\text{oil}}$ =0.2, 1 and 5. $\mu_{\text{gas}}/\mu_{\text{oil}}$ =0.01, $c_{t\text{gas}}/c_{t\text{oil}}$ =10.

Fleming et al. (1994) observed that many build-up tests from horizontal wells in a fissured reservoir with a large gas cap show several oscillations on the late time derivative response. They explain this phenomenon by the changes of saturation as the gas recedes during shut-in. The gas movement within the fracture network can be stepping, with intermittent liberation of gas pockets. Horizontal wells in double porosity reservoirs are further discussed in Section 4.1.4. In addition, multiphase reservoirs are presented in Chapter 8.

3.6.12 Multilateral horizontal wells

In single layer homogeneous reservoirs, the behavior of wells with multiple horizontal drain-holes follows a logic similar to partially open and multi-fractured horizontal wells, discussed in previous sections:

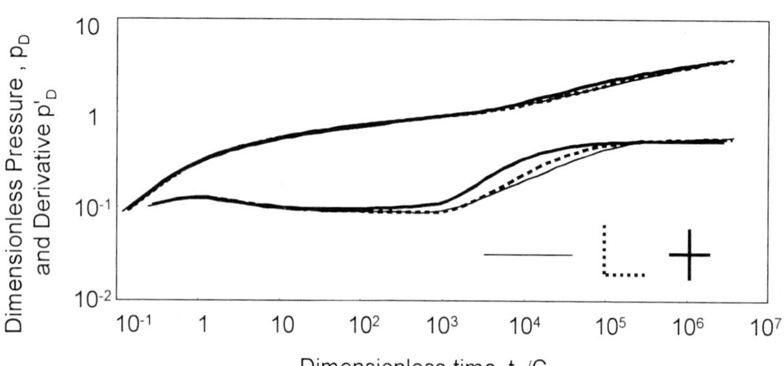

Figure 3.44. Multilateral horizontal well. Log-log scales, p_D versus t_D/C_D. C_D =100, L =1000ft (500+500 and 250+250+250+250), S_{wi} =0, h =100ft, r_w =0.25ft, k_V / k_H =0.1, z_w / h =0.5.

- At early time, the different branches produce *independently* and, when the different drain-holes have the same skin, the behavior is equivalent to a single horizontal well with a total effective length defined as the sum of the lengths of all branches.
- Later, the response deviates due to *interference effects* between the different horizontal sections. The flow geometry is a function of both horizontal and vertical distances between the branches, and orientation. An analytical simulator is required to properly interpret the well response.
- Finally, *pseudo radial flow* towards the multilateral horizontal well can develop.

On Figure 3.44, two examples of multilateral horizontal well responses are compared to the horizontal well of similar total length. The drain-hole sections are perpendicular with two and four branches (L and + shape). At intermediate time, the interference effects produce an increase of the pressure response, and the derivative deviates above the half unit slope straight line of the single drain horizontal well curve. No mechanical skin damage is assumed on the three curves. The total skin S_{TH} of the horizontal well is S_{TH} =-6.8 whereas for the multilateral well examples S_{TH} is respectively -6.6 and -6.2 with the L and + geometries.

For a given total effective length, increasing the number of intersecting branches does not improve the productivity of horizontal wells in reservoirs with isotropic horizontal permeability (Larsen, 1996 a; Salas et al., 1998). When the horizontal perforated segments do not intersect, Larsen shows that the total skin S_{TH} can be expressed as a function of the dimensionless distance r_D between the segments, with a decreasing function of ln r_D. On the examples Figure 3.45 where the distance between the two producing segments is large enough, the response becomes independent of the orientation of the branches and the total skin of the two multilateral horizontal wells is S_{TH}=-7.1 (more negative than S_{TH}=-6.8 with one branch). The responses Figure 3.45 tend to be equivalent to the example with two segments of Figure 3.37.

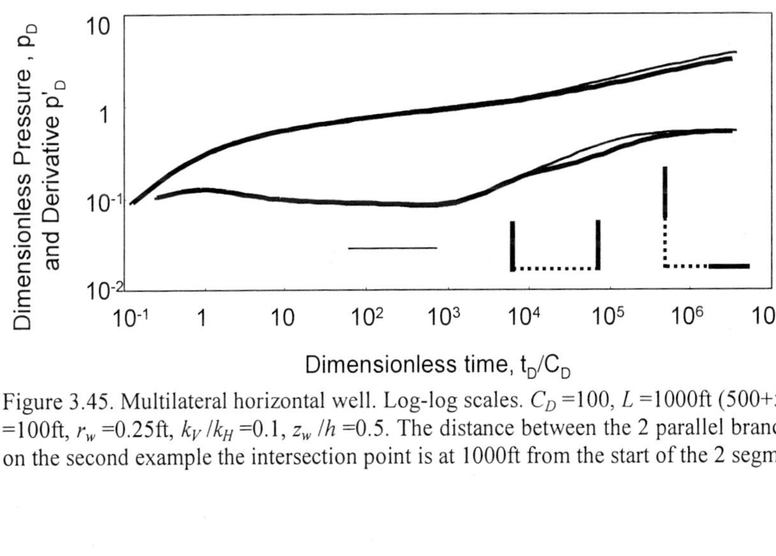

Figure 3.45. Multilateral horizontal well. Log-log scales. C_D =100, L =1000ft (500+500), S_{wi} =0, h =100ft, r_w =0.25ft, k_V/k_H =0.1, z_w/h =0.5. The distance between the 2 parallel branches is 2000ft, on the second example the intersection point is at 1000ft from the start of the 2 segments.

3.7 SKIN FACTORS

3.7.1 Components of the total skin

The different *components* contributing to the total skin S_T measured on well test responses are summarized on table 3.7 below.

Table 3.7. Components of the total skin S_T

Name	Description	Type
S_w	Infinitesimal skin at the wellbore.	Positive or negative
S_G	Geometrical skin due to the streamline curvature (fractured, partial penetration, slanted or horizontal wells).	Positive or negative
S_{ani}	Skin factor due to the anisotropy of the reservoir permeability.	Negative
S_{RC}	Skin factor due to a change of reservoir mobility near the wellbore (permeability or fluid property, radial composite behavior).	Positive or negative
$S_{2\phi}$	Skin factor due to the fissures in a double porosity reservoir.	Negative
$D.q$	Turbulent or inertial effects on gas wells.	Positive

The geometrical skin S_G has been discussed in previous Sections for various well configurations. In the following, the relationship between S_G and *derivative curves* is demonstrated by comparing three simple example responses. Negative skin produced by natural fissures is discussed in the double porosity Section 4.1.5, and turbulence effects are described in the gas well Chapter 7.

3.7.2 Geometrical skin and derivative curves

The magnitude of the geometrical skin is easy to visualize when the derivative response is considered. This can be illustrated by the theoretical response of three wells of radius r_w, producing in the same homogeneous reservoir (Figure 3.46). Well **A** is a fully penetrating *vertical well*, well **B** is in *partial penetration*, and well **C** is a *horizontal well*. For the three wells, the infinitesimal skin S_w is set to 0.

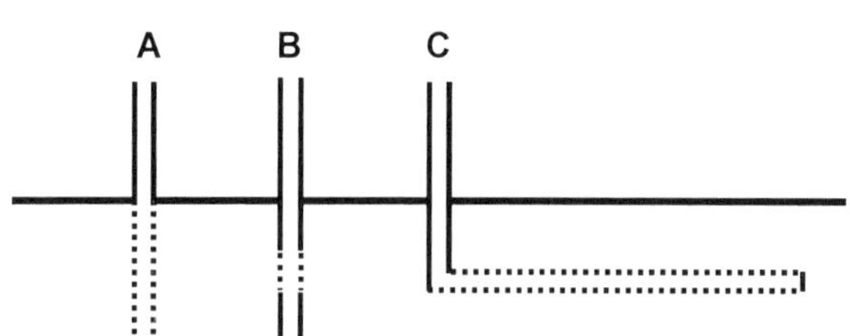

Figure 3.46. Configuration of well **A**, **B** and **C**.
A = fully penetrating vertical well, **B** = well in partial penetration and **C** = horizontal well.

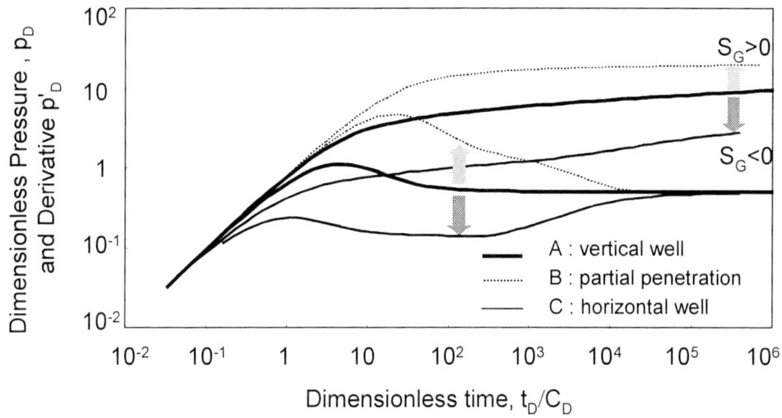

Figure 3.47. Pressure and derivative responses of well **A** (vertical), **B** (partial penetration) and **C** (horizontal). Log-log scales, p_D versus t_D/C_D.

In Figure3.47, the derivative response of the vertical well shows the usual stabilization when the wellbore storage is over. In the case of partial penetration well **B**, a first derivative stabilization is seen during the radial flow in front of the perforated interval. The derivative response is above that of the vertical well until $t_D / C_D = 10^4$, the *area* between the two curves is a measure of the positive geometrical skin. The larger this surface, the larger is the skin due to partial penetration. In terms of pressure response, the partial penetration curve **B** is above the curve for the vertical well.

For the horizontal well **C**, the derivative response stabilizes at a low level during the vertical radial flow and the resulting geometrical skin is negative. The longer is the horizontal well, the larger is the area below the vertical well derivative response, and the more negative is the total skin.

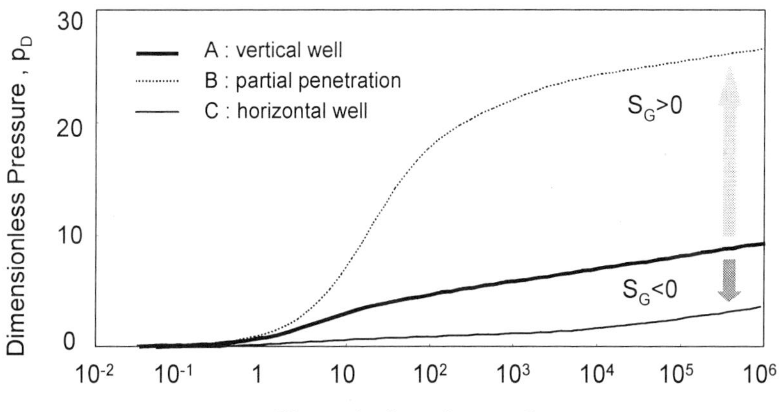

Figure 3.48. Semi-log plot of Figure 3.47 examples.

The influence of the geometrical skin on the pressure response of wells **A**, **B** and **C** is illustrated on semi-log scale Figure 3.48.

CHAPTER 4

EFFECT OF RESERVOIR HETEROGENEITIES ON WELL RESPONSES

Heterogeneous reservoir models have attracted a lot of attention in the last twenty years. The first solutions were presented in the 60's, but these only became commonly used much more recently.

Reservoir heterogeneities are identified by variations in the pressure response. Sometimes the pressure data deviates from the homogeneous behavior only during the first minutes of the test period under investigation, in other cases it takes from several hours to several days before the heterogeneity becomes evident. The introduction of high accuracy pressure measurements and computerized log-log analysis technique explains today's recent use of heterogeneous interpretation models. In addition, the derivative of pressure exaggerates the characteristic features of the response.

In this chapter, the different heterogeneous reservoir models used in well test analysis are discussed. It is assumed that the well is affected by wellbore storage and skin only, but other wellbore conditions presented in Chapter 3 can be encountered in heterogeneous formations. An infinite reservoir is considered; all outer boundary effects are presented in Chapter 5.

The basic heterogeneous solutions assume two different behaviors are combined in the reservoir response. They are described as double porosity models (restricted or unrestricted interporosity flow), double permeability models and composite systems (radial or linear interfaces). These three basic models are thoroughly presented in this chapter. The influence of the different parameters is described, and the analysis of build-up tests in heterogeneous formations is discussed.

For each model, the extension of the basic solution to a larger number of elementary behaviors is considered (multi-porosity systems for changing matrix blocks sizes, multi-layer systems and multi-composite formations). The double porosity matrix skin theory is also discussed in detail. In the final section of the Chapter, different combinations of heterogeneous solutions are presented for fissured-layered systems and composite fissured or layered formations.

4.1 FISSURED RESERVOIRS

Among the different heterogeneous interpretation models, the double porosity solutions have been the most frequently discussed in the technical literature. They assume the existence of two porous regions within the formation. One region, of *high conductivity*, is called the *fissures* whereas the other, of *low conductivity*, is called the *matrix blocks*. As described in Figure 4.1, the concept of double porosity is representative of the behavior of fissured and multiple-layer formations, when the permeability contrast between layers is high (the "fissure system" describes the high permeability layers, and the "matrix blocks" the tight zones).

The double-porosity model was first introduced by Barenblatt et al. in 1960: a low permeability porous system, the matrix blocks, is surrounded by a fissure network of high permeability. The matrix blocks are not producing to the well, but only to the fissures. Several variations of this model are available (Warren and Root, 1963; Odeh, 1965; Kazemi, 1969 a; de Swaan, 1976; Najurieta, 1980; Streltsova, 1983) for refined descriptions of the heterogeneous response. In all cases, the fissure network provides the *mobility*, and the matrix blocks supply most of the *storage capacity*. A double porosity response depends upon the storativity contrast between the two reservoir components, and the quality of the communication between them.

The basic assumptions used for the double porosity solutions are discussed in Section 4.1.1. Two types of flow from matrix to fissures are considered, depending upon the presence of minerals in the fissure network that reduce the flow from matrix to the fissures. The *restricted interporosity flow* hypothesis, also called the Warren and Root model, or pseudo-steady state interporosity flow model, was first available for transient test analysis. This model is discussed in Section 4.1.2 for a well with wellbore storage and skin. The *unrestricted interporosity flow* hypothesis is then presented in Section 4.1.3.

In Section 4.1.4, double porosity behavior is discussed in case of flow regimes other than radial flow, and extensions of the model to matrix skin and to multiple block size are considered. The double porosity models have also been extended to different well conditions presented in Chapter 4 (Housé et al., 1998, for infinite conductivity fracture; Cinco-Ley and Menh, 1988, for finite conductivity fracture and, for horizontal well, Carvalho and Rosa, 1988; Aguilera and Ng, 1991; Du. and Stewart, 1992). The effect of a sealing fault (Khachatoorian et al., 1995) and other reservoir boundaries are reviewed in Chapter 5.

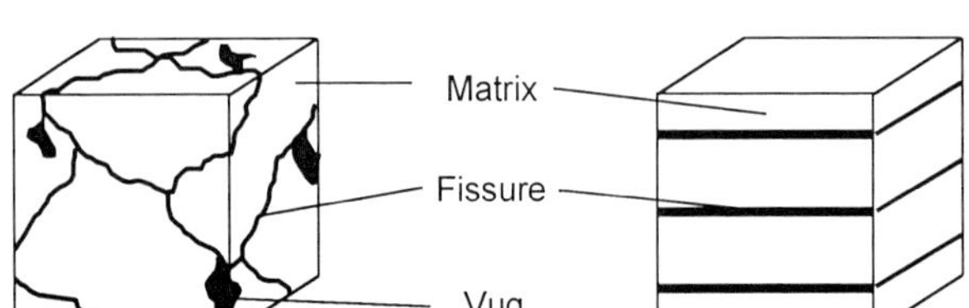

Figure 4.1. Example of double porosity reservoir, fissured and multiple-layer formations.

4.1.1 Double porosity models

Fissured reservoirs are complex. The density of the fissure network can vary with position in the reservoir, as a function of the rock stresses due to curvature of the formation. The orientation of the fissures can induce permeability anisotropy. The mathematical models for fissured reservoirs use a simplified description of the heterogeneous system. The parameters resulting from the interpretation define the idealized model, but they do not describe the geological configuration in detail. In the following, we summarize the different assumptions used in the equations for the models, and present the resulting dimensionless variables.

Basic assumptions

1 - The dimensions of the matrix blocks are *small* compared to the reservoir volume involved in the test. Each point in the reservoir is associated with two pressures, namely p_f the pressure of the fluid in the fissures, and p_m, the pressure of the fluid in the matrix pore volume.

2 - The fluid flows to the well through the fissure system only; the matrix blocks are *not connected* (or the radial permeability of the matrix system is negligible, $k_{m,\ \mathrm{rad.}} = 0$). The isolated matrix blocks are described as source terms in the fissure element, and the mobility measured during the test corresponds to the fissure system alone.

3 - Most of the reservoir fluid is *stored* in the matrix blocks porosity, the storage of the fissure network is only a small fraction of the reservoir storage.

4 - Three matrix block *geometries* are usually considered, depending upon the number n of fissure plane directions.

For $n = 3$, the matrix blocks are cubes (spheres are also described with three directions of fissure planes) but $n = 2$ (cylinder matrix blocks) and $n = 1$ (slab matrix blocks) can also be envisaged.

5 - Two different types of *matrix to fissure flow* have been considered:

In the first solution, as described by Barenblatt et al. in 1960, it is assumed that the flow of fluid from blocks to fissures occurs under *pseudo-steady state* conditions. The model was extended in the present form by Warren and Root (1963). Moench (1984) and Cinco-Ley et al. (1985) demonstrated later that it describes a *restricted* interporosity flow condition, when there is a skin effect between the matrix and the fissures, making the pressure gradient in the matrix blocks negligible.

The second type of interporosity flow described by several authors (Kazemi, 1969 a; de Swaan, 1976; Najurieta, 1980; Streltsova, 1983), considers *transient flow* in the matrix blocks. There is *no flow restriction* at the matrix - fissure interface, and the matrix blocks response starts earlier.

6 - In the double porosity models, all matrix blocks are *homogeneous*, and they have the same size. Other multiple porosity solutions consider different matrix block sizes, either uniformly distributed in the reservoir, or organized according to several possible geometries. They are discussed in Section 4.1.4 and 4.4.

Behavior

When a well is opened in a fissured reservoir, a rapid pressure response occurs in the fissure network due to its high diffusivity. A *pressure difference* is created between matrix and fissure, and the matrix blocks start to produce into the fissures. The pressure of the matrix blocks p_m decreases as flow progresses and, finally, tends to equalize with the pressure of the surrounding fissures p_f.

Definitions

In the permeability thickness product *kh*, an equivalent permeability is used. From condition 1, the fissure system is assumed to be uniformly distributed in all the reservoir thickness but, in practice, the fissures involve only a fraction of the pay zone thickness *h*. The equivalent distributed permeability (*bulk fissure permeability*) k_f is a function not only of the actual fissures thickness and intrinsic permeability, but also of the fissure network characteristics (such as tortuosity and fissure connectivity when material separates individual fractures).

$$kh = k_f h_f \tag{4.1}$$

Two porosities are defined in double porosity systems. We call ϕ_f and ϕ_m the ratio of pore volume in the fissures (or in the matrix), to the total volume of the fissures (of the matrix). V_f is the ratio of the total volume of the fissures to the reservoir volume, and V_m that of the total volume of the blocks to the reservoir volume ($V_f + V_m = 1$). The average reservoir porosity ϕ is given by:

$$\phi = \phi_f V_f + \phi_m V_m \tag{4.2}$$

In fissured formations, both ϕ_f and V_m are close to 1. The average porosity of Equation 4.2 can be simplified as:

$$\phi = V_f + \phi_m \tag{4.3}$$

Frequently, V_f is called the fissure porosity.

The storativity ratio ω expresses the contrast between the two porous systems:

$$\omega = \frac{(\phi V c_t)_f}{(\phi V c_t)_f + (\phi V c_t)_m} = \frac{(\phi V c_t)_f}{(\phi V c_t)_{f+m}} \tag{4.4}$$

ω defines the contribution of *the fissure system to the total storativity*. Usual values for ω are in the order of 10^{-1} for multiple-layer systems down to 10^{-2} or 10^{-3} for fissured formations: the fissures provide only a fraction of the total storativity.

In case of multiple-layer systems, matrix blocks and fissures are represented by horizontal slabs. h_m and h_f being the cumulative thickness of the "matrix" and the "fissure" layers, the volume ratios of Equation 4.2 are $V_f = h_f / (h_f + h_m)$ and $V_m = h_m / (h_f + h_m)$. In Equation 4.1, the equivalent permeability is expressed as $k = k_f V_f$.

A second heterogeneous parameter, called interporosity flow coefficient λ, is used to describe the ability of the matrix blocks to *flow* into the fissures. λ, as expressed by Warren and Root (1963), is a function of the matrix blocks geometry and permeability k_m.

$$\lambda = \alpha r_w^2 \frac{k_m}{k} \tag{4.5}$$

where α is related to the geometry of the fissure network. It is a function of the number n of families of fissure planes:

$$\alpha = \frac{n(n+2)}{r_m^2} \tag{4.6}$$

r_m is the characteristic size of the matrix blocks. It is defined as the ratio of the volume V of the matrix blocks, to the surface area A of the blocks with:

$$r_m = n\frac{V}{A} \tag{4.7}$$

If the matrix blocks are spheres (of radius r_m) or cubes (of side $2r_m$), n=3, and α is expressed:

$$\alpha = \frac{15}{r_m^2} \tag{4.8}$$

For n = 2, the matrix blocks are cylindrical of radius r_m, and:

$$\alpha = \frac{8}{r_m^2} \tag{4.9}$$

for n=1, the matrix blocks are slabs of thickness $2r_m$, and α is:

$$\alpha = \frac{3}{r_m^2} \tag{4.10}$$

λ defines the *communication* between the matrix blocks and the fissures. When λ is small, the fluid transfer from matrix to fissure is difficult, and it takes a long time before the double porosity model behaves like the equivalent homogeneous total system. Such behavior is obtained for example, when the matrix is tight, and the permeability k_m is small. Low density of fissures is another example of poor matrix communication: the characteristic block size r_m is large, and α is small. Usual values for λ are in the range of 10^{-4} to 10^{-10}.

In the definition of λ Equation 4.5, the matrix skin is not considered. In case of restricted interporosity flow, λ does not describe completely the matrix flow condition and an *effective interporosity flow parameter* λ_{eff} should be used (Equation 4.39 of Section 4.1.4).

Dimensionless variables

The definition of the dimensionless pressure is the same as for homogeneous reservoirs (Equation 2.3). The reference thickness corresponds to the total zone, and the permeability is the equivalent permeability k introduced in Equation 4.1:

$$p_D = \frac{kh}{141.2qB\mu}\Delta p \tag{2.3}$$

The porosity is included in the definitions of the dimensionless time Equation 2.4 and the dimensionless wellbore storage of Equation 2.5. Depending upon the reference selected, two definitions of these dimensionless parameters are used.

When the dimensionless time is expressed with the fissure system permeability and storativity, t_{Df} is:

$$t_{Df} = \frac{0.000264k}{(\phi V c_t)_f \mu r_w^2}\Delta t \tag{4.11}$$

When the reference is the total system, the dimensionless time t_{Df+m} is:

$$t_{Df+m} = \frac{0.000264k}{(\phi V c_t)_{f+m} \mu r_w^2}\Delta t \tag{4.12}$$

For the dimensionless wellbore storage coefficient, the two definitions of the storativity give respectively:

$$C_{Df} = \frac{0.8936C}{(\phi V c_t)_f h r_w^2} \tag{4.13}$$

$$C_{Df+m} = \frac{0.8936C}{(\phi V c_t)_{f+m} h r_w^2} \tag{4.14}$$

ω correlates the definitions of dimensionless time and wellbore storage:

$$t_{Df+m} = \omega t_{Df} \tag{4.15}$$

$$C_{Df+m} = \omega C_{Df} \tag{4.16}$$

4.1.2 Double porosity behavior, restricted interporosity flow model (Wellbore storage and skin)

In 1963, Warren and Root presented the double porosity solution described in this section. The flow from matrix blocks to the fissures is assumed to be *pseudo-steady state* regime.

In the original Warren and Root model, no wellbore storage effect was considered. It was introduced by Mavor and Cinco-Ley in 1979 and, Bourdet and Gringarten (1980) presented a pressure type-curve expressed in terms of independent variables (Figure 4.2), extended on Figure 4.6 to the derivative approach (Bourdet et al, 1983 b, 1989).

Moench demonstrated in 1984 that the apparent pseudo-steady state flow regime in the matrix blocks is the result of damage at the surface of the blocks. The fissures are partially plugged by mineral deposition or by chemical precipitation, but they include some channels allowing the fluid to flow towards the well. The matrix feed the channels, but the flow first has to cross the thin low permeability deposit layer on the walls of the fissures.

The *matrix skin* theory (also called interporosity skin) provides a link between the pseudo-steady state matrix flow condition discussed here and the transient interporosity solution presented in Section 4.1.3: the different mathematical approaches describe two limiting cases of the same reservoir configuration. The influence of the matrix skin S_m is further discussed in Section 4.1.4 of this chapter. It is shown that, for large interporosity skin S_m, the pseudo-steady state hypothesis of Warren and Root's (1963) is a realistic approximation of the matrix flow condition (Moench, 1984; Cinco-Ley et al., 1985; Stewart and Ascharsobbi, 1988).

In the following section, we assume strictly pseudo-steady state matrix flow: the corresponding restricted interporosity flow model provides an *effective interporosity flow parameter* λ_{eff} given in Equation 4.39, and not λ of Equation 4.5. (note that the effective interporosity flow parameter λ_{eff} is called λ for purpose of conciseness).

Behavior

With restricted interporosity flow, three different regimes can be observed on a producing well response:

1. First a *fissure flow*, when the matrix contribution is negligible. This corresponds to a homogeneous behavior, where only the fissure system is producing.
2. At intermediate times, during a *transition* regime, the response deviates from the fissure homogeneous behavior as the matrix blocks start to produce into the fissures. The pressure tends to stabilize to a constant value.
3. Later, the pressure of the matrix blocks equalizes with the pressure of the surrounding fissures. A new homogeneous behavior is reached, called the *total system flow* regime.

All the fluid flows to the wellbore through the fissures alone: the two homogeneous behaviors are characterized by the permeability thickness product kh of the fissure system (Equation 4.1). The first homogeneous regime corresponds to the fissure storativity, whereas the second involves the total storativity of Equation 4.2. The transition between the two homogeneous behaviors describes an increase of storativity (see also Sections 4.3.2 and 10.2.2), the pore volume of the fissures being a small fraction of the total.

During the two homogeneous regimes, the pressure response can exhibit a straight-line behavior on semi-log scale. The first straight line corresponding to fissure flow, the second to the total system regime. The permeability thickness kh being the same during the two homogeneous regimes, the lines are parallel (Figure 4.4).

More frequently, tests in fissured reservoirs do not show the characteristic "*two parallel semi-log straight lines*": either the first line is masked by wellbore storage effect, or the test period is too short to show the second. Many examples of analysis with double porosity type curves show that the occurrence of parallel semi-log straight lines is in fact exceptional. Furthermore, the characteristic features of double porosity responses can be identified in other regimes than the infinite acting radial flow (see discussion Section 4.1.4).

Pressure type-curve

The Bourdet and Gringarten type-curve of Figure 4.2 (1980) describes drawdown responses for a well in a reservoir with double porosity behavior, restricted interporosity matrix to fissure flow. The well can be damaged (positive skin) or stimulated (using the equivalent wellbore radius concept) and it is affected by a constant wellbore storage effect.

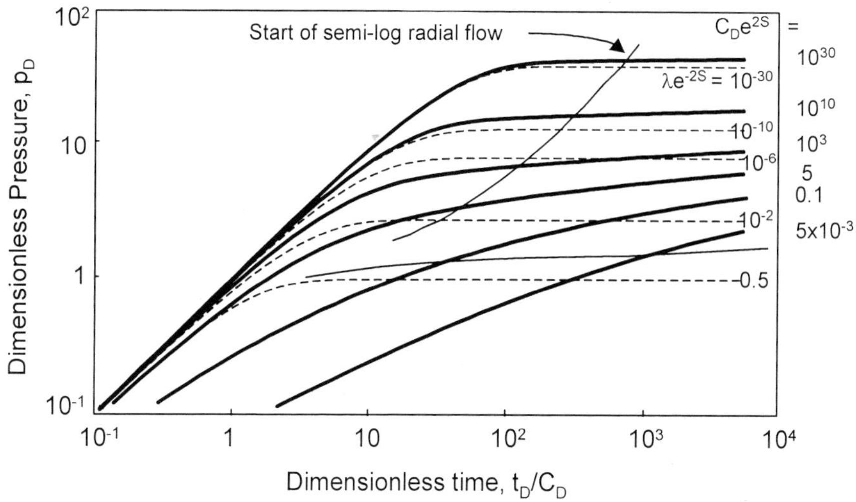

Figure 4.2. Pressure type-curve for a well with wellbore storage and skin in a double porosity reservoir, pseudo steady state interporosity flow. Log-log scales, p_D versus t_D/C_D.

Independent variables

On the type-curve of Figure 4.2, the same groups of independent variables as on homogeneous reservoir type-curve (Figure 2.2) are used. The dimensionless pressure p_D of Equation 2.3 is expressed versus the dimensionless time group t_D/C_D:

$$\frac{t_D}{C_D} = \frac{t_{Df}}{C_{Df}} = \frac{t_{Df+m}}{C_{Df+m}} = 0.000295 \frac{kh}{\mu} \frac{\Delta t}{C} \tag{4.17}$$

The pressure match and the time match are independent of the storativity. On a log-log plot, the same match point describes the two homogeneous regimes, fissure and total system flow.

Three curves are needed to define the different regimes of a double porosity response. The corresponding independent parameters are:

1 - $(C_D e^{2S})_f$ at early time, during *fissure flow*

$$\left(C_D e^{2S}\right)_f = \frac{0.8936 Ce^{2S}}{(\phi Vc_t)_f hr_w^2} \tag{4.18}$$

2 - λe^{-2S} during *transition* regime, between the two homogeneous behaviors

3 - $(C_D e^{2S})_{f+m}$ at late time, when *total system flow* behavior is reached

$$\left(C_D e^{2S}\right)_{f+m} = \frac{0.8936 C e^{2S}}{\left(\phi V c_t\right)_{f+m} h r_w^2} \tag{4.19}$$

Description

On the pressure type-curve of Figure 4.2, *two families of curves* are superimposed: the $C_D e^{2S}$ curves describe the fissure and the total system homogeneous regimes, the dashed λe^{-2S} curves describe the transition.

From the definition of the $C_D e^{2S}$ parameter (Equations 4.18 and 4.19), a double porosity response goes from a high value $(C_D e^{2S})_f$ when the storativity corresponds to fissures, to a lower value $(C_D e^{2S})_{f+m}$ when total system is acting.

The type-curve of Figure 4.2 describes the *components* of a double porosity response, not the actual curve. For illustration, two examples of double porosity responses are presented on the type curve Figure 4.3. With example **A**, the response follows first the fissure homogeneous curve $(C_D e^{2S})_f = 1$, then a transition on $\lambda e^{-2S} = 3\text{x}10^{-4}$ and finally it reaches the total system homogeneous behavior on $(C_D e^{2S})_{f+m} = 10^{-1}$.

On the pressure type-curve, the limit "approximate start of the semi-log straight line" is shown by a dashed line. Figure 4.3 indicates that example **A** shows a semi-log straight line during each of the two homogeneous behaviors as illustrated on the semi-log plot Figure 4.4. The two semi-log straight lines are parallel.

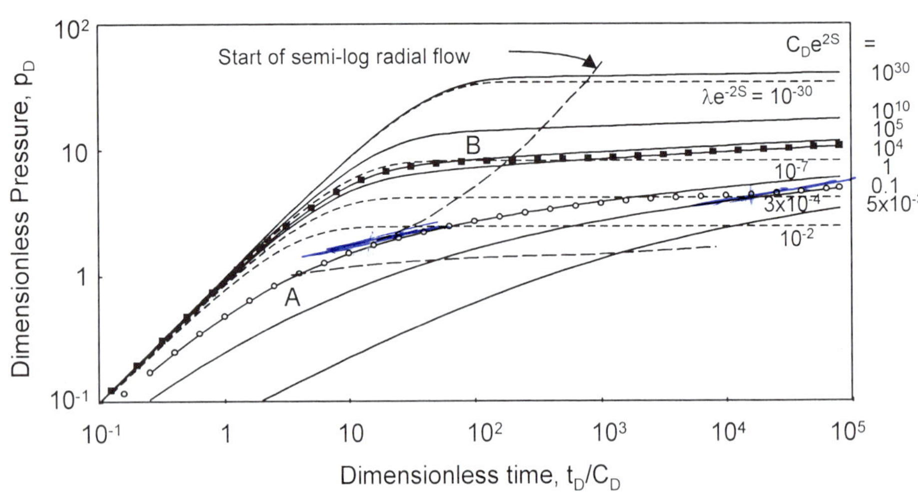

Figure 4.3. Examples for a well with wellbore storage and skin in a double porosity reservoir, pseudo steady state interporosity flow. Log-log scales, p_D versus t_D/C_D.
A : $(C_D e^{2S})_f = 1$, $(C_D e^{2S})_{f+m} = 0.1$, $\omega = 0.1$, $\lambda e^{-2S} = 3\text{x}10^{-4}$.
B : $(C_D e^{2S})_f = 10^5$, $(C_D e^{2S})_{f+m} = 10^4$, $\omega = 0.1$, $\lambda e^{-2S} = 10^{-7}$.

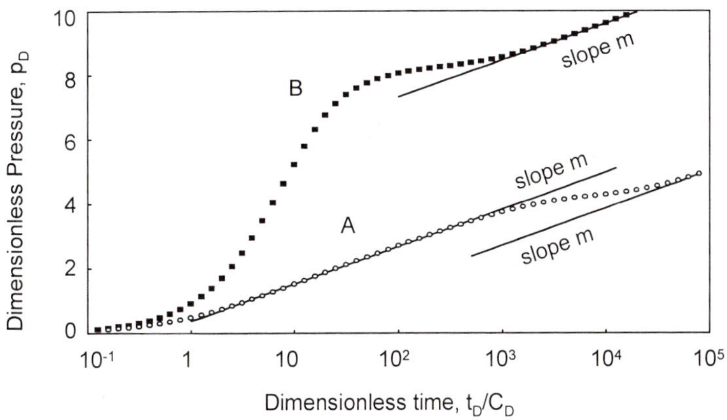

Figure 4.4. Semi-log plot of Figure 4.3 examples.

On example **B**, the pressure response goes from $(C_D e^{2S})_f = 10^5$ to $(C_D e^{2S})_{f+m} = 10^4$ through a transition on $\lambda e^{-2S} = 10^{-7}$. When the fissure regime ends, the storage effect is still present: response **B** shows only one semi-log straight line, during the second homogeneous regime when the total system is acting (Figure 4.4).

Pressure type-curve analysis

As for the homogeneous reservoir type-curve of Figure 2.2, analysis with the double porosity type-curves of Figure 4.2, yields:

1. the permeability thickness product kh from the pressure match,

$$kh = 141.2 \Delta q B \mu (\text{PM}) \tag{2.9}$$

2. the wellbore storage coefficient C from the time match,

wellbore storage coefficient

$$C = 0.000295 \frac{kh}{\mu} \left(\frac{1}{\text{TM}} \right) \tag{2.10}$$

3. the skin factor S from one of the two homogeneous curves. In general, the total system regime is used:

$$S = 0.5 \ln \frac{\left(C_D e^{2S} \right)_{f+m}}{C_{Df+m}} \tag{4.20}$$

The two heterogeneous parameters are estimated from the curve match.

4. comparing the two $C_D e^{2S}$ values gives the storativity ratio ω:

$$\omega = \frac{\left(C_D e^{2S}\right)_{f+m}}{\left(C_D e^{2S}\right)_f} \tag{4.21}$$

5. during transition, the λe^{-2S} curve describing the pressure stabilization defines the interporosity flow parameter.

$$\lambda = \left(\lambda e^{-2S}\right) e^{2S} \tag{4.22}$$

Frequently, the transition between the two homogeneous $C_D e^{2S}$ curves is too short to reach a stabilized pressure behavior. The choice of the λe^{-2S} curve is then defined on the middle point of the transition.

Derivative type-curve

The two double porosity examples **A** and **B** are presented on Figure 4.5 with the pressure and derivative. During the fissure flow, this homogeneous regime is described on the derivative response by a $C_D e^{2S}$ curve and, when semi-log radial flow is reached, the derivative stabilizes on 0.5 in dimensionless terms. At transition time, the flattening of the pressure curve is changed into an obvious *valley* on the derivative response. Later, the derivative returns to the 0.5 stabilization during the total system equivalent homogeneous behavior.

- With example **A**, the wellbore storage effect ends during fissure regime, and a first radial flow is seen before the start of transition. Two parallel semi-log straight lines are present on the semi-log plot Figure 4.4. On Figure 4.5, the derivative reaches the 0.5 line, both before and after the transition valley.

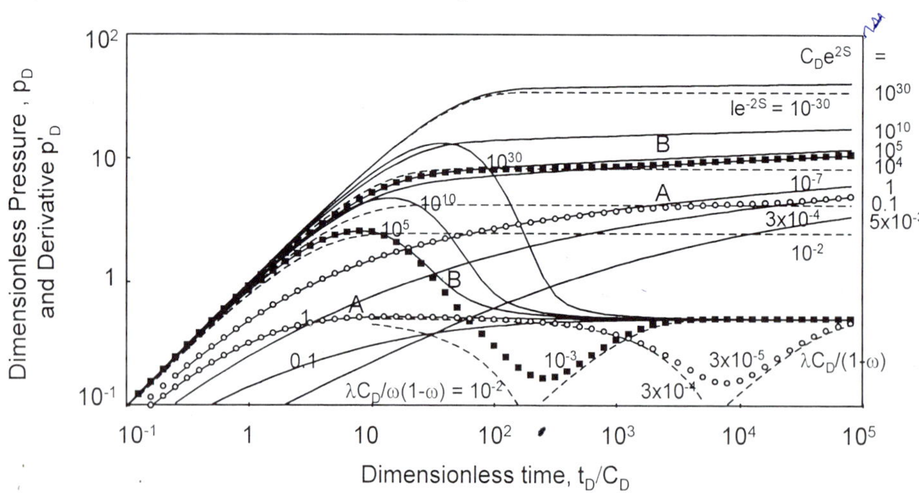

Figure 4.5. Pressure and derivative examples of Figure 4.3 for a well with wellbore storage and skin in a double porosity reservoir, pseudo steady state interporosity flow. Log-log scales.

- With example **B**, storage is still present when the transition starts: the semi-log curve of Figure 4.4 shows only one straight line, during total system flow. On Figure 4.5, the derivative response goes directly from the wellbore storage hump to the transition valley, and the first 0.5 plateau is not seen.

The two examples of Figure 4.5 illustrate that, as opposed to the log-log pressure curves, the derivative emphasizes the small variations of behavior characterizing double porosity responses. The pressure and derivative type curve for a well with wellbore storage and skin in a double porosity reservoir, restricted interporosity flow, is presented Figure 4.6 (Bourdet et al., 1983 b).

Description

With the pressure type curve of Figure 4.2, a double porosity response is defined by three components: $(C_D e^{2S})_f$, λe^{-2S} and $(C_D e^{2S})_{f+m}$. On the type curve Figure 4.6, the derivative requires *four components*: the $(C_D e^{2S})_f$ curve and the 0.5 line are used for the homogeneous behaviors but, at transition time, two other curves are needed.

After wellbore storage, the derivative valley during transition can be expressed as:

$$p'_D(t_D/C_D) = 0.5\left\{ 1 + e^{-\left[\frac{\lambda C_{Df+m}}{\omega(1-\omega)}\frac{t_D}{C_D}\right]} - e^{-\left[\frac{\lambda C_{Df+m}}{(1-\omega)}\frac{t_D}{C_D}\right]} \right\} \tag{4.23}$$

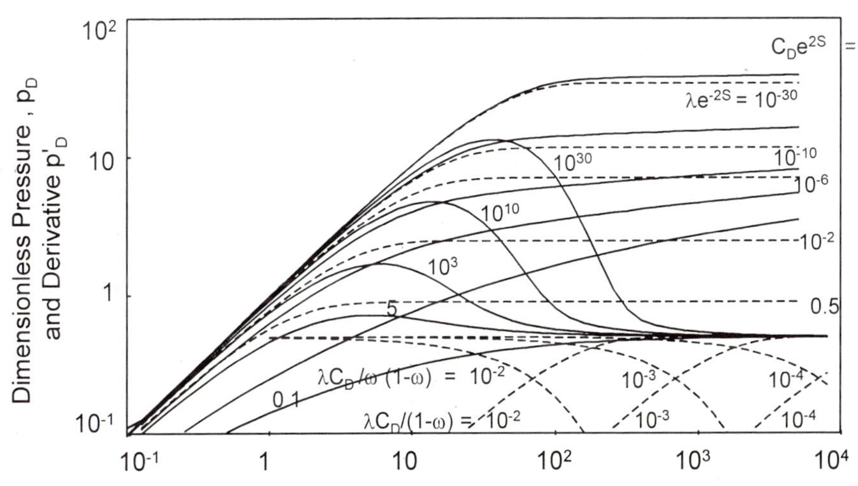

Figure 4.6. Pressure and derivative type-curve for a well with wellbore storage and skin in a double porosity reservoir, pseudo steady state interporosity flow. Log-log scales, p_D versus t_D/C_D.

The *early transition curve*, corresponding to the decreasing derivative trend, is labeled $(\lambda C_{Df+m})/[\omega(1-\omega)]$ on Figure 4.6. When the derivative returns to 0.5, the *late transition curve* is defined with the $(\lambda C_{Df+m})/(1-\omega)$ group.

With the example **B** of Figure 4.5, the transition starts when the response is still under the influence of wellbore storage and the double porosity curve deviates from the early transition curve $(\lambda C_D)/[\omega(1-\omega)]$ (at 1.11×10^{-2}). Storage being over at late transition times, the example matches on the $(\lambda C_D)/(1-\omega)$ curve (1.11×10^{-3}).

Matching procedure with the pressure and derivative type curve

As for the type-curve for homogeneous system with storage and skin, the derivative straight lines are used to select the match point on the type-curve Figure 4.6: the unit slope line during wellbore storage, and the 0.5 line during radial flow. From the pressure and time matches, *kh* and *C* are estimated (Equations 2.9 and 2.10).

A type-curve match is defined by six components curves: three components for the pressure response [$(C_D e^{2S})_f$, λe^{-2S} and $(C_D e^{2S})_{f+m}$], and three derivative curves: $(C_D e^{2S})_f$, $(\lambda C_{Df+m})/[\omega(1-\omega)]$ and $(\lambda C_{Df+m})/(1-\omega)$.

The match is adjusted by trial and errors, until consistency is achieved between pressure and derivative results:

- A first check is made on the early time fissure flow analysis: the $(C_D e^{2S})_f$ parameter of the pressure and derivative curves must be the same.

$$\left(C_D e^{2S}\right)_{f\ \text{PRESSURE}} = \left(C_D e^{2S}\right)_{f\ \text{DERIVATIVE}}$$

- The storativity ratio is estimated from the two transition curves of the derivative match. It is compared to ω found with Equation 4.21.

$$\omega = \text{late transition}/\text{early transition} = \frac{\lambda C_D}{(1-\omega)} \bigg/ \frac{\lambda C_D}{\omega(1-\omega)} \tag{4.24}$$

- The interporosity flow parameter is preferably calculated from the late transition curve. It has to be in agreement with λ estimated from pressure response (Equation 4.22).

$$\lambda = \frac{\lambda C_{Df+m}}{(1-\omega)} \frac{(1-\omega)}{C_{Df+m}} \tag{4.25}$$

In general, double porosity responses do not follow perfectly the component curves, except for short periods of time: the responses show *transitional behavior* between one component curve to the next. Matching on the pressure and derivative type-curve of Figure 4.6 is performed by interpolation between component curves, and frequently reiterations are needed before the final solution is reached. With some experience, the adjustment converges quickly, and no more than two or three tentative are necessary.

Discussion of double-porosity parameters

In this section, we describe the influence of the two heterogeneous parameters ω and λ on log-log and semi-log curves. This shows how the shape of a double porosity response curve can be adjusted on actual data, when the match is made by computer.

Influence of ω

The storativity ratio defines the contrast between the fissure regime and the total system regime. It relates the two homogeneous pressure curves $(C_D e^{2S})_f$ and $(C_D e^{2S})_{f+m}$. With small ω values, the two curves are very different and, on pressure curves, the transition regime from $(C_D e^{2S})_f$ to $(C_D e^{2S})_{f+m}$ is long and flat.

Figure 4.7 and 4.8 presents on log-log and semi-log scales three examples of curves generated for different values of ω (10^{-1}, 10^{-2} and 10^{-3}), C_{Df+m}, S and λ being the same.

On the derivative responses of Figure 4.7, the depth of the characteristic valley is a function of the transition duration. For small ω values, *long transition* regimes correspond *to deep valleys* on derivative. The minimum of the valley is given, from Equation 4.23, by:

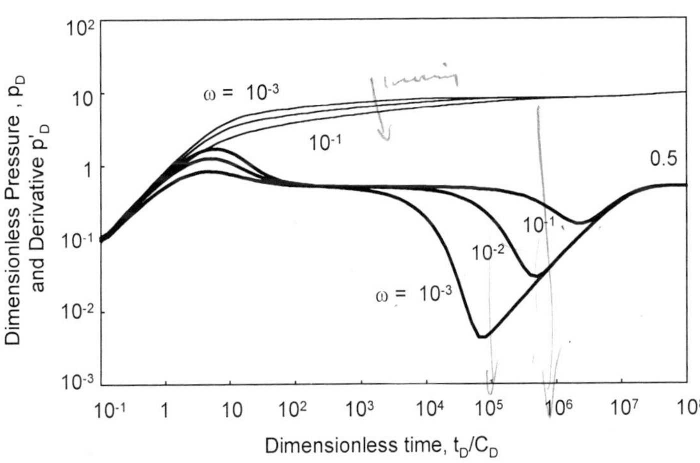

Figure 4.7. Influence of ω on pressure and derivative log-log curves, p_D versus t_D/C_D. $C_{Df+m}=1$, $S=0$, $\lambda=10^{-7}$, $\omega=10^{-1}$, 10^{-2} and 10^{-3}

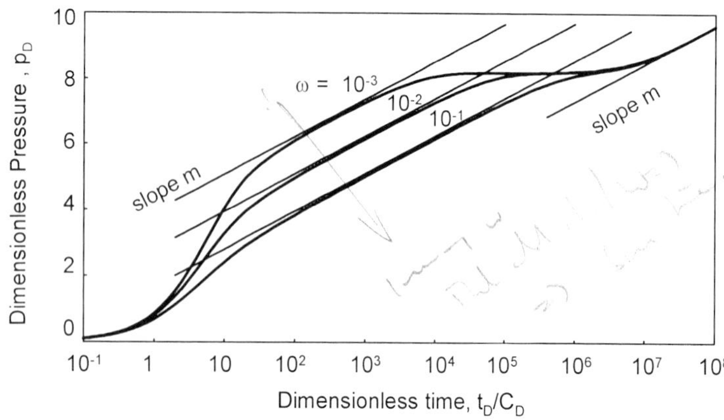

Figure 4.8. Influence of ω on semi-log plot of the Figure 4.7 examples.

$$p'_D\left(t_D/C_D\right)=0.5\left[1+\omega^{\left(\frac{1}{1-\omega}\right)}-\omega^{\left(\frac{\omega}{1-\omega}\right)}\right] \tag{4.26}$$

For a given total system curve $(C_D e^{2S})_{f+m}$, a decrease of ω value produces an increase of the fissure curve parameter $(C_D e^{2S})_f$. On the semi-log plot Figure 4.8, the first semi-log straight line is displaced upwards, the horizontal transition between the two parallel lines is longer.

Influence of λ

The interporosity flow parameter defines the ability of the matrix blocks to produce into the fissure system. The previous example Figures 4.7 and 4.8 are generated for the same value of λ: as shown by the log-log and semi-log curves, λ determines *the time of end* of the transition, and the start of the equivalent homogeneous total system flow regime.

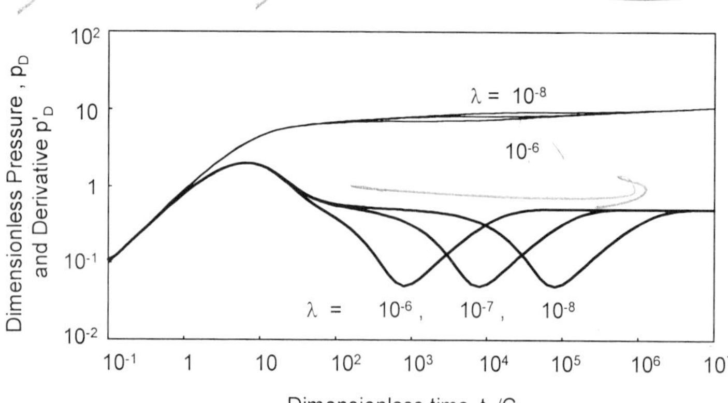

Figure 4.9. Influence of λ on pressure and derivative log-log curves, p_D versus t_D/C_D. $C_{Df+m}=100$, $S=0$, $\omega=0.02$, $\lambda=10^{-6}$, 10^{-7} and 10^{-8}.

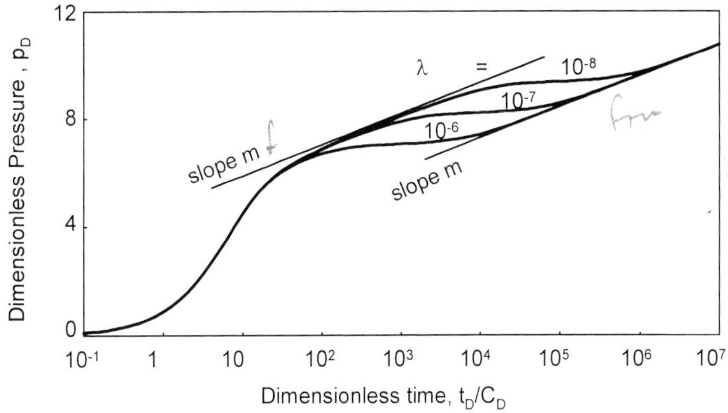

Figure 4.10. Influence of λ on semi-log plot of the Figure 4.9 examples.

On Figures 4.9 and 4.10, the three curves are generated for $\lambda = 10^{-6}$, 10^{-7} and 10^{-8}. The smaller is λ, the later the start of total system flow. On the pressure curves, the transition regime occurs at a higher amplitude and, on the derivative responses, the transition valley is displaced towards late times.

Associated specialized plot straight lines

Once the match is defined, the different flow regimes are identified, and specialized analysis can be carried out on selected time intervals.

Wellbore storage analysis

A Cartesian scale is used for wellbore storage analysis (Section 1.2.2): the slope of the early time straight line gives, by Equation 1.10, an estimate of the wellbore storage coefficient.

As for homogeneous reservoirs discussed in Chapter 3, wellbore storage analysis in fissured formations is in general feasible only on damaged wells, when data points match the unit slope log-log straight line.

Radial flow analysis

On semi-log scale, the presence of two parallel straight lines has been considered as the characteristic feature of double porosity responses (Figure 4.4, example **A**). In fact, experience shows that this configuration is exceptional, the wellbore storage effect frequently masks the first semi-log straight line during fissure flow (Figure 4.4, example **B**). When the first line is present, the fissure flow regime lasts a relatively long period, and the test often stops before the second radial flow regime is reached.

Even when the first semi-log straight line is not present, a semi-log plot exhibits a characteristic "S" shape (Figures 4.4, example **B** and 4.10, high λ value), indicating a heterogeneous response. This characteristic shape is illustrated on Figure 4.11: the double porosity curve crosses the semi-log straight line, as opposed to the curve for a homogeneous response.

The semi-log analysis of two parallel straight lines is valid only when both radial flow regimes are clearly identified from the log-log analysis. During fissure flow, the first line is expressed, from Equation 1.15:

$$\Delta p = 162.6 \frac{qB\mu}{kh}\left[\log \Delta t + \log \frac{k}{(\phi V c_t)_f \mu r_w^2} - 3.23 + 0.87S\right] \tag{4.27}$$

And the second line, for the total system regime:

$$\Delta p = 162.6 \frac{qB\mu}{kh}\left[\log \Delta t + \log \frac{k}{(\phi V c_t)_{f+m} \mu r_w^2} - 3.23 + 0.87S\right] \tag{4.28}$$

The vertical distance δp between the two lines gives ω (Warren and Root, 1963):

$$\omega = 10^{-\frac{\delta p}{m}} \tag{4.29}$$

where m is the semi-log slope. In the definition of δp, the first straight line should not be simply defined as a tangent to the data curve in the early time region, drawn parallel to the second line. Such an approximation can give an under estimated value of ω.

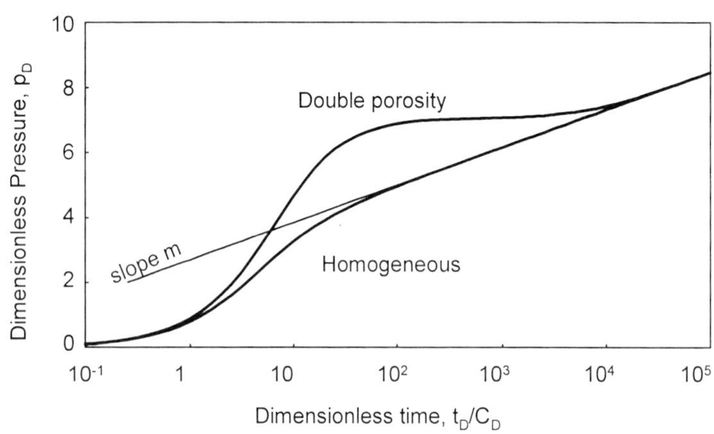

Figure 4.11. Semi-log plot of homogeneous and double porosity examples. $C_D = C_{Df+m} = 100$, $S = 0$, $\omega = 0.01$ and $\lambda = 10^{-6}$.

The interporosity flow coefficient λ is not accessible with accuracy from semi-log analysis.

When only the first semi-log straight line of fissure regime is present, the double porosity nature of the response can be overlooked by mistake. By using the total storativity instead of that of the fissure system, the calculation of the skin gives an over estimated value S_f. From Equations 4.27 and 4.28,

$$S_f = S + 0.5 \ln \frac{1}{\omega} \tag{4.30}$$

Build-up analysis

Drawdown periods are difficult to analyze and, frequently, only build-up data is used for interpretation. Build-up analysis in double porosity reservoirs, however, is a lot more complicated than for homogeneous formations. The main reason is that the behavior changes in the course of the response: for different production times, the shape of the build-up curves can show different characteristic features.

Log-log pressure build-up analysis

Build-up pressure type-curves for a well with wellbore storage and skin in a double porosity reservoir can be constructed as the drawdown type-curve of Figure 4.2: the family of $C_D e^{2S}$ component curves is replaced by build-up curves generated for the appropriate production history; the λe^{-2S} transition curves are not changed (Bourdet and Gringarten, 1980).

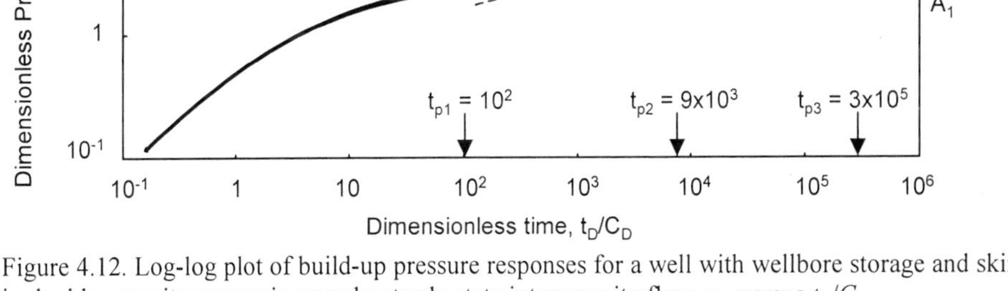

Figure 4.12. Log-log plot of build-up pressure responses for a well with wellbore storage and skin in double porosity reservoir, pseudo-steady state interporosity flow. p_D versus t_D/C_D. $C_{Df+m} = 0.1$, $S = 0$, $\omega = 0.1$ and $\lambda = 3\text{x}10^{-4}$. $t_{pD}/C_D = 100$ (**A₁**), 9000 (**A₂**), 300000 (**A₃**).

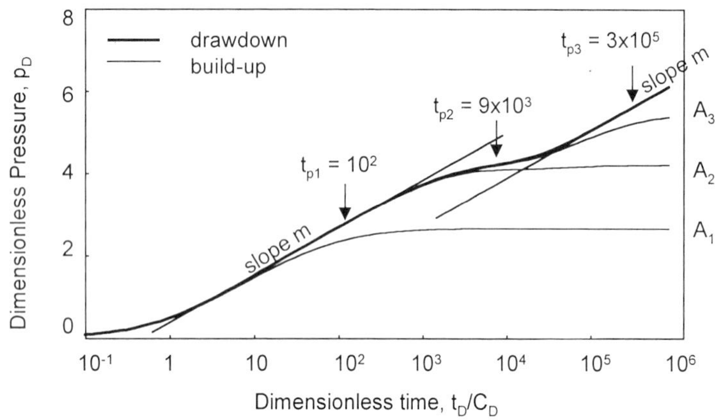

Figure 4.13. Semi-log plot of drawdown and build-up pressure responses of Figure 4.12.

On the log-log plot Figure 4.12, the drawdown double porosity example **A** of Figure 4.4 is compared to three build-up responses generated for different production times t_p. With example $\mathbf{A}_1$, $(t_p/C)_D = 100$, the drawdown prior to shut-in stopped in fissure flow, with example $\mathbf{A}_2$ $((t_p/C)_D = 9000)$ it ended during transition, and in example $\mathbf{A}_3$ $((t_p/C)_D=3\text{x}10^5)$ in total system flow.

For clarity, the same drawdown and build-up curves are presented on semi-log scale in Figure 4.13, similar to the build-up discussion of Figure 2.7. However long the three build-ups are, only example $\mathbf{A}_3$ exhibits a clear double porosity pressure response. On example $\mathbf{A}_1$, the drawdown stops in fissure flow, before the λe^{-2S} transition curve is reached. The build-up curve $\mathbf{A}_1$ appears to show only the build-up response of the fissures, and not a double porosity behavior. In example $\mathbf{A}_2$, the build-up curve flattens at the same Δp level as the λe^{-2S} transition and there is no evidence of total system flow regime.

Figure 4.12 illustrates the lack of definition of log-log pressure analysis. The three build-up curves are generated by superposition of drawdown solutions, and all characteristic features of the double porosity model are theoretically present in the responses. Due to the build-up effect, they cannot always be identified on a log-log pressure plot.

Horner & superposition analysis

Figure 4.14 is a Horner plot of the three build-up examples of Figure 4.12. With example $\mathbf{A}_1$, only *one semi-log straight line* is obtained during shut-in. It represents the fissure behavior. The straight line can be used to compute kh and, if the fissure porosity is known, the skin factor S. The *extrapolation* at infinite shut in time gives the *correct* $p^*=p_i$. When the total system storativity is used, the semi-log straight-line analysis provides only a maximum value of the skin, S_f of Equation 4.30. In the test sequence of example $\mathbf{A}_1$, no double porosity effect is apparent from either drawdown or build-up when only the pressure data is considered for analysis.

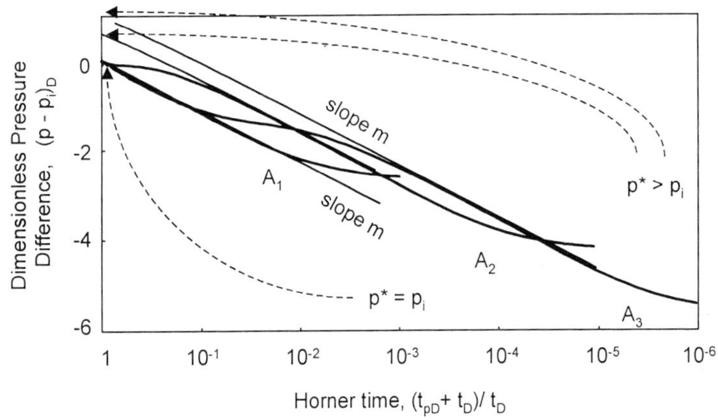

Figure 4.14. Horner plot of the three build-up examples of Figure 4.12. $t_{pD}/C_D = 100$ ($\mathbf{A_1}$), 9000 ($\mathbf{A_2}$), 300000 ($\mathbf{A_3}$).

If the drawdown test is stopped during the transition, (example $\mathbf{A_2}$), only *one semi-log straight line* develops on the Horner plot. This straight line corresponds to the fissure regime. It has the proper slope but does *not extrapolate* to p_i. The intercept at infinite shut-in time is between p_i and $p_i + m.\log(1/\omega)$, depending upon t_p. On Figure 4.14, the response leaves the Horner straight line at late time, and *flattens* to reach p_i. As for curve $\mathbf{A_1}$, the use of the total system storativity for the fissure regime semi-log straight-line analysis leads to an over estimated skin value, between S and S_f.

Finally, in example $\mathbf{A_3}$, the total system radial flow regime is reached at shut-in time and, provided the build-up period lasts long enough, *two parallel straight lines* are present on the Horner plot. These can be used to estimate kh, S and ω (Equation 4.29). p_i is obtained by extrapolation of the *second straight line*, the first extrapolates to $p_i + m \log(1/\omega)$.

Figure 4.14 illustrates this: if the heterogeneous response has not been clearly identified, Horner or superposition analysis of double porosity build-up responses can be misleading. When the double porosity nature of the response is ignored, not only the calculation of the skin can be wrong but, more importantly, *the extrapolated pressure* to infinite shut-in time can be *over estimated.* In addition, the shape of the semi-log superposition curve, where the pressure flattens at late time (examples $\mathbf{A_2}$ or $\mathbf{A_3}$ if the build-up stops during the transition), can be interpreted by mistake as the effect of a depleted closed system (see Figure 5.25, Section 5.4.7).

Derivative build-up analysis

The derivative using the Horner / superposition time corrects the influence of the previous production history, except in the case when the drawdown response changes during the extrapolation into the build-up period (see Section 2.3.4). In double porosity responses, when the well is closed before the total system flow, the drawdown response changes after shut-in from fissure to total system regime, and the derivative method does not correct perfectly the build-up curve as illustrated Figure 4.15.

In Figure 4.15, the derivative responses of the three build-up examples of Figure 4.12 are compared to the original drawdown derivative curve. Example $\mathbf{A_3}$ matches the drawdown response but not the two other build-up curves. For example $\mathbf{A_2}$ *the transition happens later,* and the deviation is more pronounced and even more so during the build-up $\mathbf{A_1}$, which corresponds to a short production history limited to fissure system.

The derivative curve $\mathbf{A_1}$ reveals a double porosity behavior whereas the same data plotted in terms of pressure, either on log-log or semi-log scale (Figure 4.12 and 4.14) suggests a homogeneous behavior. The significance of the signal is absorbed by the compression effect of the build-up correction methods (see Section 2.2.2 Figure 2.6 for log-log analysis, and Figure 2.7 for semi-log analysis).

In practice, build-up derivative data after a short drawdown does not always display a full double porosity response as on synthetic example of curve $\mathbf{A_1}$. With a small production time, the pressure builds up quickly to the initial pressure, and the response is barely changing at late times. The derivative then becomes scattered and a clear diagnosis is difficult. However, when a late time downwards trend of the derivative is observed on test data, the hypothesis of a double porosity response should not be neglected.

Practice of build-up tests in double porosity reservoirs

The previous discussion clearly demonstrates the importance of a careful test design in double porosity systems. As discussed with the pressure examples in Figure 4.3, two test conditions have to be satisfied in order to display a full double porosity response during drawdown:

1. the early time fissure flow regime should not be *masked by wellbore storage*,
2. the *analyzed period has to be long enough* to show the late time total system flow.

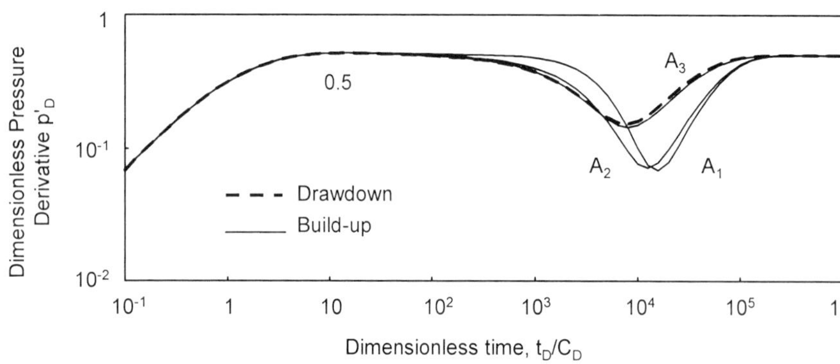

Figure 4.15. Log-log plot of drawdown and build-up derivative responses of Figure 4.12 examples. Log-log scales, p_D versus t_D/C_D.

For build-up responses, not only these two conditions must be met but, as illustrated on Figures 4.12 to 4.15, the *previous period* should ideally have been also *long enough* to reach the final total system regime.

4.1.3 Double porosity behavior, unrestricted interporosity flow model (Wellbore storage and skin)

In this section, the effect of transient flow from blocks to fissures is considered. As opposed to the pseudo steady state interporosity flow model presented in Section 4.1.2, there is no skin effect at the surface of the blocks. The transient interporosity flow solution is also called unrestricted matrix flow.

Transient matrix flow has been studied by several authors, and two matrix blocks geometries, slab or sphere, are usually considered. Following the theory developed by de Swaan in 1976, Bourdet and Gringarten (1980) presented a pressure type-curve for a well with wellbore storage and skin in a double porosity reservoir with unrestricted interporosity flow. The type curve was later extended to describe derivative responses (Bourdet et al., 1984).

Behavior

In the case of unrestricted interporosity flow, the matrix blocks react almost immediately to any change of pressure in the fissures: the transition starts earlier than in case of restricted flow, and the fissure flow regime is generally not seen. Only two flow regimes are observed with this double porosity solution:

1. At early time, both fissure and matrix are producing, but the rate of change of pressure is faster in the fissure system than in the matrix blocks. The first response observed is in *transition* regime.
2. Later, the *homogeneous* behavior corresponding to the *total system* is reached.

Pressure type-curve

The pressure type-curve of Figure 4.16 describes drawdown responses for a well with wellbore storage and skin in a double porosity reservoir with unrestricted interporosity flow. As on the type-curve of Figure 4.2 for restricted interporosity flow, the well can be damaged or stimulated. The same dimensionless terms are used (Section 4.1.1).

Independent variables

On the pressure type-curve of Figure 4.16, the dimensionless pressure p_D is expressed versus the dimensionless time group t_D/C_D. The dimensionless pressure and time are defined with respect to the equivalent permeability (Equations 2.3 and 4.17).

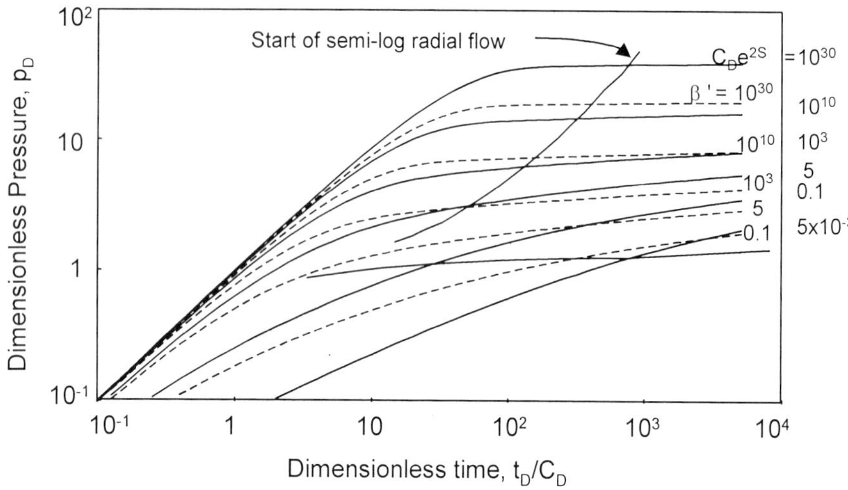

Figure 4.16. Pressure type-curve for a well with wellbore storage and skin in a double porosity reservoir, transient interporosity flow. Log-log scales, p_D versus t_D/C_D.

The homogeneous regime is described with a family of $(C_D e^{2S})_{f+m}$ curves defined in Equations 4.19, and the early time transition regime by a second set of curves, labeled β'. The β' dimensionless group is expressed as:

$$\beta' = \delta' \frac{\left(C_D e^{2S}\right)_{f+m}}{\lambda e^{2S}} \qquad (4.31)$$

The constant δ' is related to the matrix system geometry. For *slab* matrix blocks,

$$\delta' = 1.89 \qquad (4.32)$$

and for *sphere* matrix blocks:

$$\delta' = 1.05 \qquad (4.33)$$

Description

On the pressure type-curves of Figure 4.16, *two families of curves* are superimposed: the β' curves for transition regime, and the $(C_D e^{2S})_{f+m}$ curves for the total system homogeneous regime. The two families of curves have the same shape: the β' transition curves are equivalent to $C_D e^{2S}$ curves whose pressure and time are divided by a factor of two (Bourdet and Gringarten, 1980).

As shown on Figure 4.17, the match of a double porosity response with transient interporosity flow is adjusted on two component curves: at early time, a β' curve describe the transition behavior; at late time, the total system homogeneous regime matches on a $(C_D e^{2S})_{f+m}$ curve. With example **A**, the response follows first the transition curve on $\beta'=10^6$ and finally reaches the total system homogeneous behavior on $(C_D e^{2S})_{f+m}=10$. With example **B**, the pressure response starts on $\beta'=10^{10}$ and reaches later $(C_D e^{2S})_{f+m}=6\text{x}10^3$.

Since the shape of the β' transition curves is similar to that of the $C_D e^{2S}$ homogeneous curves, a semi-log straight line can develop during the transition. The β' curves are equivalent to $C_D e^{2S}$ curves displaced by a factor of two: the first line, in transition, has a *slope half* of the second. The slope of the second gives the proper estimate of the equivalent permeability thickness product *kh*.

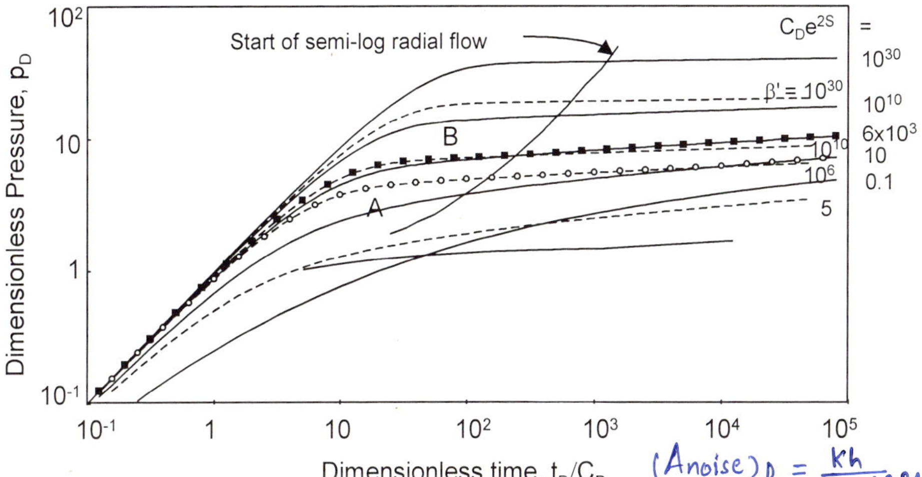

$(A_{noise})_D = \frac{kh}{141.2qB\mu}(A_{noise})$

Figure 4.17. Examples for a well with wellbore storage and skin in a double porosity reservoir, transient interporosity flow, slab matrix blocks. Log-log scales, p_D versus t_D/C_D.
A : $(C_D e^{2S})_f = 10^4$, $(C_D e^{2S})_{f+m} = 10$, $\omega = 0.001$, $\beta' = 10^6$, $\lambda e^{-2S} = 1.89\text{x}10^{-5}$.
B : $(C_D e^{2S})_f = 6\text{x}10^6$, $(C_D e^{2S})_{f+m} = 6\text{x}10^3$, $\omega = 0.001$, $\beta' = 10^{10}$, $\lambda e^{-2S} = 1.13\text{x}10^{-6}$.

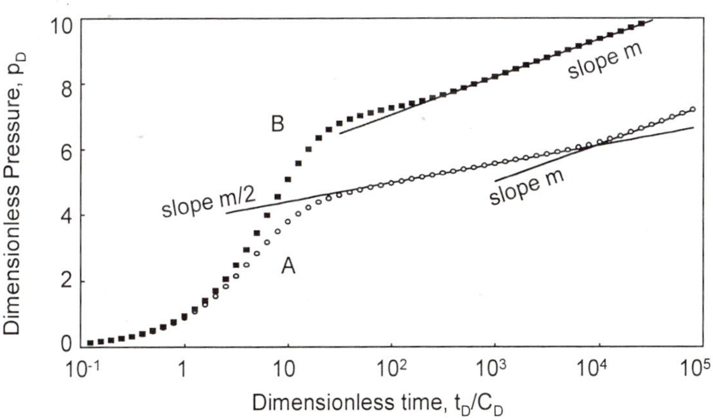

Figure 4.18. Semi-log plot of Figure 4.17 examples.

On the semi-log plot, Figure 4.18, example **A** exhibits a semi-log straight line of slope $m/2$ during the transition before the radial flow straight line of slope m. With example **B**, the first semi-log straight line of the transition regime is masked by wellbore storage, and only the final straight line of slope m is present.

Pressure type-curve analysis

As on the double porosity type-curve of Figure 4.2, pressure match and time match are used to estimate the equivalent permeability thickness product kh and the wellbore storage coefficient C (Equations 2.9 and 2.10).

The skin factor is estimated from the late time match, on the total system homogeneous curve $(C_D e^{2S})_{f+m}$ (Equation 4.20).

λ is estimated from the β' curve:

$$\lambda = \delta' \frac{\left(C_D e^{2S}\right)_{f+m}}{\beta' e^{2S}} \tag{4.34}$$

The fissure flow regime is not identified on the pressure type-curve of Figure 4.16, ω is *not accessed* from the match parameters.

Derivative type-curve

The derivative response of examples **A** and **B** are presented Figure 4.19.

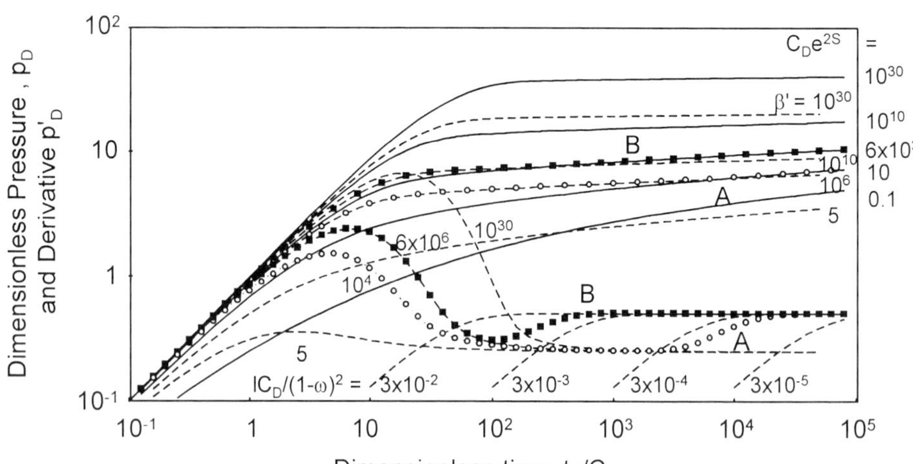

Figure 4.19. Pressure and derivative examples of Figure 4.17 for a well with wellbore storage and skin in a double porosity reservoir, transient interporosity flow. Log-log scales, p_D versus t_D/C_D.

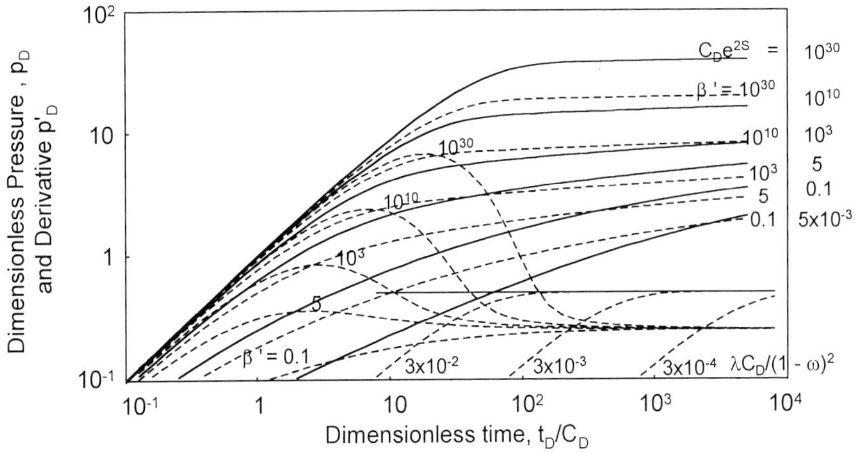

Figure 4.20. Pressure and derivative type-curve for a well with wellbore storage and skin in a double porosity reservoir, transient interporosity flow. Log-log scales, p_D versus t_D/C_D.

With example **A**, *two derivative plateaus* are evident: the first during transition is at 0.25 (semi-log straight line of slope $m/2$ on Figure 4.18) and the second, during the total system homogeneous behavior, is on the usual 0.5 line (slope m). With example **B**, the transition is descried by a short derivative valley before the stabilization at 0.5.

On drawdown responses, the main difference with the restricted matrix flow solution is in the transition regime: with the examples Figure 4.19, the derivative does *not drop below* 0.25 but tends to stabilize. It is a flat bottomed valley rather than a deep rounded valley

Description

Figure 4.20 presents the pressure and derivative type curve for transient interporosity flow (Bourdet et al., 1984). The dimensionless pressure and time are the same as on the type-curve Figure 4.16.

Derivative responses are also described by component curves. At late time, the 0.5 line defines the infinite acting radial flow regime. Before the total system homogeneous regime, *two transition curves* are used.

The early transition is described by β' derivative curves. As on the pressure type-curve of Figure 4.16, the derivative β' curves are obtained by displacing derivative $C_D e^{2S}$ curves by a factor of two along the pressure and time axes. During transition, the 0.5 line plateau of the $C_D e^{2S}$ curves is replaced by a constant derivative 0.25 line.

During late transition, the derivative is expressed as:

$$p'_D\left(t_D/C_D\right)=0.5\left\{1-e^{-\left[\frac{\lambda C_D}{(1-\omega)^2}\frac{t_D}{C_D}\right]}\right\} \tag{4.35}$$

The late transition curves, labeled $(\lambda C_D)/(1-\omega)^2$, describe the end of transition and the start of the total system homogeneous regime.

Sphere and slab matrix blocks

Figure 4.21 presents the drawdown response for the same parameters $(C_D e^{2S})_{f+m}=1$, $\beta'=10^4$ and $\omega=10^{-2}$ produced by the two types matrix geometry, namely slabs and spheres. Though the pressure curves look identical, the derivatives are different:

- The sphere model response hardly reaches the 0.25 straight line but remains above it.
- The curve generated for *slab* matrix is *tangential* to the 0.25 line, and at late transition time, the change from 0.25 to the 0.5 level is *steeper*.

Figure 4.21 illustrates that the matrix geometry has only a limited influence on double porosity responses. In practice, when the analysis is made by hand, it is not possible to differentiate between the two solutions, and the same type curve is used. When a computer is used, and provided the quality of the data is good, one of the two solutions is sometimes preferred because the derivative match appears slightly better. The choice of the matrix geometry does not influence the numerical results of analysis.

Matching procedure with the pressure and derivative type curve

The match point is fixed from the derivative 0.5 line and the early time unit slope line. The pressure curve is used to identify the total system curve $(C_D e^{2S})_{f+m}$, and a comparison between the pressure and the derivative matches defines the appropriate β' curve.

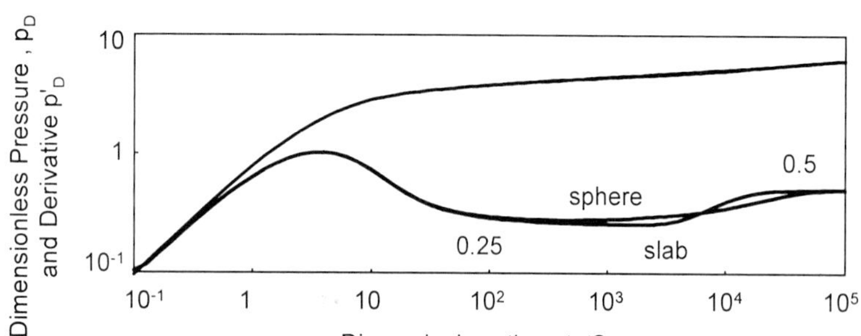

Figure 4.21. Slab and sphere matrix blocks. Log-log scales, p_D versus t_D/C_D. $(C_D e^{2S})_{f+m}=1$, $\beta'=10^4$, $\omega=0.01$, slab: $\lambda e^{-2S}=1.89\text{x}10^{-4}$, sphere: $\lambda e^{-2S}=1.05\text{x}10^{-4}$.

The permeability thickness and the wellbore storage constant are estimated from the pressure match and the time match (Equations 2.9 and 2.10). The skin is obtained from $(C_D e^{2S})_{f+m}$ (Equation 4.20), and λ from β' with Equation 4.34. At late transition, the group $(\lambda C_D)/(1-\omega)^2$ of the derivative match is not really sensitive to ω. It is not used to estimate the storativity ratio.

Influence of double porosity parameters

In this section, we discuss the influence of the two heterogeneous parameters ω and λ on double porosity responses with unrestricted matrix flow.

Influence of ω

The storativity ratio ω defines the contrast between the two homogeneous curves $(C_D e^{2S})_f$ and $(C_D e^{2S})_{f+m}$. In case of unrestricted interporosity flow, the fissure regime is short lived: after the wellbore storage effect, the response is usually already in the transition regime and the $(C_D e^{2S})_f$ curve is not seen. In such cases, the storativity ratio has *no influence* on the model response.

The influence of ω can only be demonstrated on responses generated with a very low wellbore storage coefficient. Figure 4.22 presents three double porosity curves for unrestricted flow from slab matrix blocks. All parameters are the same as in Figure 4.7: $C_{Df+m}=1$, $S=0$, $\lambda=10^{-7}$ and $\omega=10^{-1}$, 10^{-2} and 10^{-3}.

Due to the very small wellbore storage, the three derivative curves of Figure 4.22 exhibit the early time fissure regime: the response follows a $(C_D e^{2S})_f$ curve, then the β' transition curve and later the $(C_D e^{2S})_{f+m}$ curve. The examples in Figure 4.22 show that, when the wellbore storage does not mask the early time response, the influence of ω is the same as in the restricted interporosity flow response: decreasing ω increases the fissure curve parameter $(C_D e^{2S})_f$.

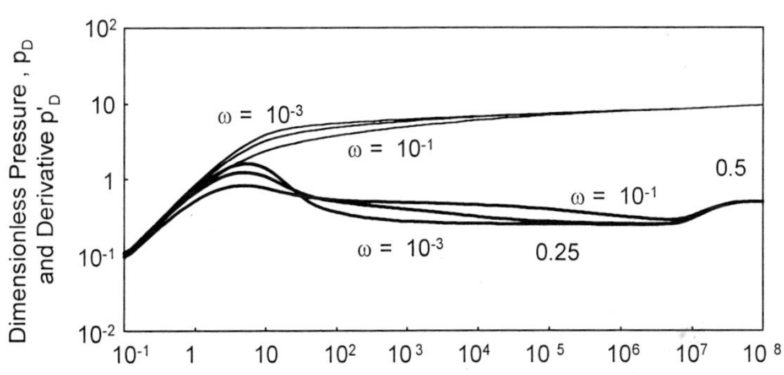

Figure 4.22. Influence of ω on pressure and derivative log-log curves, p_D versus t_D/C_D. Slab matrix blocks. $C_{Df+m}=1$, $S=0$, $\lambda=10^{-7}$, $\omega=10^{-1}$, 10^{-2} and 10^{-3}.

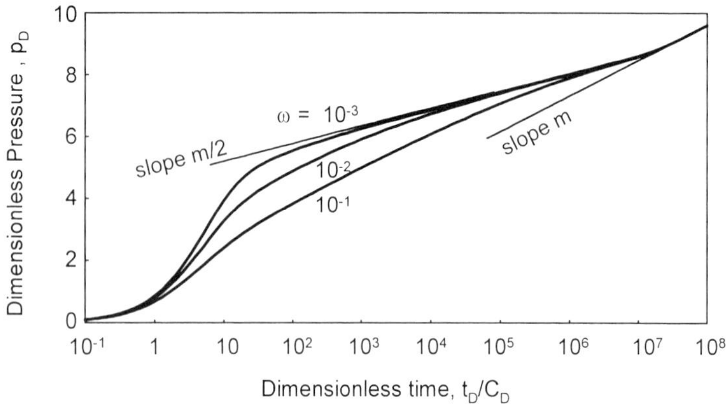

Figure 4.23. Influence of ω on semi-log plot of the Figure 4.22 examples.

On the semi-log plot Figure 4.23 for the same three examples, the curve generated for ω=10^{-3} first shows the straight line of slope $m/2$ during transition, and later the usual semi-log straight line of slope m. When the storativity ratio is high (ω =10^{-1}), a first semi-log straight line of slope m may be seen during the fissure regime, before transition (Streltsova, 1983).

The curves on two log-log plots of Figures 4.7 and 4.22 are generated for restricted and unrestricted flow with the same parameters. In the case of restricted matrix flow, the fissure regime lasts longer (t_D/C_D = 10^4 for ω =10^{-2}) than in the case of unrestricted interporosity flow (t_D/C_D = 10^2): the matrix skin *delays* the start of transition by more than two log-cycles. This confirms that with the unrestricted matrix flow solution, when the matrix blocks are not damaged, the fissure regime is generally not seen and ω cannot be accessed from pressure and derivative analysis.

Influence of λ

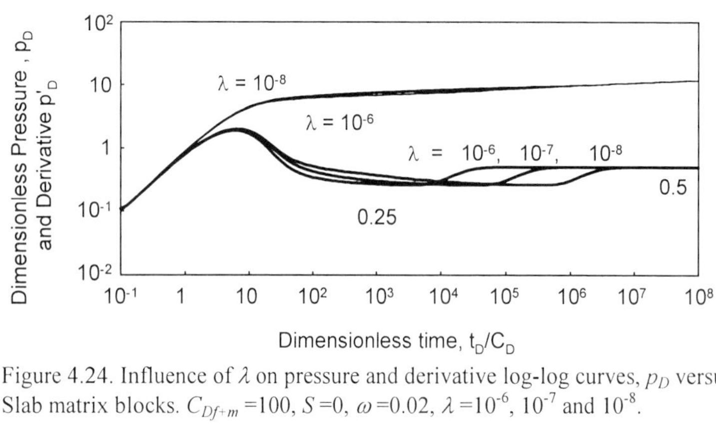

Figure 4.24. Influence of λ on pressure and derivative log-log curves, p_D versus t_D/C_D. Slab matrix blocks. C_{Df+m} =100, S =0, ω =0.02, λ =10^{-6}, 10^{-7} and 10^{-8}.

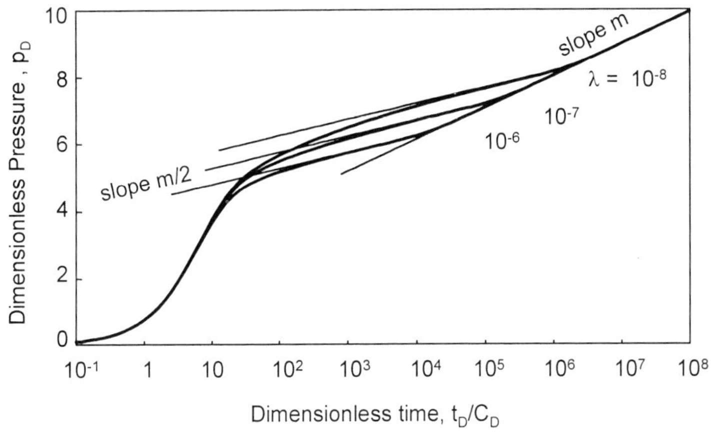

Figure 4.25. Influence of λ on semi-log plot of the Figure 4.24 examples.

Figures 4.24 and 4.25 present three double porosity curves generated for unrestricted flow with different interporosity flow parameters λ. The curves correspond to slab matrix blocks. The parameters are the same as on Figure 4.9: $C_{Df+m}=100$, $S=0$, $\lambda=10^{-6}$, 10^{-7}, 10^{-8} and $\omega=0.02$.

As in the restricted flow solution, λ defines the end of transition and the *start* of the total system flow. On derivative curves, a low λ corresponds to a long transition regime on 0.25, before the homogeneous behavior (Equation 4.35). On the semi-log plot of Figure 4.25, decreasing λ moves the transition upwards.

Associated specialized plot straight lines

The two usual specialized analyses can be used for interpretation: wellbore storage on early time data, and semi-log analysis once the storage effect has subsided. With the unrestricted matrix flow solution, three straight lines are in theory possible on a semi-log plot. Practically, the first one during fissure flow is short lived, and it is masked by wellbore storage.

Wellbore storage analysis

The analysis of wellbore storage is carried out on a linear scale (Section 1.2.2). Pure wellbore storage regime is generally seen only on damaged wells, when the skin is positive.

Radial flow analysis

As shown by the example **A** of Figure 4.18, two straight lines can be observed on a semi-log plot of double porosity response, unrestricted flow. The first, during the transition regime, has a slope half of the second, corresponding to the total system homogeneous behavior (Figure 4.23, example $\omega = 10^{-3}$ and Figure 4.25). The curve $\omega = 10^{-1}$ of Figure 4.23 shows that, at very early time, a first semi-log straight line can, in theory, demonstrate the fissure flow. The sequence is then, after wellbore storage, a first line of slope m, a second transition straight line of slope $m/2$ and a third line, slope m, parallel to the first.

In practice, the occurrence of three semi-log straight lines has not been clearly demonstrated on actual data, for example with three long stabilizations for confirmation on a log-log derivative plot such as on the theoretical example $\omega = 10^{-1}$ of Figure 4.22. Even the transition half slope semi-log straight line appears to be rather exceptional and, as discussed with Figure 4.11 of Section 4.1.2, most semi-log plots of actual field data only show the characteristic double porosity "**S**" shape.

When the radial flow of the total system has been reached, the last semi-log straight line is defined in Equation 4.28. Slope and intercept at Δt=1hr give the equivalent permeability thickness product kh and the skin coefficient S.

Build-up analysis

Build-up analysis of double porosity unrestricted interporosity flow is more difficult than for a homogeneous reservoir response. The superposition method used to take into account the influence of the production prior to shut-in can introduce a distortion on the curves (see Section 2.3.4).

Log-log analysis

For build-ups, an approximation of the pressure response is obtained by replacing the drawdown homogeneous curve $(C_D e^{2S})_{f+m}$ by the corresponding build-up curve, and keeping the β' transition unchanged. A build-up response displays the full double porosity behavior only when the production time t_p has been long enough before shut-in.

In Section 4.1.2, it was shown that for restricted interporosity flow, the derivative with respect to Horner or superposition time does not always correct perfectly the shape of the build-up curve during the transition. The same limitation is observed with the unrestricted flow models.

On Figure 4.26, three build-up examples of a double porosity response are compared to the original drawdown solution. Unrestricted flow from slab matrix blocks is assumed, the parameters are the same as for the restricted flow build-up examples of Figure 4.15, the three production times are also the same t_{pD}/C_D = 100 (Curve $\mathbf{A}_1$), 9000 (Curve $\mathbf{A}_2$), 3 10^5 (Curve $\mathbf{A}_3$).

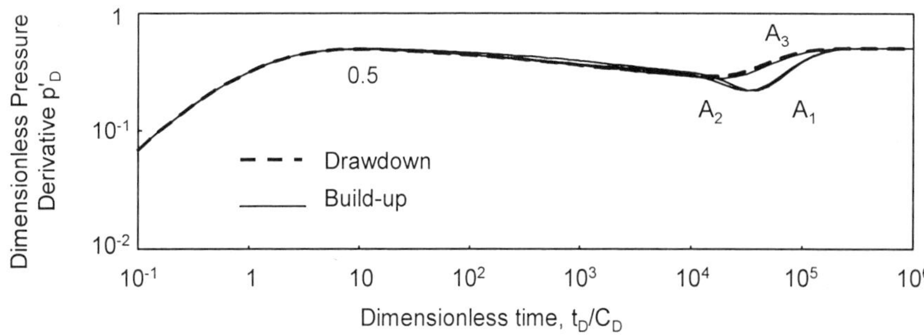

Figure 4.26. Log-log plot of build-up derivative responses for a well with wellbore storage and skin in double porosity reservoir, unrestricted interporosity flow, slab matrix blocks. C_{Df+m} =0.1, S =0, ω =0.1 and λ =3x10^{-4}. t_{pD}/C_D = 100 (**A$_1$**), 9000 (**A$_2$**), 300000 (**A$_3$**).

If the homogeneous total system flow was reached during the drawdown (curve **A$_3$**), the derivative of the build-up matches on the original drawdown response. However, if the previous drawdown stopped in transition regime, the derivative of the build-up deviates as shown in examples **A$_1$** and **A$_2$** of Figure 4.26: during transition regime, the build-up derivative drops *below* 0.25, down to 0.21 on Figure 4.26.

When the early time data is affected by wellbore storage, derivative curves can exhibit, on build-up data, a valley shape during transition, with a *minimum below 0.25*. In some cases, a match can be obtained with the two matrix flow solutions, unrestricted or restricted, the second generally with a high value of storativity ratio ω.

Semi-log analysis

As discussed in Section 4.1.2 for the restricted flow model, Horner and superposition analysis of unrestricted matrix flow double porosity data is used to estimate the *kh* product, the skin and the extrapolated pressure p^* provided that the correct total system semi-log straight line is used.

4.1.4 Extension of the double porosity models

In previous discussions, radial flow is assumed. Double porosity responses can be observed during other flow regimes encountered on well responses, due to well conditions (Chapter 3) or boundary effects (Chapter 5). In the following, the resulting shape of double porosity responses is briefly reviewed.

Next, several variations in the matrix properties are discussed: in the basic assumptions of double porosity models presented in Section 4.1.1, matrix blocks are homogeneous and of constant dimension (condition 6). A reduction of permeability at the surface of the blocks is introduced with the interporosity skin concept, and changes of block size or fissured matrix blocks are considered with the multiple porosity models. The behavior of layered and radial composite fissured reservoirs is discussed in Section 4.4.

Other flow regimes in case of double porosity behavior

In their study of horizontal wells in double porosity reservoir, Du and Stewart (1992) illustrated that the double porosity transition can occur in any of the three flow regimes characteristic of this well configuration (namely vertical radial flow, linear flow and horizontal radial flow, see Section 3.6). The derivative double porosity transition is simply superimposed onto the derivative trend describing the flow regime acting at the same time. Interestingly, they demonstrated that when unrestricted interporosity flow transition occurs during the linear flow regime, it is characterized by a quarter slope log-log straight line, typical of a bilinear flow response.

In the case of boundary effects in fissured reservoir, the derivative double porosity transition is also superimposed onto the boundary characteristic derivative shape (see Section 5.7). Channel reservoirs with layered deposits having a high contrast of permeability can also produce a bilinear flow regime, because of linear flow in an unrestricted double porosity system (see Figure 5.42).

Matrix skin

When the surface of the matrix blocks is damaged, the interporosity skin S_m is defined, in dimensionless terms, as (Moench, 1984):

$$S_m = \frac{k_m}{r_m}\frac{h_d}{k_d} \qquad (4.36)$$

where h_d and k_d are respectively the damaged zone thickness and permeability (Figure 4.27). As already mentioned, the matrix skin term is not present in the Warren and Root's (1963) definition of λ Equation 4.5. For high S_m (>10), several authors have proposed a correction to the interporosity flow parameter.

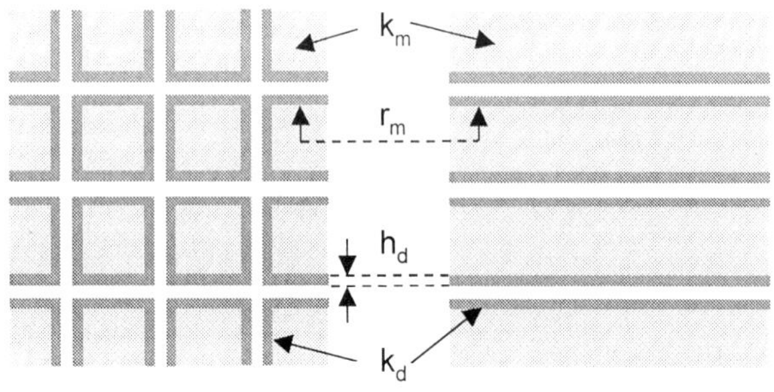

Figure 4.27. Matrix skin. Slab and sphere matrix blocks.

For slab matrix blocks of thickness $2r_m$, Cinco et al. (1985) define the effective interporosity flow parameter λ_{eff} as:

$$\lambda_{\text{eff}} = \frac{\lambda}{3(1-\omega)S_m} \tag{4.37}$$

Stewart and Ascharsobbi (1988) propose an effective interporosity flow parameter as a function of the number n of families of fissure planes :

$$\lambda_{\text{eff}} = \frac{\lambda}{1+(n+2)S_m} \tag{4.38}$$

In Figures 4.28 and 4.30, transient interporosity double porosity responses are presented for different values of the matrix skin S_m of Equation 4.36. Slab and sphere matrix block geometries are considered.

- When S_m =0, the responses correspond to the unrestricted interporosity model of Section 4.1.3.

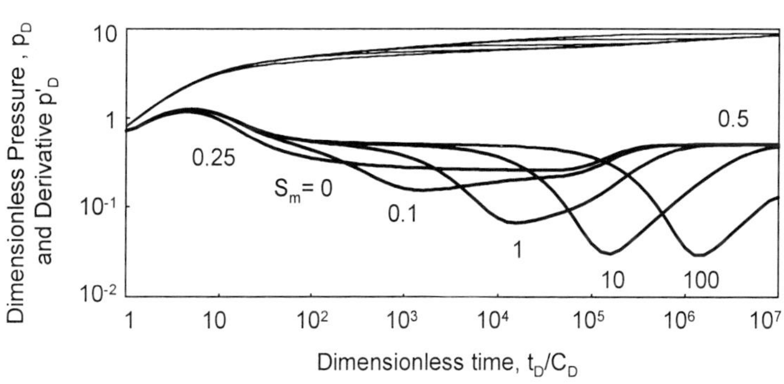

Figure 4.28. Double porosity reservoir, transient interporosity flow, slab matrix blocks with interporosity skin. Log-log scales. C_{Df+m} =1, S =0, ω =0.01 and λ =10^{-5}. S_m = 0, 0.1, 1, 10, 100.

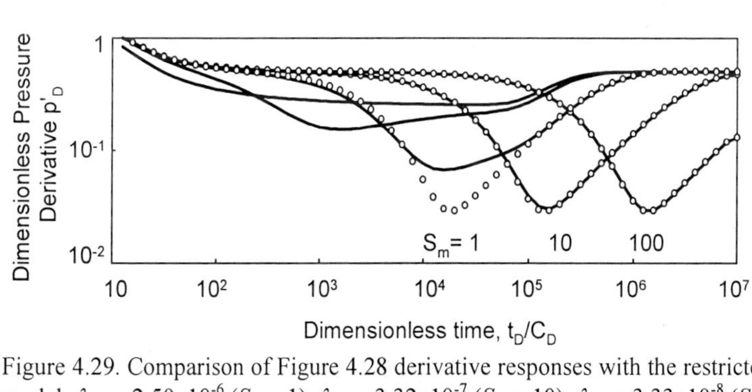

Figure 4.29. Comparison of Figure 4.28 derivative responses with the restricted interporosity flow model. λ_{eff} = 2.50x10^{-6} (S_m = 1), λ_{eff} = 3.32x10^{-7} (S_m = 10), λ_{eff} = 3.33x10^{-8} (S_m = 100).

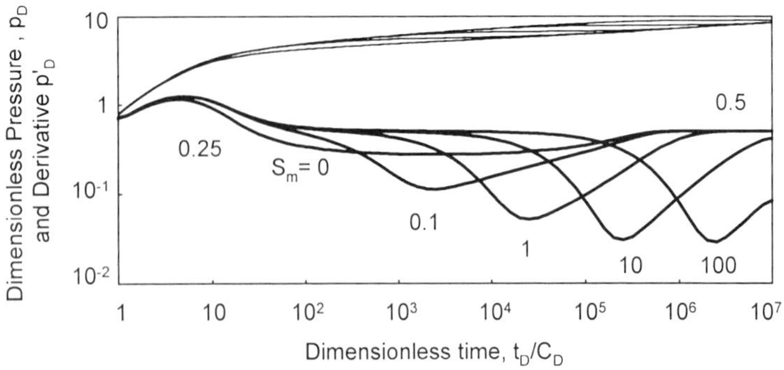

Figure 4.30. Double porosity reservoir, transient interporosity flow, sphere matrix blocks with interporosity skin. Log-log scales. $C_{Df+m}=1$, $S=0$, $\omega=0.01$ and $\lambda=10^{-5}$. $S_m = 0, 0.1, 1, 10, 100$.

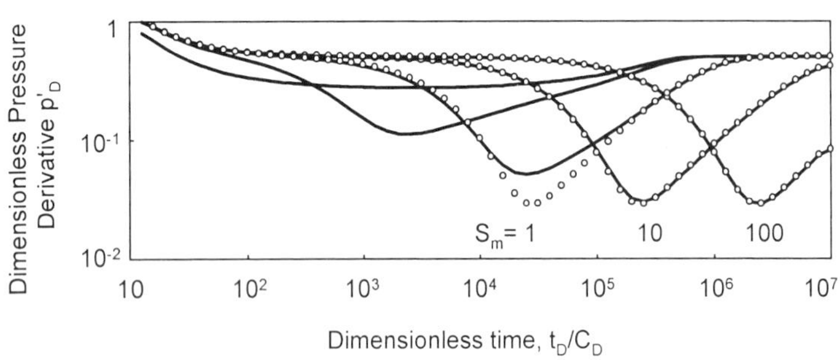

Figure 4.31. Comparison of Figure 4.30 derivative responses with the restricted interporosity flow model. $\lambda_{\text{eff}} = 1.66\text{x}10^{-6}$ ($S_m = 1$), $\lambda_{\text{eff}} = 1.96\text{x}10^{-7}$ ($S_m = 10$), $\lambda_{\text{eff}} = 2.00\text{x}10^{-8}$ ($S_m = 100$).

- When $S_m = 1$, a first derivative stabilization at 0.5 is observed before the transition. The effect of matrix skin is to *delay* the flow from blocks to fissures, and a fissure regime is identified at early time. During transition, the derivative drops *below* the usual 0.25 plateau of the unrestricted interporosity transition ($S_m = 0$).

- With larger matrix skins ($S_m = 10$ or 100), the early time fissure regime lasts longer. Later, a *valley shaped transition*, similar to the restricted interporosity curves of Section 4.1.2, is observed and the start of the total system equivalent homogeneous behavior is delayed.

On Figures 4.29 and 4.31, the derivative curves of Figures 4.28 and 4.30 are compared to restricted interporosity responses generated with the effective interporosity flow parameter λ_{eff} of Equation 4.38. When $S_m = 1$, the equivalent restricted interporosity solution describes correctly the fissure and total system flow regimes but, during the transition, the valley drops to a deeper level than on the transient interporosity response with matrix skin. When the matrix skin is larger ($S_m = 10$ or more), the two models produce equivalent derivative curves.

Figures 4.29 and 4.31 illustrate that, when the Warren and Root (1963) restricted interporosity flow solution is used for analysis, the match provides the effective interporosity flow parameter λ_{eff}, and not λ of Equation 4.5. When $S_m \geq 10$, λ_{eff} of Stewart and Ascharsobbi (1988) can be approximated by the modified pseudo steady state interporosity flow parameter of Moench (1984):

$$\lambda_{\text{eff}} = n \frac{r_w^2}{r_m h_d} \frac{k_d}{k} \tag{4.39}$$

Interestingly, λ_{eff} of Equation 4.39 is *independent* of the matrix block permeability k_m. Moench (1984) showed that when the matrix skin S_m of Equation 4.36 is large, most of the pressure change within the blocks occurs across the damaged zone of thickness h_d, and the pressure gradients in the block become small. This explains Equation 4.39, and thus justifies the assumption of the Warren and Root (1963) model where the pressure gradients in the matrix are neglected.

If the average size of the matrix blocks r_m is known, λ_{eff} can be used to estimate the permeability thickness ratio k_d / h_d of the damaged region on the fissure walls.

The theory of matrix skin unifies the two double porosity models presented in previous Sections. For small values of the matrix skin (S_m.=0.1 on Figures 4.28 and 4.30), none of the two limiting solutions, for unrestricted and restricted interporosity flow, describes the response. A transient interporosity model with matrix skin should then be used for the analysis of such data.

Multiple porosity systems

Several authors have considered changing block sizes in fissured reservoirs. Abdassah and Ershaghi (1986) proposed a triple porosity model assuming two families of matrix blocks with different characteristics. Cinco-Ley et al. (1985) and Belani and Yazdi (1988) extended the triple porosity model to multiple block size with a frequency function defining the probability of blocks of a given size. With these models, the matrix blocks are uniformly distributed in the reservoir. Al-Ghamdi and Ershaghi (1994) envisaged a different configuration, where matrix blocks are fissured with pseudo steady state interporosity flow. In such case, the matrix produces into the micro fissures, which feed a network of macro fissures producing to the well. A schematic of the two possible triple porosity configurations is presented on Figure 4.32.

Other configurations have been proposed when the density of the fissure network is not uniform. These solutions combine a double porosity response with double permeability or radial composite configurations. They are briefly discussed at the end of this chapter (Section 4.4). In the following, only the triple porosity solution is presented for illustration, and the main conclusions concerning the multiple block size configuration are summarized.

Figure 4.33 presents a triple porosity reservoir response with pseudo steady state interporosity flow. The fissure network interacts with two groups of matrix blocks. For each group, the interporosity flow is defined with a specific λ_i (i =1, 2). In the storativity ratio $\omega = 10^{-2}$, the total storage of Equation 4.4 involves the fissure and the two groups of matrix blocks. δ_i defines the contribution of the group to the total matrix storage ($\delta_1 + \delta_2 = 1$):

$$\delta_i = \frac{(\phi V c_t)_{mi}}{(\phi V c_t)_{m1} + (\phi V c_t)_{m2}} = \frac{(\phi V c_t)_{mi}}{(\phi V c_t)_m} \tag{4.40}$$

With group 1, the blocks are small and the interporosity flow coefficient is large ($\lambda_1 = 10^{-5}$) but the group represents only 10% of the matrix storage ($\delta_1 = 0.1$). The curve Figure 4.33 shows response of the fissure alone with a first derivative *stabilization* at 0.5, then a *valley* transition when the matrix blocks of group 1 start to produce, followed by a new *stabilization* at 0.5. An intermediate homogeneous radial flow behavior is seen at times 10^4 to 10^5, describing the fissures and group 1 matrix blocks. After $t_D/C_D = 10^5$, the second group enters into production ($\lambda_2 = 5x10^{-7}$) and a second derivative *valley* develops. During this transition, the storage contrast between (fissures and group 1) and the total system is 0.01+10%x0.99=0.109, and the depth of the second valley is about the same as the first. The response ends in radial flow for the total system equivalent homogeneous behavior.

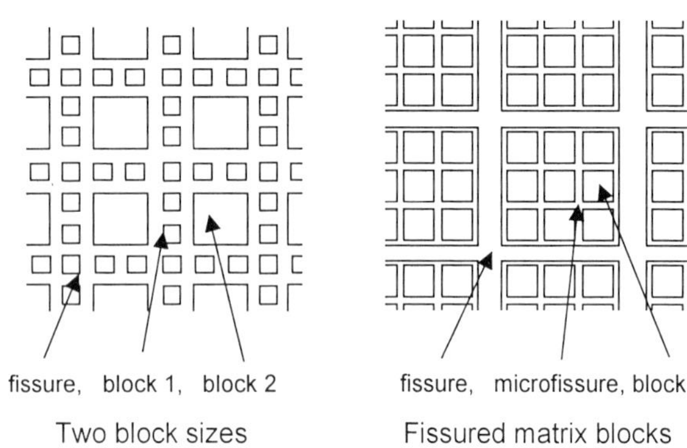

Figure 4.32. Triple porosity systems

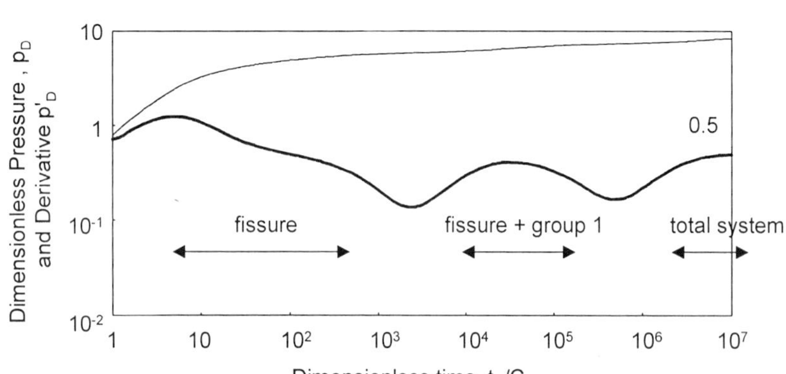

Figure 4.33. Triple porosity reservoir, pseudo steady state interporosity flow, two sizes of matrix blocks, different λ. $C_{Df+m} = 1$, $S = 0$, $\omega = 0.01$, $\lambda_1 = 10^{-6}$, $\delta_1 = 0.1$ and $\lambda_2 = 5x10^{-7}$, $\delta_2 = 0.9$.

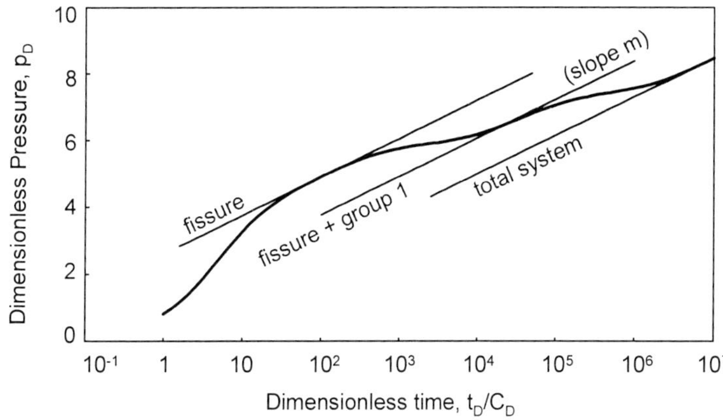

Figure 4.34. Semi-log plot of Figure 4.33 example.

The same triple porosity example is presented on semi-log scales in Figure 4.34. *Three straight lines* are present: the first for fissure flow, the second for (fissures + group 1) and finally the third for (fissures + group 1 + group 2). This sequence of regimes is theoretically possible but, more frequently, a triple porosity response is characterized by a *non-symmetric* transition valley as illustrated on Figure 4.35.

In this example, the interporosity flow coefficient is the same for the two groups ($\lambda_1=\lambda_2=10^{-6}$) and the two transitions start at about the same time. The two dashed curves of Figure 4.35 present the individual double porosity responses produced each groups: the transition for group 1 (δ_1 =0.1) ends earlier than for group 2. The resulting triple porosity transition can be described as the sum of the two valley shaped transitions:

- At early transition time, the drop of derivative can be identified earlier than on the two individual double porosity response curves.
- At late transition time, an intermediate derivative plateau is observed when the group 1 transition is finished but not for the group 2.

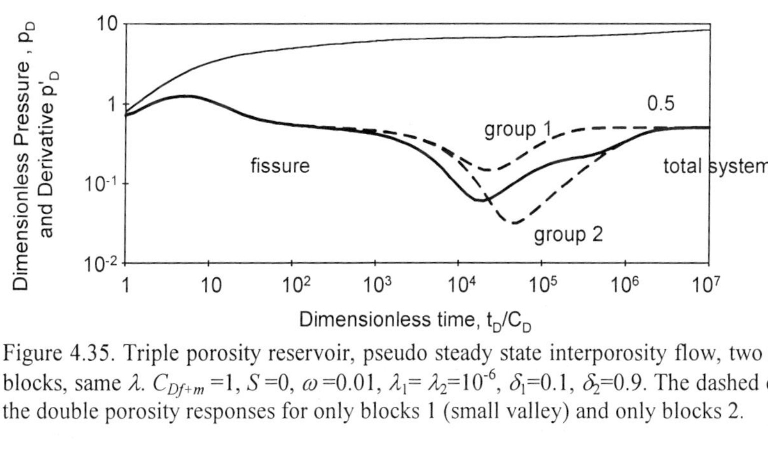

Figure 4.35. Triple porosity reservoir, pseudo steady state interporosity flow, two sizes of matrix blocks, same λ. C_{Df+m} =1, S =0, ω =0.01, $\lambda_1=\lambda_2=10^{-6}$, δ_1=0.1, δ_2=0.9. The dashed curves describe the double porosity responses for only blocks 1 (small valley) and only blocks 2.

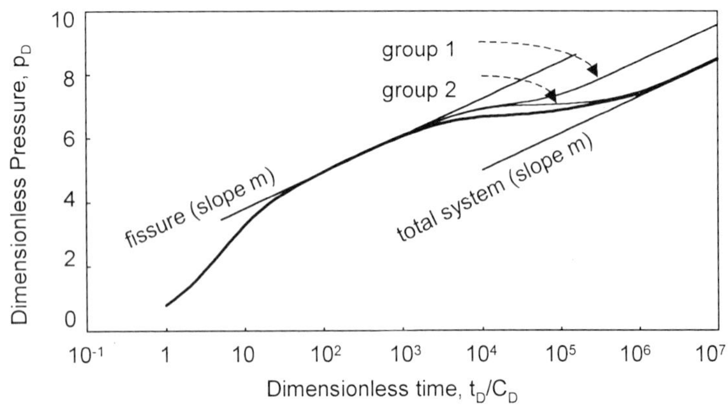

Figure 4.36. Semi-log plot of Figure 4.35 example. The thin curves describe the double porosity responses for only blocks 1 (final semi-log straight line for fissures + blocks 1) and only blocks 2 (final semi-log straight line close to the total system).

The three examples of Figure 4.35 are presented on semi-log scale on Figure 4.36. On the triple porosity response, only two straight lines are present, for fissure flow and for (fissures + group 1 + group 2). The double porosity response for group 1 alone, shown as a thin curve, describes the theoretical (fissures + group 1) semi-log straight line.

In the case of unrestricted interporosity flow, triple porosity responses deviate from the typical behavior presented in Section 4.1.3 at transition time and the derivative curves can exhibit an intermediate plateau between 0.25 and 0.5, or even an oscillation (Cinco-Ley et al., 1985). With multiple porosity systems, the authors conclude that the size of the matrix blocks r_m estimated with a double porosity model is the *harmonic weighted average* of the different blocks sizes.

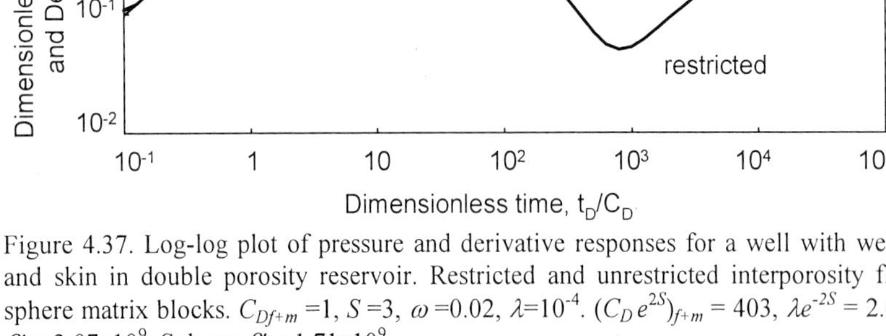

Figure 4.37. Log-log plot of pressure and derivative responses for a well with wellbore storage and skin in double porosity reservoir. Restricted and unrestricted interporosity flow, slab and sphere matrix blocks. $C_{Df+m}=1$, $S=3$, $\omega=0.02$, $\lambda=10^{-4}$. $(C_D e^{2S})_{f+m} = 403$, $\lambda e^{-2S} = 2.48\times10^{-7}$. Slab: $\beta' = 3.07\times10^9$, Sphere: $\beta' = 1.71\times10^9$

4.1.5 Discussion of double porosity analysis results

In this section, we present specific features concerning results of analyses with the double porosity solutions. We examine the question of the uniqueness of the solution, and we discuss the numerical values of the different parameters obtained from analysis. Multiple porosity systems and transient interporosity responses with matrix skin are seldom used as they require parameters that are usually not accessible. Test interpretation with these models is not considered in the following discussion.

Restricted and unrestricted interporosity flow models

Figure 4.37 compares the two types of double porosity responses generated for the same parameters: $(C_D e^{2S})_{f+m}$ =403, λe^{-2S} =2.48x10^{-7} and ω=2x10^{-2}. In the case of unrestricted interporosity flow with slab or sphere matrix blocks, the derivative is flat during transition. With the restricted interporosity flow solution, the transition is characterized by a valley in the derivative.

The choice of the appropriate double porosity solution is in general unique, the two models correspond to different sets of data. The rule is to use the model that provides the best match: with the restricted interporosity flow solution, the depth of the derivative valley during transition is a direct function of the transition duration. With long transition regimes, corresponding to small ω values, the derivative drops below the practical 0.25 limit of the transient interporosity flow solution.

An ambiguity may occur when the transition regime is of very short duration after the wellbore storage effect. In such cases, pseudo steady state curves (generated with a high ω value) can produce a shape very similar to transient solutions, generated with a ω value much smaller (of the order of 10^{-2} or less). The two resulting fissure storativity values are very different.

Some wells have been reported to change from a restricted to an unrestricted interporosity flow behavior after acid stimulation (see discussion of the radial composite double porosity model Section 4.4.2). In most cases, the type of interporosity flow regime does not change in the course of the well history. Furthermore, it is in general similar for all wells in the same formation.

Fissured reservoir versus sealing fault

The shape of the two curves of Figure 4.21 is similar to the response of a well in a homogeneous reservoir bounded by one sealing fault (Chapter 5). On the corresponding example Figure 5.1, the derivative stabilizes first at 0.5 and later at 1, but the shape of this response shows the same characteristic as a double porosity curve with unrestricted matrix flow. Using a homogeneous bounded model to interpret such pressure response gives a permeability thickness product twice the *kh* obtained with a double porosity match.

When the two alternative solutions seem equally appropriate to describe the pressure response, the analysis of the results shows frequently that only one is applicable. For example, in the hypothesis of boundary effect, the distance of the sealing fault could be unrealistic, or the same behavior may be repeated on several wells in the same reservoir etc. A double porosity model may then be preferred for interpretation. In the next Section, we show that the numerical values of the well parameters (wellbore storage and skin) can also be used in the diagnosis of double porosity behavior.

Results of double-porosity analysis

In the following, we discuss the significance of the interpretation results obtained with the double porosity model. The equivalent permeability thickness product has been discussed in Section 4.1.1. The interporosity flow parameter λ, presented in Equation 4.5 of Section 4.1.1 for the unrestricted interporosity flow condition, has been modified into Equation 4.39 of Section 4.1.4 in the case of restricted interporosity flow. In the build-up discussion of Section 4.1.2, the concept of extrapolated pressure was shown to be in some cases misleading. The next paragraphs present other characteristic features of double porosity interpretations.

Reservoir parameters

The discussion of double porosity behavior was based so far on the assumption that the fluid is single phase, and the compressibility is constant. When free gas is present in the formation, the *gas saturations* can be different in the fissures and in the matrix blocks, and they are both changing during the well production history. As the storativity ratio of Equation 4.4 depends on the total compressibility of the two phases $(c_t)_f$ and $(c_t)_m$, ω can change when the fluid characteristic changes. In the same way, λ depends on k_m or S_m, which are very sensitive to gas saturation. Some wells, after several tests at different times, have shown a change in the double porosity behavior because of the variation in ω and λ (Gringarten, 1984). This is in agreement with the Camacho-V. and Raghavan (1994) simulation results.

Well parameters

In some cases, fissured reservoir responses show a very *high value of wellbore storage* constant, associated to a *negative skin* factor, even when the well has not been stimulated.

Common values for the wellbore storage constant in homogeneous reservoirs are in the order of 10^{-2} Bbl/psi or less for an oil well. In the case of fissured reservoirs, it is not exceptional to find wellbore storage effects of 0.1 Bbl/psi or more, 10 or 100 times greater than the wellbore storage constant calculated from the completion.

Double porosity reservoirs can also show negative pseudo-skins. A skin of around -3 can be encountered in non-stimulated wells, acidized wells may have skins as low as -7, and a zero skin can correspond to a slightly damaged well (Gringarten, 1984). Following the Stewart and Ascharsobbi (1988) theory, the equivalent bulk fissure permeability concept k of Equation 4.1 is not valid for near wellbore description. At small scale, the dimension of the matrix blocks is not negligible (condition 1 of Section 4.1.1) and the flow is not radial in the vicinity of the well. The fluid flows through the fissure in direct contact with the well and, as a result, the resistance to flow is less than in the homogeneous radial flow based on k. For cubes matrix blocks, they express the corresponding negative pseudo-skin as:

$$S_{2\phi} = \frac{\pi}{2} - \ln \frac{2r_m}{r_w} \tag{4.41}$$

A high value of wellbore storage, associated with an apparent negative skin, is an indication of fissured formation, even when the shape of the data curves does not suggest any heterogeneity. This specific influence of natural fissures on the wellbore parameters has been observed in many fissured reservoirs but not all. Many double porosity reservoirs, such as multiple-layer systems, are also known to have no effect on the wellbore parameters (Gringarten, 1984).

4.1.6 Field examples

In the following section, two published double porosity examples are presented. The manual analysis of the first example (Bourdet et al., 1983 b), with the double porosity type-curve of Figure 4.6 for restricted interporosity flow is summarized. For the second example (Bourdet et al., 1984), the pressure and derivative responses are briefly discussed for a match with the double porosity unrestricted interporosity flow model. A possible triple porosity response is described in Section 5.7.5 (Figure 5.45).

Restricted interporosity flow example

After one day of production, an oil well is shut in for an 18 hour build-up test. The log-log plot of pressure and derivative (Figure 4.38) suggests a heterogeneous behavior: after an initial hump at early times, the derivative drops slowly until the third hour of shut-in, then it increases during the 15 remaining hours of build-up. The derivative does not stabilize on this build-up response, a radial flow semi-log straight-line analysis is not justified.

A first match is attempted on the homogeneous model (type-curve Figure 2.22), assuming that the lower part of the derivative plot corresponds to the 0.5 line of the radial flow regime. The later upward trend would then be possibly explained by boundary effects.

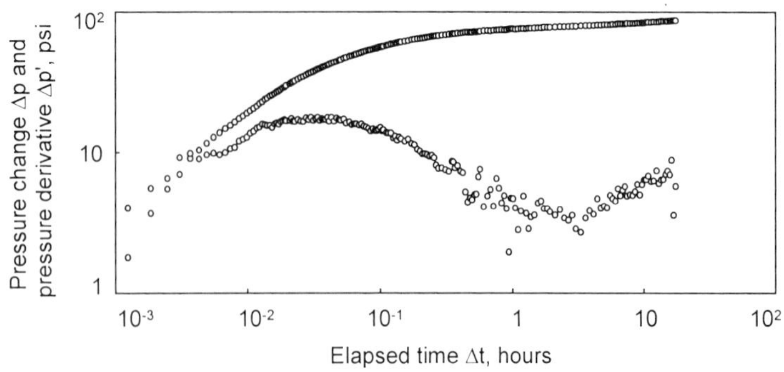

Figure 4.38. Log-log plot of the double porosity (pseudo steady state interporosity flow) field example.

No satisfactory solution is found to match the first part of the build-up: after the derivative maximum, the downward branches of the homogeneous type curves are much steeper than on the data plot.

A second attempt is made with the double porosity solution, restricted interporosity flow. The definition of the flow regimes is as follows: the derivative displays a hump corresponding to wellbore storage during the first 10 minutes of shut-in. Then, from 0.2 hours to the end, a long transition behavior is observed, the radial flow for the total system is not reached after 18 hours of shut-in.

An initial match is performed on the pressure and derivative type-curve of Figure 4.6. As radial flow has not been reached, the last derivative points are placed slightly below the 0.5 line, and then the early time data is matched on the unit slope straight line. Both pressure and derivative data curves appear to match a fissure curve in the region of $(C_D e^{2S})_f = 50$. The pressure transition period is matched against $\lambda e^{-2S} = 10^{-4}$, and the total system curves seems to be close to $(C_D e^{2S})_{f+m} = 10$.

The match is checked against the derivative transition response. With $\omega = 0.2$, the early and late transition curves are $\dfrac{\lambda C_D}{\omega(1-\omega)} = \dfrac{10^{-4} \text{x} 10}{0.2 \text{x} 0.8} = 6 \text{x} 10^{-3}$ and $\dfrac{\lambda C_D}{(1-\omega)} = 1.2 \text{x} 10^{-3}$ respectively.

Pressure and derivative give consistent results. The match can be refined by generating the complete double porosity multiple-rate response by computer (Figure 4.39). Results are:

- Pressure match: $p_D / \Delta p = 0.06$ psi-1
- Time match: $(t_D/C_D)/\Delta t = 161$ hrs-1
- Fissure curve: $(C_D e^{2S})_f = 50.$
- Total system curve: $(C_D e^{2S})_{f+m} = 8$
- Transition curve: $\lambda e^{-2S} = 1.01 \text{ x } 10^{-4}$

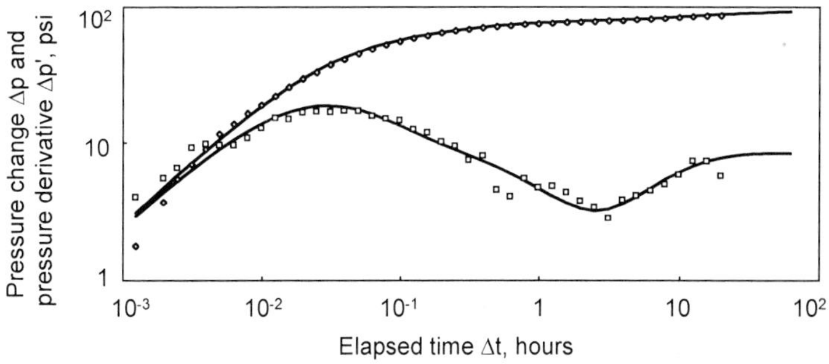

Figure 4.39. Log-log match of the field example on double porosity model, restricted interporosity flow. $(C_D e^{2S})_{f+m} = 8$, $S = -4.1$, $\omega = 0.16$, $\lambda = 2.8\text{x}10^{-8}$.

Unrestricted interporosity flow example

The second example is a short three hours build-up test, after two days of production. On the log-log plot shown in Figure 4.40, the derivative clearly describes two stabilizations, with the second approximately twice above the first. On late time data, the pressure is less than one log-cycle above the derivative response: it can be concluded that the skin is negative. This is confirmed by the early time response, where the wellbore storage effect ends before the first point, recorded 15 seconds after shut-in.

Two models can be used to match the data: a double porosity model with unrestricted interporosity flow, or a homogeneous reservoir with sealing boundaries. Both models are equally applicable. With the first hypothesis, the radial flow derivative stabilization corresponds to the second plateau, and a slab matrix block geometry is found to provide the best match. With the second hypothesis, the radial flow regime is seen during the first derivative stabilization, and two intersecting faults with an angle larger than 90° have to be used.

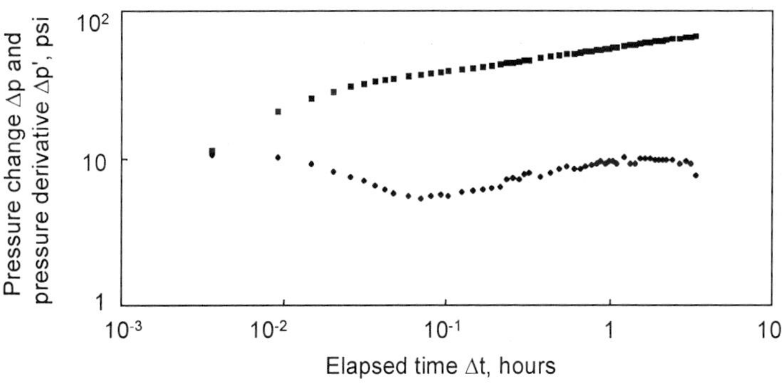

Figure 4.40. Log-log plot of the double porosity (transient interporosity flow) field example.

4.2 LAYERED RESERVOIRS WITH OR WITHOUT CROSSFLOW

As an extension of the double porosity models presented in Section 4.1, the double permeability solution considers also two distinct media in the reservoir. The two elements are defined as layers, with a porosity and a permeability. In each layer, the flow is radial, and they can both produce directly into the well. In the reservoir, a cross flow may be established when one layer flows into the other (Figure 4.41).

The double permeability behavior is observed in stratified reservoirs, when the permeability of the different layers is participating to the response, or in fissured reservoirs, when the matrix blocks are connected. In systems with a large number of layers, the high permeability layers are grouped by convention into the "Layer 1", and the "Layer 2" describes the tight zones (see discussion in multiple-layer Section 4.2.5).

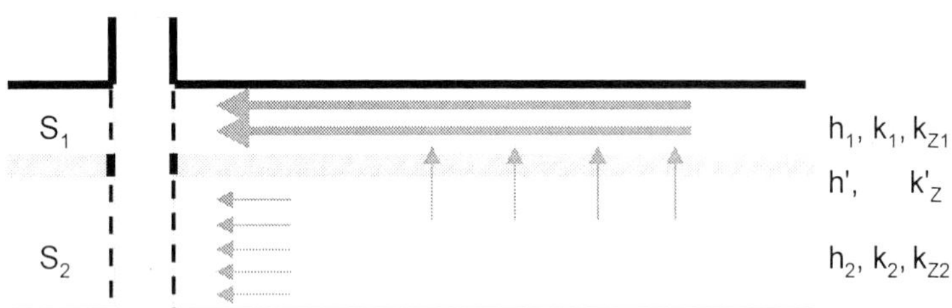

Figure 4.41. Model for double permeability reservoir

Layered reservoir have been studied either for the case of commingled systems (Lefkovits et al., 1961; Tariq and Ramey, 1978), when the layers, separated by impermeable barriers, can only communicate through the well, and also when a reservoir cross flow is possible (Jacquard, 1960; Russell and Prats, 1962; Polubarinova-Kocina, 1962; Gao, 1984; Wijesinghe and Culham, 1984; Bourdet, 1985; Prijambodo et al., 1985). In 1984, Wijesinghe and Culham presented an analytical model for transient interlayer cross flow. Pseudo steady state cross flow between the layers was envisaged for two layers by Bourdet in 1985. This type of reservoir cross flow has also been considered by Liu et al. (1987) and Liu and Wang (1993) with similar conclusions. Chen et al. (1990) derived a relatively simple transient cross flow solution. The pseudo steady state interlayer cross flow solution was extended to any number of layers (Ehlig-Economides and Joseph, 1985; Larsen, 1988; Park and Horne, 1989) and to different well and boundary conditions (Joseph et al., 1986; Larsen, 1989; Suzuki and Nanba, 1991; Bidaux et al., 1992; Kuchuk and Habashy, 1996). Limited entry wells in layered reservoir with transient cross flow have also been considered (Shah and Thambynayagam, 1992; Abbaszadeh et al., 1993).

The double permeability model described in this section corresponds to the analytical solution proposed by Bourdet (1985). It includes the wellbore storage and skin effects, and the reservoir cross flow is in pseudo steady state condition. The basic assumptions are discussed in Section 4.2.1. Two different types of well configuration are considered.

In Section 4.2.2, the model response is presented when the two layers produce into the well as in the example Figure 4.41. In Section 4.2.3, the response is examined when only one of the two layers is perforated, such as in Figure 4.47. Extensions of the model to multiple-layer reservoir, and transient reservoir cross flow are described in Section 4.2.6. Responses of commingled systems are further discussed in Section 4.2.7 for equal or unequal initial pressure.

4.2.1 Double permeability model

The basic assumptions of the double permeability solution and the definition of dimensionless parameters are based on the same concepts as in the double porosity model of Section 4.1. Again, the parameters resulting from interpretation define the idealized mathematical model used for description of the layered reservoir. In particular, the vertical permeability estimated from a match with the double permeability model depends upon the choice of a two layers simplified model to describe a complex reservoir configuration. The influence of layer refinement on the results is discussed in the multiple-layer Section 4.2.6.

Double permeability assumptions

- The well, intercepting two *homogeneous layers*, is affected by wellbore storage. At each layer, a skin defines the communication between the well and the formation.

- The *initial pressure* is the same in the two layers.

- After some production, a difference of pressure is established between the two layers and a *cross flow* takes place in the reservoir. As for the double porosity model, two different types of cross flow have been considered. In the case, discussed in the following, of pseudo steady state flow between the layers (Bourdet, 1985), the same assumption as in the Warren and Root (1963) model is made: the vertical pressure gradients in the layers are neglected and the resistance to vertical flow is described with the semi-permeable wall model of Gao (1984). With the hypothesis of transient flow between the layers (Chen et al., 1990), the cross flow is unsteady because of the vertical pressure gradients in the low permeability layer.

Other multiple-layered reservoir configurations have been considered with any number of layers, homogeneous or not. These solutions are discussed in Section 4.4.

Definitions

In the following, subscripts 1 and 2 refer respectively to layer 1 and 2 (see Figure 4.41). The total permeability thickness product is expressed as:

$$kh_{\mathrm{TOTAL}} = k_1 h_1 + k_2 h_2 \tag{4.42}$$

and the reservoir total storativity:

$$(\phi c_t h)_{\mathrm{TOTAL}} = (\phi c_t h)_1 + (\phi c_t h)_2 \tag{4.43}$$

The kh contrast between the two layers is expressed by the mobility ratio κ:

$$\kappa = \frac{k_1 h_1}{k_1 h_1 + k_2 h_2} = \frac{k_1 h_1}{kh_{\mathrm{TOTAL}}} \tag{4.44}$$

κ defines the contribution of the *high permeability layer* 1 to the total *permeability thickness product*. When κ is close to 1 (0.999 or more), the mobility of layer 2 becomes negligible. The double permeability model then tends to exhibit the same configuration as the double porosity solution defined in Section 4.1 when applied to multiple-layer systems: the double porosity model is a limiting form of double permeability responses.

The contrast of storage between the layers is expressed by the storativity ratio ω. It defines the contribution of the *high permeability layer*, to the *total storativity*:

$$\omega = \frac{(\phi c_t h)_1}{(\phi c_t h)_1 + (\phi c_t h)_2} = \frac{(\phi c_t h)_1}{(\phi c_t h)_{\mathrm{TOTAL}}} \tag{4.45}$$

The definition of the storativity ratio is the same as in the double porosity model: the volume ratios V_f and V_m of Equation 4.2 are replaced by the thickness ratios $h_1/(h_1 + h_2)$ and $h_2/(h_1 + h_2)$ in Equation 4.45. Practical values for ω can be in the same range, in the order of 10^{-1} or less.

The reservoir cross flow is defined by the *interlayer cross flow coefficient* λ: the smaller is λ, the more difficult is communication between the layers and λ=0 corresponds to two layers without cross flow, also called commingled system.

With the semi-permeable wall resistance hypothesis of Gao (1984), λ is expressed as:

$$\lambda = \frac{r_w^2}{k_1 h_1 + k_2 h_2} \frac{2}{2\dfrac{h'}{k'_z} + \dfrac{h_1}{k_{z1}} + \dfrac{h_2}{k_{z2}}} \tag{4.46}$$

As depicted on Figure 4.41, λ is a function of the vertical permeability k'_z in the low permeability "wall" of thickness h' between the layers and, by extension, of vertical permeability in the two layers k_{z1} and k_{z2}.

If the vertical resistance is mostly due to the "wall", a simplified λ can be used to characterize this interlayer skin:

$$\lambda = \frac{r_w^2}{k_1 h_1 + k_2 h_2} \frac{k'_z}{h'} \tag{4.47}$$

Equation 4.47 is equivalent to the effective interporosity flow parameter λ_{eff} of Equation 4.39 (Moench, 1984).

When there is no skin at the interface and the vertical pressure gradients are negligible in the high permeability layer 1, λ is equivalent to the interlayer flow parameter of the transient double permeability model of Chen et al. (1990):

$$\lambda = \frac{r_w^2}{k_1 h_1 + k_2 h_2} \frac{k_{z2}}{h_2/2} \tag{4.48}$$

Dimensionless variables

All dimensionless variables are expressed with reference to the total system parameters defined in Equations 4.42 and 4.43. The dimensionless pressure is based on the permeability thickness product of the total system:

$$p_D = \frac{k_1 h_1 + k_2 h_2}{141.2 qB\mu} \Delta p \tag{4.49}$$

and the dimensionless time is:

$$t_D = 0.000264 \frac{k_1 h_1 + k_2 h_2}{\left[(\phi c_t h)_1 + (\phi c_t h)_2\right] \mu r_w^2} \Delta t \tag{4.50}$$

The well condition is defined by two skins S_1 and S_2, and by the dimensionless wellbore storage:

$$C_D = \frac{0.8936 C}{\left[(\phi c_t h)_1 + (\phi c_t h)_2\right] r_w^2} \tag{4.51}$$

In the following sections, the double permeability curves are presented with the dimensionless pressure p_D of Equation 4.49 versus the dimensionless time group t_D/C_D:

$$\frac{t_D}{C_D} = 0.000295 \frac{k_1 h_1 + k_2 h_2}{\mu} \frac{\Delta t}{C} \tag{4.52}$$

4.2.2 Double permeability behavior when the two layers are producing in the well (Wellbore storage and two skins)

The response of a well with wellbore storage and skins is defined, in a double permeability reservoir, by six dimensionless parameters: C_D, S_1 and S_2 for the well, κ, ω and λ for the reservoir. No log-log type-curve has been found practical to describe double permeability responses, the analysis is performed on computer and the parameters are adjusted with pressure and derivative data.

Behavior

In a two-layer reservoir with cross flow, three different regimes can be identified in the response of a well with wellbore storage and skins :

1. First, the behavior of *two layers without cross flow* is seen: with the semi-permeable wall assumption, the reservoir cross flow is negligible at early time.
2. At intermediate times, the response deviates from the "two layer no cross flow" behavior and reaches a *transition* regime, as the fluid transfer between the layers starts in the reservoir.
3. Later, the pressure of the two layers equalizes, the equivalent *homogeneous behavior* of the total system becomes evident.

A typical double permeability response is presented in Figure 4.42. The example corresponds to a well with wellbore storage, the two layers are producing into the well and a cross flow is established in the reservoir. The derivative follows the unit slope straight line at early time, reaches a maximum, then drops below the 0.5 line in a long transition, and finally reaches the 0.5 stabilization when the radial flow in the equivalent homogeneous total system is reached. During transition, the shape of the derivative valley is a function of the contrast of storativity and permeability between the two reservoir elements, as opposed to a double porosity response such as on Figure 4.7, where only the storativity ratio influences the response.

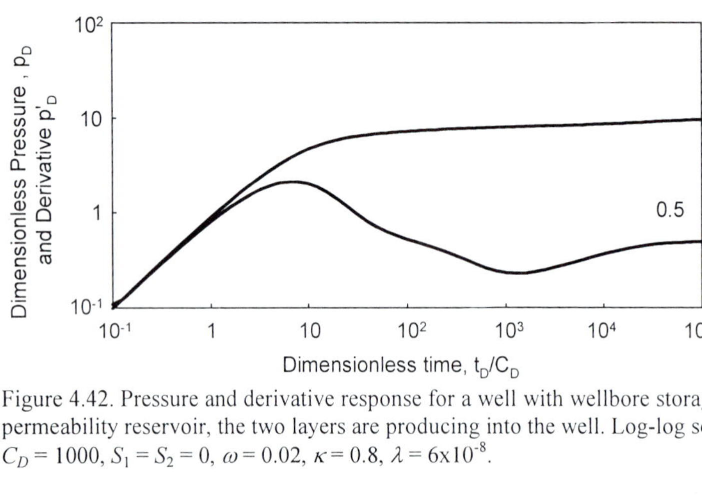

Figure 4.42. Pressure and derivative response for a well with wellbore storage and skins in double permeability reservoir, the two layers are producing into the well. Log-log scales, p_D versus t_D/C_D. $C_D = 1000$, $S_1 = S_2 = 0$, $\omega = 0.02$, $\kappa = 0.8$, $\lambda = 6\text{x}10^{-8}$.

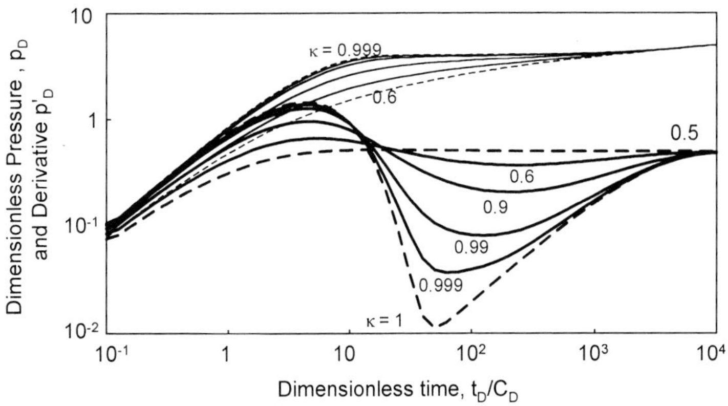

Figure 4.43. Log-log plot of double permeability responses, the two layers are producing into the well. p_D versus t_D/C_D. $C_D = 1$, $S_1 = S_2 = 0$, $\omega = 0.001$, $\lambda = 4\text{x}10^{-4}$. $\kappa = 0.6$, 0.9, 0.99 and 0.999. The two dashed curves correspond to the homogeneous reservoir response ($C_D e^{2S} = 1$) and double porosity response ($\kappa = 1$).

Discussion of double permeability parameters

In the following, it is assumed that the respective layer skins are equal: $S_1 = S_2$. The number of parameters is reduced to 5, and the influence of the three reservoir parameters κ, ω and λ is demonstrated with several log-log and semi-log examples.

Influence of κ *and* ω

Figure 4.43 presents pressure and derivative examples of double permeability responses for different values of κ (0.6, 0.9, 0.99, 0.999). The curves are generated with $\omega = 10^{-3}$, and λ, C_D, S_1 and S_2 are the same on all responses. The upper and lower dashed curves are the two limiting cases: the homogeneous reservoir ($\kappa = \omega$) and the double porosity ($\kappa = 1$) responses.

When $\kappa = 1$, the maximum value of the mobility ratio, layer 2 has no radial permeability and the model corresponds strictly to a double porosity configuration: the derivative valley has the characteristic shape of a double porosity transition with restricted interporosity flow. When $\kappa = 0.999$, layer 2 provides 0.1% of the mobility, and the derivative curve deviates from the double porosity shape. With lower values of κ, the *depth* of the transition valley is reduced and for $\kappa = 0.6$ the derivative hardly drops below the 0.5 line.

The Figure 4.43 double permeability examples are presented on semi-log scales in Figure 4.44. The dashed curves describe the equivalent homogeneous behavior ($C_D e^{2S} = 1$) and the double porosity response ($\kappa = 1$). The thin curves illustrate two examples of the "*two layers no-cross flow*" responses corresponding to $\kappa = 0.99$ and 0.6: the interlayer flow parameter is set to $\lambda = 0$, other parameters are unchanged. When the mobility contrast is large (curve κ = 0.99), the two layer no-cross flow early time response is very different from the final total system equivalent homogeneous regime $C_D e^{2S}=1$ (see discussion of Equation 4.74, Section 4.2.6), and the double permeability transition is long and flat. The "S" shape of the semi-log curves is reduced when the contrast in layer permeability is smaller, such as with the curve for $\kappa = 0.6$.

On the Figures 4.45 and 4.46, the storage contrast between the layers is reduced: the storativity ratio is $\omega = 10^{-1}$, the other parameters are the same as on the responses Figures 4.43 and 4.44. When the storage of layer 1 becomes significant, the two characteristic regimes "two layers no-cross flow" and "total system homogeneous" are closer, and the double permeability transition is *shorter*.

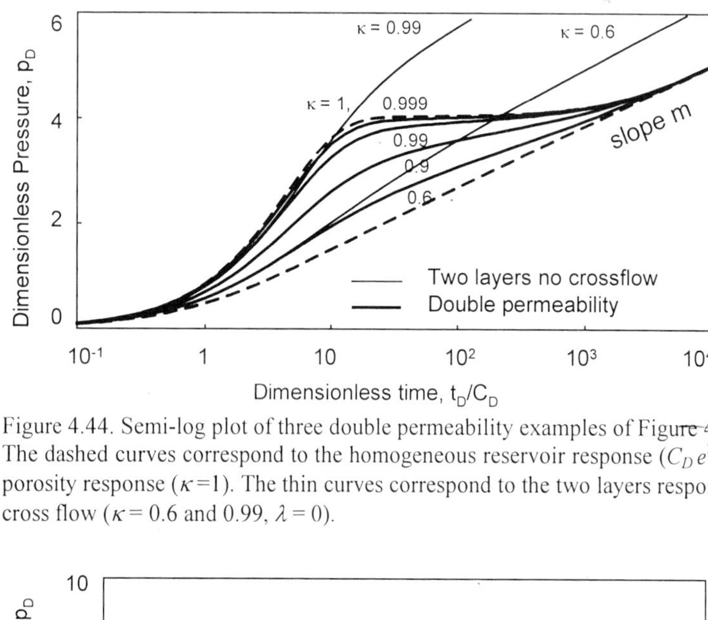

Figure 4.44. Semi-log plot of three double permeability examples of Figure 4.43.
The dashed curves correspond to the homogeneous reservoir response ($C_D e^{2S} = 1$) and the double porosity response (κ=1). The thin curves correspond to the two layers responses with no reservoir cross flow (κ = 0.6 and 0.99, λ = 0).

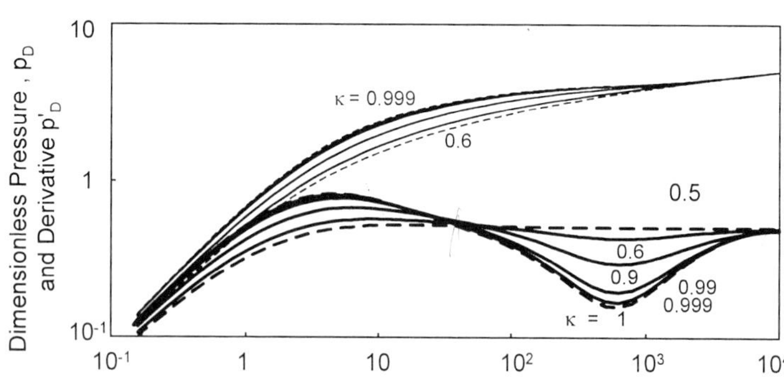

Figure 4.45. Log-log plot of double permeability responses, the two layers are producing into the well. p_D versus t_D/C_D. $C_D = 1$, $S_1 = S_2 = 0$, $\omega = 0.1$, $\lambda = 4\mathrm{x}10^{-4}$. κ = 0.6, 0.9, 0.99 and 0.999.
The two dashed curves correspond to the homogeneous reservoir response ($C_D e^{2S} = 1$) and the double porosity response (κ = 1).

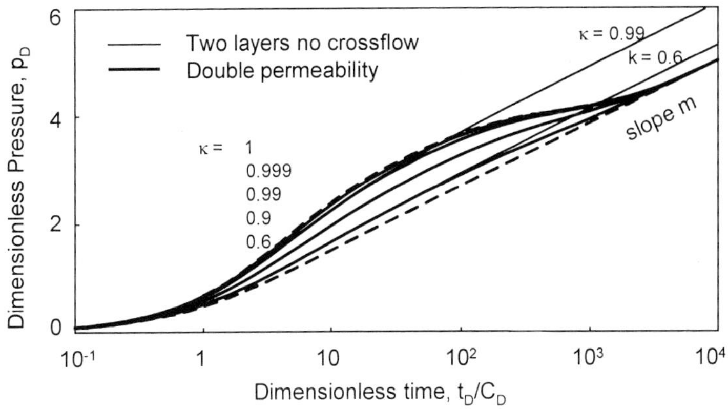

Figure 4.46. Semi-log plot of three double permeability examples of Figure 4.45.
The dashed curves correspond to the homogeneous reservoir response ($C_D e^{2S} = 1$) and the double porosity response (κ=1). The thin curves correspond to the two layers responses with no reservoir cross flow (κ= 0.6 and 0.99, $\lambda = 0$).

On a double permeability response, the shape of the transition is a function of the two ratios ω and κ. The storativity ratio ω defines the *duration* of the transition, but the *depth* of the derivative valley is also related to the mobility ratio κ. For example, the transition is deeper on the curve κ = 0.99 of Figure 4.45 ($\omega=10^{-1}$) than on the curve κ= 0.6 of Figure 4.43 ($\omega=10^{-3}$).

Influence of λ

The interlayer flow parameter λ depends upon the ability of the vertical cross flow to be established between the layers. When λ= 0, the response corresponds to the "two layers no-cross flow" solution. In other cases, a cross flow is established between the layers, and λ defines the *time of start* of the equivalent homogeneous total system flow regime like in double porosity responses. The smaller is λ, the later the start of total system flow.

Matching procedure with the pressure and derivative data

As for all responses affected by wellbore storage effect, the match point is fixed by the two derivative straight lines: at early time, the wellbore storage unit slope line, and during total system radial flow, the 0.5 line. The kh_{TOTAL} and C are estimated from the pressure and the time matches (Equations 4.49 and 4.52):

$$k_1 h_1 + k_2 h_2 = 141.2 qB\mu(\mathrm{PM}) \tag{4.53}$$

$$C = 0.000295 \frac{k_1 h_1 + k_2 h_2}{\mu} \left(\frac{1}{\mathrm{TM}} \right) \tag{4.54}$$

The time of the end of transition and start of the total system homogeneous behavior is used to estimate the interlayer cross flow parameter λ. The two other heterogeneous parameters are adjusted on the derivative transition data: a long transition valley suggests a small storativity ratio ω, and the depth below the 0.5 line indicates the range of the mobility ratio κ.

ω and κ define the storativity and the mobility distribution between the two layers from Equations 4.45 and 4.44, λ is used to provide an estimate of the vertical permeability. If the vertical resistance to the flow between the layers is concentrated at the wall interface, the vertical permeability of the "wall" k'_Z is obtained from Equation 4.47:

$$k'_z = (k_1 h_1 + k_2 h_2) \frac{\lambda}{r_w^2} h' \tag{4.55}$$

If no semi-permeable wall is present between the layers and the vertical pressure gradients are negligible in the high permeability layer 1, k_{Z2} can be expressed from Equation 4.48:

$$k_{z2} = (k_1 h_1 + k_2 h_2) \frac{\lambda}{r_w^2} \frac{h_2}{2} \tag{4.56}$$

When the same vertical permeability k_Z is assumed in the two layers and there is no skin at the interface:

$$k_{z1,2} = (k_1 h_1 + k_2 h_2) \frac{\lambda}{r_w^2} \frac{h_1 + h_2}{2} \tag{4.57}$$

In practice, when the storativity ratio ω is small, $h_1 << h_2$ and Equations 4.56, 4.57 provide similar results. The estimation of vertical permeability from λ is further discussed in the Multiple layer Section 4.2.6.

Once the derivative is matched and the three heterogeneous parameters are defined, the pressure response is used to evaluate the skins S_1 and S_2. If the two skin effects are different, the well condition influences the shape of the derivative transition, and it is difficult to conclude a unique match.

In the next Section 4.2.3, only one of the two layers is producing into the well: this configuration corresponds to the highest contrast between the two skins. It can be applied when a layered system is tested on a selected interval only.

4.2.3 Double permeability behavior when only one of the two layers is producing in the well (Wellbore storage and skin)

The well is perforated in one layer only: in the model, a very high skin is given to the non-communicating zone. Even though the well configuration corresponds to partial penetration, no spherical flow regime is produced. With the semi-permeable wall assumption of the double permeability model, the vertical pressure gradient is concentrated at the interface between the two layers.

The response of a partial completion well with wellbore storage and skin in a double permeability reservoir is defined by five dimensionless parameters: κ, ω, λ, C_D, and S_1 or S_2, depending upon the well configuration.

Behavior

Three different regimes are observed in the response of a partially completed well with wellbore storage and skin in a layered reservoir:

1. First, the perforated layer response is seen alone, and the behavior is *homogeneous*. The level of the first derivative stabilization depends upon the permeability of the perforated interval.
2. When the second layer starts producing into the perforated zone, the response deviates in a *transition* regime and the derivative drops.
3. Later, the pressure of the two layers equalizes, the equivalent *homogeneous* behavior of the total system is seen and the derivative stabilizes at 0.5, as the total *kh* is acting.

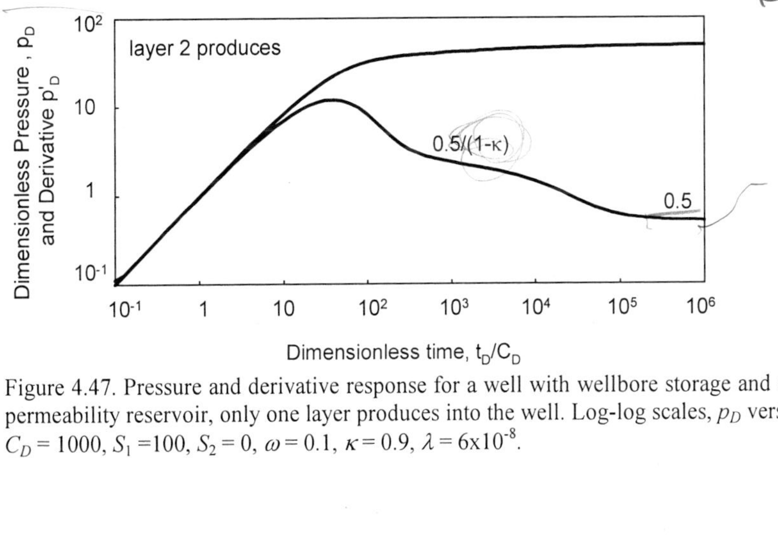

Figure 4.47. Pressure and derivative response for a well with wellbore storage and skin in double permeability reservoir, only one layer produces into the well. Log-log scales, p_D versus t_D/C_D. $C_D = 1000$, $S_1 = 100$, $S_2 = 0$, $\omega = 0.1$, $\kappa = 0.9$, $\lambda = 6\text{x}10^{-8}$.

In Figure 4.47, only layer 2 is producing in the well. After the wellbore storage hump, the derivative shows a first plateau corresponding to radial flow in the perforated zone, until the curve decreases to reach a second plateau, when the interlayer cross flow is established, and the total system produces. The double permeability behavior is then similar to the response of a well in partial penetration, such as Figure 3.15.

Discussion of double permeability parameters

The shape of the transition between the two homogeneous regimes depends upon the permeability ratio between the perforated layer and the second layer. In Figure 4.48, the two hypotheses are presented for the same layered system. The reservoir parameters are defined as $\omega = 0.2$, $\kappa = 0.9$ and $\lambda = 10^{-4}$. The wellbore storage coefficient is $C_D = 1$ and, for one of the curves $S_1 = 100$, $S_2 = 0$ and for the other $S_1 = 0$, $S_2 = 100$. The thin curve describes the double permeability response when both layers are perforated (the two skins are set to 0). Figure 4.49 presents the same response curves on semi-log scales.

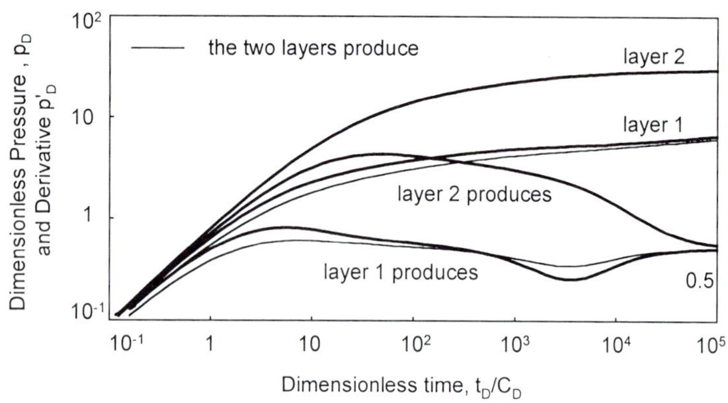

Figure 4.48. Log-log plot of double permeability responses, only one layer is producing into the well. $C_D = 1$, $\omega = 0.2$, $\kappa = 0.9$, $\lambda = 10^{-4}$. $S_1 = 100$, $S_2 = 0$ and $S_1 = 0$, $S_2 = 100$. The thin curves correspond to the double permeability response with no skin when both layers are perforated.

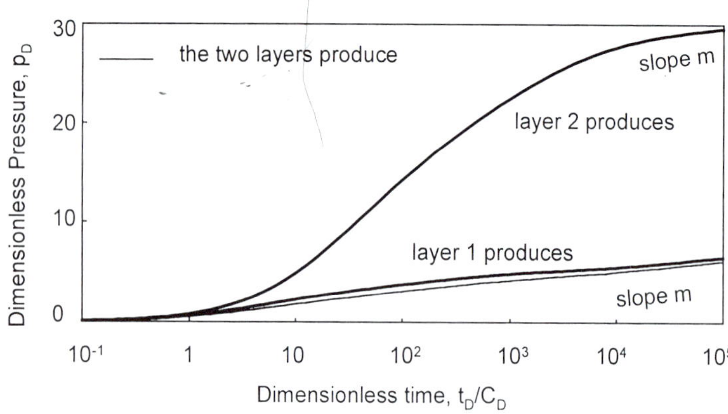

Figure 4.49. Semi-log plot of Figure 4.48 double permeability examples.

Taking only the low permeability layer producing into the well, we will examine first the response shown as "layer 2 produces". During the initial homogeneous regime, the derivative tends to show a stabilization at $0.5/(1-\kappa)$. On the example of Figure 4.48, the first radial flow regime is masked by wellbore storage, and the derivative plateau (at 5 with κ=0.9) is not seen. The curve displays a long hump before the *drop of derivative* in transition, and the second *stabilization* on 0.5 for the total system. During this final radial flow regime, the amplitude of the response is large, as illustrated on the semi-log plot Figure 4.48. This configuration produces, when the well is perforated on the low permeability layer, a large positive geometric skin.

The response is very similar to the behavior of a well in partial penetration as described in Section 3.4; the two models exhibit a different behavior only during the transition. In a homogeneous reservoir, partial penetration is characterized by a spherical flow regime with a negative half unit slope derivative straight line. In case of double permeability response, the vertical pressure gradient is modeled with the semi-permeable wall interface, and the transition between the two radial flow regimes is *steeper*.

Conversely, when only the high permeability layer 1 is producing to the well, the permeability of the first homogeneous regime is, with κ=0.9, 90% of the total system permeability, and the two derivative stabilizations are almost at *same level*: $0.5/\kappa$ for the first (in the example: 0.55) and 0.5 for the second. At transition time, the derivative *valley* is deeper than on the thin curve, when the two layers produce into the well.

With this second type of double permeability partial completion, the response tends to show the same behavior as a double porosity with restricted interporosity flow: introducing a high skin at the low permeability layer produces a similar effect as a reduction of permeability of the layer.

The influence of the double permeability parameters on a partial completion well can be summarized as follows:

- When only one layer is communicating with the well, κ defines the level of the first derivative plateau, $0.5/\kappa$ or $0.5/(1-\kappa)$, thus the type of response. When the perforated layer is of high permeability, the first derivative stabilization is close to 0.5, and the response tends to a double porosity behavior. Determining whether the reservoir behavior is double porosity or double permeability is difficult. In the opposite case, the response shows a first derivative plateau above 0.5, and tends to a partial penetration response.
- λ indicates the *time of transition* between the two homogeneous behaviors. A small λ corresponds to a long early time "one layer" regime, and a late total system regime. ω can influence the shape of the transition when the derivative drops.

Matching procedure on the pressure and derivative data

In the case of a partially completed well, frequently only one derivative stabilization is observed on the log-log data plot. When radial flow is reached for the late time total system homogeneous behavior, the second stabilization is evident and the match point is fixed by the two usual derivative straight lines: the wellbore storage unit slope line and the 0.5 line during total system radial flow. The total permeability thickness product kh_{TOTAL} and the wellbore storage coefficient C are estimated with Equations 4.53 and 4.54. The curve match defines the other parameters: the skin of the perforated layer, and the ratios of diffusivity and mobility.

When the test stops before total system radial flow, only the first line can be expected on the response curve and the total permeability thickness product cannot be accessed. In such case, the static parameters used for analysis correspond generally to the open interval, as discussed next.

In the following definition of the dimensionless pressure, time and wellbore storage, the open interval is referenced with subscript i (i =1 or 2):

$$p_D = \frac{k_i h_i}{141.2 qB\mu} \Delta p \tag{4.58}$$

$$t_D = \frac{0.000264 k_i h_i}{(\phi c_t h)_i \mu r_w^2} \Delta t \tag{4.59}$$

$$C_D = \frac{0.8936 C}{(\phi c_t h)_i r_w^2} \tag{4.60}$$

$$\frac{t_D}{C_D} = 0.000295 \frac{k_i h_i}{\mu} \frac{\Delta t}{C} \tag{4.61}$$

The 0.5 derivative stabilization corresponds to radial flow in the open interval, shown by the first plateau before the drop of derivative response. From the pressure and time matches, $k_i h_i$ and C are estimated:

$$k_i h_i = 141.2 qB\mu (\mathrm{PM}) \tag{4.62}$$

$$C = 0.000295 \frac{k_i h_i}{\mu} \left(\frac{1}{\mathrm{TM}} \right) \tag{4.63}$$

The skin S_i is estimated from the homogeneous $C_D e^{2S}$ curve matching the early time data before the drop of the derivative, with Equation 4.60 and Equation 2.11.

When the second derivative stabilization is not reached, the parameters of the complete system are not well defined. The time of end of the first homogeneous behavior and start of the transition is used to estimate the interlayer cross flow parameter λ, and possibly the vertical permeability from Equations 4.55 to 4.57. Depending on how well defined the shape of the derivative curve during transition is, the two other heterogeneous parameters κ and ω can be approximated or not. In general, a unique match is difficult to obtain.

4.2.4 Associated specialized plot straight lines

Wellbore storage effects and radial flow regimes can be identified on double permeability responses. The specialized analysis of the wellbore storage period is not affected by the layered nature of the reservoir but, for the radial flow regimes, the analysis is adapted to the number of producing zones during the selected time interval, as diagnosed with the log-log analysis.

Wellbore storage analysis

On a Cartesian scale, the slope of the early time straight line provides an estimate of the wellbore storage coefficient with the relationship Equation 1.10.

Radial flow analysis

When the two layers are producing into the well, as described in Section 4.2.2, the sequence of flow regimes is "two layers without cross flow", transition and "total system equivalent homogeneous". On a semi-log scale, the first regime is not characteristic and only one straight line can be analyzed at late time (Figures 4.44 and 4.46).

With the other well configuration presented in Section 4.2.3, only one layer produces into the well and a first radial flow regime can be seen before the total system response. Two semi-log straight lines are then possible, the mobility during the first radial flow being a fraction of the total; the first slope is higher than the second is. Frequently, only one of the two lines is observed: the wellbore storage effect can cover the first as in Figure 4.49, or the test is not long enough to reach the second. The semi-log analysis of the two straight lines is valid only when the log-log analysis confirms the presence of two radial flow regimes.

When one of the two layers (called layer *i*) starts to produce alone, the first line is expressed as (from Equation 1.15):

$$\Delta p = 162.6\frac{qB\mu}{k_i h_i}\left[\log\Delta t + \log\frac{k_i}{(\phi c_t)_i \mu r_w^2} - 3.23 + 0.87S_i\right] \tag{4.64}$$

This first line gives $k_i h_i$ and S_i. The second line, for the total system regime, gives the total mobility:

$$\Delta p = 162.6\frac{qB\mu}{kh_{\mathrm{TOTAL}}}\left[\log\Delta t + \log\frac{kh_{\mathrm{TOTAL}}}{(\phi c_t h)_{\mathrm{TOTAL}}\,\mu r_w^2} - 3.23 + 0.87S\right] \tag{4.65}$$

When only one layer is producing into the well, the global skin S measured on the total system semi-log straight line includes a geometrical skin due to the curvature of the flow lines. This is equivalent to the geometrical skin of partially penetrating wells. When the two layers are producing into the well, the global skin S, measured on the total system semi-log straight line of Equation 4.65, is a function of the two layers skins S_1 and S_2, with (Prijambodo et al., 1985; Park and Horne, 1989):

$$S = S_1\, q_{1D}^{\mathrm{LT}} + S_2\, q_{2D}^{\mathrm{LT}} = \frac{1}{q_1 + q_2}\left(S_1\, q_1^{\mathrm{LT}} + S_2\, q_2^{\mathrm{LT}}\right) \tag{4.66}$$

where q_{iD}^{LT} are the late time stabilized fractional rates of the two layers. When production-logging data is not available, q_{iD}^{LT} can be expressed (Park and Horne, 1989) as a function of k, S_i and λ (see discussion for multiple layer in next Section).

4.2.5 Field example

On the build-up example presented in Figure 4.50 (published in 1985, Bourdet), the derivative response shows a long transition valley before a final stabilization, from the sixth hour to the end of the test. When the final stabilization is assumed to describe the infinite acting radial flow regime, it is possible to match the data with a double porosity solution, restricted interporosity flow. The match is slightly improved with the double permeability model, when 2.5 % of the permeability thickness product is attributed the low permeability layer (κ=0.975 with the same negative skin on both layers S_1=S_2=-4).

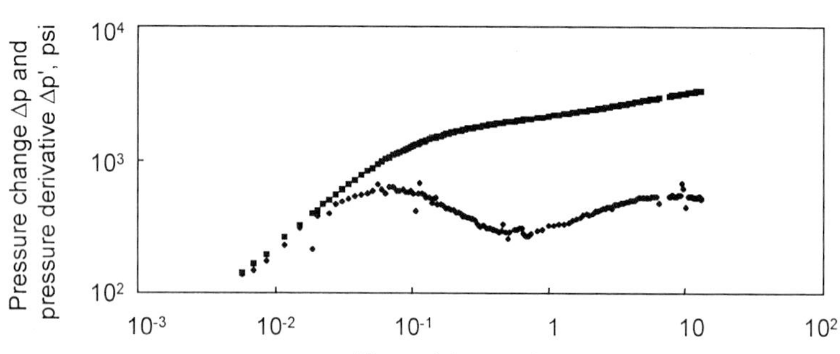

Figure 4.50. Field example build-up test. Log-log scale.

Other models, such as a radial composite system (Section 4.3.2) or a homogeneous reservoir with intersecting sealing faults (Section 5.3) can be used, but the radial flow regime ends before the first hour of shut-in. The derivative stabilization must be located a little below the bottom of the valley, the permeability thickness product is almost three times larger and the skin is less negative than with previous double porosity or permeability matches.

4.2.6 Extensions of the double permeability model

In the following, several variations of the double permeability behavior are reviewed. First, the extension of the model to multiple layer system is considered. The analysis is more complex and the uniqueness of the solution seems frequently to be not clearly established. Transient reservoir cross flow is then presented. These two developments are not currently used for the time being and they are only briefly discussed.

Multiple layer

The double permeability model can be extended to an n-layer system (Ehlig-Economides and Joseph, 1985; Larsen, 1988; Bidaux et al., 1992). For layer i, the n-permeability parameters are defined as:

$$\kappa_i = \frac{k_i h_i}{\sum_{j=1}^{n} k_j h_j} = \frac{k_i h_i}{kh_{\text{TOTAL}}} \tag{4.67}$$

$$\omega_i = \frac{(\phi c_t h)_i}{\sum_{j=1}^{n} (\phi c_t h)_j} = \frac{(\phi c_t h)_i}{(\phi c_t h)_{\text{TOTAL}}} \tag{4.68}$$

λ_i is a function of the vertical permeability k'_{iz} in the low permeability "wall" of thickness h'_i between the layers i and $i+1$ and, of vertical property of the two layers $(h/k_z)_{i,\,i+1}$.

$$\lambda_i = \frac{r_w^2}{\sum_{j=1}^{n} k_j h_j} \frac{2}{2\dfrac{h'_i}{k'_{iz}} + \left(\dfrac{h}{k_z}\right)_i + \left(\dfrac{h}{k_z}\right)_{i+1}} \tag{4.69}$$

If there is no cross flow between layers i and $i+1$, $k'_{iz}=0$ and $\lambda_i=0$.

As for the double permeability configuration, when all layers of an n-layer system are communicating in the reservoir, the response is identical to that of the commingled system at early time and to that of the equivalent homogeneous total system at late time (with the average horizontal permeability of Equation 3.55).

Each layer is defined with 4 parameters (Equations 4.67 to 4.69 and the layer skin S_i) and, with the wellbore storage coefficient, the total number of parameter required to define the response is $4n$. In order to reduce the non-uniqueness of the solution, several authors suggest to incorporate layer rate measurements into multiple layer analysis. Transient layer rate responses can be summarized as follows: during the commingled regime, the fractional flow rate of each layer is expressed at very *early time* as a function of the *layer skin factors* S_i (Ehlig-Economides and Joseph, 1985; Park and Horne, 1989):

$$\lim_{t_D \to 0} q_{iD} = \frac{k_i / S_i}{\sum_{j=1}^{n} k_j / S_j} \tag{4.70}$$

When there is no skin at the layers, the fractional flow rate is expressed:

$$\lim_{t_D \to 0} q_{iD} = \frac{\sqrt{\omega_i k_i}}{\sum_{j=1}^{n} \sqrt{\omega_j k_j}} \tag{4.71}$$

When there is no cross flow in the reservoir, the *late time rates* of the commingled system are independent of the layer skin S_i with (Park and Horne, 1989):

$$\lim_{t_D \to \infty} q_{iD} = \frac{k_i h_i}{\sum_{j=1}^{n} k_j h_j} = \kappa_i \tag{4.72}$$

In case of formation cross flow, the layer rates become *constant* after the transition, during the equivalent total system homogeneous behavior, whereas those of the commingled system do not stabilize as fast. This stabilized rate after transition is a function of the permeability k_i and the skin S_i of the layers, with several groups such as $k_i S_i$ and λ_i / k_i, but also with a term $k_i(S_i - S_j)$ defining the contrast between the skin factors (Park and Horne, 1989).

- In the high permeability layers, the rate is higher than that of the commingled system of Equation 4.70 if the skin is lower than in the low permeability layers.
- Conversely, when the high permeability layers are the most damaged, the rate is lower because the reservoir cross flow is established from the layers with large skin to the layers with smaller skin.

With the current down-hole rate measurement technology, transient layer rate is frequently noisy and only stabilized rate data is available for the analysis. Caution must be exercised in interpreting multiple layer systems. A regression algorithm can be used to match pressures and layer rates, but the question of the uniqueness of the solution must be examined carefully. Larsen (1994 b) investigated the non-uniqueness question versus layers skin and refinement in the discretization of the layered reservoir (see following discussion). He concluded that layer flow rates are not strongly influenced by layer refinement but, with large contrast in skin values between the layers, the analysis become highly uncertain.

Interpretation of the interlayer cross flow coefficients λ_i is the next step for characterizing the multiple layer reservoir. When there is no low permeability "wall" separating the layers, Equation 4.69 can be used to express the vertical permeability (Boutaud de la Combe et al., 1996). With $k'_{iZ} = \infty$,

$$k_{zi} = h_i \frac{k_{zi+1}\lambda_i}{k_{zi+1}\left(2r_w^2 \Big/ \sum_{j=1}^{n} k_j h_j\right) - h_{i+1}\lambda_i} \tag{4.73}$$

The different vertical permeability k_{iZ} can be expressed as a function of one of the vertical permeability k_{jZ} . A graph of the k_{iZ} versus k_{jZ} defines n branches of hyperbola that indicate the different minimum and maximum k_Z values.

In his discussion of n permeability model, Larsen (1988) observed that by dividing a layer into several smaller layers with similar characteristics, the vertical pressure gradients due to transient vertical flow are reproduced. Increasing the number of layers tends to produce a transient interlayer flow response (see next Section): the cross flow begins earlier and the bottom of the derivative valley is moved upward. Boutaud et al. reached similar conclusion for the low permeability interval of a two layers reservoir but, in the high permeability layer, the discretization has little effect. When only the high permeability layer of a two layers system is perforated (see Section 4.2.3), the discretization of the non-perforated interval into several layers tends to exhibit the half unit slope derivative straight line characteristic of the spherical flow regime of a well in partial penetration presented in Section 3.4 (Larsen, 1988; Boutaud et al., 1996).

Conversely, the double permeability model can be used for the analysis of multiple layer reservoirs. Assuming an n layers system is described as a sequence of alternating high and low permeability layers, the response can be approximated by that of one element of the series (a two layers reservoir) produced at rate $2q/n$. In the analysis of the multiple layer reservoir, it is therefore possible to group the $n/2$ high permeability layers into "Layer 1" and the $n/2$ low permeability layers into "Layer 2". Such approximation provides an accurate estimate of the total mobility and storage ratios κ and ω. In the interpretation of λ to estimate the vertical permeability with Equation 4.46, the layer thickness h_1 and h_2 should not be defined as the total thickness of the high and low permeability layers. In his discussion of layer refinement, Larsen (1988) illustrated the difficulty in the interpretation of the cross flow parameter λ with a three layers example.

Transient reservoir cross flow

Following Chen et al. model (1990), transient reservoir cross flow is defined, as for the unrestricted double porosity model of Section 4.1.3, by introducing vertical pressure gradients in the low permeability layer. The vertical pressure gradients in the high permeability layer are neglected.

With transient reservoir cross flow, the low permeability layer reacts quickly to any change of pressure in the high permeability drain. The transition starts earlier than in case of the semi-permeable wall model: the two layers no cross flow regime is short lived and the transition is longer. During the transition, the derivative tends to *stabilize* at a level between 0.5 (homogeneous system) and 0.25 (unrestricted double porosity model, $\kappa=1$). The level of the intermediate stabilization is a function of κ.

When a skin is introduced between the layers, the Chen et al. model (1990) reproduces the double permeability transition of Section 4.2.2 for the semi-permeable wall. For large skin between the layers, the effective interporosity flow parameter λ_{eff} is given by Equation 4.47.

When there is no skin between the layers, the two models are clearly different provided the contrast between the two layers is very high ($\kappa > 0.99$ and $\omega < 0.01$), making the transition characteristic. In such case, the unrestricted double porosity model with slab matrix block can be used as an approximation. In other cases, Equation 4.48 seems to provide acceptable results (Boutaud de la Combe et al., 1996).

4.2.7 Commingled systems with equal or unequal initial pressure

Many authors have discussed pressure responses from commingled systems (Lefkovits et al., 1961; Tariq and Ramey, 1978; Larsen, 1981; Bourdet, 1985; Ehlig-Economides and Joseph, 1985; Joseph et al., 1986; Larsen, 1989; Park and Horne, 1989) considering infinite or closed system behavior. Different initial pressure between the layers has also been envisaged (Larsen, 1981; Agarwal et al., 1992; Aly et al., 1994). In the following, transient pressure analysis in an infinite multiple layer reservoir without cross flow is briefly reviewed.

Same initial pressure

Commingled responses have been illustrated on the semi-log plots Figures 4.44 and 4.46 for a two layers system with no skin at the wellbore. When there is no reservoir cross flow, the amplitude of the response is larger than that of the equivalent homogenous system. Furthermore, the semi-log straight line is not accurately defined until long elapsed time.

In multiple layer systems, Larsen (1981) defines the *pseudo-skin* factor induced by layering as:

$$S_L = \frac{1}{2}\sum_{i=1}^{n} \frac{k_i h_i}{kh_{\text{TOTAL}}} \ln \frac{(kh/\phi c_t h)_i}{(kh/\phi c_t h)_{\text{TOTAL}}} \quad (4.74)$$

The layering pseudo-skin factor is relatively small in magnitude. In the case of large contrast such as on the example $\kappa = 0.999$ and $\omega = 0.001$ of Figure 4.44, the pseudo-skin is estimated at S_L=3.5. For the curve $\kappa = 0.9$ and $\omega = 0.1$ of Figure 4.46, S_L is only 0.9.

When in addition the layers have different mechanical skin factors, Larsen shows that the semi-log slope is larger at early time, and it decreases slowly with increasing time to reach finally the equivalent total system slope of Equation 4.65: $162.6 qB\mu / kh_{\text{TOTAL}}$. For a two layers reservoir, the apparent early time semi-log slope is a function of κ and ω, but also of the skin contrast (S_1- S_2). As the reduction of slope is slow, the author indicates that the analysis of an approximate semi-log straight line on early time data could yield a total *kh* up to 50% below the correct value. This error can be larger when both positive and negative skins are present in systems with more than two layers.

When the semi-log approximation is valid, the global skin has two components: S_L of Equation 4.74 and the average mechanical skin $\overline{S}$, resulting from the skin S_j of the different layers. When the layer skins S_j are not considerably different, the author proposes to approximate $\overline{S}$ with:

$$\overline{S} = \sum_{i=1}^{n} \frac{k_i h_i}{kh_{\text{TOTAL}}} S_i = \sum_{i=1}^{n} \kappa_i S_i \quad (4.75)$$

On build-up responses, the extrapolated semi-log straight line to infinite shut-in time can be significantly lower than the true build-up pressure at $\Delta t = \infty$. As discussed in Section 5.8.3, some layers can show boundary effects when other are still infinite acting at time of shut-in. Such configuration produces a late time rise of the build-up pressure above the semi-log straight line.

Unequal initial pressure

In case of different initial pressure, a cross flow *through the wellbore* is established as soon as the well is perforated. If the well is not opened to surface production, the bottom hole pressure tends asymptotically towards the *average* initial pressure. For an infinite system, $\overline{p_i}$ is defined as the *kh* average of the different initial pressures with (Larsen, 1981):

$$\overline{p_i} = \sum_{j=1}^{n} \frac{k_j h_j}{kh_{\text{TOTAL}}} p_{ij} \quad (4.76)$$

The author developed a technique to analyze this initial transient regime with a plot of the reciprocal of a pressure difference versus the logarithm of time.

It is interesting to note that, if the non-producing commingled reservoir is closed, the final average reservoir pressure $\bar{p}$ can be greater or smaller than the "infinite" average initial pressure $\bar{p}_i$ of Equation 4.76 (Boutaud de la Combe et al., 1996). The average reservoir pressure $\bar{p}$ is defined as the "pore volume" average of the initial pressures (Larsen, 1981):

$$\bar{p} = \sum_{j=1}^{n} \frac{V_j c_{t\,j}}{V c_{t\ \mathrm{TOTAL}}} p_{i\,j} \tag{4.77}$$

where V_j is the pore volume $\phi\, hA$ of the closed layer j.

When the well is opened to production, the effect of the initial transient due to unequal pressures is inversely proportional to the production rate. Larsen concludes that a high flow rate is desirable to reduce this adverse pressure trend. Before the test, a long shut-in period allowing cross flow through the well also tends to minimize the effect of unequal pressures (Agarwal et al., 1992).

During shut-in periods, the transient effect of unequal pressures ultimately dominates the build-up response. On early time data, the semi-log straight line does not in general extrapolate to the average initial pressure $\bar{p}_i$ of Equation 4.76, and analysis of the late time data is uncertain.

4.3 COMPOSITE RESERVOIRS

The composite reservoir models, like all basic heterogeneous reservoir solutions presented in this chapter, consider two distinct media in the reservoir. Each component is defined by a porosity and a permeability, and they are located in different reservoir regions. Two geometries are considered for the interface between the reservoir areas.

Radial composite systems have been studied from the early 1960's (Hurst, 1960; Loucks and Guerrero, 1961; Carter, 1966; Satman, 1980; Olarewaju, 1989): it is assumed that the well is at the center of a circular zone, the outer reservoir structure corresponds to the second element (Figure 4.51). This geometry is used to describe a radial change of properties, resulting from a change of fluid or formation characteristic. Such change can be man-induced in case of injection wells and in some cases of damaged or stimulated wells. It can also be observed when oil and gas saturations vary around the wellbore, for example when the reservoir produces below bubble point or dew point (see Section

8.2.2). Radial composite solutions have also been used in some cases of sparsely fractured reservoirs, when the well is in a block. The radial composite model presented in this section is an extension of the analytical solution of Satman (1980).

With the linear composite model, a vertical plane interface is assumed between the two reservoir regions (Bixel et al., 1963; Streltsova and McKinley, 1984 b; Ambastha et al., 1987): the reservoir is divided into two semi-infinite zones, the well is located in one of them (Figure 4.51). This composite configuration can be observed for example when a linear fault separates two different reservoir elements with different characteristics, or when a water drive is active in one direction of the producing zone. The linear composite model presented in the following corresponds to the solution of Ambastha et al. (1987), completed by the effects of wellbore storage and skin.

The hypotheses used for the description of composite reservoirs are presented in Section 4.3.1, the models behaviors are discussed in Section 4.3.2 for the radial model, and in Section 4.3.3 when the interface is linear. In Section 4.3.4, extensions of the models to multiple composite systems are presented (Barua and Horne, 1987; Acosta and Ambastha, 1994; Abbaszadeh and Kamal, 1989; Bratvold and Horne, 1990; Kamal et al., 1992; Bourgeois et al., 1996 b; Kuchuk and Habashy, 1997), and tests in reservoirs with complex changes of permeability (Levitan and Crawford, 1995; Oliver, 1990; Thompson and Reynolds, 1997; Oliver, 1992; Feitosa et al., 1993; Yeh and Agarwal, 1989; Kuchuk et al., 1993; Tauzin and Horne, 1994). Practical interpretation of injection wells has been discussed by several authors (Abbaszadeh and Kamal, 1989; Bratvold and Horne, 1990; Yeh and Agarwal, 1989), the main conclusions are summarized in Section 4.3.5.

In this chapter, only the response of a well with wellbore storage and skin is envisaged, but the radial composite solution has been also extended to the case of a fractured well (Chu and Shank, 1993, Chen and Raghavan, 1995).

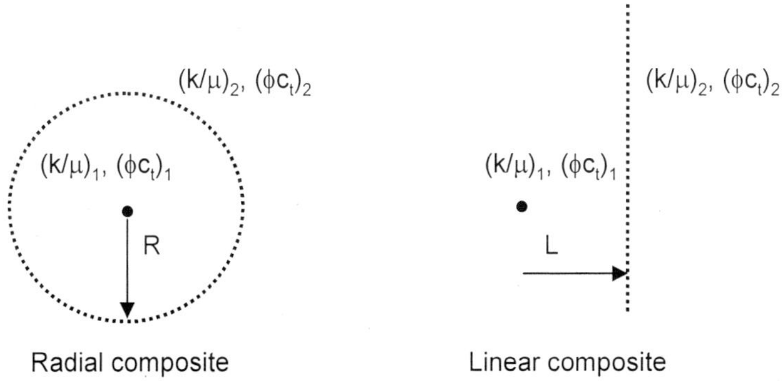

Figure 4.51. Models for composite reservoirs.

4.3.1 Composite reservoir models

Composite reservoir assumptions

A discontinuity defines two distinct homogeneous regions in the infinite reservoir. The interface is *stationary* and it has *no thickness*. The mobility (k/μ) and storativity (ϕc_t) are different on each side, but the reservoir thickness h is constant. The change of reservoir properties is abrupt, and there is *no resistance to flow* .between the two reservoir regions.

The well, affected by wellbore storage and skin, is located in the region 1: with the radial composite model, it is at the center of a circular zone of radius R, with the linear composite, the interface is at a distance L (Figure 4.51). The characteristics parameters of the second region are defined with a subscript 2.

Definition

The changes of reservoir mobility (k/μ) and storativitystorativity (ϕc_t) are expressed with the mobility M and storativity F ratios, defined as region 1 compared to region 2:

$$M = \frac{(k/\mu)_1}{(k/\mu)_2} \tag{4.78}$$

$$F = \frac{(\phi c_t)_1}{(\phi c_t)_2} \tag{4.79}$$

A mobility ratio M greater than 1 indicates a decrease of mobility from region 1 to region 2. A decrease of the storage is expressed with the ratio F greater than 1.

Dimensionless variables

All dimensionless variables are expressed with reference to the parameters of the region 1 around the well. The dimensionless pressure is:

$$p_D = \frac{k_1 h}{141.2 q B \mu_1} \Delta p \tag{4.80}$$

the dimensionless time:

$$t_D = \frac{0.000264 k_1}{(\phi \mu c_t)_1 r_w^2} \Delta t \tag{4.81}$$

and the dimensionless wellbore storage:

$$C_D = \frac{0.8936C}{(\phi c_t)_1 h r_w^2} \tag{4.82}$$

The composite reservoir response curves are presented with the dimensionless pressure p_D of Equation 4.80 versus the dimensionless time group t_D/C_D:

$$\frac{t_D}{C_D} = 0.000295 \frac{k_1 h}{\mu_1} \frac{\Delta t}{C} \tag{4.83}$$

The skin factor expresses a dimensionless pressure change around the wellbore (Equation 1.11). As the dimensionless pressure of Equation 4.80, it is defined with respect to the inner zone parameters.

$$S = \frac{k_1 h}{141.2 q B \mu_1} \Delta p_{\text{skin}} \tag{4.84}$$

The distance between the well and the interface is expressed in dimensionless terms as R_D or L_D, depending upon the model geometry. For radial and linear systems, the distances are respectively:

$$R_D = \frac{R}{r_w} \tag{1.21}$$

$$L_D = \frac{L}{r_w} \tag{1.34}$$

4.3.2 Radial composite reservoir

Olarewaju and Lee (1989) presented a pressure and derivative type-curve for a well with wellbore storage in a radial composite reservoir. They use a dimensionless time group based on the distance R to the interface, and the curves are defined in term of $C_D R_D^2$.

In the following, the mechanical skin of Equation 4.84 is included, and the usual t_D/C_D time group is used (Equation 4.83). Five dimensionless parameters are used to define the response of a radial composite system: C_D, S for the well and R_D, M, and F for the reservoir. No type-curves are available for this configuration, a log-log match of the complete response is made on computer-generated pressure and derivative curves. When the analysis is performed by hand, the different regimes of a radial composite response can also be described individually, with the homogeneous log-log type-curve and the specialized semi-log scale.

Behavior

With the radial symmetry of the system, the two reservoir regions are seen in sequence:
1. First, the pressure response depends upon the *inner zone* characteristics, and the well behavior corresponds to a homogeneous reservoir response.
2. When the circular interface is reached, a second homogeneous behavior, corresponding to the *outer region*, is observed.

Influence of M

In Figure 4.52, derivative responses are presented for different values of the mobility ratio M: the parameters of the well and of the inner zone are constant, the two reservoir regions have the same storativity (F=1).

In Equation 4.80, the dimensionless pressure is defined with respect to the inner zone parameters, the first derivative plateau corresponds then to 0.5. Before the different curves diverge, the early time data is described by the type-curve for a homogeneous reservoir, Figure 2.22.

For large values of mobility ratio (M =2 and 10 on Figure 4.52), the mobility of the outer region is *reduced*, and the second derivative plateau is displaced *upwards* (to 0.5M). The dotted derivative curves show the drawdown response of a well in a closed circle of same radius R_D: it illustrates the limiting case of a zero mobility in the outer zone.

In the same Figure 4.52, *low values* of the mobility ratio are shown by a *reduction* of the derivative amplitude (curves M=0.5 and 0.1). The corresponding limiting case, illustrated by the second dotted derivative curve, is a circle at constant pressure, when the mobility of the outer region is infinite.

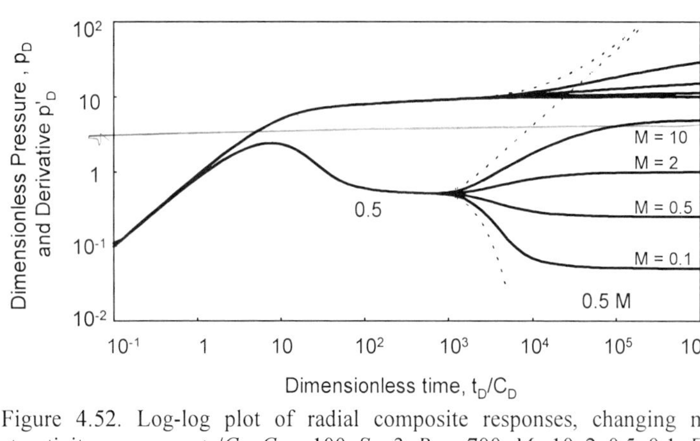

Figure 4.52. Log-log plot of radial composite responses, changing mobility and constant storativity. p_D versus t_D/C_D. C_D = 100, S = 3, R_D = 700, M =10, 2, 0.5, 0.1, F =1.
The two dotted curves correspond to the closed and the constant pressure circle solutions.

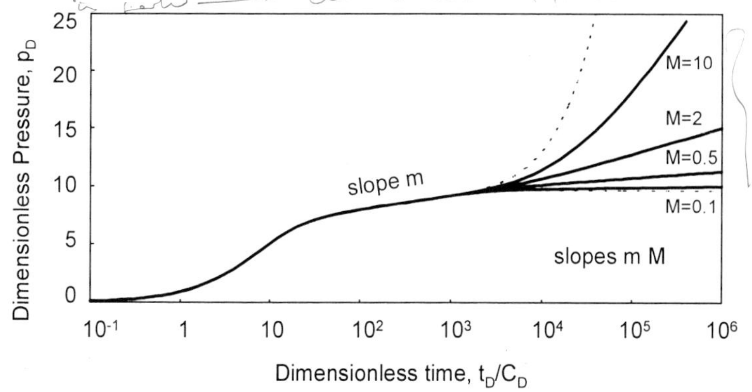

Figure 4.53. Semi-log plot of Figure 4.52 radial composite examples.
The two dotted curves correspond to the closed and the constant pressure circle solutions.

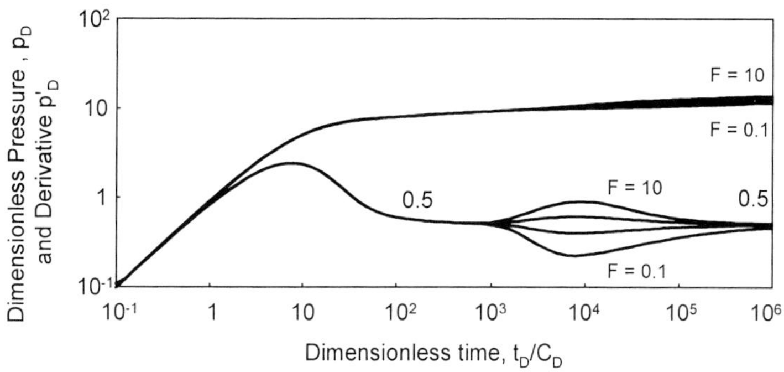

Figure 4.54. Log-log plot of radial composite responses, constant mobility and changing storativity. p_D versus t_D/C_D. $C_D = 100$, $S = 3$, $R_D = 700$, $M = 1$, and $F = =10, 2, 0.5, 0.1$
The two pressure curves correspond to $F = 10$ and $F = 0.1$.

On semi-log scale, the radial composite examples of Figure 4.52 exhibit two straight lines as illustrated on Figure 4.53. During the initial radial flow in the inner zone, the slope is m (Equation 4.88). The slope of the second line at late time (mM, or Equation 4.89) defines the mobility of the outer region. For a mobility ratio M higher than 1, the semi-log straight line slope is increased, and for M values lower than 1, the second line tends to flatten, towards a constant pressure behavior for very low values of M.

Influence of F

For the examples chosen in Figure 4.54, it is assumed that the two regions have the same mobility: M=1. The well and inner zone parameters of Figure 4.52 examples are used, several pressure and derivative responses are presented for different storativity ratios F.

When F=0.1, the storage of the outer zone is 10 times larger than the storage of the inner zone (with a constant mobility through the reservoir). The response corresponds to an *increase of storativity*. A similar effect is observed in the double porosity model for restricted interporosity flow, when the response moves from the fissures to the total system flow regime (see Section 4.1.2). The shape of the curve for F=0.1 is in fact close to a double porosity response: the derivative shows a *valley* between the two plateaus at 0.5.

Conversely, when F is greater than unity (and M=1), the storage of the outer zone is reduced, and the response shows a *decrease of storativity* from early time to late time: the transition on derivative curves exhibits a *hump* above the two 0.5 plateaus, as on the curve F=10 of Figure 4.54.

Figure 4.55 is a semi-log plot of the four radial composite log-log curves of Figure 4.54. The mobility is constant in the reservoir: during the two homogeneous behaviors and the two semi-log straight lines have the same slope.

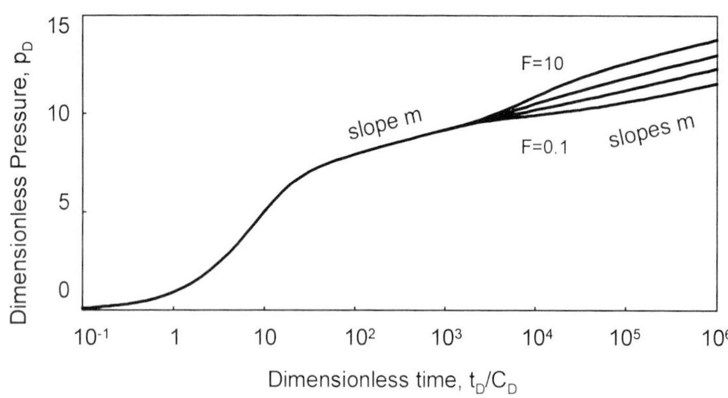

Figure 4.55. Semi-log plot of Figure 4.54 radial composite examples.

When the storativity increases (curve F=0.1 of Figure 4.55), the transition between the two parallel semi-log straight lines tends to the horizontal as for a double porosity response. By symmetry, when the storativity decreases (curve F=10), the transition between the two parallel lines tends to the vertical, suggesting then the opposite reservoir configuration.

Radial composite responses

The duration of the first homogeneous regime is a function of the inner region radius: with a large R_D, the transition occurs later. Before the transition, the early time response corresponds to the behavior of a well with wellbore storage and skin in a homogeneous reservoir.

The shape of the transition is a function of M and F. When both the mobility and the storativity change, the two transitions illustrated on Figure 4.52 and 4.54 examples are superimposed on the response. For example, both low mobility and low storage in the outer zone tend to increase the rate of change of pressure during transition and, when the outer zone storage is very small, the derivative curve can show a hump during transition, before the second plateau. For a given mobility ratio M, the shorter transition is obtained for constant storativity.

Matching procedure with the pressure and derivative log-log data curves

Pressure and time matches are defined by the 0.5 line and the early time unit slope wellbore storage line. The permeability thickness product k_1h of the inner region is estimated from Equation 4.80:

$$k_1 h = 141.2\, qB\mu_1 (\text{PM}) \tag{4.85}$$

with Equation 4.83, the wellbore storage coefficient is obtained:

$$C = 0.000295 \frac{k_1 h}{\mu_1} \left(\frac{1}{\text{TM}} \right) \tag{4.86}$$

At early time, the homogeneous $C_D e^{2S}$ curve matching the data before the start of transition is used, with Equation 2.11 and Equation 4.82, to estimate the skin factor. The skin, defined in Equation 4.84 with the inner zone parameters, corresponds to the wellbore skin.

The mobility ratio M is obtained by comparing the level of the two derivative stabilizations:

$$M = \frac{\Delta p_{\text{2nd stab.}}}{\Delta p_{\text{1st stab.}}} \tag{4.87}$$

The storativity ratio F is in general difficult to access. When the match is performed on a complete radial composite response generated by computer, F is adjusted from the derivative transition.

Semi-log analysis

The first semi-log straight line describes the inner zone response. From Equation 1.15, the pressure is expressed as:

$$\Delta p = 162.6 \frac{qB\mu_1}{k_1 h} \left(\log \Delta t + \log \frac{k_1}{(\phi\mu c_t)_1 r_w^2} - 3.23 + 0.87S \right) \tag{4.88}$$

The analysis of the first semi-log line provides the mobility of the inner zone, and the wellbore skin factor (also called S_w).

The second line, for the outer zone, is defined by:

$$\Delta p = 162.6\frac{qB\mu_2}{k_2 h}\left(\log \Delta t + \log\frac{k_2}{(\phi\mu c_t)_2 r_w^2} - 3.23 + 0.87 S_T\right) \tag{4.89}$$

From the second straight line, the outer zone mobility is estimated and, if the first line is also present on the response, the mobility ratio M is defined.

The total skin S_T is calculated from the second line and includes two components: the wellbore skin factor S and a radial composite *apparent skin* effect S_{RC}, function of the mobility ratio M and the radius R_D of the circular interface, as shown in Equation 1.13. In Equation 4.89, S_T is defined with respect to the outer zone storativity and mobility. The corresponding pressure change $\Delta p_{Skin\ T}$ is expressed:

$$S_T = \frac{k_2 h}{141.2 qB\mu_2}\Delta p_{skin T} \tag{4.90}$$

The two components of S_T are defined as:

$$S_T = \frac{1}{M}S + \left(\frac{1}{M} - 1\right)\ln R_D \tag{4.91}$$

The second term of Equation 4.91 is the radial composite apparent skin effect S_{RC} discussed in Section 1.2.3. It describes the influence of the inner zone during the late time homogeneous response. When the near wellbore mobility is higher than in the outer zone (M >1), the inner zone appears as a negative skin. In the opposite case (M<1), a reduced mobility around the wellbore is equivalent to a well damage, and the apparent radial composite skin is positive. Equation 4.91 is further discussed in the multiple composite model, Section 4.3.4.

When the two semi-log straight lines are clearly defined, the analysis provides M, S and, if the outer zone storativity is known, S_T can be calculated. The radius R_D of the circular interface between the two reservoir regions can then be estimated from Equation 4.91.

When the storativity of the outer zone is not known and $(\phi c_t)_1$ is used in Equation 4.89 for the calculation of S_T, the global skin is wrong by $0.5\ln(F)$. If $(\phi c_t)_1 > (\phi c_t)_2$, S_T is over estimated (as suggested for example by the increased amplitude of the curve F=10 on Figure 4.55). A similar effect of the storativity on the skin factor calculated from semi-log straight line has been already discussed for double porosity reservoir responses with Equation 4.30.

Build-up analysis

With the radial composite model, the pressure behavior changes in the course of the response. As for other heterogeneous models, the superposition method used in build-up analysis can introduce distortions on the derivative response curve (see Section 2.3.4).

In Figure 4.56, a radial composite build-up curve is compared to the original drawdown model response. At the time of shut-in, the production period had just reached the start of the transition. During the build-up, the extrapolated drawdown has passed beyond the early time behavior of the inner zone: as long as the shut-in period has not reached the second regime, the derivative of the build-up curve with respect to the superposition time does not follow exactly the original drawdown behavior. It drops slightly below the 0.5 line, and the transition towards the second homogeneous regime appears delayed.

The curves of Figure 4.56 shows also that, on a Horner plot of the radial composite build-up example, the first semi-log straight line can be distorted and the resulting slope is too low (the mobility of the inner zone (k_1h/μ_1) would be then over-estimated). The extrapolated pressure to infinite shut-in time is taken from the second line to provide the correct $p^* = p_i$.

When, such as in some lens type reservoirs, the mobility of the outer reservoir region is significantly smaller than in the inner zone, the build-up effect can produce a valley shaped response, as illustrated on Figure 4.57. In this example, the mobility of the outer reservoir region is reduced by a factor of 100, the dimensionless production time prior to shut-in is t_{pD}/C_D=3200. If the duration of the shut-in period is of the same order of magnitude, only the downward trend of the derivative is recorded. Both the drawdown and the build-up responses show the characteristic behavior of a closed depleted system (See Sections 5.4 and 5.9.2), and the late time contribution of the infinite outer reservoir region can be overlooked.

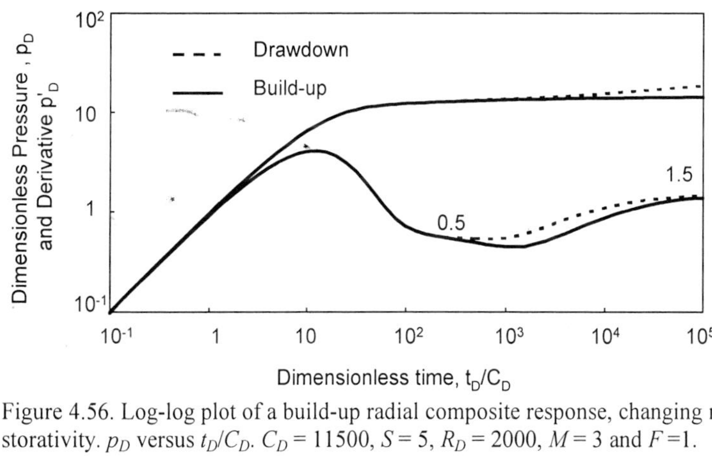

Figure 4.56. Log-log plot of a build-up radial composite response, changing mobility and constant storativity. p_D versus t_D/C_D. C_D = 11500, S = 5, R_D = 2000, M = 3 and F =1.
The dotted pressure and derivative curves correspond to the drawdown solution.

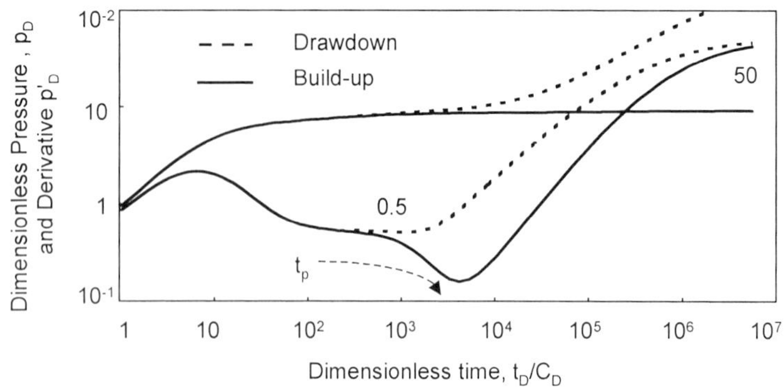

Figure 4.57. Log-log plot of a build-up radial composite response, changing mobility and constant storativity. p_D versus t_D/C_D. $C_D = 1000$, $S = 0$, $R_D = 10000$, $M = 100$, $F = 1$ and $t_p/C_D = 3200$. The dotted pressure and derivative curves correspond to the drawdown solution.

The build-up data plot Figure 4.50, of the double permeability field example Section 4.2.5, illustrates a possible radial composite test response.

4.3.3 Linear composite reservoir

For linear composite responses, the five dimensionless parameters are: C_D, S, and L_D, M, and F. As for the radial composite solution, no log-log type curve is available, the match is performed on computer-generated pressure and derivative curves.

Behavior

Two homogeneous regimes are seen on linear composite reservoirs responses but the second homogeneous regime describes an equivalent total system, not the reservoir region far from the well as it is in case of radial symmetry:

1. First, the region around the well is producing alone, and the pressure behavior corresponds to a *homogeneous* reservoir.
2. When the linear interface is reached, the two regions are producing together. A second homogeneous behavior is observed, the corresponding equivalent homogeneous system is defined by the *average properties* of the two regions.

Influence of M

During the second homogeneous regime of linear composite reservoirs, the two reservoir regions are participating to the production. The two limiting cases are defined as an infinite, or a zero mobility in the external region. In Figure 4.58, the two limit behaviors, shown as dotted curves, correspond respectively to a constant pressure and a sealing linear boundary (see Chapter 5).

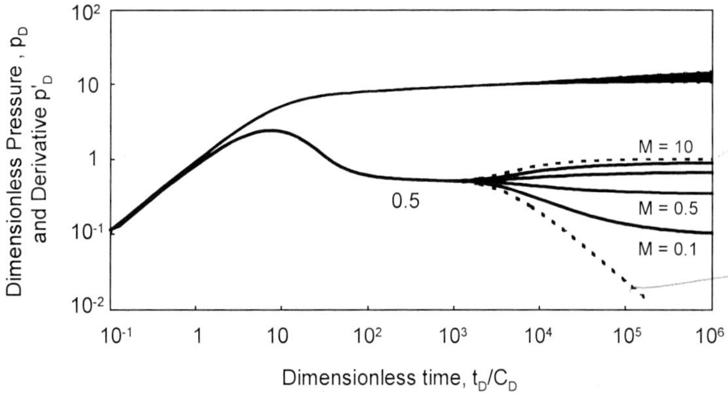

Figure 4.58. Log-log plot of linear composite responses, changing mobility and constant storativity. p_D versus t_D/C_D. $C_D = 100$, $S = 3$, $L_D = 700$, $M = 10, 2, 0.5, 0.1$, $F = 1$.
The two dotted curves correspond to the sealing and the constant pressure boundary solutions.

Figure 4.58 presents, between the two limiting cases, several linear composite examples generated for different values of the mobility ratio M: the two reservoir regions have the same storativity ($F = 1$), the mobility ratio varies from 10 to 0.1. The same well parameters as on the radial composite examples of Figure 4.52 are used.

As for the radial composite responses discussed in Section 4.3.2, the early time data is described by the type-curve for a well with storage and skin in a homogeneous reservoir, of Figure 2.22. With the dimensionless variables defined in Section 4.3.1 for the inner region, the first derivative plateau corresponds to 0.5.

The level of the second plateau indicates the apparent mobility of the equivalent homogeneous regime. This is obtained as the average of the two regions mobility:

$$\left(\frac{k}{\mu}\right)_{APPARENT} = \frac{(k/\mu)_1 + (k/\mu)_2}{2} = 0.5\left(1 + \frac{1}{M}\right)\left(\frac{k}{\mu}\right)_1 \tag{4.92}$$

When the mobility of the second region is small, the second derivative plateau is above the first. For example, with the curve M=10 on Figure 4.58, the apparent mobility of the equivalent homogeneous regime is only 55% of $(k/\mu)_1$, and the second plateau corresponds to a dimensionless pressure of 0.91. As opposed to radial composite responses, the level of the second stabilization is *limited* to twice the level of the first plateau, illustrated by the sealing fault dotted curve.

With values of the mobility ratio less than 1 (curves M=0.5 and 0.1), the mobility of the second region is larger, and the level of the final derivative stabilization is less than 0.5.

On semi-log scale, as illustrated on Figure 4.59 with the same example responses, linear composite systems exhibit two straight lines. The slope of the second line at late time defines the apparent mobility of Equation 4.92.

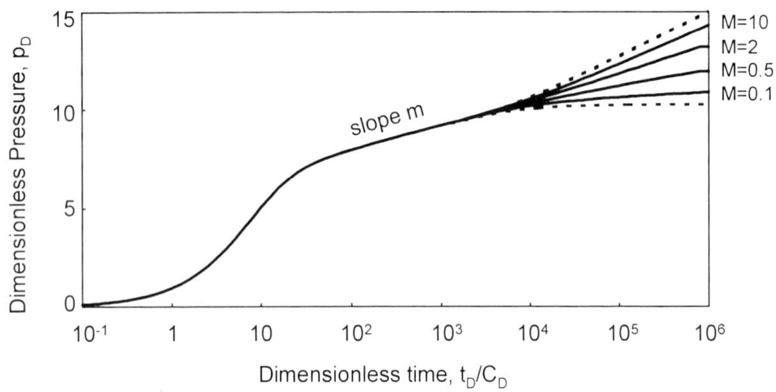

Figure 4.59. Semi-log plot of Figure 4.58 linear composite examples.

Linear composite responses

The duration of the first homogeneous regime is a function of the distance between the well and the interface: with a large L_D, the transition occurs late. The shape of the transition before the second plateau is a function of M and F. The transition is in general not well defined with linear composite responses, F is frequently difficult to determine. The final derivative stabilization defines the apparent mobility, and therefore M.

On Figure 4.60, a radial and a linear composite response are compared. The parameters have been adjusted so that the two models show the same derivative stabilizations during both homogeneous behaviors, the distance of the interface being also the same: L_D=R_D=300. The apparent mobility of the second regime gives, with the linear composite solution, a mobility ratio M=5 whereas, on the radial composite curve, it is only M=1.667 (Equation 4.92).

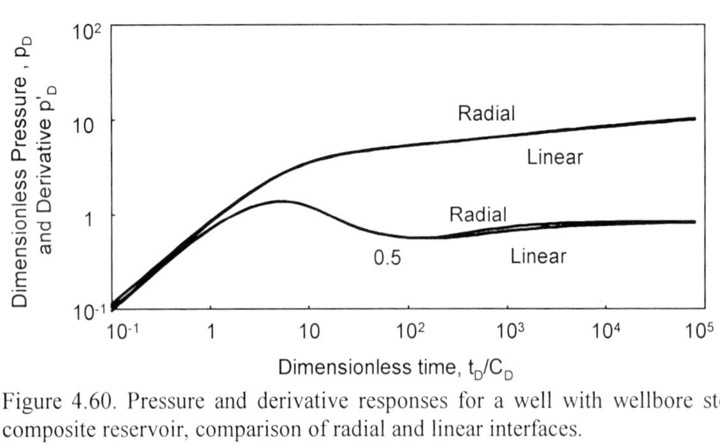

Figure 4.60. Pressure and derivative responses for a well with wellbore storage and skin in a composite reservoir, comparison of radial and linear interfaces.
Log-log scales, p_D versus t_D/C_D. C_D = 200, S = 0.
Linear composite: M = 5, F =1, L_D = 300. Radial composite M = 1.667, F =1, R_D = 300

Matching procedure with the pressure and derivative log-log data curves

The matching procedure is the same for both types of composite reservoirs, with a linear or radial interface: pressure and time matches are defined by the unit slope and the 0.5 straight lines. The permeability thickness product k_1h of the inner region is estimated from Equation 4.85, the wellbore storage coefficient from Equation 4.86. The wellbore skin is estimated from the early time $C_D e^{2S}$ curve match with Equation 2.11 and Equation 4.82.

The two derivative stabilizations are used to estimate the mobility ratio M. From the definition of the apparent mobility Equation 4.92,

$$M = \frac{\Delta p_{\text{2nd stab.}}}{2\Delta p_{\text{1st stab.}} - \Delta p_{\text{2nd stab.}}} \tag{4.93}$$

The storativity ratio F is in general difficult to access, even when the match is performed on a computer generated response.

4.3.4 Extensions of the composite reservoir models

In the following section, several variations of the composite reservoir models presented above are reviewed. For multiple composite solutions, the radial and linear composite models consider n regions with concentric or parallel interfaces. With these models, it is possible to refine the abrupt change of properties assumed with the two region models. In another approach, the mobility and storativity are not changing step-wise but defined as a function of the distance. Finally, the estimation of permeability distribution from well test data is discussed.

Multiple composite systems

Several authors considered multiple composite radial systems. The extension of the radial composite model to three regions was considered by Barua and Horne (1987) and an n regions radial composite model has been envisaged for the analysis of falloff tests by Acosta and Ambastha (1994), Abbaszadeh and Kamal (1989), and Bratvold and Horne (1990). The multiple composite model has also been used for the identification of reservoir damage in producing wells from a water-flooded field (Kamal et al., 1992) (see Section 4.3.5).

Figure 4.61 presents an example of 4 regions radial composite response. The circular interfaces are defined at R_{1D} = 1000, R_{2D} = 2500 and R_{3D} = 50,000, the mobility is respectively 1.5, 5 and 10 times larger than in the inner region and the storativity is constant (F_i=1). The thin curves describe 2 zones radial composite responses with the same mobility in the infinite outer reservoir region. The three inner regions are changed into one circular zone with respectively R_{1D}, M=0.1; R_{2D}, M=0.15 and R_{3D}, M=0.5.

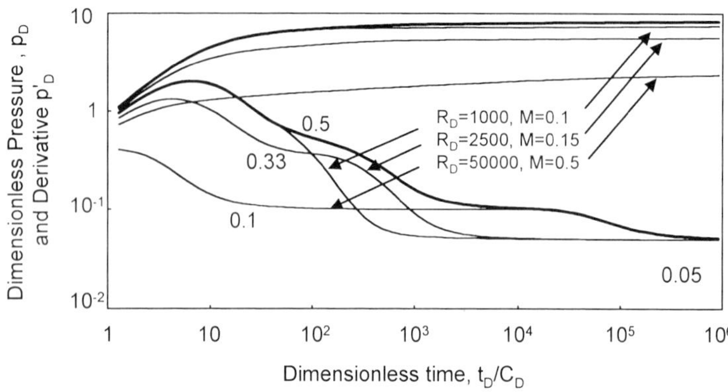

Figure 4.61. Pressure and derivative responses for a well with wellbore storage and skin in a 4 zones radial composite reservoir. Log-log scales, p_D versus t_D/C_D.
$C_D = 5440$, $S = 0$, $F = 1$. $R_{1D} = 1000$, $k/\mu_2 = 1.5\ k/\mu_1$, $R_{2D} = 2500$, $k/\mu_3 = 5\ k/\mu_1$, $R_{3D} = 50{,}000$, $k/\mu_4 = 10\ k/\mu_1$. The thin curves correspond to radial composite responses with only one zone ($R_D = 1000$, $M = 0.1$, $R_D = 2500$, $M = 0.15$, $R_D = 50{,}000$, $M = 0.5$).

The first and the third radial composite thin derivative curves on Figure 4.61 match partially the multiple composite response but, at intermediate time, the curve for the second zone does not. At time t_D/C_D=250, the approximation of a homogeneous inner region is not applicable to the multiple composite response. When the second interface R_{2D} is reached, the central region of radius R_{1D} delays the transition compared to the two region radial composite thin curve with M=0.15.

In terms of pressure behavior, the curve (R_{1D}=1000, M=0.1) is close to the multiple composite response on Figure 4.61. This suggests that the large global skin of the multiple composite curve is mainly produced by the small low permeability region close to the well. As suggested by Equation 4.91, the large region with a moderately reduced permeability has less effect on the global skin.

The two regions radial composite configuration is a flexible model that offers a large range of derivative signatures. This flexibility is higher with the multiple composite model and, for a realistic match, the distribution of the different radial zones characteristics should be supported by geological or petrophysical considerations.

In a multiple composite system, the radius of investigation r_i can be expressed as a function of the permeability distribution. Assuming a constant storativity, when the radius of investigation r_i lies between the interfaces R_n and R_{n+1}, r_i is defined as:

$$r_i = R_n\left(1-\sqrt{\frac{k_{n+1}}{k_n}}\right)+\sqrt{k_{n+1}}\left[\frac{R_{n-1}}{\sqrt{k_n}}\left(1-\sqrt{\frac{k_n}{k_{n-1}}}\right)+\ldots+\frac{R_i}{\sqrt{k_{j+1}}}\left(1-\sqrt{\frac{k_{j+1}}{k_j}}\right)+\ldots\right.$$
$$\left.\ldots+\frac{R_1}{\sqrt{k_2}}\left(1-\sqrt{\frac{k_2}{k_1}}\right)\right]+0.029\sqrt{\frac{k_{n+1}\Delta t}{\phi\mu c_t}} \tag{4.94}$$

Bourgeois et al. (1996 b) extended the linear composite solution of Ambastha et al. (1987) to a three zone composite system with parallel boundaries (see discussion of channel reservoir Section 5.9.1). Kuchuk and Habashy (1997) developed a multi-linear-composite solution to describe reservoirs with lateral changes of properties, such as in deltaic sandstone deposit, or when a network of partially communicating faults improves the permeability at some distance away from the well (see Section 5.6).

Unbroken change of properties in composite systems

In a different approach, Levitan and Crawford (1995) envisaged gradational changes of properties, instead of the piecewise sharp interface models used by other authors. They considered both radial and linear geometries, for a symmetric system. The storativity ($\phi c_t h$) and transmissibility (kh/μ) are function of the distance from the well.

Figure 4.62 presents an example of radial composite response when the permeability increases linearly from R_D=1000 to R_D=10,000, the storativity is assumed to be constant (F=1). The two thin curves describe the radial composite responses with R_D=1000 and R_D=10,000. With a linear increase of mobility, the transition between the two derivative plateaus is smoother than in the case of a sharp change of properties.

Discussion of permeability distribution

Oliver (1990) investigated the perturbation produced by radial changes of permeability. He showed that the derivative of the wellbore pressure is a weighted average of the inverse permeability inside the drainage radius. The weighting function, expressed as a function of time and distance, shows a maximum and drops to 0 beyond the radius of investigation. The maximum indicates that the wellbore pressure responds primarily to the permeability at a distance very similar to the radius of investigation of Equation 1.22:

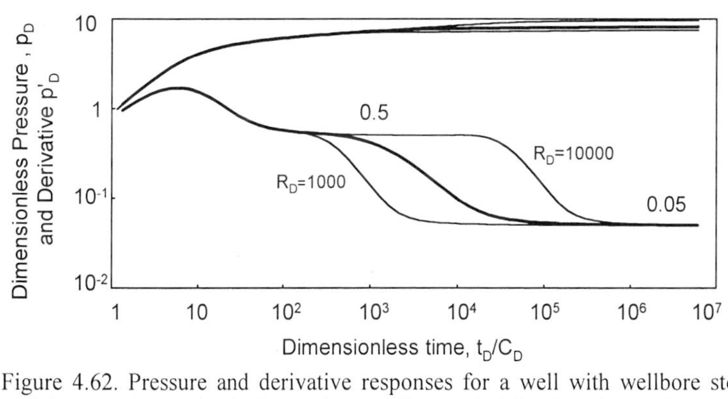

Figure 4.62. Pressure and derivative responses for a well with wellbore storage and skin in a radial composite reservoir, linear change of transmissivity. Log-log scales, p_D versus t_D/C_D. C_D =1000, S =0, F =1. From R_{1D}=1000 to R_{2D} = 10,000, M decreases linearly from 1 to 0.1. The thin curves correspond to radial composite responses (R_D =1000, R_D =10,000).

$$r_D = 0.92\sqrt{t_D} \tag{4.95}$$

Thompson and Reynolds (1997) correlated the weighting function with reservoir rate versus the radial distance. In a circular region near the well, the rate is equal to the sand face flow rate and this inner region has not effect on the derivative response. Only the next circular region where the rate changes until the drainage radius is reached, influences the wellbore pressure derivative.

Oliver's theory has been used in a tentative way to determine the permeability profile from wellbore pressure data, assuming radial symmetric systems and constant porosity (Oliver, 1992; Feitosa et al., 1993). Algorithms are proposed to construct a radial permeability profile, but the solution is non-unique. Levitan and Crawford (1995) investigated similar problem and they conclude that a heterogeneity function can be defined from the analysis of well test data, but not the true permeability and porosity distribution.

For water injection wells, Yeh and Agarwal (1989) propose a simpler approach to convert the derivative curve into a mobility profile versus the radial distance. For each point ($\Delta p' = d\Delta p / d\ln\Delta t,\ \Delta t$), they use Equation 2.24 and the radius of investigation r_i of Equation 1.23. Such procedure gives a volumetric average mobility profile:

$$\lambda_t = k\left(\frac{k_{ro}}{\mu_o} + \frac{k_{rw}}{\mu_w}\right)_r = 0.5\frac{141.2qB}{h\Delta p'} \tag{4.96}$$

and

$$r = 0.024\sqrt{\frac{\lambda_t}{\phi c_t}\Delta t} = 0.024\sqrt{\frac{141.2qB}{\phi c_t h}\frac{\Delta t}{2\Delta p'}} \tag{4.97}$$

Non-radial geometry has also been considered for arbitrary heterogeneous reservoirs (Kuchuk et al., 1993; Tauzin and Horne, 1994). The solution of the inverse problem becomes problematic.

4.3.5 Injection wells

In the following section, the applicability of the radial composite solution to water injection wells is discussed. With a two region radial composite system, the change of mobility and compressibility is abrupt and stationary (See Section 4.3.1). This is not the case with water injection wells :

- The water saturation is constant at S_w=1-S_{or} only in the region near the well from which oil has been flushed. Afterwards, it decreases in a *transition zone* until the flood front is reached. At the injection front, water saturation drops abruptly to the initial water saturation S_{wi} of the outer unflooded region.

- The pressure behavior of an injection well is different during injection and falloff. During the injection period, the flooded region expands with the cumulative injection volume, and a *moving interface* model has to be used. During a falloff, if the injection time is large compared to the shut-in period, the *stationary interface* hypothesis is applicable (Abbaszadeh and Kamal, 1989; Bratvold and Horne, 1990; Yeh and Agarwal, 1989).

Abbaszadeh and Kamal (1989) presented injection and falloff solutions considering the effect of saturation gradient, Bratvold and Horne (1990) introduced non-isothermal effects due to injection of cold water in a hot oil reservoir.

In the following, the reservoir is assumed homogeneous. When the injection well intercepts several layers, a multiple layer composite reservoir model can be used (see Section 4.4.3).

Injection period

After enough injection time, when a constant water saturation region is established around the well, the injection pressure response follows a semi-log straight line whose slope yields the water region mobility. The apparent skin estimated from injection data is made of two components: the wellbore skin and a skin effect due to the two phases fluid condition when the water displaces the oil. It is defined as a function of the rock and fluids characteristics and, in practice, the injection skin is difficult to interpret.

Falloff

As for a build-up period after a production, a falloff response depends upon the duration of the injection prior to shut-in, and a build-up or multiple-rate analysis method should be used. Due to the different behavior during injection and falloff, the superposition method is in theory not applicable to generate the shut-in response. In practice, it is found that, when a radial composite solution with a stationary front is used to analyze the falloff, the superposition described in Section 2.2.2 does not introduce a significant error. Falloff responses can be approximated with a standard build-up or multiple-rate radial composite type curve (Abbaszadeh and Kamal, 1989).

On a log-log derivative plot of falloff data, the difference between the first and the final derivative stabilizations define in theory the mobility ratio M (Equation 4.87). The skin is estimated from the early time homogeneous match for the water zone response, as described in Section 4.3.2. Between the two plateaus, the derivative response describes the changes of saturation in the transition zone separating the inner water region and the original reservoir fluid region.

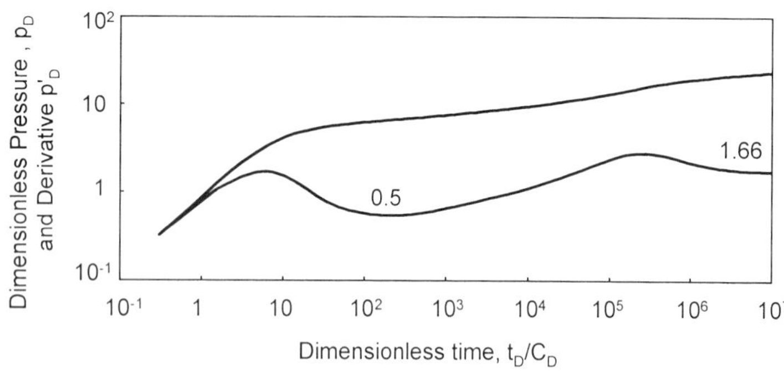

Figure 4.63. Pressure and derivative responses for a falloff. Well with wellbore storage and skin. Log-log scales, p_D versus t_D/C_D. $C_D = 1000$, $S = 0$, $F = 1$.
From $R_{1D} = 1000$, M increases until $R_{2D} = 10{,}000$ and, after the flood front, $M = 3.33$.

In practice, the inner region with constant water saturation $S_w=1-S_{or}$ is generally small (if not absent), and most of the flooded region is *transition*, the first derivative stabilization is frequently not seen. After wellbore storage, the response shows the reduced mobility of the transition, including possible temperature effects (Bratvold and Horne, 1990). When the mobility ratio M is not significantly greater than unity, the transition is described by a hump such as on the synthetic falloff example of Figure 4.63 (this example has been generated with a multiple composite model, for a given radial distribution of the total fluid mobility). If the typical falloff of Figure 4.63 is matched on a the radial composite model, the reduction of mobility is ignored and the match is frequently poor at transition time. In such case, both the mobility of the inner region and the radius R_D of the injection front are underestimated (Abbaszadeh and Kamal, 1989).

Multiple composite radial models have been developed to describe the saturation gradients in the reservoir: the circular interfaces are moving during injection, and stationary after shut-in (Abbaszadeh and Kamal, 1989; Bratvold and Horne, 1990). Such refined models require the relative permeability curves for the reservoir to be known, and therefore they are reservoir dependent.

When Equations 4.96 and 4.97 are used to estimate the average mobility profile, Yeh and Agarwal (1989) conclude that the flood front radius corresponds to the minimum of the average mobility curve.

4.4 COMBINED RESERVOIR HETEROGENEITIES

The response of fissured reservoirs with changing block sizes has been presented in Section 4.1.4 for a uniform fissure network. In the following, the fissured layered and the fissured radial composite models are considered to describe respectively a vertical or a radial change of fissure characteristics. Next, radial composite behavior in layered formations is presented and several examples of layered radial composite responses are discussed.

4.4.1 Fissured layered reservoirs

Several authors have envisaged presence of fissures in layered reservoirs (Al-Ghamdi and Ershaghi, 1994; Larsen, 1988). The two different types of heterogeneities are combined on the response and several sequences of flow regimes are possible, depending which system reacts first. In the following, two examples are briefly discussed to illustrate fissured layered reservoir responses.

On Figure 4.64, the thin curves describe a double permeability response with low contrast (ω = 0.1, κ = 0.7) and an early cross flow ($\lambda = 10^{-3}$). When the two layers are fissured, and assuming restricted interporosity flow from matrix to fissure, the derivative exhibits *two valleys*. For each layer, the storativity ratio between the fissure and the layer storage (Equation 4.4) is $\omega_1 = \omega_2 = 0.01$ and the interporosity flow coefficient are $\lambda_1 = 10^{-5}$ and $\lambda_2 = 5\text{x}10^{-7}$ respectively, as on the triple porosity example of Figure 4.33. With this layer configuration, all parameters are equivalent to those of Figure 4.33 and the fissured layered response matches the triple porosity example (shown with circles on Figure 4.64). When the vertical communication is good in a fissured layered reservoir, grouping of matrix size by layers has no effect on the response.

When reservoir cross flow between layers is not allowed, the response is different as shown by the second thick curves of Figure 4.64. The first transition is observed earlier and, especially during the second transition, the shape of derivative valley is smoother. In the absence of vertical cross flow, the time of start of the matrix blocks production is different from that of the triple porosity example.

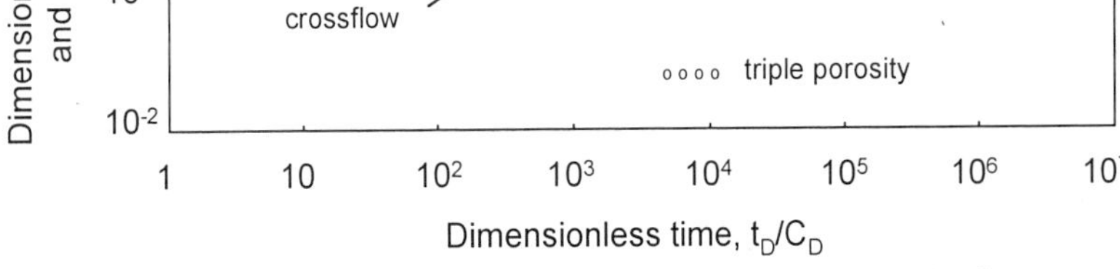

Figure 4.64. Fissured layered reservoir, pseudo steady state interporosity flow, two sizes of matrix blocks, different λ in each layer. Log-log scales, p_D versus t_D/C_D. $C_{Df+m} = 1$, $S_1 = S_2 = 0$, $\omega = 0.1$, $\kappa = 0.7$, $\lambda = 10^{-3}$ or $\lambda = 0$. $\omega_1 = 0.01$, $\lambda_1 = 10^{-5}$, $\omega_2 = 0.01$, $\lambda_2 = 5\text{x}10^{-7}$.
The (o) dotted curves correspond to the triple porosity response of Figure 4.33.

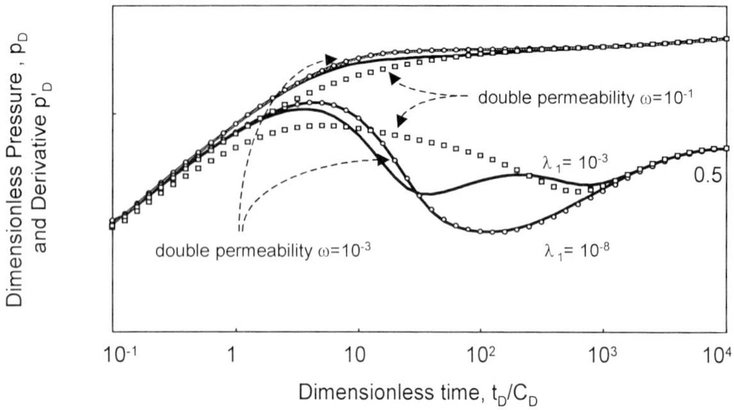

Figure 4.65. Fissured layered reservoir, pseudo steady state interporosity flow, only layer 1 is fissured. Log-log scales, p_D versus t_D/C_D. C_{Df+m} = 1, S_1= S_2= 0, ω = 0.1, κ = 0.99, λ =10^{-3}. ω_1 =0.01, λ_1 =10^{-3} or λ_1 =10^{-8}. The dotted curves correspond to the double permeability responses with κ= 0.99 and λ=4x10^{-4} (o : ω= 10^{-3} , Figure 4.43, and □ : ω= 10^{-1}, Figure 4.45).

When the high permeability layer 1 is fissured and not the other, the response depends upon which transition, the double porosity or the double permeability transition, is seen first. On Figure 4.65, the two layers double permeability system is defined with C_{Df+m}=1, S_1=S_2=0, ω=0.1, κ=0.99, λ=4x10^{-4}. Layer 1 is fissured with ω_1=0.01 and λ_1=10^{-3} or λ_1=10^{-8}.

When the interporosity flow parameter is small (λ_1=10^{-8}), layer 1 is in fissure regime when the double permeability transition starts. The reservoir cross flow is established between the layer 2 and the fissure network of layer 1. The storativity ratio between the two reservoir components is therefore ω x ω_1=10^{-3} and the response becomes equivalent to the double permeability response κ=0.99 of Figure 4.43 (shown with circle on Figure 4.65). The double porosity transition is not seen on this response, it would show at later time, after t_D/C_D=10^4.

Conversely, if layer 1 is in total system flow (λ_1=10^{-3}) at start of the double permeability transition, the double porosity transition in layer 1 is first seen during the two layers no cross flow regime. Later, the double permeability transition corresponds to a storativity ratio of ω=10^{-1} and the second derivative valley tends to be similar to that of the double permeability response κ= 0.99 of Figure 4.45 (square symbols on Figure 4.65).

4.4.2 Radial composite double porosity reservoirs

In a fissured reservoir, a radial composite behavior is sometimes observed because of change of fissure density in the reservoir. Injection wells, and some acid stimulated wells in fissured reservoir also correspond to this geometry. Radial composite double porosity responses have been first considered by Poon (1984) for restricted interporosity flow. Satman (1991) studied unrestricted matrix flow and Kikani and Walkup (1991) presented a general model, including the effect of matrix skin.

Figure 4.66 presents two examples of radial composite responses when the inner region is fissured but not the outer region. Well and reservoir parameters are the same as on curve M=10 of Figure 4.52 (shown with circle on Figure 4.66) with C_D=100, S=3, F=1 and R_D=700. The inner region is double porosity restricted interporosity flow with ω_1=0.01, λ_1=10^{-4} or λ_1=10^{-6}.

With λ_1=10^{-4}, the response shows the double porosity behavior of the inner zone and, after the characteristic valley, the derivative follows the 0.5 stabilization for the total system matrix + fissure flow, until the radial composite interface is reached at t_D/C_D=10^3. Then, the response becomes equivalent to the radial composite with a homogeneous inner region, shown Figure 4.52.

When λ_1=10^{-6}, the inner region is still in fissure regime when the interface is seen. Since the diffusivity is high during fissure flow, the interface is felt earlier, and the derivative deviates from the double porosity transition valley at t_D/C_D=10^2 (the infinite reservoir double porosity response is shown as a dashed curve on Figure 4.66). The transition towards the final derivative stabilization for the outer region combines the double porosity and the radial composite responses, the derivative increase is delayed compared to the radial composite curve.

When the outer zone is fissured, the double porosity derivative valley transition can be observed after the radial composite derivative deviation, provided $\lambda_2 R_D^2$< 10 (Kikani and Walkup, 1991). When both regions are fissured, the authors conclude that only one transition is likely to be observed as a contrast of five order is needed between the two interporosity flow parameters $\lambda_{1,2}$ in order to display two double porosity transitions.

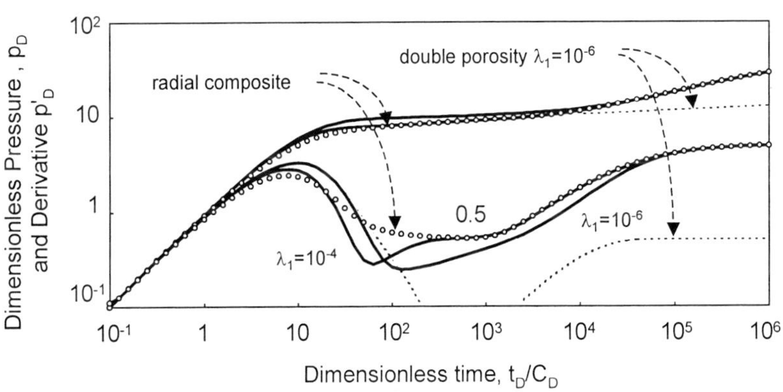

Figure 4.66. Radial composite reservoir, the inner region is fissured, pseudo steady state interporosity flow. Log-log scales. C_D =100, S =3, M =10, F =1 R_D =700, ω_1 =0.01, λ_1= 10^{-4} or λ_1= 10^{-6}. The (o) dotted curves correspond to the radial composite response of Figure 4.52 with M=10, the thin (·) dotted curves describe the double porosity response ω_1 =0.01, λ_1= 10^{-6}.

Kikani and Walkup (1991) presented an example of a well response in a fissured reservoir before and after acid stimulation. As the acid penetrates mostly the fissures, it can travel a long distances from the well, resulting frequently in a radial composite behavior (as opposed to homogeneous reservoir where acid stimulation is a near wellbore process). More interestingly, the authors observed that the acid treatment not only removes the wellbore damage, but also changes the type of interporosity flow. Before acid, the response corresponds to restricted matrix to fissure flow whereas after acid, it is unrestricted interporosity flow. In such a case, the acid removes the matrix skin in the invaded region.

4.4.3 Radial composite double permeability reservoirs

Radial composite behavior can be observed in layered formations. Hatzignatiou et al. (1987) investigated interference responses in a two-layer composite reservoir with the same interface distance R_D in the layers. Larsen (1988) described the extension of the model to an n-layer multiple composite system. Anbarci et al. (1989) used a similar solution to investigate the location of the flood fronts for injection wells in a two-layer reservoir. Gomes and Ambastha (1993) suggest using an n-layer multiple composite model to describe tilted or irregularly - shaped fronts.

In the following, a two-layer, two-region, radial composite solution is used to illustrate several variations of double permeability model presented Section 4.2.1. Extensions to n layers or multiple composite configurations are not envisaged here.

Frequently, layered reservoirs combine homogeneous and lenticular layers, commingled or not. The examples shown in Figure 4.67 illustrate the response of a two-layer reservoir without cross flow when one layer is radial composite with a strong reduction of mobility at R_{2D}=100. Near the well, the two layers have equal mobility (κ=0.5) and the storativity ratio is ω=0.1. After the circular interface, the mobility of layer 2 is reduced with M_2=10, 100 and 1000 and the storativity ratio is constant in layer 2 (F_2=1).

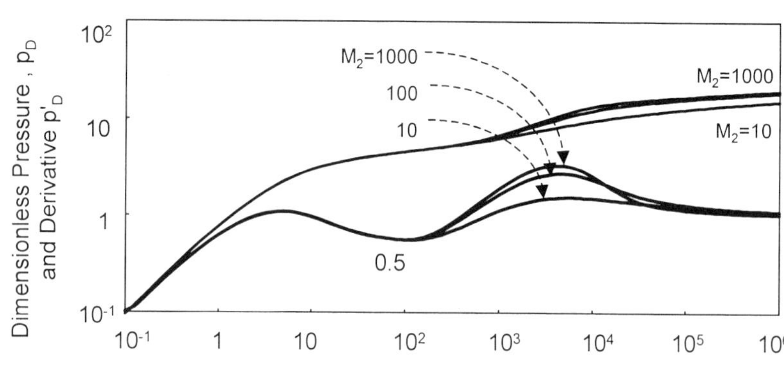

Figure 4.67. Layered reservoir, no cross flow, layer 1 homogeneous, layer 2 radial composite. Log-log scales. C_D = 30, S_1 = S_2 =0, ω=0.1, κ=0.5, λ=0. R_{2D} = 100, M_2 = 10, 100, 1000, F_2 = 1.

After wellbore storage, the response describes the commingled two layers behavior until the interface of layer 2 is reached. The derivative increases and, when the mobility of the outer region is low (M_2=100 or 1000), it tends to follow a *unit slope* straight line and produces a *hump*. Later, a new two layers commingled infinite reservoir response is seen with, in layer 2, the outer region mobility. The derivative tends to stabilize at 0.5 $M_2/[1+(M_2-1)\kappa]$, respectively 0.90, 0.990 and 0.999 on Figure 4.67.

The hump of the derivative examples M_2=100 or 1000 describes the storage effect of the inner region of layer 2. Reservoir storage is further discussed in the chapter on boundary effects, Section 5.8.3 for closed circle double permeability reservoir. When the reduction of mobility is large in the radial composite layer, the responses of Figure 4.67 tend to be similar to the first part of a closed system layered reservoir response, as illustrated on Figure 5.48.

On the radial composite double permeability examples Figures 4.68 and 4.69, only the cross flow parameter λ changes between the inner regions and the outer regions. The two layers have constant properties ($M_{1,2}=F_{1,2}=1$) and several distances R_D are considered for the change of cross flow parameter λ.

When no cross flow is allowed in the inner region and the layers are communicating only beyond the radius R_D, the valley shaped derivative transition is *delayed* and it tends to be *steeper* than the double permeability response with a constant interlayer cross flow parameter throughout the reservoir. In Figure 4.68, the dotted curve describe the double permeability response and the radial composite double permeability curves are generated for R_D =30, 100 and 300. In the inner region, λ_1=0, and in the outer region λ_2=4 10^{-4}.

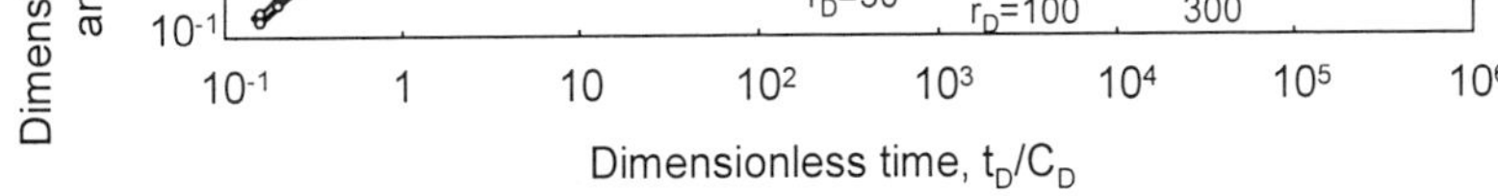

Figure 4.68. Layered reservoir, no cross flow in the inner region. Log-log scales, p_D versus t_D/C_D. C_D =1, $S_1=S_2$ =0, ω=0.1, κ=0.9, $M=F$=1. λ_1=0, λ_2=4x10^{-4}, R_D =30, 100, 300.
The (o) dotted curves correspond to the double permeability response of Figure 4.45 with κ=0.9.

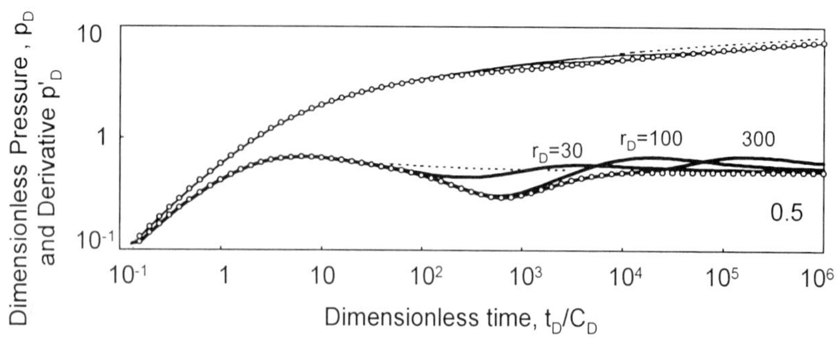

Figure 4.69. Layered reservoir, no cross flow in the outer region. Log-log scales, p_D versus t_D/C_D. C_D =1, S_1= S_2 =0, ω=0.1, κ=0.9, M = F =1. λ_1=4x10^{-4}, λ_2=0, R_D =30, 100, 300.
The (o) dotted curves correspond to the double permeability response of Figure 4.45 with κ =0.9 and the thin (·) dotted curves to the commingled reservoir (λ=0).

The opposite configuration is presented in Figure 4.69 with the same parameters. The reservoir cross flow is only possible in the inner region of radius R_D, the layers are not communicating in the outer region. When the R_D interface is reached, the responses change to the two layers without cross flow behavior and the derivative deviates above the 0.5 stabilization (see Section 4.2.7). The introduction of a permeability barrier between the layers produces a smooth *hump* on the derivative response. The dotted curve of Figure 4.69 describes the double permeability response (same as on Figure 4.68), and the dashed curve corresponds to the commingled configuration.

CHAPTER 5

EFFECT OF RESERVOIRS BOUNDARIES ON WELL RESPONSES

The effect of boundaries has been considered from the earlier studies of pressure transient analysis. In 1951, when presenting his historic paper, Horner (1951) discussed the response due to a single linear sealing fault on a build-up example. Today, complex boundary systems are used in well test interpretation, with sealing or constant pressure conditions. Partially sealing and conductive linear boundaries can also be identified and interpreted on well pressure responses.

In this chapter, the different boundary conditions used in well test analysis are described. First, a homogeneous reservoir is assumed and the identification of sealing boundaries is demonstrated for drawdown and build-up responses. When several boundaries are present, the effect of the different limits appears as a function of the different distances. It is shown that the late time response is influenced by the time when the first boundaries are reached. The question of the extrapolated shut-in pressure is discussed in detail for semi-infinite and closed systems. The effect of constant pressure and communicating boundaries is then described. Boundaries in heterogeneous double porosity, double permeability and composite reservoirs are considered in later sections of this chapter. A constant formation thickness is assumed except in the final section 5.10, where complex boundary configurations are described.

In the following, the distance from the producing well to the boundary is called L. The dimensionless distance L_D is defined as:

$$L_D = \frac{L}{r_w} \qquad (1.34)$$

5.1 SINGLE SEALING FAULT IN A HOMOGENEOUS RESERVOIR

With the sealing fault model, a linear no-flow boundary closes the reservoir in one direction. Such a configuration is encountered in faulted reservoirs but it can also be considered, as an extension of the linear composite solution presented in Section 4.3.3, when the reservoir flow capacity kh becomes zero. A pinch-out for example is sometimes analysed using the sealing fault solution.

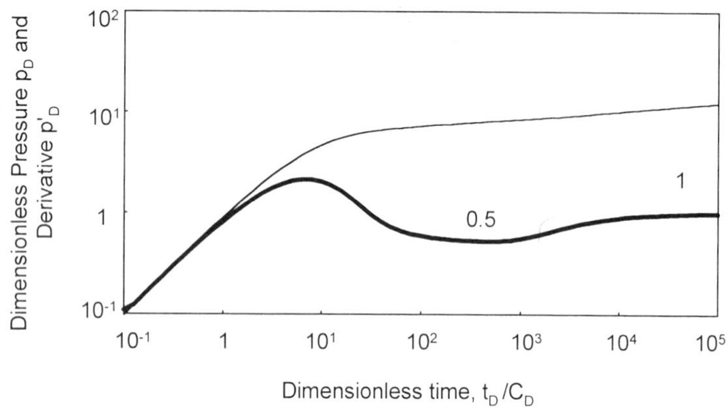

Figure 5.1 Pressure and derivative responses for a well with wellbore storage near one sealing fault in a homogeneous reservoir. Log-log scales, p_D versus t_D/C_D. $C_D = 10^4$, $S = 0$, $L_D = 5000$.

The effect of a sealing fault has been presented in Section 1.2.7 (Figures 1.21 to 1.23) and, for shut-in periods, in Sections 2.2.2 and 2.3.4 (Figures 2.10 and 2.20). A typical drawdown response is presented on Figure 5.1 for a well with wellbore storage and skin in a homogeneous reservoir limited by a sealing fault.

- The early time part of the well response corresponds to the *infinite reservoir* behavior. During radial flow, the pressure response follows the first semi-log straight line as illustrated on the semi-log plot Figure 1.23 and, on Figure 5.1, the derivative follows the first stabilization.
- When the influence of the sealing fault is felt, the flow becomes *hemi-radial*, and the apparent mobility is reduced by a factor of two. On semi-log scale, the slope of the straight line doubles and, with the derivative, the curve follows a second stabilization at a level twice the first. In dimensionless terms, the first derivative plateau is at 0.5 and the second at 1.

5.1.1 Sealing fault model

With analytical well test interpretation models, the image well method is used to produce the effect of a no-flow barrier: a imaginary second well, at a distance $2L_D$ from the active well, is assumed to be produced with the same flow rate history. The symmetry condition of the image method requires, in theory, the use of the same wellbore condition for the two wells. In most cases, however, the influence of the image well can be simplified using a line source response (see Chapter 6.1.1), expressed with the exponential integral function of Equation 6.1: $-0.5\,\mathrm{Ei}\left[-(2L_D)^2/4t_D\right]$.

5.1.2 Log-log behavior

Figure 5.2 presents several examples of the influence of a sealing fault on a well pressure response. Different distances L_D are considered: 100, 300, 1000 and 3000. The time Δt_x at *midpoint* of transition between the first derivative plateau on 0.5, and the second at 1, is given by (Larsen, 1983) :

$$L = 0.01217\sqrt{\frac{k\Delta t_x}{\phi\mu c_t}} \tag{1.33}$$

Δt_x is proportional to the *square of the distance*: when L_D is increased by a factor of ten, the end of the early time infinite acting behavior is displaced by two log-cycles along the time scale.

The curve $L_D = 100$ shows a limiting case of the behavior of a well near a sealing fault. When the distance L_D to the linear fault is small and the wellbore storage coefficient is high, the fault influence can start during the wellbore storage dominated regime: after the derivative hump, the curve stabilizes directly on 1, and does not show the first plateau at 0.5. In such a case, the sealing fault is difficult to identify, and the response can be misinterpreted with an infinite reservoir solution (the resulting permeability is then half the true reservoir permeability).

5.1.3 Matching procedure with the pressure and derivative data

The derivative response curves of Figure 5.2 are characterized by a doubling of the derivative plateau. As discussed in Sections 4.1.3 and 4.1.5, the same behavior is observed on double porosity responses with transient interporosity flow condition. The responses of the two solutions show a similar shape, but the matching procedure is different.

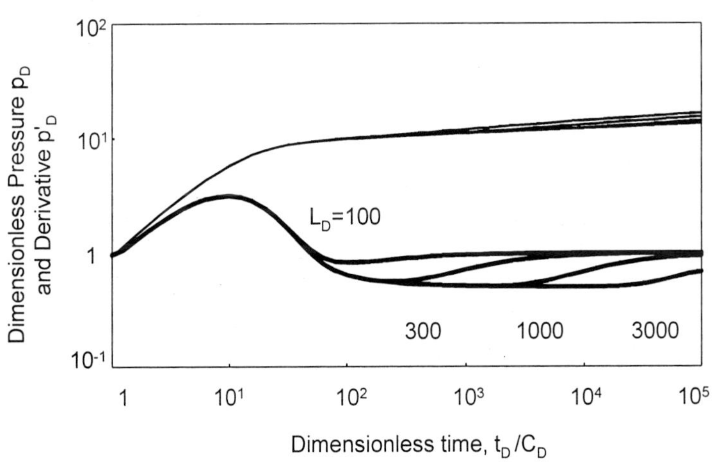

Figure 5.2. Pressure and derivative responses for a well with wellbore storage and skin near one sealing fault in a homogeneous reservoir. Log-log scales, p_D versus t_D/C_D.
$C_D = 100$, $S = 5$. Several distances $L_D = 100, 300, 1000, 3000$.

For a well with wellbore storage and skin near a sealing fault, a $C_D e^{2S}$ curve describes the early time response before the start of the boundary influence. The pressure match is defined by positioning the 0.5 line on the first derivative stabilization, and the time match is fixed by the wellbore storage unit slope line, as for the homogeneous type-curve of Figure 2.22. The permeability thickness product *kh* is estimated with Equation 2.9, the wellbore storage coefficient *C* with Equation 2.10 and the skin factor with Equation 2.11.

The distance *L* of the sealing fault can be estimated from the time Δt_x of transition between the two derivative stabilizations with Equation 1.33, or by matching the data on a computer generated response.

5.1.4 Semi-log analysis

On Figure 5.3, the four examples of Figure 5.2 are plotted on a semi-log scale: the duration of the infinite acting radial flow regime, described by the first straight line of slope *m*, is a function of L_D. When the distance of the sealing fault is increased, the second straight line of slope 2*m* is displaced towards late times.

The first semi-log straight line of slope *m* is used to estimate the reservoir permeability thickness product *kh*, and the skin *S*:

$$kh = 162.6 \frac{qB\mu}{m} \tag{1.16}$$

$$S = 1.151 \left(\frac{\Delta p_{1\,\mathrm{hr}}}{m} - \log \frac{k}{\phi \mu c_t r_w^2} + 3.23 \right) \tag{1.17}$$

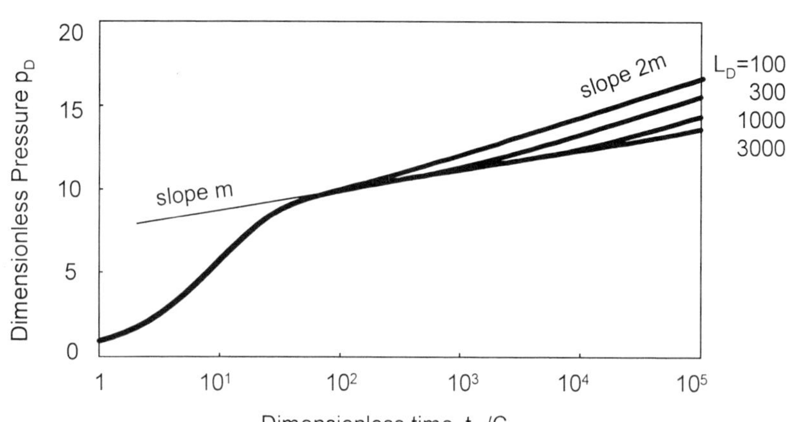

Figure 5.3. Semi-log plot of Figure 5.2 examples. p_D versus t_D/C_D.

The time of intercept Δt_x between the two semi-log straight lines, is used to estimate the distance between the well and the sealing fault (Gray, 1965) with Equation 1.33.

On the four examples of Figure 5.3, the transition between the two straight lines is smooth and lasts more than one log-cycle on the time scale. Unless the two lines are well defined such as on the curve $L_D = 300$, the intercept time Δt_x is difficult to define with accuracy. In practice, when the analysis is made by computer, the distance is estimated from the log-log derivative match rather than with Equation 1.33.

5.1.5 Build-up analysis

Log-log pressure and derivative build-up analysis

As discussed in the Section 2.3.4 of the pressure derivative discussion, if the drawdown behavior changes when extrapolated in the course of the build-up, the multiple rate build-up derivative curve does not match perfectly on the original drawdown response.

With the sealing fault model, the flow behavior changes from a radial to a hemi-radial geometry: a build-up derivative curve can deviate from the drawdown behavior. This is illustrated by the build-up curve presented Figure 2.20. The well is near a sealing fault, and the drawdown before shut-in has stopped before, or during transition between the radial and the hemi-radial regimes (Figure 2.10), when the derivative is between the 0.5 line and the stabilization on 1.

At early shut-in time, the drawdown period extrapolated with the superposition principle at $(t_p+\Delta t)$ is under the influence of the fault whereas the shut-in period, generated for the elapsed time Δt, is still in an infinite reservoir regime. The two superposed periods are not on the same semi-log straight line and, on Figure 2.20, the build-up derivative first drops below the 0.5 line. The curve deviates upwards towards the second derivative stabilization at 1 only after the two superposed periods follow the same behavior, when Δt has passed the time of start of the fault influence. On the build-up example Figure 2.20, the sealing fault appears to be slightly delayed compared to the original drawdown response. The effect of the production history on build-up derivative responses has been also illustrated with similar conclusions in Sections 4.1.2 and 4.1.3 for heterogeneous double porosity reservoirs (Figures 4.15 and 4.26).

Semi-log build-up analysis

For semi-log analysis of build-up periods, the Horner or multiple rate superposition time is used (Section 2.2.2). The first semi-log straight line gives access to kh and S and, provided only one sealing fault is present in the reservoir area investigated during the test, the behavior is semi-infinite and the extrapolation of the *second line* to infinite shut-in time gives the initial reservoir pressure.

The extrapolated pressure of the first line is not used, except in bounded systems with the Matthews-Brons-Hazebroek (1954) method, to predict the average reservoir pressure (see Section 5.4.7). The M.B.H. method, designed for closed reservoirs, requires the shape of the drainage area to be known.

5.2 TWO PARALLEL SEALING FAULTS IN HOMOGENEOUS RESERVOIR

With this boundary solution, the well is located between two parallel sealing faults. Even though this type of configuration is encountered in faulted systems, frequently it corresponds to long narrow reservoirs such as channel sands.

Two examples of parallel sealing faults responses are presented on Figures 5.4 and 5.5 for a well with wellbore storage and skin in a homogeneous reservoir.

- On the log-log plot Figure 5.4, the derivative describes first the wellbore storage effect, then it follows the 0.5 line.
- Later, when the two reservoir boundaries have been reached, the flow lines become parallel to the reservoirs limits, and a linear flow regime is established (Miller, 1962). The pressure changes proportionally to $\sqrt{\Delta t}$, and the derivative follows a *half unit slope* straight line.
- The shape of the transition between radial and linear flow is a function of the well location in the channel. When the well is equidistant from the two boundaries such as on curve **A** of Figure 5.4, the transition between radial and linear flow regimes is short. If the well is closer to one of the two boundaries, the characteristic behavior of *one sealing fault* is seen before the linear flow. The derivative stabilizes first at 0.5, then 1 and finally it reaches the half unit slope straight line (Figure 5.4, **B**).

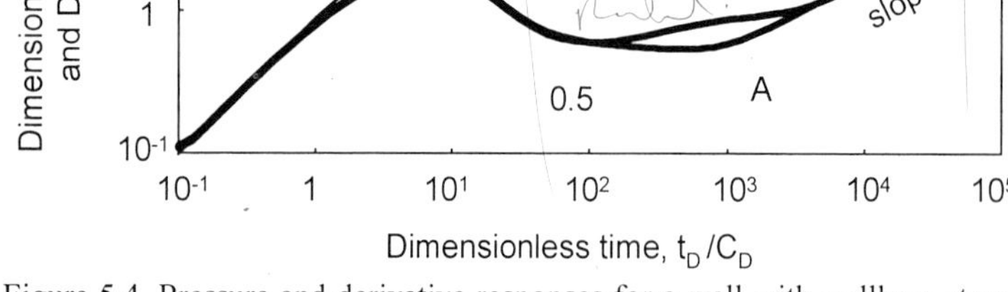

Figure 5.4. Pressure and derivative responses for a well with wellbore storage in a homogeneous reservoir limited by two parallel sealing faults. Log-log scales, p_D versus t_D/C_D.
$C_D = 3000$, $S = 0$, $L_{1D} = L_{2D} = 3000$ (curve **A**) and $L_{1D} = 1000$, $L_{2D} = 5000$ (curve **B**).

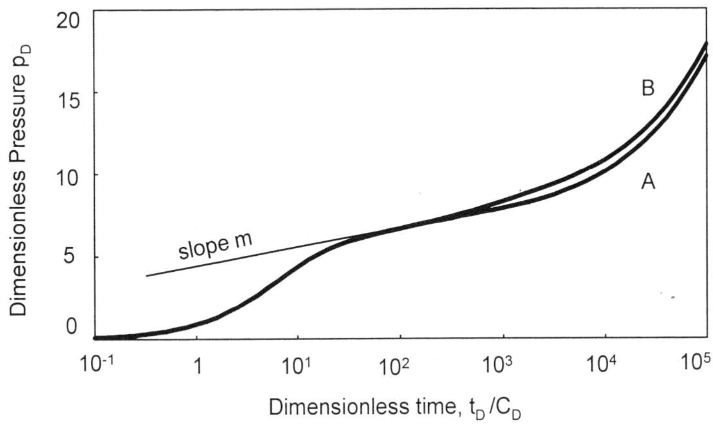

Figure 5.5. Semi-log plot of Figure 5.4 examples. p_D versus t_D/C_D.

On a semi-log plot of the same example (Figure 5.5), only one straight line is present, describing the infinite acting radial flow regime. During linear flow, the pressure response deviates in a curve above the semi-log straight line.

5.2.1 Parallel sealing fault model

The well is at a dimensionless distance L_{1D} and L_{2D} from the two sealing boundaries. The channel, of dimensionless width of $L_{1D}+L_{2D}$, is of infinite extension on both directions.

The behavior of the pressure derivative response for parallel sealing faults was originally considered in 1980 (b) by Tiab and Kumar. They used the image well method to generate the response of a well between two parallel sealing faults. Considering a graph of the pressure derivative with respect to time, as did van Poollen (1965) for intersecting faults, Tiab and Kumar identified a late time straight line with a negative slope of 0.5 in case of parallel sealing faults. In 1986, the technique of source and Green's function was considered by Wong et al. for long rectangular reservoir systems. Using the logarithm derivative, the linear flow regime was evident with a positive 0.5 slope.

5.2.2 Linear flow regime

During the late time linear flow regime, the dimensionless pressure is expressed as (Miller, 1962; Nutakki. and Mattar, 1982; Wong et al., 1986):

$$p_D = 2\sqrt{\pi t_{DL}} + S + S_{\rm ch} \tag{5.1}$$

where the dimensionless time t_{DL} is defined with respect to the channel width:

$$t_{DL} = \frac{0.000264k}{\phi\mu c_t (L_1 + L_2)^2}\Delta t \tag{5.2}$$

S is the wellbore skin coefficient and $S_{\rm ch}$ expresses a geometrical skin component, due to the convergence of the linear flow lines towards the well. Larsen and Hovdan (1987) related $S_{\rm ch}$ to the offset of the well in the channel:

$$S_{\rm ch} = \ln\frac{L_1 + L_2}{2\pi r_w} - \ln\left(\sin\frac{\pi L_1}{L_1 + L_2}\right) \tag{5.3}$$

The geometrical skin $S_{\rm ch}$ of Equation 5.3 has a form very similar to the skin S_z of Equation 3.34, for linear flow towards an horizontal well. The dimensional Equation 5.4 is obtained by combining Equation 5.1 and Equation 5.2.

$$\Delta p = 8.133\frac{qB}{h(L_1 + L_2)}\sqrt{\frac{\mu\Delta t}{k\phi c_t}} + 141.2\frac{qB\mu}{kh}(S_{\rm ch} + S) \tag{5.4}$$

5.2.3 Log-log behavior

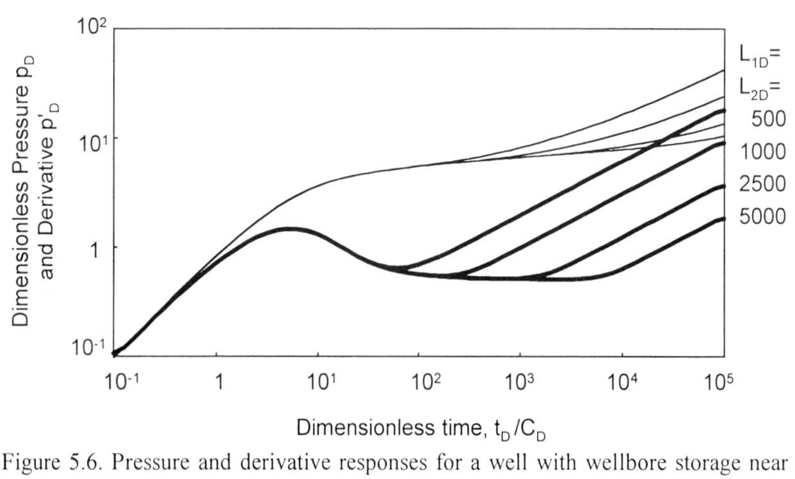

Figure 5.6. Pressure and derivative responses for a well with wellbore storage near two parallel sealing faults in a homogeneous reservoir. The well is located midway between the two boundaries, several distances between the two faults are considered.
Log-log scales, p_D versus t_D/C_D. $C_D = 300$, $S = 0$, $L_{1D} = L_{2D} = 500$, 1000, 2500 and 5000.

The linear flow regime was discussed earlier in Sections 1.2.4 and 3.2, for high conductivity fractured wells. The response is then characterized by linear flow at the start of the period, and the derivative response follows a half-unit slope straight line before the radial flow stabilization (Figure 3.5). With parallel sealing faults, the sequence of flow regimes is inverse and the radial flow regime is observed before linear flow.

On Figure 5.6, four examples of responses for different channel width are presented on log-log scale, with the dimensionless pressure p_D and derivative p'_D versus the dimensionless time t_D/C_D. The well is located midway between the two parallel sealing boundaries, the curves are generated for $L_{1D} = L_{2D} = 500$, 1000, 2500 and 5000.

The time of start of the linear flow regime is proportional to the square of the channel width. For large $(L_{1D}+L_{2D})$, a long infinite acting radial flow regime is needed before the boundaries are reached, the 1/2 slope line is displaced towards late times. Conversely, in case of a narrow channel, the radial flow regime is short and can be masked by wellbore storage. For example, on the curve $L_{1D}=L_{2D} = 500$ the derivative follows the half unit slope straight line immediately after the wellbore storage effect.

On Figure 5.4, the two examples are generated for the same channel width ($L_{1D}+L_{2D}=$ 6000) and the derivative curves merge on the same linear flow half unit slope line. The well **A** is centered and for well **B** the offset is defined with $L_{1D}/(L_{1D}+L_{2D})=1/6$. The geometrical skin S_{ch} of Equation 5.3 is higher on well **B** response and, after radial flow, the pressure curve **B** is above the curve **A**. This effect is more evident on the specialized scale plots presented in later sections.

5.2.4 Matching procedure with the pressure and derivative data

The early time response is analyzed with the infinite reservoir model. For a well with wellbore storage and skin, a $C_D\, e^{2S}$ curve is selected, *kh*, *C* and the skin coefficient *S* are evaluated as described in Section 2.2.1.

When the match is made on a computer generated curve, the derivative half unit slope line is used to estimate the channel width $(L_{1D}+L_{2D})$. The distance of the closest boundary, and therefore the well position in the channel, can be estimated from the shape of the derivative transition between the 0.5 line of radial flow, and the half unit slope line of linear flow. On the pressure curve, the well position is shown when the effect of the geometrical skin S_{ch} becomes evident, during the linear flow regime.

5.2.5 Semi-log analysis

Figure 5.7 is a semi-log plot of the examples of channel responses presented on log-log scales in Figure 5.6: at late time, during linear flow, the responses *curve upwards* from the semi-log straight line. For small channel width, the deviation occurs early, and the radial flow straight line may not be seen.

Figure 5.5 illustrates on semi-log scale the influence of the geometrical skin effect S_{ch} with the two examples of Figure 5.4: the channel width is $L_{1D}+L_{2D} = 6000$, the well is centered (**A**) or close to one of the boundaries with $L_{1D} = 1000$ (**B**). When one of the boundaries is seen before the other, the geometrical skin S_{ch} of Equation 5.3 is larger and, after the infinite acting radial flow regime, the amplitude of the response is higher than when the well is equidistant from the two limits. The influence of the well position in multiple boundary systems will be discussed more generally in Section 5.4.6, with the shape factor concept.

For the radial flow analysis of actual data on a Δp versus Δt semi-log plot, the semi-log straight line before linear flow provides the permeability thickness product kh and wellbore skin coefficient S (Equations 1.16 and 1.17).

5.2.6 Linear flow analysis

Figure 5.8 is a plot of the four examples of parallel sealing fault responses Figures 5.6 and 5.7, with the dimensionless pressure p_D versus the square root of the dimensionless time t_D/C_D.

During the late time linear flow regime, the response follows a straight line behavior (Equation 5.4) and the slope of the straight line is inversely proportional to the *channel width* ($L_{1D}+L_{2D}$). For small distances between the two boundaries, the rate of change of pressure is faster during linear flow than when the channel is large, and the straight line slope on Figure 5.8 is higher.

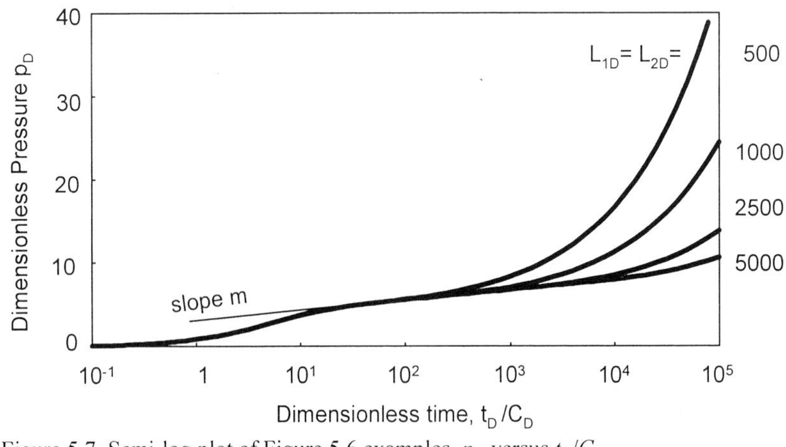

Figure 5.7. Semi-log plot of Figure 5.6 examples. p_D versus t_D/C_D.

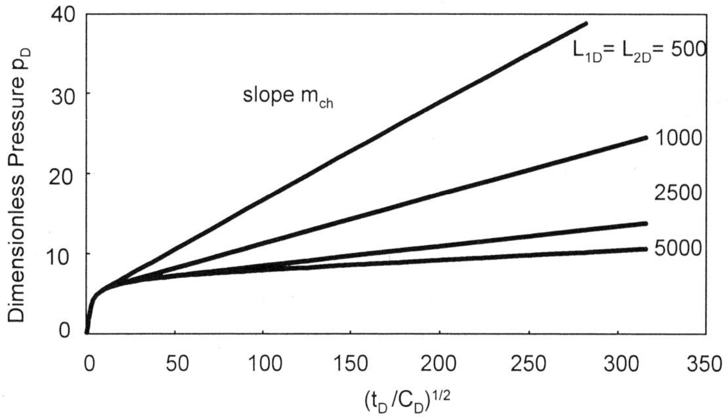

Figure 5.8. Linear flow analysis of Figure 5.6 examples. p_D versus $(t_D/C_D)^{1/2}$.

With real data, the specialized analysis of linear flow is made on a plot of the pressure change Δp versus the square root of the elapsed time Δt. The slope m_{ch} and the intercept Δp_{chint} of the linear flow straight line are used to estimate the channel width and the well location.

$$m_{ch} = 8.133 \frac{qB}{h(L_1 + L_2)} \sqrt{\frac{\mu}{k\phi c_t}} \tag{5.5}$$

Knowing the permeability from radial flow analysis, the channel width is obtained by:

$$L_1 + L_2 = 8.133 \frac{qB}{hm_{ch}} \sqrt{\frac{\mu}{k\phi c_t}} \tag{5.6}$$

The straight line intercept at time 0, Δp_{chint}, gives the total skin effect ($S_{ch} + S$), defined as the sum of the wellbore skin and the linear flow geometrical skin of Equation 5.3. When the permeability and the wellbore skin coefficient are known from radial flow analysis, the geometrical skin S_{ch} is estimated with :

$$S_{ch} = \frac{kh}{141.2qB\mu} \Delta p_{ch\,int} - S \tag{5.7}$$

and it is possible to estimate the well location between the two parallel faults (Larsen and Hovdan, 1987).

$$\frac{L_1}{L_1 + L_2} = \frac{1}{\pi} \arcsin\left(\frac{L_1 + L_2}{2\pi r_w} e^{-S_{ch}} \right) \tag{5.8}$$

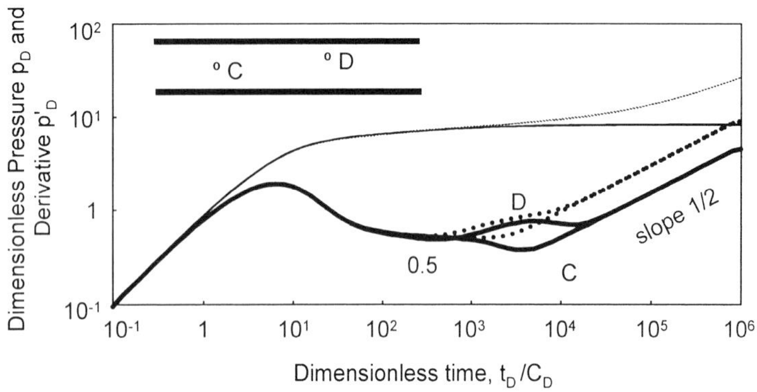

Figure 5.9. Build-up pressure and derivative responses for a well with wellbore storage in a homogeneous reservoir limited by two parallel sealing faults. Log-log scales, p_D versus t_D/C_D. $C_D = 3000$, $S = 0$, $L_{1D} = L_{2D} = 5000$ (curve **C**) and $L_{1D} = 2000$, $L_{2D} = 8000$ (curve **D**).
The dotted curves are drawdown, the build-up responses are generated for $(t_p/C)_D = 2000$.

5.2.7 Build-up analysis

Log-log pressure and derivative build-up analysis

Figure 5.9 presents on log-log scales the build-up response of two wells in a channel reservoir. The distance between the two parallel sealing faults is $(L_{1D}+L_{2D}) = 10000$, the well is centered for example **C** and off-centered with $L_{1D}/(L_{1D}+L_{2D}) = 1/5$ in example **D**. The dimensionless production time t_{pD}/C_D prior to shut-in is 2000 on both examples. The two dotted curves show the original drawdown responses of wells **C** and **D**.

The wells are closed for build-up just before the start of linear flow: at $t_{pD}/C_D = 2000$, example **C** is in infinite acting radial flow and example **D** in hemi-radial regime. On the build-up curves, the pressure flattens at late times (see Section 2.2.2) and the derivative responses display a half unit slope straight line.

For the two build-up curves of Figure 5.9, the drawdown behavior changes after extrapolation into shut-in times. The two build-up derivative curves do not match on the original drawdown responses: the half unit slope straight line is displaced towards late times and, during the transition between radial to linear flow, example **C** drops below 0.5 while example **D** shows an oscillation. For larger production times t_{pD}/C_D, the build-up derivative curves would be closer to the original drawdown responses.

As already discussed in Section 5.1.5 for a single sealing fault, build-up derivative responses in a bounded reservoir are sensitive to the production history prior to shut-in. Drawdown derivative curves are not always applicable, and caution must be exercised during the diagnosis of build-up responses. For example, if a build-up similar to example **C** of Figure 5.9 stops between the stabilization and the valley bottom, the channel response can be interpreted by error with several different models:

- increasing of mobility such as in a composite reservoir (see Section 4.3)
- effect of a constant pressure boundary (see Section 5.5)
- closed depleted reservoir (see Section 5.4.7)

The choice of the appropriate interpretation model can be frequently concluded when the results of the different interpretations are applied to flowing periods, such as on a test simulation plot. In particular, any difference between the initial pressure used for simulation, and the actual pressure before the test, must be carefully examined (see Section 10.2.3).

Semi-log analysis of build-up

The derivative distortion of the build-up curves of Figure 5.9 shows that, on a Horner or multiple rate superposition plot, the time of end of the semi-log straight line is influenced by the production history before shut-in. On example **C**, the drawdown derivative follows the 0.5 line until $t_D/C_D = 2000$ but the build-up starts to drop below 0.5 at $t_D/C_D = 1000$, the duration of the radial flow behavior is shorter.

Figure 5.10 is a Horner plot of the two build-up examples in Figure 5.9. After the radial flow straight line, curve **C** tends to flatten before the late time linear flow regime, and curve **D** shows the opposite behavior, by rising above the semi-log straight line.

The proper straight line of the Horner plot defines the permeability thickness product *kh* and wellbore skin coefficient *S* but, as shown on Figure 5.10, the extrapolation of the straight line does *not* correspond to the infinite shut-in time pressure.

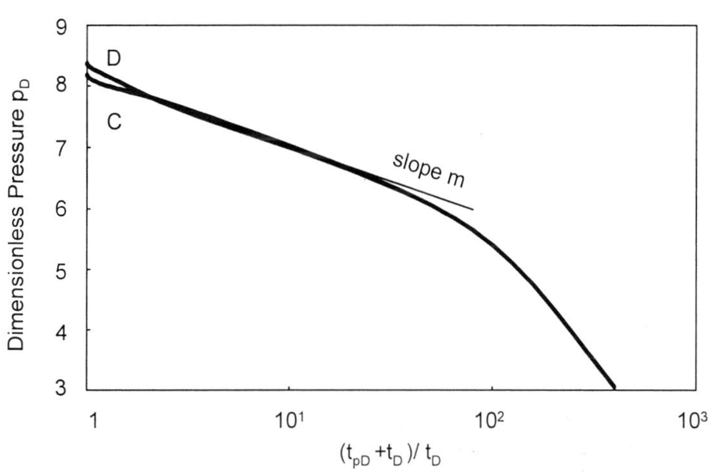

Figure 5.10. Horner plot of the build-up examples Figure 5.9. p_D versus $(t_{pD} + t_D)/t_D$.

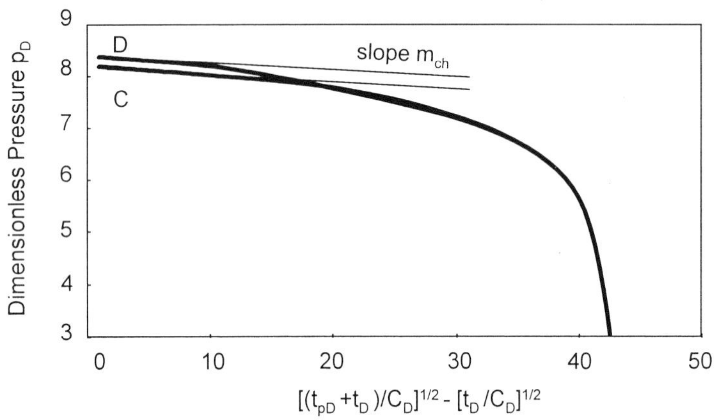

Figure 5.11. Linear flow analysis of the build-up examples Figure 5.9. p_D versus $[(t_{pD}+t_D)/C_D]^{1/2}$- $[t_D/C_D]^{1/2}$.

Linear flow specialized analysis of build-up

In a two rate test, when both the drawdown and the shut-in periods are in linear flow regime, the superposition function is expressed as $\sqrt{t_p+\Delta t}-\sqrt{\Delta t}$ (Equation 2.19 Section 2.2.2). Figure 5.11 is a plot of the two build-up examples from Figure 5.9, with the dimensionless pressure p_D versus the dimensionless linear flow superposition time $\sqrt{t_{pD}/C_D+t_D/C_D}-\sqrt{t_D/C_D}$. The two build-up curves exhibit a straight line at late times.

The slope m_{ch} of the linear flow straight line is expressed, as for drawdown responses, by Equation 5.5. On the examples **C** and **D** of Figure 5.11, the channel width is the same and the two straight lines are parallel. Well **D** is off-centered in the channel, the geometrical skin S_{ch} is larger than for well **C**: the two dimensionless responses are superimposed until the end of infinite acting radial flow, then they diverge before reaching linear flow straight lines. The amplitude of the pressure change is higher on curve **D** than on curve **C**. For infinite channels, the *extrapolation* of the linear flow straight line to $\sqrt{t_p+\Delta t}-\sqrt{\Delta t}=0$ gives the initial reservoir pressure p_i.

5.3 TWO INTERSECTING SEALING FAULTS IN HOMOGENEOUS RESERVOIR

With the intersecting sealing faults model, two linear no-flow boundaries limit the reservoir drainage area, the wedge is otherwise of infinite extension. The angle of intersection between the two faults can take any value smaller than 180°.

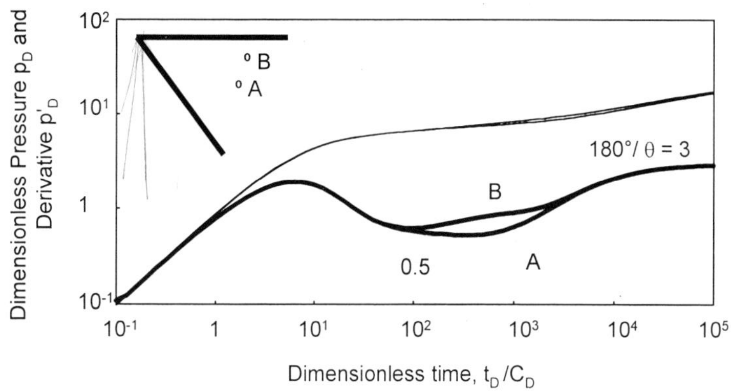

Figure 5.12. Pressure and derivative responses for a well with wellbore storage in a homogeneous reservoir limited by two intersecting sealing faults. Log-log scales, p_D versus t_D/C_D. $C_D = 3000$, $S = 0$, $L_D = 5000$, $\theta = 60°$, $\theta_w = 30°$ (curve **A**) and $\theta_w = 10°$ (curve **B**).

The effect of two intersecting sealing faults is illustrated on Figures 5.12 and 5.13 for a well with wellbore storage and skin in a homogeneous reservoir. In these examples, the angle between the faults is 60° ($\pi/3$).

- The response first describes the infinite reservoir behavior and later, when the two faults are reached, the *fraction* of radial flow limited by the wedge.

- In the case of one sealing fault, half of the plane is producing at late times and the semi-log slope doubles, the dimensionless derivative stabilizes at 1 (see Section 5.1). When two intersecting faults limit the drainage area, a smaller fraction of the plane produces: on the semi-log scale, the slope of the straight line is increased by a factor of $360/\theta$ and, with the derivative, the curve follows a *second stabilization* at a level equal to $180/\theta$. On the examples of Figure 5.12, the wedge shaped reservoir is 1/6 of the infinite plane (360°) and the dimensionless derivative stabilizes at 3, 6 times above the first 0.5 line. On semi-log scale (Figure 5.13), the second straight line slope is 6 times higher than the first.

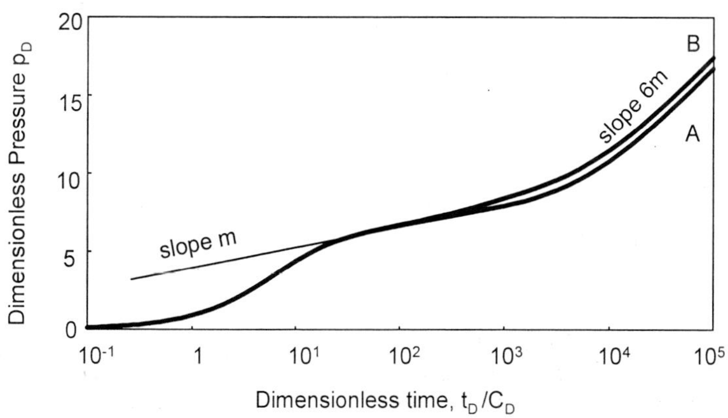

Figure 5.13. Semi-log plot of Figure 5.12 examples. p_D versus t_D/C_D.

- The shape of the transition between the two derivative plateaus depends upon the location of the well in the angle. If the well is located on the bisector, the two boundaries are equidistant from the well, and the derivative transition follows a *half unit slope* straight line (Figure 5.12, curve **A**). If the well is close to one of the two boundaries, one sealing fault is seen before the wedge response: the derivative *stabilizes* at 0.5, then 1 and finally the wedge stabilization (Figure 5.12, curve **B**). The transition between the initial and the final derivative plateaus is longer.

5.3.1 Intersecting sealing fault model

The wedge geometry is defined on Figure 5.14: θ is the angle between the faults, θ_w describes the well location in the wedge and L_D is the dimensionless distance between the well and the intercept of the two faults. The distances L_1 and L_2 between the well and the sealing faults are expressed as

$$L_1 = L_D r_w \sin\theta_w \tag{5.9}$$

$$L_2 = L_D r_w \sin(\theta - \theta_w) \tag{5.10}$$

In 1965, van Poollen used the image well method to generate the effect of intersecting sealing faults. He considered pressure and derivative curves and showed that the image well method can be used only for a restricted number of configurations: the angle of intersection must be π/n (or $2\pi/n$ if the well is located on the bisector), where n is an integer. The first possible configuration, for n=2, is then a right angle, when the reservoir is limited by two perpendicular sealing faults.

A more general solution, using Green's function, was presented in 1975 by Prassad. It allows any angle of intersection, for any location of the producing well. This solution is used in the following discussion.

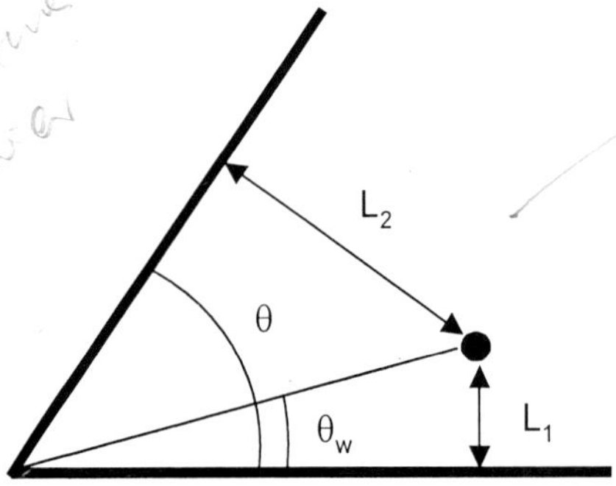

Figure 5.14. Two intersecting sealing faults in a homogeneous reservoir.

5.3.2 Log-log behavior

On Figure 5.15, the angle of intersection θ varies from 10° to 180°, the producing well is on the bisector.

The distances L_{1D} and L_{2D} between the well and the two faults are fixed at 1000, the distance to the fault's intercept changes with the angle. From Equations 5.9 and 5.10, for small angles the intercept is far from the producer, for example with θ=10°, L_D=11473 but, with θ=135°, it is only 1082.

The level of the second derivative plateau is defined by the angle θ, namely $180/\theta$ in dimensionless terms. For a single linear sealing fault (θ= 180°) the second plateau is at 1, for perpendicular faults (θ= 90°) at 2, and for the smallest angle presented on Figure 5.15, θ= 10°, it is at 18.

The start of the transition above the 0.5 derivative plateau, at the end of the infinite reservoir behavior, indicates the distance to the faults. On the examples of Figure 5.15, with $L_{1D} = L_{2D} = 1000$, the initial infinite behavior ends at about $t_D/C_D = 300$. When the well is not centered on the bisector, the shape of the transition is different as shown on curve **B** of Figure 5.12: the initial radial flow regime is shorter, and the transition between the two plateaus lasts longer.

The transition is in general long, even when the well is on the bisector. For small angles, the transition can last two log-cycles or more, before the final derivative stabilization. On Figure 5.15, with θ = 90°, the second stabilization is reached at t_D/C_D = 20,000 whereas when θ= 10°, it is hardly starting at $t_D/C_D = 10^6$.

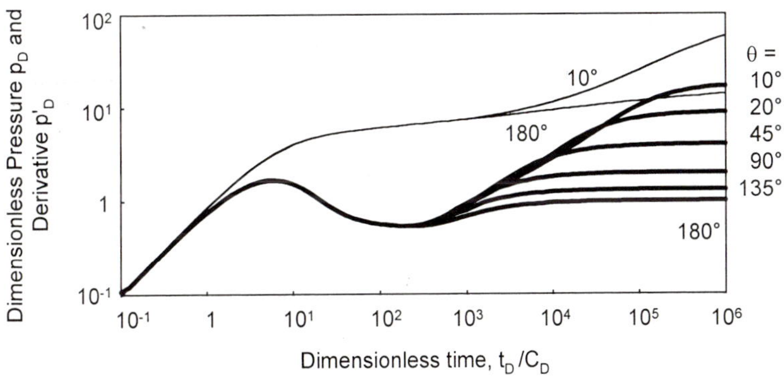

Figure 5.15. Pressure and derivative responses for a well with wellbore storage near two intersecting sealing faults in a homogeneous reservoir. Several angles of intersection θ, the well is on the bisector $\theta_w = 0.5\theta$, the distance to the two faults is constant $L_{1D} = L_{2D} = 1000$, the distance L_D to the fault intercept changes. Log-log scales, p_D versus t_D/C_D.
C_D = 1000, S = 0, θ =10°, L_D = 11473; θ =20°, L_D = 5759; θ =45°, L_D = 2613; θ =90°, L_D = 1414; θ=135°, L_D = 1082; θ=180°, L_D = 1000.

During the transition between the 0.5 and $180/\theta$ plateaus, the response tends to show a straight line of slope 1/2. When the angle of intersection θ is small, the two faults are reached a long time before the intercept is seen. Since the characteristic wedge flow behavior is not established at transition time, the behavior is similar to the response of a well between two parallel sealing faults (Section 5.2).

5.3.3 Matching procedure with the pressure and derivative data

The early time response is analyzed as usual with the infinite reservoir model and *kh*, *C* and the skin coefficient *S* are evaluated.

When the response displays a complete intersecting faults behavior, the ratio between the two derivative plateaus, at 0.5 and $180/\theta$, is used to estimate the angle of intersection θ. With actual data, the angle is expressed, in degrees:

$$\theta = 360° \frac{\Delta p_{1\text{st stab.}}}{\Delta p_{2\text{nd stab.}}} \tag{5.11}$$

The distance L_1 to the closest fault is accessed by the time of deviation from the 0.5 plateau, as for a single sealing fault. The distance L_2 to the second fault is then adjusted from the shape of the derivative transition. The distance L of the fault intercept and the angular position of the well in the wedge θ_w are obtained by solving the Equations 5.9 and 5.10.

In many field examples of intersecting faults' responses, the second plateau is not completely developed at the end of the data curve, and the response is only partially defined. When the derivative increases after the 0.5 line by a factor greater than 2, one sealing fault is not sufficient to describe the late time behavior, and the intersecting faults model has to considered. Provided the transition starts to show a decreasing curvature after the half unit slope line, the level of the second plateau is relatively easy to guess, and a good estimate of the angle of intersection θ between the two faults can be obtained.

5.3.4 Semi-log analysis

Two semi-log straight lines can be identified when the complete response of a well near two intersecting sealing faults is available. The first, of slope *m*, corresponds to the early time infinite regime shown by the 0.5 line of the derivative response. The second, with a slope $m_{\text{wedge}}=(360/\theta)m$, defines the fraction of radial flow limited by the wedge, shown on the derivative curve by the second plateau. The two semi-log straight lines can be analyzed, provided the derivative plot confirms the presence of data points during the two characteristic regimes.

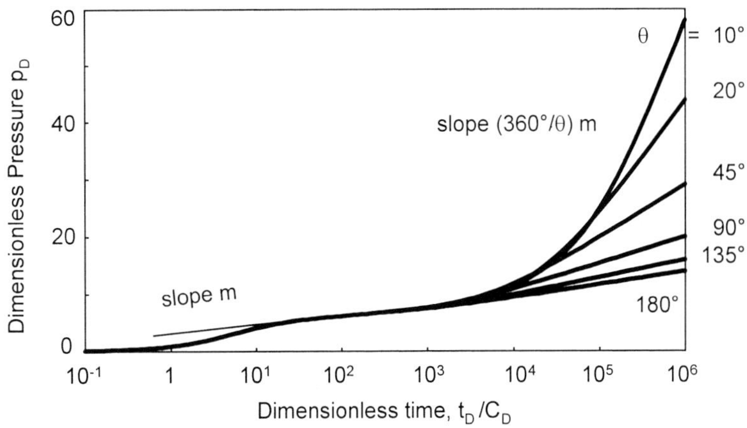

Figure 5.16. Semi-log plot of Figure 5.15 examples. p_D versus t_D/C_D.

Figure 5.16 is a semi-log plot of the six examples of wedge responses presented in Figure 5.15. The bottom curve corresponds to one sealing fault (θ=180°), the semi-log slope doubles at late times. For smaller angles, the curve deviates upwards, for example with θ =10° the second slope is 36 times greater than the first. The semi-log scale confirms that the transition between the two straight lines is long, particularly when the angle is small.

The shape of the semi-log curves for intersecting sealing faults such as on Figures 5.13 and 5.16 is not very different from the semi-log responses of Figures 5.5 and 5.7 for channel reservoirs: after the initial infinite acting regime, the curves deviates upwards. On the intersecting sealing faults examples of Figures 5.13 and 5.16, the late time responses follow a second semi-log straight line, but the channel reservoirs examples of Figures 5.5 and 5.7 are curved. In practice, the presence of the second line is difficult to ascertain on a semi-log scale, and the choice of the appropriate boundary configuration is easier when the derivative log-log data plot is considered.

With actual data, the time limits of the two semi-log straight lines on a plot of Δp versus $\log(\Delta t)$ must have been previously defined by the identification of the log-log derivative plateaus, and not decided after observation of the semi-log plot only. The first semi-log straight line of slope m is used for infinite acting radial flow analysis. The permeability thickness product kh and the skin S are estimated with Equations 1.16 and 1.17. When the two lines are present, the ratio of the first to the second slope gives the angle of intersection:

$$\theta° = 360 \frac{m}{m_{\text{wedge}}} \tag{5.12}$$

On the semi-log plot Figure 5.13 for two intersecting faults at θ=60°, well **A** is centered (θ_w=30°). Well **B** is off-centered (θ_w=10°) and one sealing fault is seen before the other.

As a result, the final semi-log straight line of well **B** is parallel to that of well **A**, but the amplitude of the response is higher. A similar geometric skin effect has been observed on Figure 5.5 of Section 5.2 for parallel sealing faults. When several boundaries are present in a pressure response, the time when the first boundary is reached influences the pressure response during all subsequent regimes.

5.3.5 Build-up analysis

Log-log pressure and derivative build-up analysis

The limitations of the multiple rate derivative, illustrated with one or two parallel sealing faults, are applicable for all boundary solutions. In the case of intersecting faults, the transition between the two derivative plateaus is long, and the build-up distortion can be significant. The match is preferably made with a computer generated build-up model, especially when the second derivative plateau, being not well defined, is estimated by extrapolation of the late transition behavior.

When several solutions are found to provide a good match on build-up data, extrapolation of the models at times larger than the shut-in duration helps reducing the number of choices. As discussed in Section 5.2.7, this is achieved with the test simulation plot. The quality of the match during the flow periods, and the accuracy of the initial pressure used for simulation, can indicate an inconsistent model.

Semi-log analysis of build-up

If only two sealing faults have been reached during the test, the wedge shaped reservoir is of infinite extension and the *second straight line* of the Horner or multiple rate superposition plot can be extrapolated to infinite shut-in time for an estimate of the initial reservoir pressure p_i.

5.3.6 Field examples

The double porosity example presented in Figure 4.40, and the double permeability example of Figure 4.50 can also be analyzed with the interesting sealing fault model in a homogeneous reservoir. With these two examples, only the final stabilization is accurately defined and the match provides the product θkh. In the case of Figure 4.40, good matches are obtained with θ ranging from 120° to 160° and, with the example of Figure 4.50, only the maximum angle can be determined with $\theta \leq 130°$. For the two tests, the reservoir permeability thickness product is not accurately defined.

5.4 CLOSED HOMOGENEOUS RESERVOIR

A closed system behavior is characteristic of limited reservoirs but it can also be encountered in developed fields, when several wells are producing and each well drains only a certain volume of the reservoir (Matthews and Russell, 1967).

The analysis of a closed reservoir has been presented in Section 1.2.8 (Cartesian scale for drawdown pseudo-steady state analysis, Figure 1.25) and in Section 2.3.3 (drawdown pressure derivative Figure 2.19). It is important to note that the responses are different for a *drawdown* and a *build up*. This is clearly illustrated in figure 5.17.

- During drawdown periods, when all boundaries have been reached after the infinite acting behavior, the reservoir starts to deplete. The response follows the *pseudo steady state* flow regime, and the well flowing pressure becomes proportional to time. Pressure and derivative log-log curves merge on a straight line of *slope unity* at late time.
- During build-ups, the shape of the well response is different. After shut-in, the pressure starts to build-up during the initial infinite regime but, later, it stabilizes and tends towards the *average reservoir pressure* $\overline{p}$.

On the linear plot Figure 1.25, a characteristic closed system response is illustrated with a very simple flow sequence. The well, at initial reservoir pressure p_i, is opened until all reservoir boundaries are reached and the pseudo steady state regime is established. This is shown on Figure 1.25 by a linear drawdown pressure trend. When the well is then closed for a shut-in period, the pressure builds up until the average reservoir pressure $\overline{p}$ is reached, and the curve flattens. The *depletion*, expressed by the difference $\left(p_i - \overline{p}\right)$ between the initial pressure and the final stabilized pressure, is proportional to the cumulative production. The longer the duration of the drawdown period, the lower is the final average reservoir pressure $\overline{p}$.

Typical pressure and derivative log-log curves are presented on Figure 5.17 for a well with wellbore storage and skin in a closed square reservoir. Well **A** is centered whereas well **B** is close to one corner of the bounded system. The dotted curves describe the drawdown responses, and the build-ups are shown with lines.

The pseudo steady state flow regime, characterized by a straight line of slope unity on the late time pressure and derivative curves, is seen only during drawdown periods (dots). If the well is not centered in the closed area, one or several boundaries can be seen during the transition between the initial radial flow and the late time pseudo steady state flow. The derivative response of example **B** shows the behavior of a 90° wedge, between the derivative 0.5 plateau and the late time unit slope line. Similarly, the pressure response **B** deviates above the response **A**, until the two curves finally reach the same asymptote.

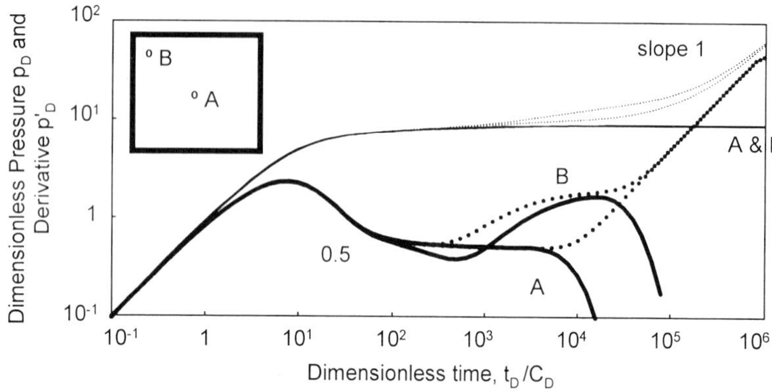

Figure 5.17. Pressure and derivative responses for a well with wellbore storage in a closed square homogeneous reservoir. The dotted curves are drawdown, the build-up responses are generated for $(t_p/C)_D = 1000$. Log-log scales, p_D versus t_D/C_D. $C_D = 25000$, $S = 0$. Curve **A**: (the well is centered) $L_{1D} = L_{2D} = L_{3D} = L_{4D} = 30000$. Curve **B**: $L_{1D} = L_{2D} = 6000$, $L_{3D} = L_{4D} = 54000$.

As opposed to drawdown responses, the build-up pressure curves of Figure 5.17 flatten, and the *derivative drops*. This illustrates the particular behavior of closed systems, where drawdown and build-up curves have totally different late time responses. Due to the presence of two boundaries close to the well, the derivative response of example **B** shows an oscillating shape.

Figure 5.18 is a semi-log plot of the two drawdown examples of Figure 5.17. The two well responses are superposed during the infinite acting radial flow regime. During pseudo steady state flow, the two responses curve above the semi-log straight line. At intermediate time, well **B** shows the behavior of two perpendicular sealing faults with an increase of the semi-log straight line slope by a factor of 4.

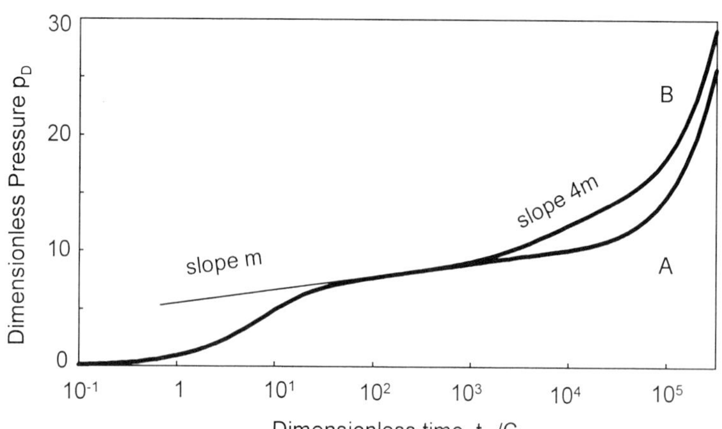

Figure 5.18. Semi-log plot of Figure 5.17 drawdown examples. p_D versus t_D/C_D.

5.4.1 Closed reservoir model

With the analytical solutions used for well test interpretation, several reservoir geometries can be envisaged but, for most practical purposes, only circular and rectangular reservoir shapes are considered. This last boundary model is presented in the following text. It is generated with the image well method as described by Matthews et al. (1954), the well can have any position in the bounded rectangular area.

The reservoir is homogeneous. The well, with wellbore storage and skin, is at dimensionless distances L_{1D}, L_{2D}, L_{3D}, and L_{4D} from the four sealing boundaries of the rectangle. The actual distances are defined by equation 1.34, and the area of the closed reservoir is expressed as :

$$A = \left(L_{1D} + L_{3D}\right)\left(L_{2D} + L_{4D}\right)r_w^2 \tag{5.13}$$

5.4.2 Drawdown behavior

During the pseudo steady state flow regime, the drawdown pressure drop is a linear function of time (Equation 1.35 of Section 1.2.8). Brons and Miller (1961 b) expressed the dimensionless pressure as:

$$p_D = 2\pi t_{DA} + \frac{1}{2}\left(\ln\frac{A}{r_w^2} + \ln\frac{2.2458}{C_A}\right) + S \tag{5.14}$$

where the dimensionless time t_{DA} is defined with respect to the drainage area,

$$t_{DA} = \frac{0.000264k}{\phi\mu c_t A}\Delta t \tag{5.15}$$

and S is the dimensionless wellbore skin coefficient.

The shape factor C_A characterizes the geometry of the reservoir and the well location. Brons and Miller (1961), Dietz (1965) and Earlougher (1971) have presented tables of shape factors for various drainage configurations. As shown in Table C.1 of Earlougher, Monograph (1977), the term 1/2 $\ln(2.2458/\ C_A)$ of the pseudo steady state Equation 5.14 is as low as -1.31 for a well at the center of a closed square such as in example **A** of Figure 5.17. In long narrow rectangular systems, it can be greater than 1, in particular when the well is near one or two reservoir boundaries. On the pressure curves, this constant geometrical factor tends to increase the amplitude of the response at late times, as shown for example on curve **B** of Figure 5.18.

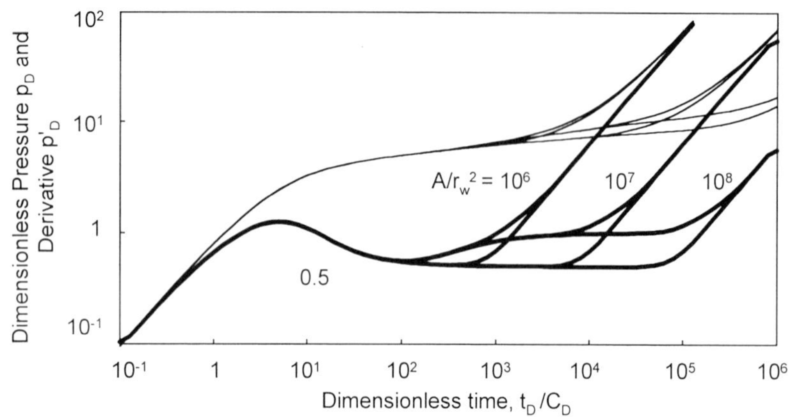

Figure 5.19. Pressure and derivative drawdown responses for a well with wellbore storage in a closed square homogeneous reservoir. Three reservoir sizes, the well is centered or near one of the boundaries ($L_{1D} = 200$). Log-log scales, p_D versus t_D/C_D. $C_D = 100$, $S = 0$, $A/r_w^2 = 10^6, 10^7, 10^8$.

The two dotted derivative curves of Figure 5.17 reach the unit slope straight line at least one log cycle earlier than the pressure responses. During the pseudo steady state flow regime, the log-log pressure curve starts to follow the straight line only when the time group $2\pi t_{DA}$ of Equation 5.14 is large compared to the three constant terms. With the derivative presentation, all constants are eliminated and the diagnosis of pseudo steady state is easier.

The location of the late time unit slope straight line is a function of the reservoir area. On Figure 5.19, several drawdown examples are presented on log-log scale, with p_D and p'_D versus the dimensionless time t_D/C_D. Three closed square reservoirs are considered, the areas are $A/r_w^2 = 10^6, 10^7, 10^8$. When the size is increased by a factor of 10, the line is displaced by one log-cycle towards late times.

For each square, the well is either centered, or near one boundary with the distance L_{1D}=200. In the latter cases, the derivative shows an intermediate plateau at 1, before reaching the closed system straight line of unit slope. In the following Sections 5.4.4 and 5.4.6, the examples of Figure 5.19 are further used to illustrate the influence of the reservoir geometry and well location on the shape factor C_A.

The two drawdown examples of Figure 5.20 correspond to a long narrow rectangular reservoir. With curve **C** the well is centered and with curve **D** it is close to one end of the channel. In both cases, the derivative responses show a half unit slope straight line at intermediate time, before the start of the closed system flow regime. A 1/2 slope is characteristic of the linear flow condition, as discussed in Section 5.2 when the well is between two parallel sealing faults. For example **D**, the channel produces in only one direction and the flow is in fact hemi-linear. The derivative half unit slope straight line of curve **D** is double that of example **C**. The analysis of the hemi-linear flow regime is discussed in Section 5.4.5.

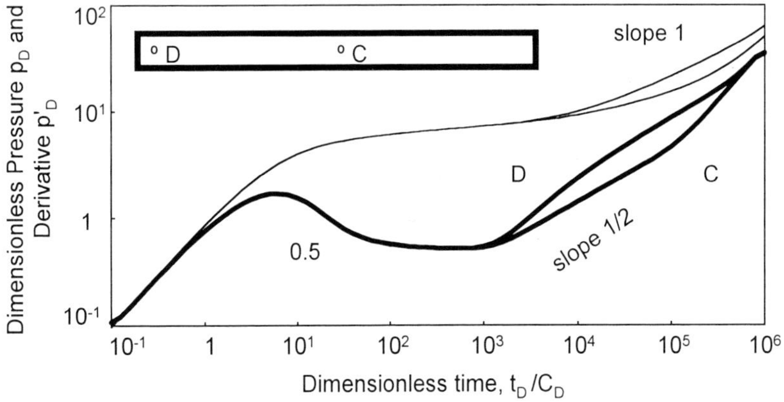

Figure 5.20. Pressure and derivative drawdown responses for a well with wellbore storage in a closed channel. Log-log scales, p_D versus t_D/C_D.
$C_D = 1000$, $S = 0$. Curve **C**, the well is centered: $L_{1D} = L_{3D} = 20000$, $L_{2D} = L_{4D} = 2000$. Curve **D**, the well is close to one end of the channel: $L_{1D} = L_{2D} = L_{3D} = 2000$, $L_{4D} = 38000$.

5.4.3 Matching procedure with pressure and derivative drawdown data

The early time response is matched on the homogeneous infinite reservoir model. For a well with wellbore storage and skin, *kh*, *C* and *S* are estimated from Equations 2.9, 2.10 and 2.11.

The late time derivative unit slope straight line defines the reservoir area. When the type-curve is generated by computer, the shape of the derivative transition between the 0.5 line and the unit slope line is used to estimate the distance of the different boundaries. Assuming a rectangular geometry, it can be possible to identify one sealing fault (as on Figure 5.19), two perpendicular sealing faults (Figure 5.17) and also two parallel sealing faults (Figure 5.20).

In practice, the derivative transition is seldom characteristic of a clear boundary configuration, and frequently several solutions are possible for the shape of the rectangular reservoir and the well location. The pressure response is also used for adjusting the geometry, with the shape factor effect.

5.4.4 Semi-log analysis

Figure 5.21 is a semi-log plot of the six examples of closed square responses presented on log-log scale Figure 5.19: at late time, during pseudo steady state flow, the curves deviate above the semi-log straight line. Before, the three examples generated with the well near one of the boundaries show a typical sealing fault behavior at transition time, with a doubling semi-log straight line slope.

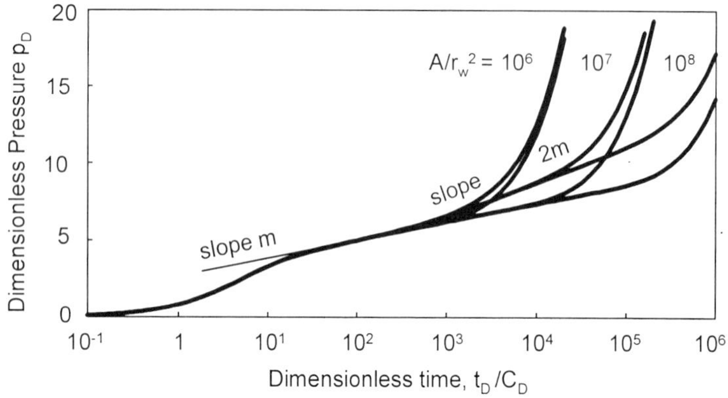

Figure 5.21. Semi-log plot of Figure 5.19 drawdown examples. p_D versus t_D/C_D.

On these three examples, the distance to the first boundary is constant with L_{1D}=200 and the duration of the semi-infinite regime is a function of the reservoir area. The curve generated for $A/r_w^2=10^8$ shows the longest hemi-radial flow regime: at time of start of the pseudo steady state flow, this curve is significantly higher than the curve generated for the well at the center of the square. Conversely, the transition on the example $A/r_w^2=10^6$ is short, and the two responses are relatively close. Figure 5.21 illustrates on semi-log scale the influence of the geometrical effect of boundaries on pseudo steady state pressure responses: the shape factor is high when the hemi-radial regime is long.

On a Δp versus Δt semi-log plot of actual data, provided the correct semi-log straight line has been identified from log-log analysis, the permeability thickness product *kh* and skin coefficient *S* are estimated with Equations 1.16 and 1.17. During the transition before pseudo steady state flow, if the derivative shows a second stabilization such as on the examples Figures 5.17 and 5.19, the intermediate semi infinite regime also displays a semi-log straight line (Figures 5.18 and 5.21). It is then possible to estimate the distance of one or two boundaries (Section 5.1.4 for one sealing fault and Section 5.3.4 for two intersecting sealing faults).

If the log-log analysis shows the behavior of two parallel sealing faults before the pseudo steady state regime as on Figures 5.20, the linear flow transition regime can be analyzed with a plot of the pressure change versus the square root of time.

5.4.5 Linear and hemi-linear flow analysis

Figure 5.22 is a plot of the two examples of closed channel reservoir of Figure 5.20, with the dimensionless pressure p_D expressed versus the square root of the dimensionless time t_D/C_D (see Section 5.2.6). With well **C**, the flow is linear at intermediate time but, on the second example, well **D** is close to one end of the channel and the flow is hemi-linear.

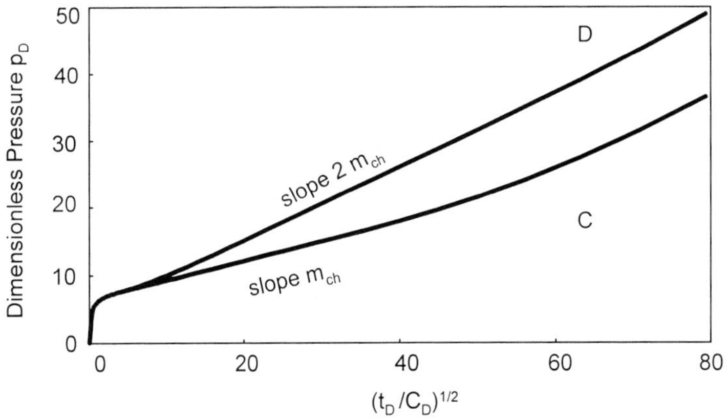

Figure 5.22. Linear flow analysis of Figure 5.20 drawdown examples. p_D versus $(t_D/C_D)^{1/2}$.

The two responses show a straight line on the linear flow plot of Figure 5.22. Example **C** corresponds to an infinite channel, the slope m_{ch} is expressed in Equation 5.5. In the case of semi-linear flow curve **D**, the slope m_{hch} is double:

$$m_{\mathrm{hch}} = 16.27 \frac{qB}{h(L_1 + L_2)} \sqrt{\frac{\mu}{k\phi c_t}} \tag{5.16}$$

If the semi-infinite nature of the linear flow regime is ignored and Equation 5.5 is used in place of Equation 5.16, the resulting calculated width is half of the true channel width.

This illustrates the frequent non-uniqueness of the solution when several boundaries are acting during a well response. For a closed system, the reservoir area is well defined from the pseudo-steady state analysis but, in general, several combinations of shape and well position can be used to describe the transition between the infinite radial flow regime and the pseudo-steady state flow.

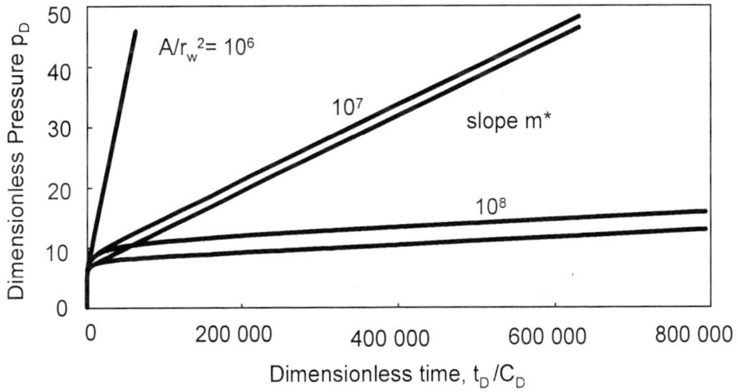

Figure 5.23. Pseudo steady state flow analysis of Figure 5.19 drawdown examples. p_D versus t_D/C_D.

5.4.6 Pseudo-steady state analysis

Figure 5.23 is a cartesian plot of the six examples of closed reservoir responses presented Figure 5.19 and 5.21: three square areas are considered and two well locations are assumed, with the well either centered in the square or near one of boundaries.

On a linear plot, the responses follow a straight line during pseudo steady state regime. With the dimensionless variables p_D and t_D/C_D, the slope is inversely proportional to the reservoir area A and the six straight lines of Figure 5.23 can be grouped into three pairs. With the smallest area $A/r_w^2=10^6$, the two lines are almost superimposed but, as discussed in the previous section 5.4.4, when the area is larger and the well is off-centered, a long semi-infinite regime is seen before pseudo steady state flow and the shape factor C_A becomes large. For $A/r_w^2=10^8$, the two pseudo steady state straight lines are clearly different.

With real data, the pressure is expressed in Equation 1.35 during the pseudo steady state flow regime:

$$\Delta p = 0.234\frac{qB}{\phi c_t hA}\Delta t + 162.6\frac{qB\mu}{kh}\left[\log\frac{A}{r_w^2} - \log(C_A) + 0.351 + 0.87S\right] \tag{1.35}$$

The slope $m*$ of the pseudo-steady state straight line provides the reservoir connected pore volume:

$$\phi hA = 0.234\frac{qB}{c_t m*} \tag{1.36}$$

When kh and S are known from semi-log analysis of the early time response, the shape factor C_A can be estimated from the time zero intercept $\Delta p*_{\text{int}}$ of the pseudo-steady state straight line with Equation 5.17, or 5.18 (Earlougher, 1971) :

$$C_A = 2.2458e^{2.303\left[\left(p_i - p_{\text{int}}^*\right)/m - \log\left(A/r_w^2\right) - 0.87S\right]} \tag{5.17}$$

$$C_A = 5.456\frac{m}{m*}e^{-\left[2.303\left(p_i - p_{\text{int}}^*\right)/m\right]} \tag{5.18}$$

The reservoir shape is deduced from a table of shape factors such as Table C.1 in Earlougher (1977).

5.4.7 Build-up analysis

As already shown on Figures 1.25 and 5.17, the pressure behavior of closed systems is different during drawdown and build-up periods. At late time, the build-up pressure stabilizes at the average reservoir pressure $\bar{p}$, the difference ($p_i - \bar{p}$) defines the depletion.

Log-log pressure and derivative build-up analysis

Figure 5.24 presents three build-up examples for a well in a closed rectangle. The length of the rectangle is twice the width, the distance between the well and one of the small sides is 1/8 of the length, the dimensionless rectangle area is A/r_w^2=8*10^6. This configuration is described in the Shape Factors Table C.1 of Earlougher (1977): C_A is 0.5813, the start of pseudo steady state t_{pssD} is at t_{DA}=2 and, at t_{DA}=0.6, the error in drawdown pseudo-steady state analysis (Section 5.4.6) is less than 1%.

In the log-log plot Figure 5.24, the dimensionless pressure and derivative are expressed versus the dimensionless time t_D/C_D. The dimensionless wellbore storage coefficient is C_D = 292 and the dimensionless time t_{DA} of Equation 5.15 can be evaluated as:

$$t_{DA} = \frac{C_D}{A/r_w^2} \frac{t_D}{C_D} = 3.65\text{x}10^{-5} \frac{t_D}{C_D} \tag{5.19}$$

Three production times are considered:
1: The well is closed at start of the pseudo steady state regime (within 1% error: t_{pDA}=0.6; t_{pD}/C_D=16,400),
2 & 3: The well is closed during pure pseudo steady state flow (t_{pDA}=2 (=t_{pssD}) and 10; t_{pD}/C_D=54,600 and 273,000).

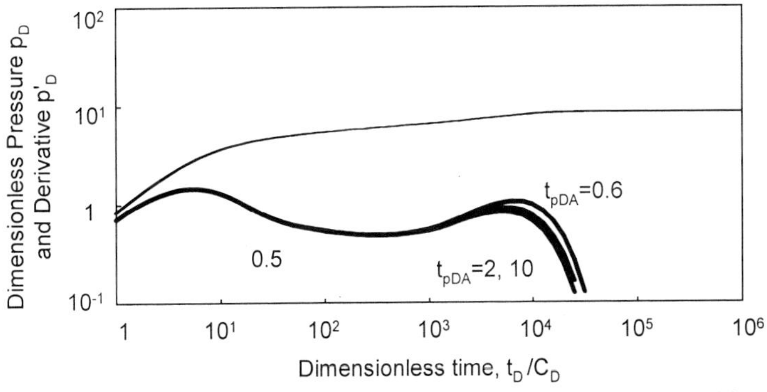

Figure 5.24. Pressure and derivative build-up responses for a well with wellbore storage in a closed rectangle. The well is close to one boundary. Three production times are considered. Log-log scales, p_D versus t_D/C_D. C_D = 292, S = 0, L_{1D} = 500, L_{2D} = 1000, L_{3D} = 3500, L_{4D} = 1000. t_{pD}/C_D (t_{pDA}) = 16400 (0.6), 54600 (2), 273000 (10)

In Figure 5.24, the difference between the pressure responses appears negligible and the three curves merge into one single build-up response. When all reservoir boundaries have been reached during drawdown, the late time dimensionless build-up pressure stabilizes at:

$$\overline{p_{\mathrm{BU}_D}} = 1.151\left(\log\frac{A}{r_w^2} - \log C_A + 0.351 + 0.87S\right) \tag{5.20}$$

The examples are generated with a skin factor S =0. From Equation 5.20, the dimensionless stabilized pressure at late time is $\overline{p_{\mathrm{BU}_D}} = 8.62$.

Interestingly, when Equation 5.20 is changed to dimensional data, it has the same form as the constant term of the drawdown Equation 1.35. This shows that the amplitude of the pressure change during a complete build-up, $\overline{\Delta p} = \overline{p} - p_{(\Delta t\,=\,0)}$, is equal to $\Delta p^*_{\mathrm{int}} = p_i - p^*_{\mathrm{int}}$. The intercept $\Delta p^*_{\mathrm{int}}$ of the pseudo-steady state straight line of the drawdown data (Δt=0 in Equation 1.35) indicates the amplitude of the subsequent build-up period at infinite shut-in time, and the depletion is expressed by the first term of Equation 1.35:

$$p_i - \overline{p} = 0.234\frac{qB}{\phi c_t hA}\Delta t \tag{5.21}$$

When the pressure stabilizes, the derivative drops towards zero. The three derivative curves are estimated with respect to the natural logarithm of the Horner time (see Section 2.3.4), and three different dimensionless production times are used: t_{pD}/C_D=16,400, 54,600 and 273,000. Even though the pressure curves appear similar, the late time derivative responses are different on Figure 5.24.

Before the final derivative drop, the build-up responses show the influence of the closest reservoir boundary and the derivative curves deviate above the 0.5 line, producing a hump. The examples of Figure 5.24 define the following sequence of flow behaviors: after the initial wellbore storage effect, the infinite acting regime is followed by a semi-infinite behavior and finally by the closed system stabilized pressure.

With t_{pDA}=0.6, the derivative is above the two other build-up curves during the hump, and the final drop is delayed. When the pseudo steady state has been reached during drawdown ($t_{pDA} \geq 2$ in the examples Figure 5.24), the shape of the build-up curve becomes independent of t_p.

Other boundary configurations can generate a similar shape, with a derivative response curve going up and then down. For example when the reservoir is limited by a constant pressure boundary (see Section 5.5), the pressure also stabilizes at late time, and the derivative drops. If sealing boundaries are seen before the constant pressure support, the derivative increases before falling to zero (Figure 5.31). A semi permeable fault (Figure

5.33), and double porosity systems with sealing boundaries (Figure 5.41), can also generate a hump on derivative responses. In such cases, the derivative does not drop to zero at late time. These different boundary configurations are discussed in later sections.

Semi-log analysis of build-up

The three dimensionless production times t_{pD}/C_D=16,400, 54,600 and 273,000 used for Horner analysis are considerably larger than the time of end of the semi-log regime, estimated at t_D/C_D=1000 on Figure 5.24. As discussed in Section 2.2.2, when $t_p >> \Delta t$, the Horner time can be simplified with Equation 2.16:

$$\log \frac{t_p + \Delta t}{\Delta t} \approx \log t_p - \log \Delta t \tag{2.16}$$

and the semi-log straight line slope m becomes independent of t_p.

When the analysis plots were made by hand, the M.D.H. semi-log scale was sometimes used for build-ups after long drawdown periods, since it required less work. With computing facilities available today, the Horner or superposition methods are preferred.

When the production time t_p is greater than the time of start of the pseudo-steady state t_{pss}, two different definition of Horner time have been proposed (Earlougher, 1977), using the true production time t_p as on Figure 5.25, or t_{pss}. The build-up pressure response being independent of t_p, the Horner time should ideally also be independent of the production history, and therefore calculated with a fixed t_{pss}.

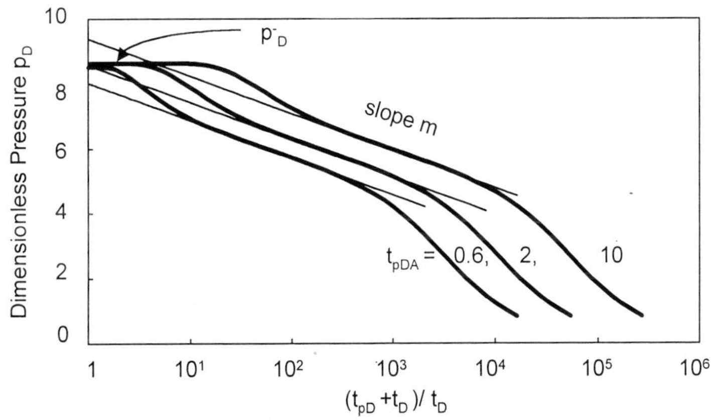

Figure 5.25. Horner plot of Figure 5.24 build-up examples. p_D versus $(t_{pD} + t_D)/t_D$. The Horner production time is t_{pD}/C_D = 16400, 54600 and 273000 (t_{pDA} =0.6, 2 and 10).

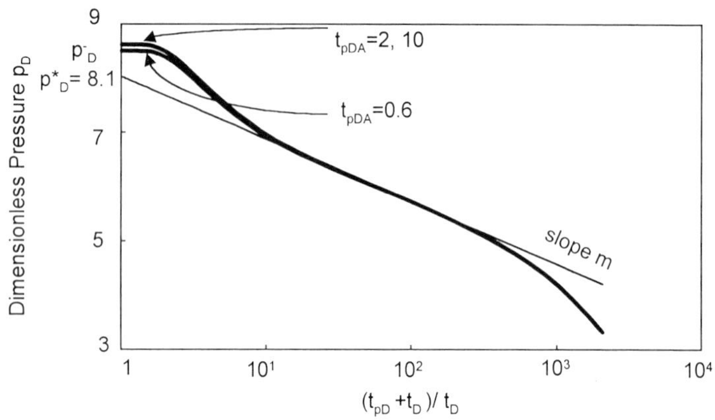

Figure 5.26. Horner plot of Figure 5.24 build-up examples. p_D versus $(t_{pD} + t_D)/t_D$. For the three examples, the Horner production time is t_{pD}/C_D = 16400 (t_{pDA} =0.6).

Figure 5.26 is the Horner plot of the dimensionless log-log curves of Figure 5.24 where the Horner dimensionless production time is fixed at t_{pDA} = 0.6 for the three responses. This corresponds to the 1% error criteria of the pseudo steady state drawdown straight line and to t_{pD}/C_D=16400. From Equation 2.16, the choice of the production time, t_p or t_{pss}, does not change the slope, but only displaces the Horner curve along the time scale. The two definitions of Horner time give the same permeability thickness product *kh* and skin coefficient *S* with Equations 1.16 and 2.15, the only difference is in the extrapolation of the straight line to "infinite shut-in time", p^*.

With a fixed t_{pss} as on Figure 9.26, the same semi-log straight line describes the three examples during radial flow. When pseudo steady state has been reached during drawdown (curves t_{pDA}=2 and 10), the two Horner curves are superimposed whereas with t_{pDA} = 0.6, the curve deviates and flattens at $\overline{p_D} = 8.50$, slightly below the theoretical dimensionless stabilized pressure $\overline{p_D} = 8.62$. For the three curves, the semi-log straight line extrapolated pressure p_D*=8.1 is lower than the final stabilized pressure.

The Horner curve in Figure 5.26 being independent of t_p, the relative position of the extrapolated pressure p^* compared to the stabilized pressure $\overline{p}$ is only a function of the well location in the rectangle. The final stabilized build-up pressure is higher than p^* when a long semi-infinite regime is seen after the radial flow straight line, due to one or several boundaries close to the well. In addition to the build-up example Figure 5.24, the different drawdown examples Figures 5.17, 5.19 and 5.20 with a high shape factor C_A correspond to this type of behavior. Conversely, when the well is centered in the reservoir, the radial flow regime is not interrupted by early boundary effects, and the Horner curve stabilizes below the extrapolated pressure p^*.

In practice, the Horner time is naturally expressed with reference to the true production time t_p as shown on Figure 5.25. When compared with Figure 5.26, the Horner curves are simply displaced by Log[t_p/t_{pss}] along the time scale (from Equation 2.16), and therefore the straight line extrapolated pressure *changes* with t_p. For example, the curve generated for t_{pDA}=2 is displaced by log(2/0.6) to the right, the dimensionless extrapolated pressure p_D* is then 1.151x0.52=0.60 higher than on Figure 5.26, and p_D*=8.7 is almost the same as the dimensionless stabilized pressure $\overline{p_D} = 8.62$. With t_{pDA}=10, the semi-log straight line has to be extrapolated more than one log cycle further, the new p_D* is 1.40 higher, and p_D*=9.5 is above the constant stabilized pressure $\overline{p_D}$.

In producing fields, routine shut-in periods are made to monitor the depletion. On a Horner/superposition plot of real data, both the extrapolated pressure $p*$ and the average pressure $\overline{p}$ decrease when t_p increases. When the same production time, such as on Figure 5.26, is used for Horner analysis of the different build-up tests, the change of extrapolated pressure $p*$ from one build-up to the next defines the drop of average pressure $\overline{p}$ between the tests. If the actual multiple rate sequence, or the total production time as on Figure 5.25, are used, the difference $\left(\overline{p} - p*\right)$ is not a constant, and $p*$ is difficult to interpret.

The M.B.H. method

In 1954, Matthews-Brons-Hazebroek presented a method for estimating the average reservoir pressure when build-up surveys in bounded reservoirs are terminated before the final pressure stabilization. The M.B.H. method has been thoroughly discussed and illustrated in the petroleum literature. In the following, the technique is only summarized in relation to the previous discussion of semi-log build-up analysis.

The M.B.H. method uses results of the Horner analysis based on the production time t_p such as on Figure 9.25. In addition to the straight line slope m and extrapolated pressure p*, the technique requires the reservoir area A and the geometry of the well location to be known. A set of semi-log curves is presented for various reservoir shapes and well locations. The x axis expresses, in term of t_{pDA} of Equation 5.15, the production time prior to shut-in and the y scale provides the M.B.H. dimensionless pressure, defined as:

$$p_{D,\text{MBH}} = 2.303\frac{p*-\overline{p}}{m} = 2\left(p_D^* - \overline{p_D}\right) \tag{5.22}$$

From the Horner analysis, m and p* are estimated, the permeability is calculated with Equation 1.16 and, knowing the reservoir area A, the dimensionless production time t_{pDA} is evaluated from Equation 5.15. Then, by selecting the proper curve for reservoir and well location, a direct reading provides the dimensionless M.B.H. pressure and the theoretical average reservoir pressure $\overline{p}$ can be evaluated from Equation 5.22.

The M.B.H. curve corresponding to the examples in Figure 9.24 is presented in Figure 6.4 of Earlougher (1977). For the three production times t_{pDA}=0.6, 2 and 10, the dimensionless M.B.H. pressures are respectively $p_{D,MBH}$=-0.8, 0.1 and 1.8. Using the Figure 9.25 previous calculations of extrapolated pressures for the Horner plot defined with t_p (p_D*=8.1, 8.7 and 9.5), the expected dimensionless average pressures are obtained: $\overline{p_D}$ =8.5 for t_{pDA} = 0.6, and $\overline{p_D}$ =8.6 on the two other examples.

It is assumed with the M.B.H. method that the production is relatively constant during the complete flowing time. In case of multiple rate production history, since the M.B.H. pressure is not available with the superposition method (Larsen, 1983), the equivalent Horner time is used. In practice, analytical simulations are preferred today rather than the M.B.H. method for estimation of the average reservoir pressure. No simplification of the rate history has to be made, the range of reservoir systems is not limited to a defined catalog of shapes, and curve matching by simulation is less prone to error than the straight line methods.

5.5 CONSTANT PRESSURE BOUNDARY

Sealing and constant pressure linear boundaries express the two limiting cases of the linear composite solution presented in Section 4.3.3, with a zero and an infinite mobility in the outer reservoir region. A constant pressure boundary is used to describe the influence of a linear change of fluid properties, such as the presence of a gas or a water contact some distance away from an oil well (Figure 5.27). Strictly speaking, the mobility of the outer zone is not infinite but it must be very large compared to the mobility of the oil region. In the following Section 5.6 on communicating faults, it is shown that an infinite conductivity fault also acts as a constant pressure boundary, even though the reservoir properties can be constant on both sides of the fault.

Vertical changes of fluid properties, such as the influence of a gas cap or a bottom water drive for example, correspond to a different boundary system. These configurations have been introduced for partially penetrating wells in Section 3.4.7 and for horizontal wells in Section 3.6.11 (Figure 3.43). The effect of a gas cap or bottom water drive is further discussed at the end of the chapter, in Section 5.10.3.

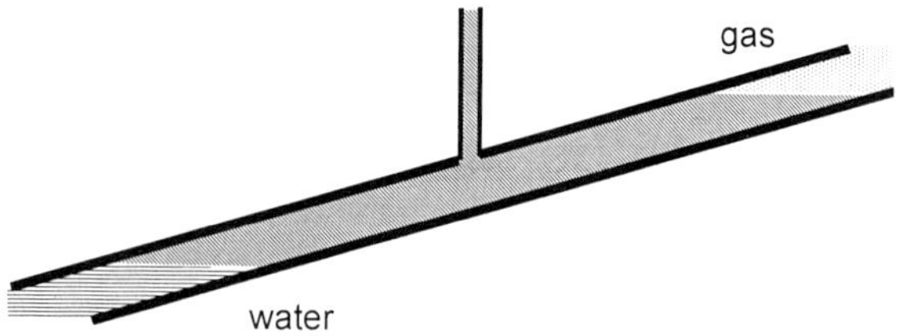

Figure 5.27. Physical model for a linear constant pressure boundary: gas or a water drive pressure support.

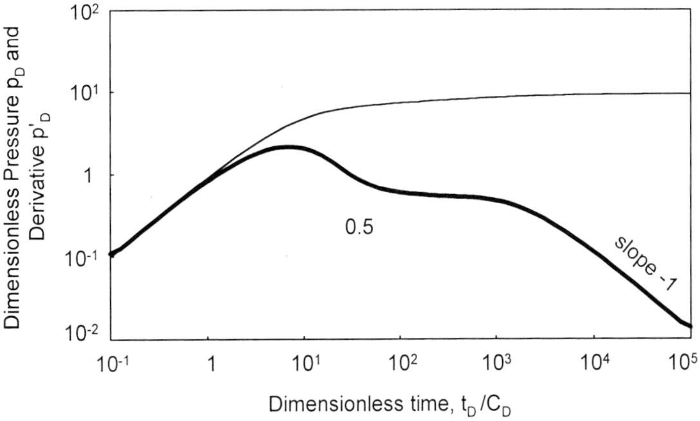

Figure 5.28. Pressure and derivative responses for a well with wellbore storage near one constant pressure linear boundary in a homogeneous reservoir. Log-log scales, p_D versus t_D/C_D. $C_D = 104$, $S = 0$, $L_D = 5000$.

A typical influence of a linear constant pressure boundary is illustrated in Figure 5.28 for a well with wellbore storage and skin in a homogeneous reservoir.

- During drawdown and shut-in periods, the *pressure stabilizes* and the *derivative tends to zero* when the influence of the constant pressure boundaries is felt.
- The rate of decline of the derivative response gives an indication of the *geometry* of the constant pressure boundaries.

In Figure 5.28, the same dimensionless parameters as on the example of a sealing fault response presented Figure 5.1 are used, only the boundary condition is changed. With the two solutions, a common infinite acting regime is seen at early time but the models diverge when the boundary effect starts to be felt, at t_D/C_D=1000. In the case of a unique linear constant pressure boundary, when the pressure stabilizes, the derivative follows a straight line with a *negative unit slope* (Abbaszadeh and Cinco-Ley, 1995).

When several constant pressure boundaries are reached, the shape of the response becomes close to that of a build-up curve in a bounded (closed) system such as in Figure 5.17 but, with a constant pressure boundary, the same stabilized pressure behavior is seen during drawdown and build-up periods.

All the configurations of linear sealing boundaries presented in previous sections can be considered with constant pressure boundaries. When several boundaries are considered, a combination of the two conditions can be also envisaged.

5.5.1 Single linear constant pressure boundary model

The constant pressure model is identical to the sealing fault analytical solution presented in Section 5.1.1, except that the producing image well at distance $2L_D$ is changed into an injection well, thus introducing the linear constant pressure condition at a distance L_D from the producing well.

When the producing well has reached the radial flow regime, the influence of the image well, described by a line source injector, is introduced as:

$$p_D = 0.5(\ln t_D + 0.80907 + 2S) + 0.5\,\mathrm{Ei}\left[-\frac{(2L_D)^2}{4t_D}\right] \tag{5.22}$$

When the radial flow condition also becomes valid for the image well response, the exponential-integral solution $\mathrm{Ei}\left[-(2L_D)^2/4t_D\right]$ can be approximated by the semi-log function $-\ln\left[t_D/(2L_D)^2\right]-0.80907$, and Equation 5.23 reduces to:

$$\overline{p_D} = \ln 2L_D + S \tag{5.23}$$

In the absence of any skin factor *S*, the amplitude of the stabilized dimensionless pressure is only function of the distance L_D between the well and the constant pressure boundary. On the example Figure 5.28, with a distance L_D = 5000 and a skin S =0, the stabilized pressure is $\overline{p_D} = 9.21$.

5.5.2 Behavior

Figure 5.29 presents four constant pressure boundary responses for the same well configuration and boundary distances as on the sealing faults examples in Figure 5.2: L_D=100, 300, 1000 and 3000.

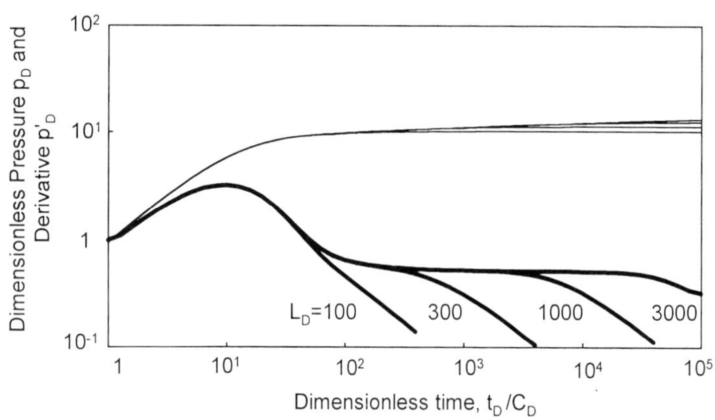

Figure 5.29. Pressure and derivative responses for a well with wellbore storage and skin near one constant pressure boundary in a homogeneous reservoir. Log-log scales, p_D versus t_D/C_D. C_D = 100, S = 5. Several distances L_D = 100, 300, 1000, 3000.

At late time, when the constant pressure is felt, the four responses become parallel in Figure 5.29: the pressure curves reach the different stabilized pressures p_D of Equation 5.24, and the derivative negative unit slope lines are displaced along the time scale proportionally to L_D^2, the square of the distance. As for the sealing fault model, the response for L_D=100 does not show any radial flow behavior.

The same examples are presented on a semi-log scales in Figure 5.30: when the constant pressure is reached, the curves deviate from the radial flow straight line and stabilize.

An example of mixed sealing and constant pressure boundaries is given in Figure 5.31: the well is near a sealing fault at L_{1D}=340, and the reservoir is under the influence of a linear constant pressure boundary perpendicular to the fault at a distance L_{2D}=940 from the well. The derivative first shows the influence of the no-flow barrier and the response deviates above the 0.5 plateau, but later the constant pressure is seen and the derivative drops. This combination of mixed boundaries produces a hump, very similar to the first part of a finite conductivity fault response as described in Section 5.6.2.

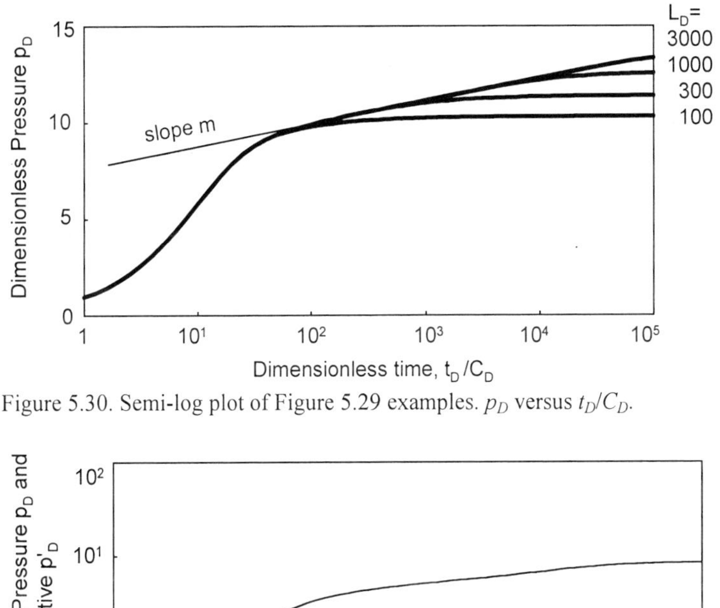

Figure 5.30. Semi-log plot of Figure 5.29 examples. p_D versus t_D/C_D.

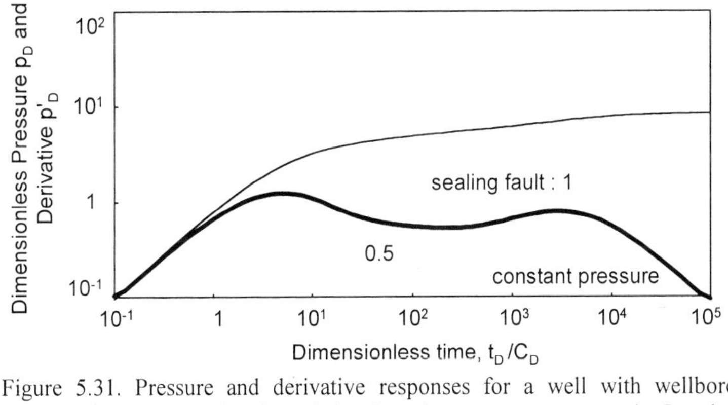

Figure 5.31. Pressure and derivative responses for a well with wellbore storage near two perpendicular intersecting boundaries in a homogeneous reservoir. Log-log scales, p_D versus t_D/C_D. C_D = 100, S = 0. The closest boundary is sealing, the second at constant pressure. The angle of intersection is θ= 90°, the well location is θ_w = 20°, the distance to the intercept is L_D = 1000.

When several constant pressure boundaries are present, or in the case of a circular constant pressure boundary, the shape of the derivative decline is steeper.

5.5.3 Drawdown and build-up analysis

Like other boundary conditions, a response influenced by constant pressure is analyzed in two steps. The early time response corresponds to the infinite reservoir solution. On log-log scales, a $C_D e^{2S}$ curve describes the first part of the response in the case of wellbore storage and skin effects. On semi-log or Horner/superposition scales, the usual straight-line methods are applicable. This first analysis provides the wellbore storage coefficient C, the permeability thickness product kh and the skin factor S.

When the constant pressure influence is seen, the distance L to the pressure support boundary can be estimated. On a log-log scale, the time of the drop of the derivative curve indicates L, in the same way as the increase of the derivative response gives the distance to a no flow barrier. With mixed boundaries, a match on a computer-generated response describes the influence of the sealing boundaries reached before the constant pressure one.

In the hypothesis of a single constant pressure boundary, the time of intercept Δt_x between the semi-log straight line and the constant pressure is used on semi-log and superposition plots. The Equation 1.33 for a sealing fault is applicable to the constant pressure boundary. The difference between the pressure at the start of the period and the stabilized pressure $\overline{\Delta p} = \overline{p} - p(\Delta t = 0)$ can also be used to estimate L. From Equation 5.24,

$$L = 0.5 r_w e^{\left[1.151\left(\overline{p} - p(\Delta t = 0)\right)/m - S\right]} \tag{5.24}$$

As the derivative follows a straight line with a negative unit slope when a constant pressure boundary is reached, Abbaszadeh and Cinco (1995) propose to estimate the distance from a Cartesian plot of the pressure derivative group $\Delta p'$ of Equation 2.23, versus the inverse of the elapsed time $1/\Delta t$. The derivative follows a straight line of slope m_{cp}, giving the distance to the linear constant pressure with:

$$L = 0.0193\sqrt{\frac{m_{\mathrm{cp}} h}{qB\phi c_t}} \tag{5.25}$$

For shut-in periods, the extrapolated pressure of the semi-log straight line is not used; the pressure at infinite shut-in time corresponds to the *stabilized pressure*.

Frequently, on build-up data, the derivative becomes noisy when the pressure stabilizes at late time and it can be difficult to differentiate a constant pressure influence from a closed system response, as illustrated in Figure 5.24 for example. The analysis of the previous drawdown and the test simulation plot are used to decide between the two alternative solutions. The comparison of the initial reservoir pressure p_i with the build-up extrapolated or stabilized pressures, p^*or $\overline{p}$, also shows if depletion has to be considered.

5.6 COMMUNICATING FAULT

In hydrocarbon bearing formations, faults are frequently non-sealing and allow communication between two reservoir regions. If the fault shows an infinite conductivity behavior, a flux parallel to the fault plane is established, improving the drainage in the reservoir region. By providing pressure support, this configuration shows the behavior of a constant pressure boundary (Abbaszadeh and Cinco-Ley, 1995; Kuchuk and Habashy, 1997).

Between the two limiting cases, namely the sealing and the infinite conductivity fault, intermediate behaviors can be encountered:
1. A *partially communicating* fault (also called semi-permeable fault) describes a reduction of permeability in the vertical plane fault.
2. With the *finite conductivity fault* model, the fault permeability is larger than the formation permeability.

In the following, partially communicating and finite conductivity fault models are presented. It is assumed that the fault, of infinite extension, is unique. For clarity of the presentation, the two reservoir regions separated by the fault are supposed to have the same characteristic. It is shown that the fault influence is expressed by a temporary deviation from the 0.5 plateau. The derivative shows a *hump* above the stabilization with the semi-permeable fault, and a *valley* below the stabilization in the case of a finite conductivity fault.

If high conductivity faults are connected and form a network in the reservoir, the behavior corresponds to the double porosity model described in Section 4.1

Cinco et al. (1976) investigated the behavior of a well near a single infinite conductivity fault of limited extension. The shape of the response is close to that of a finite conductivity fault of infinite extension. They describe the effect of the fault with a negative pseudo-skin. The fault is acting like a drain, and the pseudo-skin is more negative when the orientation of the fault is radial.

5.6.1 Semi-permeable linear boundary

The semi-permeable linear boundary is designed to describe the presence of a partially communicating fault: a flow is possible between the two reservoir regions through the fault plane, no flow is allowed along the fault. On the example of Figure 5.32, the throw of the fault plane is insufficient to cause a complete separation between the two permeable regions and a leak is possible, depending upon transmissibility in the vertical plane interface. The reservoir properties are the same on both sides; the linear interface is only acting like a restriction.

The typical influence of a semi-permeable fault is shown in Figure 5.33, the same dimensionless parameters as on the sealing fault example of Figure 5.1 are used.

- The response starts to follow the usual *infinite acting regime* and, at $t_D/C_D = 1000$, the derivative deviates above the 0.5 plateau and *tends to* 1, like for the no-flow barrier of Figure 5.1.
- When the communication starts through the fault, the derivative slowly *decreases* until a second infinite radial flow regime is established, and finally the response returns to the 0.5 stabilization line.

The influence of the semi-permeable fault is expressed by a temporary deviation above the 0.5 plateau. The duration of this hump is a function of the fault transmissibility: a long derivative transition at 1 suggests that the fault is almost sealing.

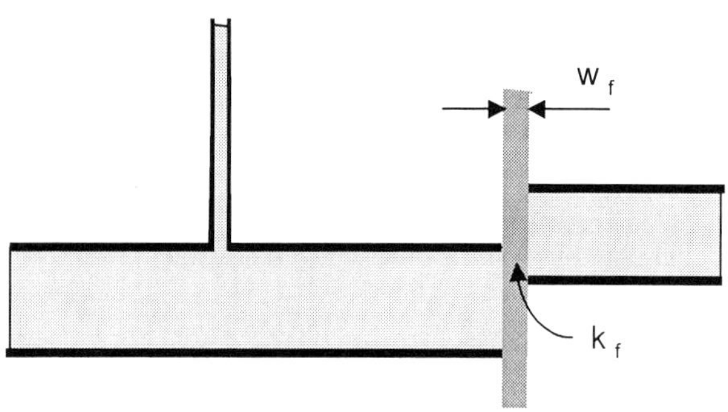

Figure 5.32. Physical model for a semi-permeable boundary.

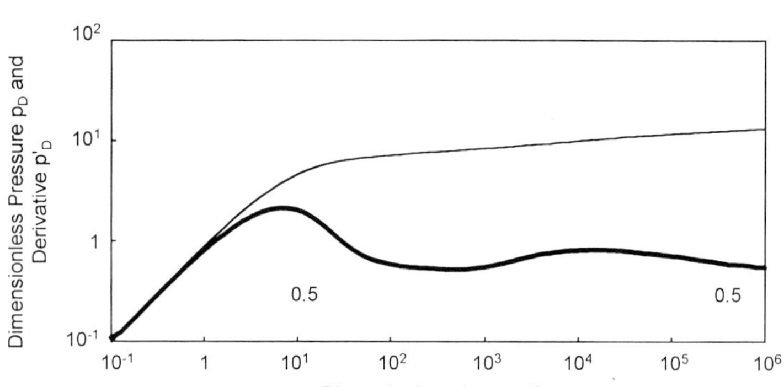

Figure 5.33. Pressure and derivative responses for a well with wellbore storage near a semi-permeable linear boundary. Log-log scales, p_D versus t_D/C_D. C_D=10000, S=0, L_D=5000, α=0.05.

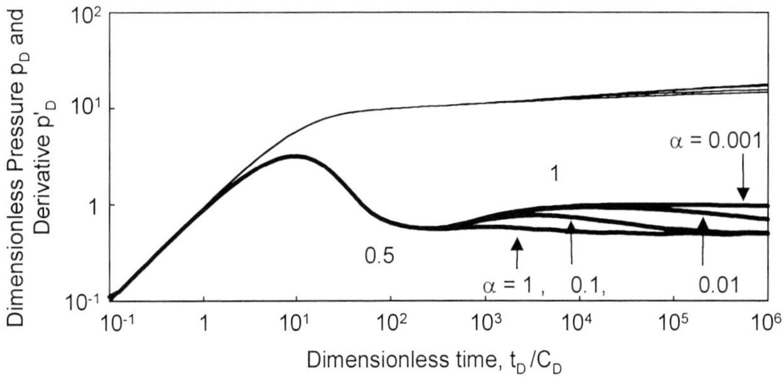

Figure 5.34. Pressure and derivative responses for a well with wellbore storage and skin near a semi-permeable linear boundary. Log-log scales, p_D versus t_D/C_D. $C_D = 100$, $S = 5$, $L_D = 300$. Several transmissibility ratios $\alpha = 1, 0.1, 0.01, 0.001$.

Semi-permeable fault model

The partially communicating fault solution presented in the following corresponds to the model of Yaxley (1985): the reservoir is homogeneous and the two regions have the same characteristics. The fault is at distance L from the well, its thickness w_f is small and the fault permeability is called k_f (Figure 5.32). Yaxley defines the dimensionless transmissibility ratio α as the *reciprocal of a skin*:

$$\alpha = \frac{k_f / w_f}{k / L} \tag{5.26}$$

With the linear composite solution presented in Section 4.3.3, Ambastha et al. (1987) have considered also the transmissibility in the interface between the two reservoir regions but they use the concept of an infinitesimal skin. When the reservoir properties are the same on both sides of the fault, their linear composite solution is equivalent to the semi-permeable fault model presented in this section.

Kuchuk and Habashy (1997) extended the model of Yaxley (1985) to a series of parallel partially communicating faults. In their examples of multiple communicating faults, the maximum of the hump above the 0.5 derivative stabilization can exceed the limit of 1 for a single fault. They also investigated a high permeability fault, allowing flow along the fault plane. They obtained results similar to Abbaszadeh and Cinco (1995) for a finite conductivity fault with no skin.

Behavior

In Figure 5.34, several transmissibility ratios α are considered for a semi-permeable fault located at L_D=300 from the well. With α=1, the linear discontinuity is hardly visible and the derivative response does not deviate much above the 0.5 line whereas with α=0.001 the fault appears almost impermeable, and the response is similar to the sealing boundary response L_D=300 of Figure 5.2 (the well parameters are the same with C_D=100 and S=5). The two intermediate values of transmissibility ratio, α=0.1 and 0.01, show the characteristic response of a leaky fault: a second radial flow regime is established at late time, and the derivative stabilizes twice on the 0.5 line.

On a semi-log scale, the two limit cases of Figure 5.34 both appear homogeneous with a single semi-log straight line of slope m for α=1 and, for α=0.001, a sealing fault response with a second line of slope $2m$ (Figure 5.35). With the two other examples, the semi-log curves show *two parallel straight lines*, but the second one is *above the first*. During the transition between the two parallel straight lines, the response follows the behavior of a sealing fault, and a temporary semi-log straight line of slope $2m$ is observed.

When the second radial flow regime is reached, the presence of a leaky fault in the reservoir simply produces an additional pressure drop, like a damage effect on the late time response. A similar semi-log response, with parallel straight lines separated by a vertical transition, can also be observed in radial composite systems (see Section 4.3.2, Figure 4.55), when the storativity decreases in the outer reservoir region. With the two solutions, the ability to flow is reduced at late time, even though the permeability is constant throughout the reservoir.

Drawdown and build-up analysis

A preliminary analysis of a semi-permeable fault response can be made with the sealing fault solution (Sections 5.1). The first infinite acting radial flow regime is analyzed on log-log and semi-log scales to provide the wellbore storage coefficient C, the permeability thickness product kh and the skin factor S. The time of the deviation at the end of the first semi-log regime is used to estimate the distance L to the semi-permeable fault, as for a no-flow barrier.

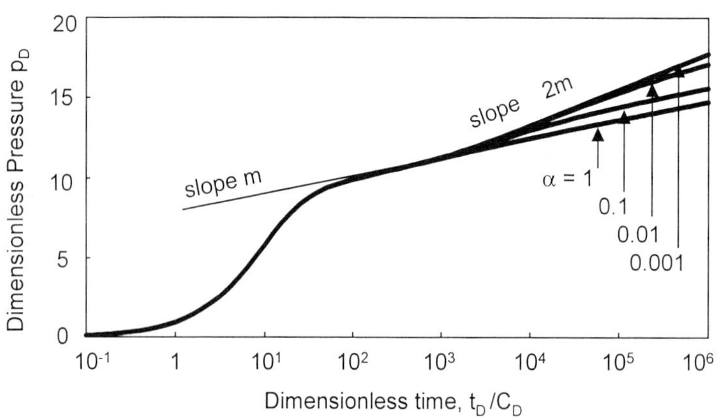

Figure 5.35. Semi-log plot of Figure 5.34 examples. p_D versus t_D/C_D.

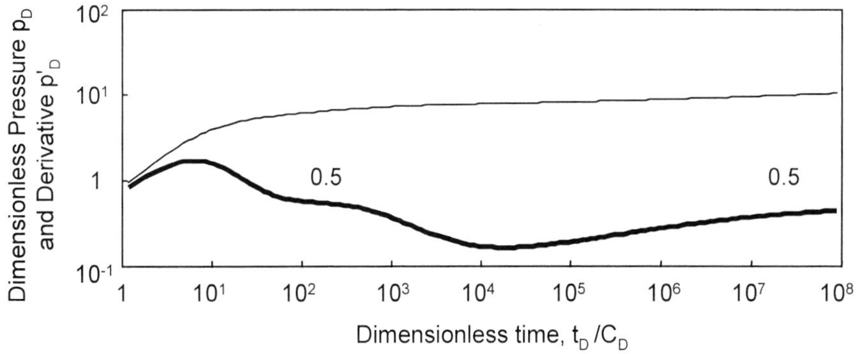

Figure 5.36. Pressure and derivative responses for a well with wellbore storage near a finite conductivity fault. No fault skin. Log-log scales, p_D versus t_D/C_D. $C_D = 10^3$, $S = 0$, $L_D = 1000$, $F_{cD} = 100$, $S_f = 0$.

The transmissibility ratio α is obtained from the transition between the two radial flow regimes. When the communication between the two reservoir regions is poor, the transition is long. On a log-log scale, the derivative hump can last more than two log-cycles, and frequently the second radial flow regime is not completely established at the end of the test period. If a complete response is available, the vertical distance between the two parallel straight lines of the semi-log plot increases as the transmissibility ratio is reduced.

For build-up responses, the extrapolated pressure of the first semi-log straight line is not used, p^* is evaluated by extrapolation of the *second line* to infinite shut-in time.

5.6.2 Finite conductivity fault

With the finite conductivity fault model, the permeability of the fault is larger than the reservoir permeability. Flow is allowed across and along the fault plane, and the fault enhances the drainage in the reservoir. In their original solution, Abbaszadeh and Cinco (1995) allow a change of mobility and storativity in the two reservoir regions. It is assumed in the following that the reservoir properties are the same on both sides of the fault.

The typical influence of a semi-permeable fault is shown in Figure 5.36 (with no fault skin S_f).

- The response starts to follow the usual *infinite acting regime*.
- When the finite conductivity fault is seen, the derivative drops along a straight line of *slope* -1. First, the fault provides a pressure support similar to a constant pressure linear boundary.
- Later, as the pressure drops in the fault, a flow is established in the thickness of the fault plane. As depicted Figure 5.37, this results in a *bi-linear flow regime*: one linear flow takes place in the reservoir when the fluid enters and exits the fault, the second

linear flow describes the flux in the fault thickness. In Figure 5.36, the derivative leaves the constant pressure negative unit slope straight line, and starts to increases along a 1/4 *slope* straight line.

- Finally, the derivative response returns to the 0.5 stabilization describing the infinite acting *radial flow* regime. At late time, the fault has no effect on the derivative response.

The bi-linear flow regime has been already discussed for well intercepting a finite conductivity vertical fracture (Sections 1.2.5 and 3.3). With the fractured well model, the reservoir flows into the fracture from the two sides of the vertical fracture plane, as opposed to the finite conductivity fault where flow both enter and exit the interface.

The effect of a finite conductivity fault is shown by a valley on the derivative curve. This shape appears similar to a double porosity response, but the two reservoir models describe different flow behavior. With the double porosity model, fluid flow through the fissure network and the derivative valley defines an increase of storativity from fissures to the total system. With finite conductivity fault model, radial flow is established in the matrix, before the conductive fault enhance the drainage in the reservoir. With the two solutions, the ability to flow is improved at late time, even though the permeability is constant throughout the reservoir.

Finite conductivity fault model

The fault thickness w_f is small, the fault permeability is called k_f and the distance from the well is L (Figure 5.37). Two dimensionless parameters are used to define the fault properties. The fault dimensionless conductivity F_{cD} describes the ability of flow in the fault plane:

$$F_{cD} = \frac{k_f w_f}{kL} \tag{5.27}$$

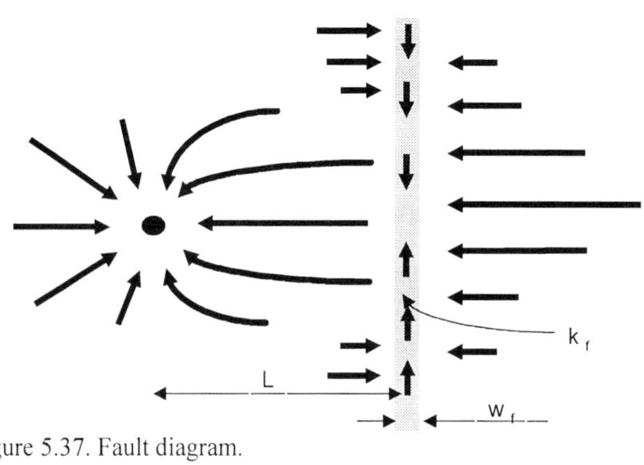

Figure 5.37. Fault diagram.

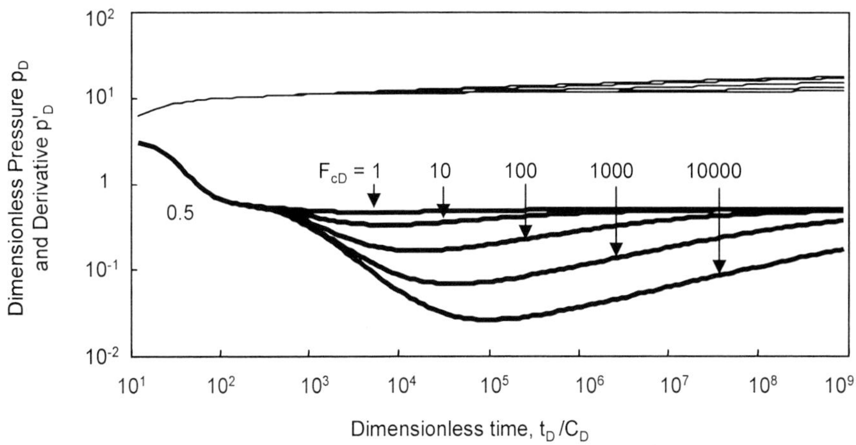

Figure 5.38. Pressure and derivative responses for a well with wellbore storage and skin near a finite conductivity fault. Log-log scales, p_D versus t_D/C_D. $C_D = 100$, $S = 5$, $L_D = 300$. No fault skin and several conductivity: $S_f = 0$, $F_{cD} = 1, 10, 100, 1000, 10000$.

The skin factor S_f describes the resistance to flow across the fault plane. In their model, Abbaszadeh and Cinco (1995) introduced a region of altered permeability k_a with an extension w_a around the fault. The dimensionless skin S_f is expressed as:

$$S_f = \frac{2\pi k}{L}\left(\frac{w_a}{k_a} + \frac{w_f}{2k_k}\right) \tag{5.28}$$

A zero fault conductivity F_{cD} corresponds to the semi-permeable fault solution of previous Section 5.6.1. The skin factor S_f is related to the transmissibility ratio a of Yaxley (1985) with:

$$\alpha = \frac{\pi}{S_f} \tag{5.29}$$

Behavior

In Figure 5.38, the fault has no skin S_f and several dimensionless conductivities F_{cD} are considered. When $F_{cD} < 0.1$, the pressure derivative stays on the 0.5 line and the fault has no effect on the well pressure response. With a low conductivity fault, there is no flow along the fault plane and the fluid transfer occurs only across the interface. The fault does not improve the drainage but, since no skin is assumed, the response is similar to the semi-permeable fault of Yaxley (1985) with $\alpha > 1$ (Figure 5.34).

For larger fault conductivity, the fault is acting like a drain and the derivative drops along a straight line of *slope* -1. The time of start of the apparent constant pressure behavior is a function of the fault distance L. The depth of the valley, and the location of the subsequent 1/4 *slope* straight line, defines the fault conductivity F_{cD}. For example with $F_{cD}=10^3$, the valley lasts 6 log cycles in Figure 5.38. This shows that, if the fault is reached after 1 hour, the second radial flow regime only starts after several years. In practice, during a standard test, the response is expected to end before the second radial flow regime.

On a plot of the pressure versus the fourth root of time, a straight line is present during the bi-linear flow regime, as shown by the dimensionless equation of Abbaszadeh and Cinco (1995):

$$p_D = C + \frac{2.45}{\sqrt{2F_{cD}\frac{\sqrt{M}+1}{\sqrt{M}}}} t_{DL}^{1/4} \tag{5.30}$$

Equation 5.31 is general; a change of mobility is possible on the two sides of the fault. When the reservoir properties are constant, the mobility ratio M, defined in Equation 4.78, is set to 1 in Equation 5.31. The constant C describes the flow restriction between the fault and the well, corresponding to the early time radial flow regime.

When a skin factor S_f reduces the ability to flow across the fault plane, the resistance to flow changes the response to that of a sealing boundary at the beginning of the fault influence. In Figure 5.39, the finite conductivity fault is affected by a skin S_f of 10, 100 and 1000. When the fault is reached, the derivative changes from the 0.5 stabilization to a plateau at 1, then it drops along a straight line of slope -1, reaches a minimum, follows the 1/4 slope straight line until the final radial flow regime is seen. The presence of a skin across the fault delays the time of the start of the apparent constant pressure behavior, with an intermediate sealing fault response. The straight line of slope -1 is delayed along the time scale, but the 1/4 slope straight line is the same for all skin factors.

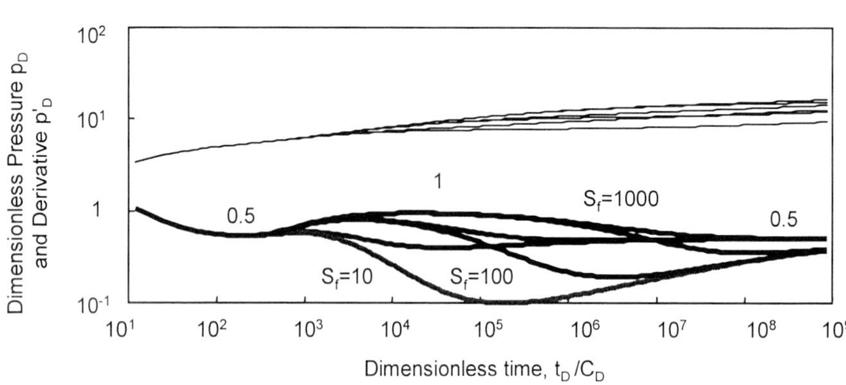

Figure 5.39. Pressure and derivative responses for a well with wellbore storage and skin near a finite conductivity fault. Log-log scales, p_D versus t_D/C_D. $C_D = 100$, $S = 5$, $L_D = 300$. Several fault skin and conductivity: $F_{cD} = 10, 1000$, $S_f = 10, 100, 1000$.

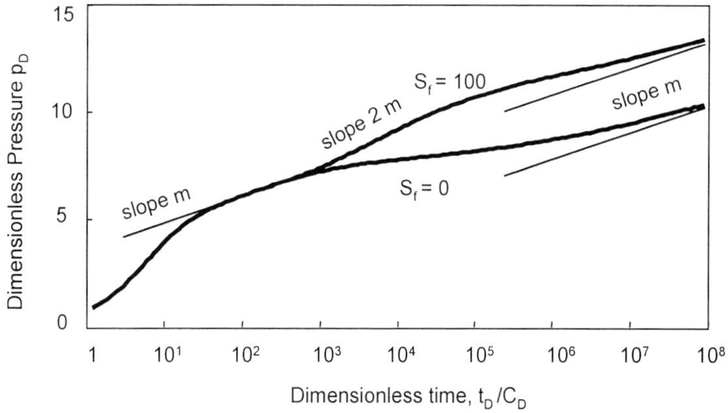

Figure 5.40. Semi-log plot for a well with wellbore storage near a finite conductivity fault. p_D versus t_D/C_D. $C_D = 10^3$, $S = 0$, $L_D = 1000$, $F_{cD} = 100$, $S_f = 0$ or 100.

On the semi-log plot Figure 5.40, the finite conductivity fault responses of Figure 5.36 with a skin S_f of 0 is compared to a similar fault configuration with a skin $S_f = 10^2$. When there is no fracture skin, the response follows the usual semi-log straight line of slope m (1.151 in dimensionless terms), flattens during the constant pressure behavior, increases during the bi-linear flow regime and, finally follows another line parallel to the first straight line. The degree of deviation below the semi-log straight line is more pronounced when the fault conductivity F_{cD} is large. When a fracture skin S_f is present, the semi-log straight line doubles (slope $2m$) before the sequence occurs of constant pressure, bi-linear flow and final radial flow regimes. The apparent flattening is delayed and the corresponding pressure change is larger than when there is no fault skin. The semi-log curves of Figure 5.40 shows that the effect of a fault skin increases the amplitude of the pressure response at late times.

Drawdown and build-up analysis

As for the semi-permeable fault, a preliminary analysis of a finite conductivity fault response can be made with a single fault solution. The analysis of the early time infinite acting regime provides the wellbore storage coefficient C, the permeability thickness product kh and the wellbore skin factor S. The time of the deviation at the end of the first semi-log regime is used to estimate the distance L to the finite conductivity fault. If the derivative drops immediately after the 0.5 stabilization, the fault has no skin S_f and a linear constant pressure boundary (Sections 5.5) is used. When the derivative increases to 1 before falling, the sealing fault solution (Sections 5.1) is used to estimate the distance L. A constant pressure boundary can then be used to match the drop of derivative. The resulting distance L_{cp} of the apparent constant pressure boundary can be used to evaluate the fault skin S_f:

$$\left(1 + S_f^2/2\pi^2\right) = \left(\frac{L_{cp}}{L}\right)^2 \tag{5.31}$$

When the bi-linear flow regime is present on the data, a plot of the pressure versus the fourth root of time exhibits a straight line whose slope, defined in Equation 5.31, is related to the fault conductivity and the mobility ratio. In field units,

$$\Delta p = C + 44.1 \frac{qB\mu}{h\sqrt[4]{\phi\mu c_t k}\sqrt{2k_f w_f \dfrac{\sqrt{M}+1}{\sqrt{M}}}} \sqrt[4]{\Delta t} \tag{5.32}$$

Equation 5.33 means that the finite conductivity fault shows the same behavior as a well intercepting a finite conductivity fracture with a flow restriction near the wellbore (Abbaszadeh and Cinco-Ley, 1995). By comparing with Equation 1.27, if the mobility ratio is assumed to be M=1, the fault behaves like a fracture of conductivity $4k_f w_f$.

As mentioned previously, the shape of the transition valley of a finite conductivity fault response is very long and actual test data generally end during the transition. The possibility to identify a change of mobility across the fault plane appears hypothetical, unless the fault skin and conductivity are low, and the system becomes equivalent to the linear composite solution of Section 4.3.3.

For build-up responses, the extrapolated pressure of the first semi-log straight line is not used. When a bi-linear flow regime has been reached during the flow and shut-in periods, the corresponding Horner or superposition times (see Equation 2.20) can be used to estimate the fracture conductivity (with the same slope as in the drawdown Equation 5.33) and p^*, by extrapolation of the *bi-linear flow straight line* to infinite shut-in time.

5.7 EFFECT OF BOUNDARIES IN DOUBLE POROSITY RESERVOIRS

When boundary effects are encountered in a double porosity reservoir, two types of transitions are present on the response. One describes the changes of flow geometry due to the boundaries, and the other corresponds to the double porosity transition (see Sections 4.1 to 4.4). Depending upon which transitions are seen first, boundary effects can be identified during the fissure flow, the double porosity transition, or during the total system flow. The number of possible pressure behaviors is large.

If the reservoir limits appear after the double porosity transition, the response is homogeneous during the total system flow regime and the effects of boundaries are the same as in a homogeneous reservoir, the different boundary models presented in the previous sections are directly applicable. Conversely, when the limits are reached during fissure flow or in transition regime, the double porosity nature of the system changes the response of boundaries. During the fissure flow regime, the storativity is only a fraction of the reservoir total storativity (see Section 4.1) and the diffusivity is high. The presence of limits appears earlier than in the case of homogeneous systems.

In the following sections, several examples are presented to illustrate the effect of boundaries during the fissure flow and the transition regimes of double porosity responses.

5.7.1 One sealing fault in double porosity reservoir, pseudo-steady state interporosity flow

In Figure 5.41 a double porosity response, pseudo-steady state interporosity flow, is influenced by a sealing fault during the fissure flow. The fault, at a distance $L_D = 5000$, is seen immediately after the wellbore storage effect and the derivative starts to increase as soon as the 0.5 line is reached at $t_D/C_D = 10^2$. The double porosity transition is then observed during the hemi-radial flow regime, and finally the response stabilizes at 1. The derivative curve presents *two valleys*: the first corresponds to the end of wellbore storage effect and the start of the fault influence; the second is the double porosity transition during the semi-infinite regime.

On this example, there is no infinite acting radial flow regime. The fault is seen at $t_D/C_D=10^2$ but, due to the double porosity transition, the semi-log straight line of slope $2m$ characteristic of the sealing fault starts after $t_D/C_D = 10^5$. A semi-log analysis of the hemi-radial flow regime can be performed only when the pressure data is available for more than three log-cycles after the end of wellbore storage.

In their study of the influence of a sealing fault in double porosity reservoir, Khachatoorian et al. (1995) indicate that a linear no-flow barrier shows during the fissure flow regime of the restricted interporosity flow model if the distance is less than a critical distance L_{cD}, defined as a function of λ. They propose:

$$L_{cD} = 0.54/\sqrt{\lambda} \tag{5.33}$$

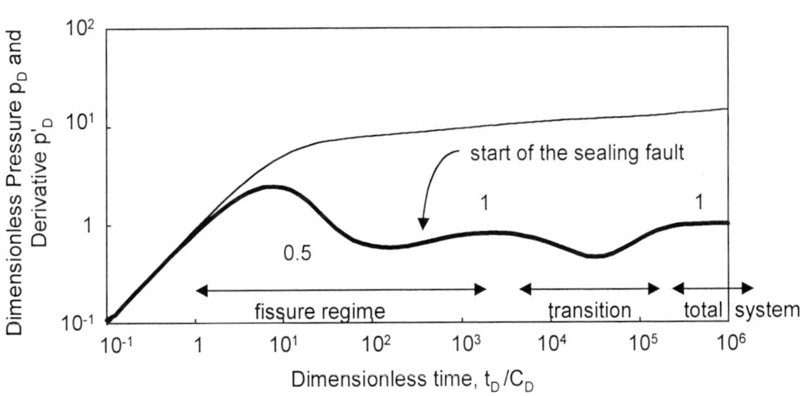

Figure 5.41. Pressure and derivative responses for a well with wellbore storage in a double porosity reservoir with a sealing fault, pseudo steady state interporosity flow.
Log-log scales, p_D versus t_D/C_D. $C_D = 104$, $S = 0$, $L_D = 5000$, $\omega = 0.2$, $\lambda = 10^{-9}$.

The critical distance L_{cD} is related to the radius of influence r_{iD} of the fissures, presented in the discussion of interference tests Section 6.4.2, Equation 6.12.

5.7.2 Parallel sealing faults in double porosity reservoir, unrestricted interporosity flow

Fluvial deposit sandstone reservoirs frequently provide examples of heterogeneous behavior with boundary effect. In river channels, the sediments are often sorted in strata, with the high permeability elements accumulated at the base of the deposit body. Depending upon the permeability contrast between the strata, channel reservoirs can show a double porosity, or a double permeability behavior. In the following, an example of double porosity channel response is presented for the unrestricted interporosity flow hypothesis. Double permeability channels are discussed in Section 5.8.2 and, in the Section 5.9.1 for composite reservoirs, a different configuration is considered: the sediments are sorted in strips parallel or perpendicular to the riverbed. Additional features, specific to channel deposits, are further reviewed in Section 5.10.1.

When the permeability of the deposit at the base of the channel sand is larger than in the higher tight sediment zone, Stewart (1997) reports that this two layers configuration can produce a *bi-linear flow* regime during the channel response. A horizontal linear flow occurs along the channel base, and a vertical linear flow is supported from the tight overlaying formation. In such a case, a *quarter slope* derivative straight line is observed on the log-log plot of the response, such as illustrated in Figure 5.42. A similar bi-linear flow behavior has been reported in the case of a horizontal well in a double porosity reservoir, if the unrestricted interporosity transition is seen when the flow is linear towards the well (Du and Stewart, 1992; see discussion Section 4.1.4).

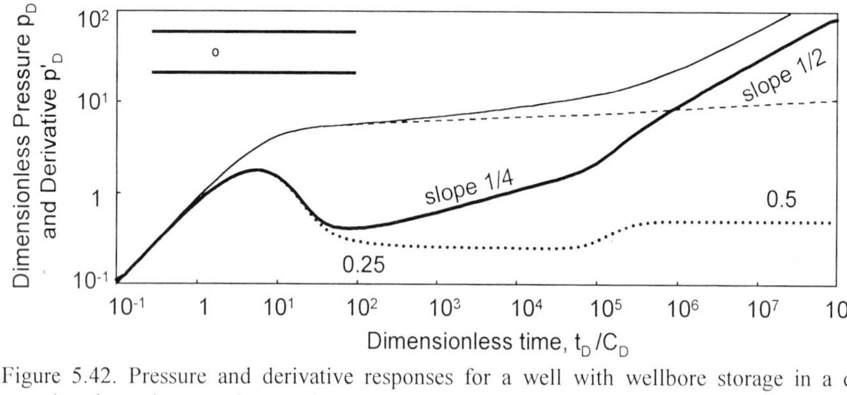

Figure 5.42. Pressure and derivative responses for a well with wellbore storage in a double porosity channel reservoir, transient interporosity flow, slab matrix blocks. The dotted curves correspond to the equivalent infinite double porosity reservoir.
Log-log scales, p_D versus t_D/C_D. $C_D = 10$, $S = 0$, $L_{1D} = L_{2D} = 300$, $\omega = 10^{-3}$, $\lambda = 10^{-6}$.

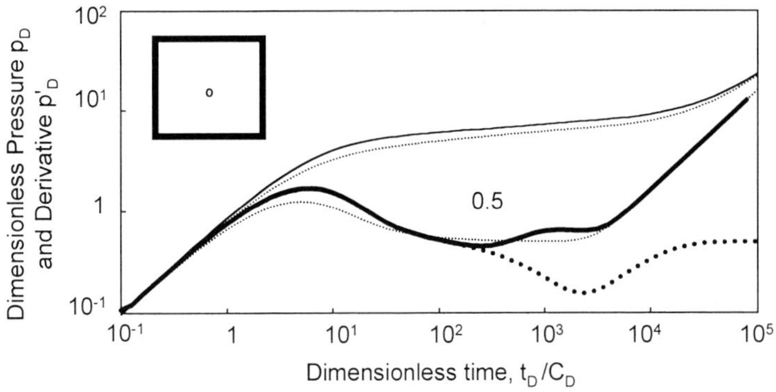

Figure 5.43. Pressure and derivative drawdown responses for a well with wellbore storage at the center of closed square double porosity reservoir, pseudo steady state interporosity flow. The thin dotted curves correspond to the equivalent homogeneous closed square reservoir. The infinite reservoir double porosity derivative response is presented by the thick dotted curve. Log-log scales, p_D versus t_D/C_D. $C_D = 100$, $S = 0$, $L_{iD} = 1000$, $\omega = 0.1$, $\lambda = 10^{-6}$.

5.7.3 Closed square double porosity reservoir, pseudo-steady state interporosity flow

On the drawdown example of Figure 5.43, the well is at the center of a closed square double porosity reservoir, pseudo-steady state interporosity flow. The thick dotted derivative curve describes the infinite double porosity reservoir response and the two thin dotted curves the equivalent homogeneous closed square reservoir.

No fissure radial flow regime can be identified on the response. After wellbore storage effects, the transition starts and the derivative drops below the 0.5 line but, quickly the boundaries are seen. As the derivative tends to increase with sealing boundaries, two opposite trends are superimposed during the double porosity transition, producing an *oscillation* on the derivative curve until the total system homogeneous behavior is reached. The response is then in a pseudo-steady state regime, and the derivative follows the characteristic unit slope straight line.

When the two transitions are superimposed on the response such as on the example of Figure 5.43, the diagnosis becomes difficult and frequently, if the data curve does not display the full response but stops during the transition, the match is not unique.

5.7.4 Square double porosity reservoir with composite boundaries, pseudo-steady state interporosity flow

In Figure 5.44, the well is near a corner of a square double porosity reservoir, pseudo-steady state interporosity flow. One of the farthest sides of the square is at a constant pressure, the three other ones are sealing. The two dotted curves describe the infinite double porosity reservoir response.

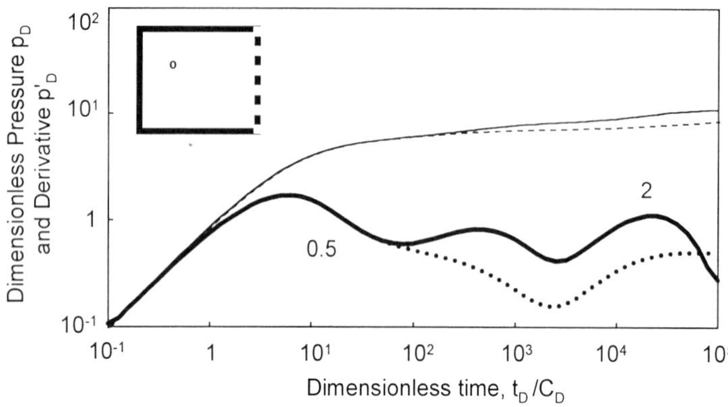

Figure 5.44. Pressure and derivative drawdown responses for a well with wellbore storage in a square double porosity reservoir with composite boundaries, pseudo steady state interporosity flow. The dotted curves correspond to the equivalent infinite double porosity reservoir.
Log-log scales, p_D versus t_D/C_D. $C_D = 100$, $S = 0$, $\omega = 0.1$, $\lambda = 10^{-6}$, $L_{1D} = L_{2D} = 500$ (sealing), $L_{3D} = 1500$ (constant pressure) and $L_{4D} = 1500$ (sealing).

On this example, the distance of the two first perpendicular sealing boundaries has been adjusted in order to produce a typical intersecting sealing fault behavior as soon as the 0.5 line is reached. The derivative increases, but the double porosity transition appears before the plateau at 2, and a second valley is observed on the response. Later, the derivative reaches the plateau at 2 and finally the constant pressure boundary produces the final drop of the derivative.

The theoretical example of Figure 5.44 is designed specifically to exhibit three consecutive *humps* on the derivative response. No radial flow is present, even though the complete curve is displayed over six log-cycles. The example is not representative of usual well pressure responses but it illustrates how the individual behaviors of the different components of the model are superimposed in the derivative response curve. As discussed in Section 5.11, by knowing the sequence of flow regimes, it is possible to predict the shape of a response such as Figure 5.44 and, reciprocally, from the observation of the shape it is possible to propose one (or several) sequence of flow regimes.

On complex heterogeneous responses, only the derivative presentation is appropriate for interpretation. Log-log derivative analysis allows the identification of the complete sequence of typical behaviors, whereas the straight-line methods are of no use.

5.7.5 Field example

On the two-day build-up test in a fissured reservoir shown in Figure 5.45, the derivative describes two humps. Several models can be used to match this data, considering that no constant pressure boundary effect is expected on the well response.

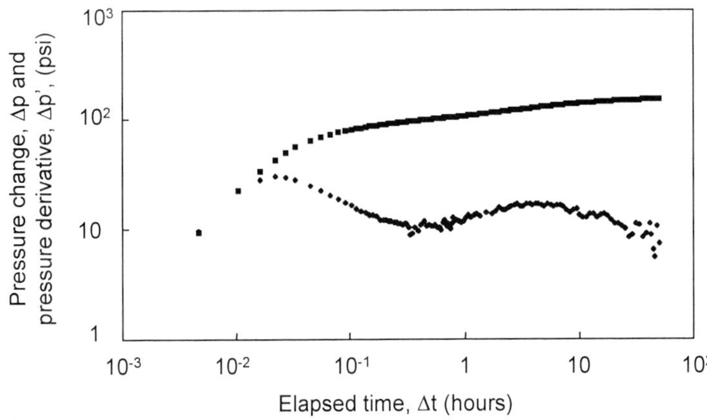

Figure 5.45. Build-up test for a well in a fissured reservoir. Log-log scales.

A triple porosity solution, similar to that in Figure 4.33, is applicable to describe the test response. A first transition valley is observed at 0.5 hours, and the test stops during the second transition, before the bottom of the valley has been reached. The derivative stabilization for radial flow is located on the top of the second hump, at 5 hours.

A double porosity model with a sealing fault can also be envisaged. The sequence of regimes is the same as in Figure 5.41, the test is too short to completely describe the double porosity transition valley. The permeability is estimated by locating the derivative stabilization on the bottom of the first valley at 0.5 hours. Since the storativity ratio ω is not defined from the truncated transition valley, the hydraulic diffusivity during the fissure flow is not known, and therefore the fault distance cannot be evaluated accurately.

5.8 EFFECT OF BOUNDARIES IN DOUBLE PERMEABILITY RESERVOIRS

With double permeability responses, two layers, each with different flow capacity are producing, first independently and later, when the reservoir cross-flow is established, as a single equivalent homogeneous layer (see Sections 4.2.2). If boundaries are reached before the final homogeneous behavior, they show first in the high diffusivity layer.

In the following sections, three examples of boundary effects in double permeability reservoir are presented. One sealing fault, two parallel sealing faults and a closed system are considered when the boundaries are seen during the two layer no cross flow regime. The sealing fault solutions are generated with a modified image well method for multi-layers reservoirs, as described by Larsen (1989).

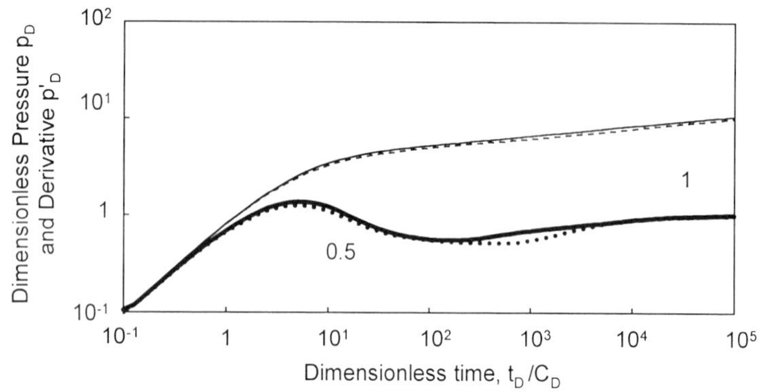

Figure 5.46. Pressure and derivative responses for a well with wellbore storage in a double permeability reservoir with a sealing fault. The dotted curves describe the sealing fault response in the equivalent homogeneous reservoir. Log-log scales, p_D versus t_D/C_D.
$C_D = 100$, $S_1=S_2 = 0$, $L_D = 500$, $\omega = 0.15$, $\kappa = 0.7$, $\lambda = 10^{-10}$.

5.8.1 One sealing fault in double permeability reservoir

Figure 5.46 presents the effect of a sealing fault in a double permeability reservoir. The sealing fault is at distance L_D=500 from the well, the interlayer cross flow parameter is low $\lambda = 10^{-10}$ and the reservoir cross flow is not established when the fault is seen. The diffusivity of Layer 1 (ω=0.15, κ=0.7) is higher than in layer 2, the fault is first reached in this layer 1 and the derivative response starts to deviate shortly after $t_D/C_D=10^2$. As long as the second layer has not reached the fault influence, the derivative tends to show an *intermediate plateau* between 0.5 and 1, describing the transition period where one layer is in a hemi-infinite regime and the other still has an infinite behavior. The plateau at 1 is reached when the two layers are influenced by the fault. On the equivalent homogeneous response described by the dotted curve, the start of the fault influence is seen almost ten times later, at $t_D/C_D = 10^3$.

The sealing fault example of Figure 5.46 shows that the analysis of a boundary effect in a layered reservoir is difficult:

- On a log-log scale, if the layered nature of the system is ignored and the start of the fault influence is analyzed with the homogeneous solution, the resulting calculated distance is *under-estimated* (one third of the true fault distance on the example in Figure 5.46).
- If a layered reservoir model is used, the time of start of the fault influence, and the shape of the transition between the derivative plateaus at 0.5 and 1, are dependent upon the refinement used for the layers description (Larsen, 1989). Increasing the number of layers delays the onset of boundary effect in the high permeability layer.
- On semi-log scales, the point of intersection between the early and late semi-log straight lines is a function not only of the fault distance, but also of the layered reservoir behavior. If the first semi-log straight line for the infinite reservoir response is drawn during the commingled regime and the second line, after the sealing fault has been reached, during the equivalent homogeneous total system response, the distance

between the well and the sealing fault can be *over-estimated* when the Equation 1.33 for a homogeneous reservoir is used. During the commingled regime, the apparent semi-log straight line is affected by the layer pseudo-skin factor of Equation 4.74 (Section 4.2.6), and the time of intercept Δt_x between the two straight lines is delayed compared to the homogeneous reservoir response.

5.8.2 Two parallel sealing faults in double permeability reservoir

Figure 5.47 presents the effect of parallel sealing faults in the same double permeability reservoir as that presented in the previous example Figure 5.46 (ω=0.15, κ=0.7 and $\lambda=10^{-10}$). The well is centered in the channel with $L_{1D}=L_{2D}=1000$. The equivalent homogeneous channel response is described by the dotted curve.

At late time, the double permeability curve apparently follows a half unit slope straight-line characteristic of linear flow, but it is earlier than the equivalent homogeneous response. When channel sand reservoirs are made of several layers of different flow capacity, the width can be *under-estimated* if a homogeneous model is used for analysis (Larsen, 1989).

5.8.3 Closed circle double permeability reservoir

Figure 5.48 presents a double permeability drawdown response for a well at the center of a closed circle. The solution for circular layered reservoir without cross flow was presented in 1978 by Tariq and Ramey, (1978). In 1986, Joseph et al. discussed the two layer double permeability closed circle solution.

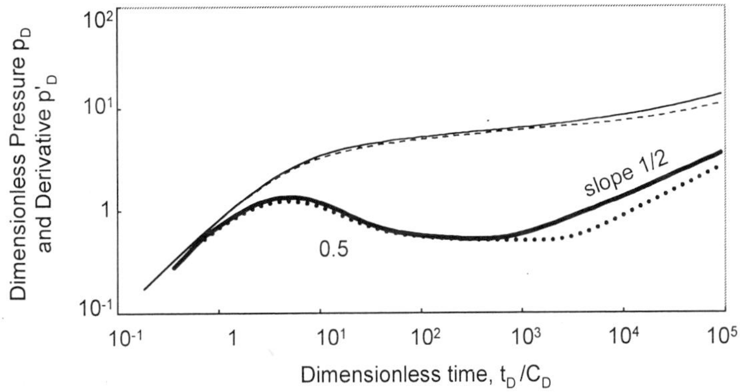

Figure 5.47. Pressure and derivative responses for a well with wellbore storage in a double permeability reservoir with two parallel sealing faults. The dotted curves describe to the channel response of the equivalent homogeneous reservoir. Log-log scales, p_D versus t_D/C_D.
$C_D = 100$, $S_1=S_2=0$, $L_{1D}=L_{2D}=1000$, $\omega=0.15$, $\kappa=0.7$, $\lambda=10^{-10}$.

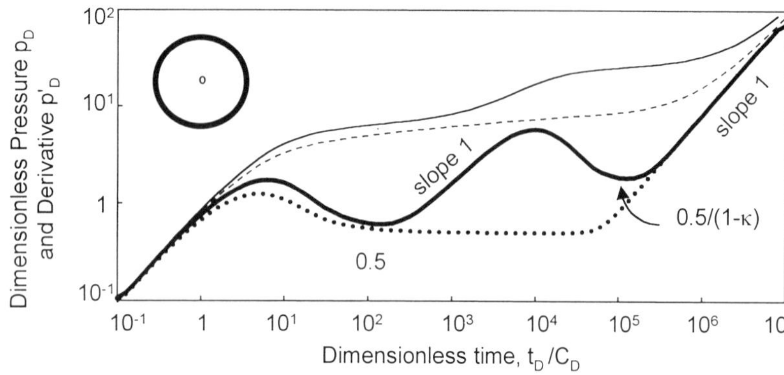

Figure 5.48. Pressure and derivative responses for a well with wellbore storage in a closed circle double permeability reservoir. The dotted curves correspond to the closed equivalent homogeneous reservoir. Log-log scales, p_D versus t_D/C_D.
$C_D = 100$, $S_1=S_2 = 0$, $r_D = 5000$, $\omega = 0.002$, $\kappa = 0.7$, $\lambda = 10^{-10}$.

As on the previous example, the interlayer cross flow parameter is $\lambda=10^{-10}$ and the double permeability response is still in the "two layers no-cross flow" regime when the closed boundary is reached at $r_D=5000$. After the early time wellbore storage hump, the derivative tends to *stabilize on* 0.5 as for an infinite commingled system but, after $t_D/C_D=10^2$, the closed boundary starts to be seen. A *second unit slope* straight line is observed, followed also by a *hump* and a new *derivative plateau*. Finally, a *third unit slope* line defines the pseudo steady state regime later. The dotted curve in Figure 5.48 describes the response of the equivalent homogeneous reservoir with the same boundary.

The shape of the second derivative hump is similar to a storage response, such as a wellbore storage displaced towards late time, or a pseudo steady state regime displaced towards early time. When the closed circular boundary is reached in layer 1 (ω=0.002, κ=0.7), this high diffusivity layer starts to be depleted while layer 2 is still in the infinite acting regime. The response then shows a storage behavior at intermediate time, but this storage corresponds to the layer 1 pore volume, a fraction of the total system volume defined by the storativity ratio ω. At the end of the layer 1 storage effect, the radial flow in layer 2 is seen (Joseph et al., 1986; Gao et al., 1994), and the derivative stabilizes at $0.5/(1 - \kappa)$, before the third unit slope straight line for the pseudo steady state regime. The location of the last unit slope straight line is a function of the *reservoir storage* $\phi c_t hA$ (A/r_w^2 in dimensionless terms, see Section 5.4.2), that of the intermediate line to the *layer* 1 *storage* ($\omega A/r_w^2$ in dimensionless terms).

For shut in periods, build-up curves show the same pattern as in Figure 5.48 until the start of the pseudo-steady state regime. As discussed in Section 5.4.7, drawdown and build-up responses diverge at late time, the build-up pressure stabilizes at the average pressure, and the derivative curve drops towards zero. The prediction of the average reservoir pressure from build-up analysis is a difficult question, in particular when non-communicating layers have different skin factors (Chen et al., 1993). If the recorded shut-in data curve is not complete and stops during the intermediate unit slope straight line, the build-up response can suggest an inexplicable pseudo-steady state behavior, when it is merely a reservoir storage effect (Joseph et al., 1986).

In Figure 5.48, the first and the last unit slope straight lines, corresponding respectively to wellbore storage and to pseudo-steady state, are separated by more than five log-cycles on the time scale. In such a reservoir configuration, a complete drawdown or build-up response requires unpractical long tests, and the analysis therefore can only be made on part of the total response.

A similar behavior has been discussed by Anisur Rahman and Ambastha (1997) for reservoirs compartmentalized in both vertical and horizontal directions. In their example, the well penetrates only one closed body but, at some distance, a communication is established with another closed compartment. Even though the flow geometry is different from that of the double permeability example of Figure 5.48, the same two pseudo-steady state straight lines can be seen on the response (see Section 5.10.1).

Layered reservoirs can also combine infinite and closed intervals. When the permeability of the closed body is large enough, the derivative response can show a hump as in Figure 5.48 but, afterwards, it stabilizes at $0.5/(1 - \kappa)$ when the radial flow regime in the infinite layer dominates. This configuration, producing a unit slope straight line at intermediate time, is a limiting case of the radial composite double permeability behavior illustrated in Figure 4.67 of Section 4.4.3. More generally, a unit slope straight line on drawdown and build-up responses is characteristic of commingled layered bounded reservoirs or radial composite systems (see Section 4.3.2 and discussion Section 5.9.2).

5.8.4 Field example

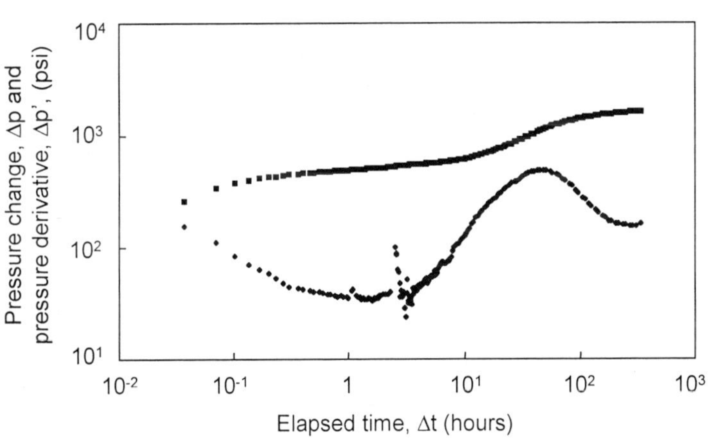

Figure 5.49. Build-up test in a multi layer reservoir. Log-log scales.

The build-up test of Figure 5.49 has been recorded in a well intercepting several layers. Some layers are small lenses of limited extension, while others show an infinite acting behavior. The resulting derivative response is very characteristic of a bounded layered system.

5.9 EFFECT OF BOUNDARIES IN COMPOSITE RESERVOIRS

In composite reservoirs with boundaries, two types of transitions are also superposed on the response. One describes the heterogeneous reservoir behavior (radial or linear composite, see Section 4.3), and the other the change from the radial to a limited flow geometry. Multiple combinations are possible, depending upon the distance of the composite reservoir interfaces, and that of the boundaries.

In the following sections, a composite channel configuration is considered for illustration of boundary effects in reservoir with changing properties. The two geometries considered by Bourgeois et al. (1996) are envisaged. The interfaces are parallel to the riverbed when the sediment properties change in the vicinity of the shorelines, producing a change of mobility from the center to the edge of the channel. With perpendicular interfaces, the properties change along the channel extension.

5.9.1 Linear composite channel

In Figures 5.50 and 5.51, the well is centered in a channel defined with $L_{1D}=L_{2D}=1000$ as in Figures 5.47. The mobility is not constant, the interfaces of the composite channel are parallel to the boundaries in Figure 5.50 and perpendicular in Figure 5.51. The mobility k/μ of the outer region is either reduced ($M=5$) or increased ($M=0.2$) by a factor of 5, the curve $M=1$ describes the homogeneous channel. The mobility ratio M is defined in Equation 4.78, and the distance of the two interfaces d_{1D} and d_{2D} are defined by Equation 1.34.

In Figure 5.50, the mobility changes near the edges of the channel. The distance to the interfaces is $d_{1D}=d_{2D}=500$: a central channel region of width 1000 is surrounded by two border channels of width 500. The three curves exhibit a half unit slope straight line at late time, but they are not superposed. When the mobility is reduced ($M=5$), the response is equivalent to a channel of *smaller* width. Conversely, when the mobility is increased ($M=0.2$), the derivative drops below the 0.5 stabilization and finally reaches a half unit slope straight line corresponding to a *larger* channel.

In the specialized plot of Δp versus $\sqrt{\Delta t}$, Bourgeois et al. (1996) found that, in the Equation 5.5 for the slope of the channel flow straight-line m_{ch}, the permeability must be weighted with the widths of each zone.

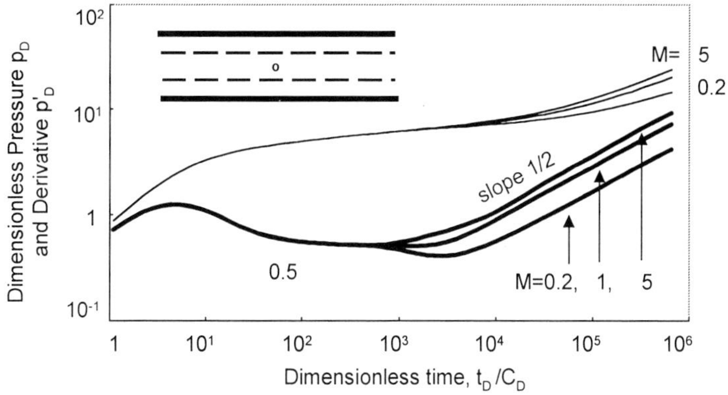

Figure 5.50. Pressure and derivative responses for a well with wellbore storage in a composite channel. The interfaces are parallel to the boundaries. Log-log scales, p_D versus t_D/C_D.
$C_D = 100$, $S = 0$, $L_{1D} = L_{2D} = 1000$, $d_{1D} = d_{2D} = 500$, $M_1 = M_2 = 0.2$, 1 and 5.

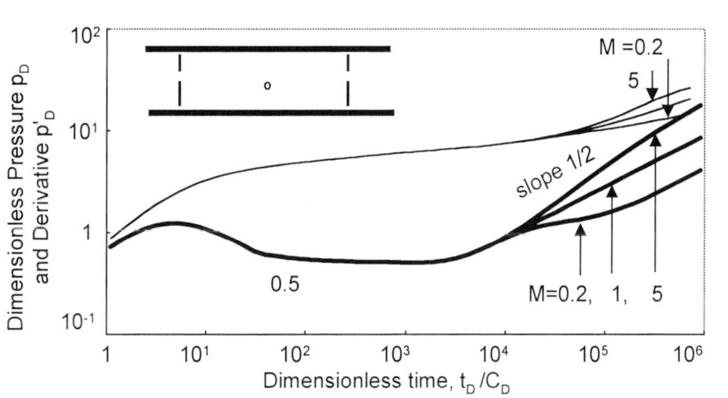

Figure 5.51. Pressure and derivative responses for a well with wellbore storage in a composite channel. The interfaces are perpendicular to the boundaries. Log-log scales, p_D versus t_D/C_D.
$C_D = 100$, $S = 0$, $L_{1D} = L_{2D} = 1000$, $d_{1D} = d_{2D} = 2000$, $M_1 = M_2 = 0.2$, 1 and 5.

When the width of the border channels become large compared to that of a center channel region of high mobility, the behavior is equivalent to a linear composite model (see Section 4.3.3) before the sealing boundaries are reached. The derivative follows two stabilizations before the late time half unit slope straight line and, at transition time, it can show an increasing trend with a slope smaller than 1/2 (Bourgeois et al., 1996).

In Figure 5.51, the mobility changes along the channel length in both directions. The distance of the interfaces are d_{1D}=d_{2D}=2000. The responses first show the typical channel behavior corresponding to the well region and, when the two interfaces are reached during the linear flow regime, the derivative curves deviate to follow a second half unit slope straight line. As in the previous example of Figure 5.50, when the mobility is reduced (M=5), the response becomes equivalent to a channel of smaller width and the second linear flow straight line is displaced upwards. This configuration can be compared to curve **D** of Figure 5.20, when the channel is limited in one direction (see Section 5.4.5), or to the double permeability channel response of Figure 5.47.

When the mobility is increased (M=0.2), the derivative flattens and then reaches a half unit slope straight line corresponding to a larger channel.

When the contrast of mobility is large, the composite channel response tends towards one of the two limiting cases of composite reservoirs, namely the closed channel and the channel with constant pressure boundaries. In Figure 5.52, the mobility changes by a factor of 50 along the channel length (M=0.02 and M=50). When the mobility is reduced (M=50), the transition between the two linear flow straight lines follows a unit slope straight line. As shown by the dotted derivative curve, this unit slope straight line corresponds to a pseudo steady state regime in a channel closed on the two interfaces at d_{1D} and d_{2D}. After the initial derivative stabilization, the response can be described with three straight-line segments, of slopes respectively 1/2, 1 and 1/2. Conversely, with a large increase of mobility (M=0.02), the behavior tends towards that of constant pressure boundaries during the transition. In the corresponding dotted derivative curve, the interfaces are changed into constant pressure boundaries. With build-up data, the effect of the time superposition can introduce a distortion on the response. Build-up analysis in composite reservoirs is discussed in the next section with the curve M=50 of Figure 5.52.

If a semi-permeable boundary closes the channel, the derivative response can exhibit a hump above the half unit slope straight line. When the reservoir properties are not changed on both sides of the fault, the same linear flow straight line describes the response before and after the fault influence. If the transmissibility ratio a is small and the fault appears to be sealing (α =0.001 or less), the hump can describe the response of a channel closed at one end (such as curve **D** of Figure 5.20), before returning to the half unit slope straight line.

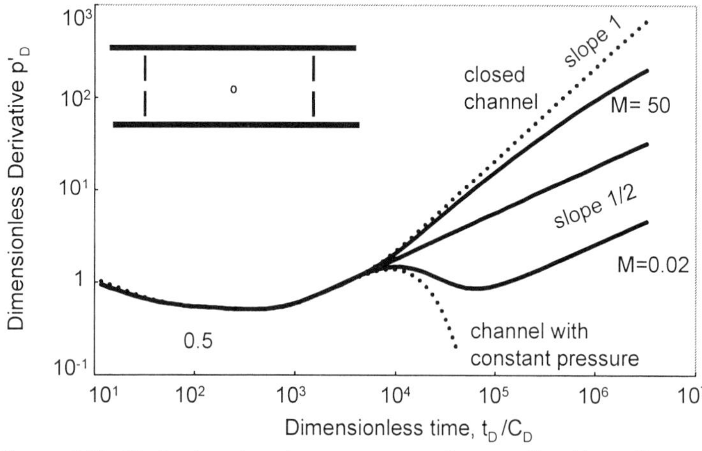

Figure 5.52. Derivative drawdown responses for a well with wellbore storage in composite channel. The interfaces are perpendicular to the boundaries. On the dotted curves, the interfaces are changed into sealing and constant pressure boundaries. Log-log scales, p_D versus t_D/C_D. C_D = 100, S = 0, L_{1D} = L_{2D} = 500, d_{1D} = d_{2D}= 1500, M_1 = M_2 =0.02, 1 and 50.

In the examples Figures 5.50 to 5.52, a simple symmetric geometry is assumed. With practical field configurations, the geometry is in general complex and, even when the match is made by computer, the results of analysis are frequently not unique. Composite channel responses are further discussed in Section 5.10.1.

5.9.2 Apparent boundary effects in composite reservoirs

Composite reservoirs with a reduction of mobility some distance away from the well can be interpreted by mistake with sealing boundary models. A no flow approximation may be acceptable to describe the response during the duration of a test but, when the sealing boundary model is used to predict longer well behavior, the result could lead to a pessimistic forecast.

For example, in the discussion of the radial composite model, it is shown in Section 4.3.2, Figure 4.57, that when the mobility of the outer region is greatly reduced, a build-up period can exhibit the characteristic response of a depleted closed reservoir under certain circumstances. After radial flow, the derivative starts falling similar to the example of Figure 5.17. Later, the infinite outer region re-compresses the depleted region, and the derivative increases to finally reach a stabilization corresponding to the reduced external mobility.

Chen et al. (1996) report similar behavior in gas reservoirs with edge-water drive. They state that, because of the large mobility contrast between gas and water, the water behaves first like an impermeable medium. Drawdown responses exhibit a unit slope straight line after the initial radial flow regime. During shut-in periods, the pressure derivative, such as in the radial composite example of Figure 4.57,demonstrates the sequence of stabilization, valley and final unit slope straight line.

The apparent *closed reservoir* behavior is illustrated as follows with a build-up example in a limited composite reservoir. In Figure 5.53, a build-up response of the example M=50 of Figure 5.52 is compared to the original drawdown response (presented with dotted curves). The well is closed at t_{pD}/C_D = 650, before the start of the two parallel boundaries influence. The derivative response shows several oscillations: after the radial flow stabilization, the build-up derivative follows a first valley, similar to curve **C** of Figure 5.9 and the 1/2 slope channel response is delayed. Later, the two composite interfaces are reached and the derivative shows a second decreasing trend, until the influence of the outer channel regions is seen and the derivative starts increasing again, to ultimately reach a 1/2 slope straight line. With such response, if the build-up data stops at $t_D/C_D = 10^4$ when the derivative is dropping, a log-log analysis with a depleted closed reservoir model would be possible. If the model can be checked over a longer time, such as in a test simulation plot with a long production history prior to shut-in, the hypothesis of a closed reservoir model will be inconsistent with the data (see Section 10.2.3).

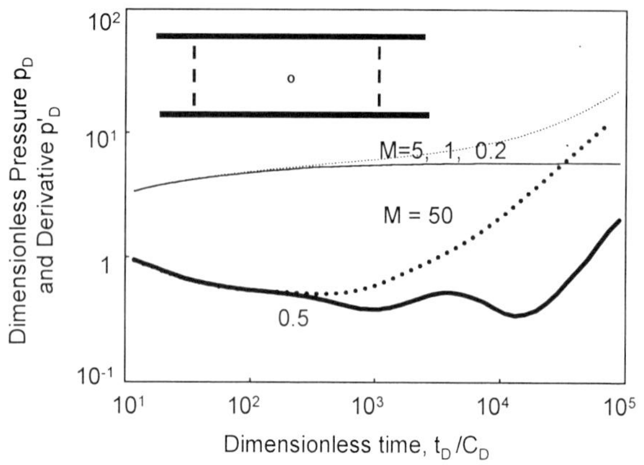

Figure 5.53. Pressure and derivative responses of curve M =50 of Figure 5.52. The two dotted derivative curves are drawdown, the build-up response (thick line) is generated for $(t_p/C)_D = 650$. Log-log scales, p_D versus t_D/C_D.

5.10 OTHER BOUNDARY CONFIGURATIONS

5.10.1 Channel reservoirs

In the previous sections, several extensions of the basic two parallel sealing fault model of Section 5.2 have been proposed for the analysis of channel reservoir responses. The behavior of a closed channel has been discussed in Section 5.4.5, double porosity, double permeability and linear composite channel models have been presented in Sections 5.7.2, 5.8.2 and 5.9. Yet, the geological description of fluvial reservoirs can be more complex than the simple assumptions used with these models. In the following sections, specific features encountered channel reservoirs are discussed in relation to transient test interpretation.

The constant thickness hypothesis is not always valid for fluvial reservoir. Frequently, the channel section is not rectangular but the shape of the sand body is parabolic (Stewart, 1997). The resulting response tends towards the linear composite channel behavior described in Figure 5.50. When the well is located in the maximum thickness region, the response corresponds to a reduction of mobility $M>1$. Conversely, when the well is close to one of the boundary, the thickness increases before the start of the linear flow regime and the channel response can show a drop of derivative before the linear flow regime, as in the curve $M<1$ of Figure 5.50.

Larsen (1993, 1996 b) investigated intersecting channel reservoirs. When two independent channels intersect, the response describes first the channel corresponding to the well location and, when the intersection is reached, a larger area contributes to the flow. By using an equivalent linear channel model, the response can be described as a stepwise increase of the channel section. The derivative curve shows two half unit slope straight lines, similar to the curve $M<1$ of the linear composite channel example of Figure 5.51.

When more than two independent channels intersect, the response shows an increase of flow area each time a new channel is reached. When a large number of channels make a connected network, the response tends to a pseudo radial flow regime and the derivative ultimately stabilizes. This flow configuration becomes similar to the fissure regime of a double porosity reservoir. Assuming the infinite channel network can be described with a regular square grid network, the pseudo radial flow regime in the channel network can be approximated by the response of a reservoir with two intersecting sealing faults. The author concludes that the angle is determined by the ratio of sand volume to bulk volume of the pattern involved.

With self-intersecting linear reservoirs, the infinite channel makes a loop. The well can be located in one of the two infinite segments, or in the loop. The loop influence can be described as an increase of the channel section, some distance away from the well and on a limited extension along an equivalent rectilinear channel. The derivative curve deviates temporarily below the half unit slope straight line corresponding to linear flow. The amplitude and the duration of this valley shaped transition is a function of the distance from the well to the channel intersection point, and to the length of the loop. At late time, the derivative returns on the early time half unit slope straight line and the loop has no effect any more on the derivative response.

The case of stacked channels has been discussed by Anisur Rahman and Ambastha (1997). They considered two intersecting closed channels when the well is located in the smallest channel sand body. A skin factor is used to define the communication at the interface between the two channels. When the flow restriction at the interface is large, a response similar to the double permeability closed reservoir example of Figure 5.48 is obtained. A first pseudo steady state regime is seen for the small volume communicating with the well and, later, a second pseudo steady state regime describes the depletion of the complete system.

5.10.2 Complex boundary systems

The identification of complex boundary systems has been the subject of many recent publications. Two examples of new approaches are briefly discussed in the following section. One aims at automatic imaging of the boundary system, the other discuss the use and limitations of numerical simulations for boundary interpretation.

Slevinsky (1997) proposes an original method for the identification of an arbitrary reservoir boundary system. The increase of radius of investigation in the reservoir is viewed as a transient wave phenomenon. In the case of a single sealing fault for example, after the circular front has reached the fault, it is reflected back towards the well (as shown in Figure 1.22). Since the radius of investigation is a function of the square root of the elapsed time, the reflected wave reaches the well four times later than when the first contact with the boundary occurred (see Section 10.3.3). With this description of the wave expansion, Slevinsky defines an "angle of view" of the boundary, which asymptotically approaches 180 degrees as radius expands.

By analyzing the "angle of view", the distance and shape of any system of boundaries can be determined. A first match of the early time test data is used to define the infinite acting reservoir model. A "boundary ratio type curve", built by dividing the actual test response by the non-boundary predicted response, defines the variation of "angle of view" as a function of time. This is used to build a diagnostic image, similar to a sonar display, of the boundary system. Indeed, the results of analysis are dependent upon the accuracy of the initial match.

For complex geological shapes, Rahon et al. (1997) proposed to use an optimization algorithm with a classical numerical simulation to match pressure data. Using a synthetic channel example with a parabolic section, they demonstrate that the algorithm converges towards the correct section shape. The sensitivity of the method to the assumed parameters used for numerical simulation, and to noise of real data is not documented.

With a numerical approach, a high degree of refinement in the results is reached to the detriment of the uniqueness of the inverse problem solution. The introduction of geostatic techniques can reduce the range of uncertainty. Schildberg et al. (1997) propose to integrate the numerical simulation of well test data as a constraint for geostatic modeling. In their study of a heterogeneous reservoir example, they use analytical simulations for a quick test of different hypothesis, and numerical simulations associated to an optimization algorithm. They conclude that, even though the results are not unique, the approach provides new information in terms of size of the reservoir bodies.

5.10.3 Effect of a gas cap or bottom water drive

In this section, a different type of boundary is considered. When the upper or lower limit of the reservoir is not sealing but at a constant pressure, such as in the case of a gas cap or a strong lower water drive, the effect of boundary is a function of the vertical permeability, not the radial permeability. The effect of a constant pressure upper or lower limit has been presented in Chapter 3 for wells in partial penetration and horizontal wells. Results are summarized in the following section for completeness of the boundary discussion.

When the well intercepts a gas cap or an aquifer, limited-entry completion or horizontal well techniques are used to prevent gas or water production. The well is opened in the oil interval away from the supporting gas or water zone and, as long as coning is not established, the high mobility of the adjacent zone maintains a fairly constant pressure at the interface with the oil-bearing interval. The resulting pressure behavior then becomes similar to a constant pressure boundary in the horizontal plane (see Section 5.5): the pressure stabilizes and the derivative drops.

A solution for limited entry wells influenced by a gas cap has been presented by Streltsova (1981) and Kuchuk derived in 1990 a solution for horizontal wells with gas a cap or aquifer. Abbaszadeh and Hegeman (1988) presented a general study of pressure and derivative responses for different well configurations under the influence of a constant pressure support, comparing the responses to the usual no-flow upper and lower limits.

In the case of a limited entry well, after a first radial flow corresponding to the opened interval h_w, a spherical flow regime is established until the upper and lower limits are reached (see Section 3.4.7, Figure 3.21). When the two limits are impermeable, a second radial flow develops in the complete reservoir thickness but, if one limit is at constant pressure, the pressure response stabilizes and the derivative drops towards zero at the end of the spherical flow regime.

For horizontal wells, the first reservoir flow regime is a vertical radial flow, followed by a linear flow when the upper and lower limits are sealing. If one of the two limits is at constant pressure, as soon as this boundary is reached the pressure becomes constant and the derivative drops (see Section 3.6.11, Figure 3.43). No linear flow regime is seen, and no horizontal radial flow.

5.11 CONCLUSION

The effect of reservoir boundaries is in general easy to identify in homogeneous systems. The different types of boundary configurations are characterized by a specific pressure behavior, well evidenced with the derivative presentation. When several boundaries are seen, the effect of the different limits is synchronized with the different distances.

Figure 5.55 illustrates, with the example of closed reservoir depicted in Figure 5.54, how it is possible to predict a derivative response from the geometry of the flow barrier and the well location. In this example, the well is in the corner of an elongated trapezoid reservoir, close to one of the system boundaries.

Figure 5.54. Well and reservoir geometry of Figure 5.55 example.

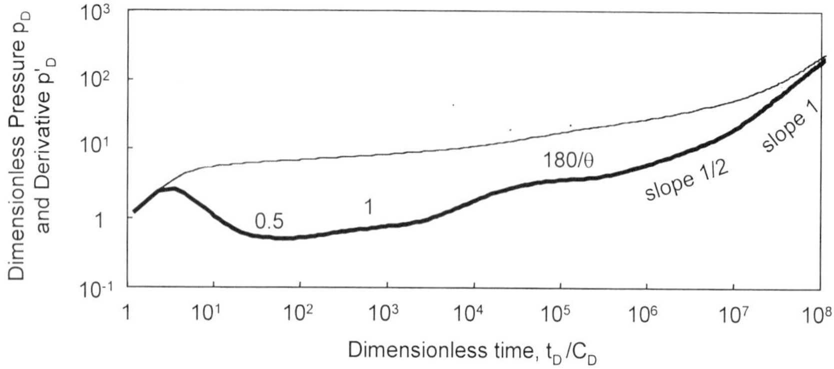

Figure 5.55. Pressure and derivative drawdown responses for a well with wellbore storage and skin in a closed trapezoid reservoir. Log-log scales, p_D versus t_D/C_D.

On this drawdown example, six characteristic flow regimes can be identified:

1. Wellbore storage with a *unit slope* straight line at early time.
2. Infinite acting radial flow with a derivative *stabilization at* 0.5.
3. Hemi-radial flow when the first sealing boundary is reached. The derivative *stabilizes at* 1.
4. After the second fault is reached, the derivative describes the wedge response with a *stabilization at 180/θ.*
5. The third boundary, parallel to the first, produces a semi-linear flow regime with a derivative *half unit slope* straight line.
6. Finally, the reservoir is closed with a fourth boundary, and the derivative follows the *unit slope* straight-line characteristic of the late time pseudo steady sate regime.

In heterogeneous systems, the characteristic derivative shapes are additive and the influence of boundaries is simply superimposed on the derivative heterogeneous behaviors. The effect of boundaries can appear on the early time response, even when the boundaries are far from the producing well. In some cases, boundary effects are identified before the heterogeneous reservoir response. With some practice, it is possible to predict derivative responses for heterogeneous bounded systems, or conversely to define the sequence of flow regimes in order to explain unusual well pressure responses.

For shut-in periods in bounded systems, the time superposition used for build-up analysis can introduce distortions on the derivative curve. Several examples have been presented where sealing boundaries produce a decrease of the derivative response, before a final upwards trend similar to that of the drawdown response. Heterogeneous reservoir with boundaries can show an oscillating derivative shape.

When several solutions are found to be applicable to match build-up data, consistency of the models during the drawdown periods, and examination of the test simulation plot, are used for reducing the number of choices. An accurate measurement of the initial pressure before the test can be the key point for the selection the appropriate boundary configuration (see discussion in Section 10.2.3).

The question of extrapolated pressure p^* from straight line analysis has been the subject of much attention in the specialized literature. In partially bounded systems, p^* gives an estimate of the initial reservoir pressure p_i, provided the correct straight line is used (versus $\log(\Delta t)$, $\sqrt{\Delta t}$ or $\sqrt[4]{\Delta t}$). In practice, the test period frequently stops in transition before the proper straight line, and p^* is difficult to estimate accurately. When the match is made using a build-up or multiple rate type-curve generated by a computer, the initial pressure p_i is implicitly defined by the match. For example in the case of a build-up type-curve, once the pressure match [PM] is defined:

$$p_i = p(\Delta t = 0) + \frac{p_D(t_{pD})}{\text{PM}} \tag{5.34}$$

This method is more general and more accurate than straight-line extrapolation. In the case of a closed system, the average pressure $\bar{p}$ is also estimated directly from the computer model.

CHAPTER 6

MULTIPLE WELL TESTING

With multiple well tests, the pressure response is measured in an observation well some distance away from the active well, which may be a producing or an injection well. The purpose is to establish communication, and to determine average reservoir properties in the area separating the wells. The analysis of the observation well response provides, in addition to the permeability thickness product kh, an estimate of the apparent storage capacity ($\phi\ c_t\ h$), which cannot be accessed from tests of active wells. When several observation wells are located in different directions, any permeability anisotropy can also be evaluated.

For interference tests, the response of the observation well corresponds either to a production period, or to a shut-in of the active well. When the interference has reached the observation well, the change of pressure can be analyzed on log-log scales and on specialized plots. In the case of a homogeneous reservoir, the log-log pressure type-curve of Figure 6.1, presented by Theis in 1935, is used. This type-curve, called the exponential integral solution, expresses the dimensionless pressure p_D versus the dimensionless time-distance group t_D/r_D^2 on a unique response curve. In the case of heterogeneous systems, the log-log analysis of the response provides a diagnosis of the reservoir behavior, and defines the choice of the appropriate interpretation model.

In 1966, Johnson et al. proposed the pulse testing method as an alternative multi-well testing procedure. A series of relatively short flow and shut-in periods is applied at the active well, and the resulting pressure oscillations in the observation well are analyzed in terms of amplitude and time lag. The measured parameters are compared to theoretical simulated responses, and the permeability and storativity are estimated.

The amplitudes of interference or pulse tests responses are small, frequently less than 10 psi and sometimes less than 1 psi, but the test duration can be as long as 2 or 3 months. Interference tests require great skill, particularly in producing fields where the general pressure trend is decreasing and operational requirements may take precedence over reservoir studies. Even though they are more difficult to interpret, pulse tests are often preferred because the oscillating response is easier to identify in a noisy reservoir environment, and it is less affected by a possible drift of the pressure gauge.

In this chapter, the interpretation of interference tests is presented first. After a review of the homogeneous reservoir behavior, the following well and reservoir configurations are considered:

- Influence of wellbore storage and skin at the two wells,
- Effects of boundaries and reservoir directional anisotropy,
- Interference tests in composite, double porosity and double permeability systems.

The interpretation and optimization of pulse tests have been explained in detail in the petroleum engineering literature (Earlougher, 1977; Bourdarot, 1998), only the recent developments in the pulse testing methods are presented in this Chapter.

6.1 INTERFERENCE TESTS IN RESERVOIRS WITH HOMOGENEOUS BEHAVIOR

6.1.1 Pressure and derivative line-source solution

The exponential integral solution of Figure 6.1 is also called the *line source* response. The effect of wellbore storage and skin are considered as negligible at the two wells: the wellbore of the active well has no volume and the well is described as a simple line intersecting the formation, the observation well has no influence on the reservoir pressure.

In dimensionless terms, the exponential integral of Equation 1.18 is expressed:

$$p_D = -\frac{1}{2}\,\mathrm{Ei}\left(\frac{-r_D^2}{4t_D}\right) \tag{6.1}$$

p_D, t_D, and r_D are defined respectively in Equations 2.3, 2.4 and 1.21. The time group t_D/r_D^2 is:

$$\frac{t_D}{r_D^2} = \frac{0.000264k}{\phi \mu c_t r^2}\Delta t \tag{6.2}$$

The interference pressure drop of Equation 1.18 is expressed versus the elapsed time by combining the dimensionless Equations 6.1, 2.3 and 6.2.

The exponential integral pressure type-curve of Theis (1935) is now associated to the pressure derivative for practical interference test analysis, as shown in Figure 6.1. (In 1980-a, Tiab and Kumar proposed an interference derivative type curve using the rate of pressure change versus time, and not the logarithm derivative. Their type curve is equivalent to the curve presented in Figure 6.1 by multiplying the time derivative by Δt). The type-curve of Figure 6.1 exhibits two important characteristics:

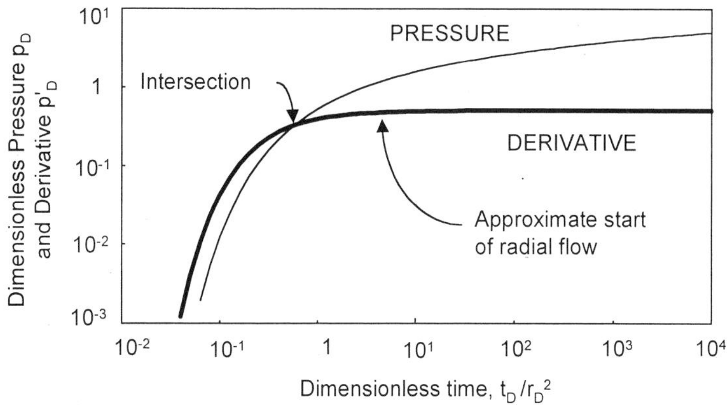

Figure 6.1. The Theis solution (exponential integral).
Log-log scales, pressure and derivative responses versus t_D/r_D^2.

1. The two curves *intersect.* At the start of the response, the amplitude of the derivative curve is higher than the pressure change but, later, the two curves intersect and the derivative stabilizes on the 0.5 line while the pressure continues to increase. Provided the interference test data also demonstrates a cross-over of pressure and derivative response curves, the match point is uniquely defined. In dimensionless terms, the intersection point is defined at $t_D/r_D^2=0.57$ and $p_D=p'_D=0.32$.
2. The start of the semi-log straight line is *late.* The semi-log radial flow behavior, characterized by the derivative stabilization, starts at approximately $t_D/r_D^2=5$ (Ramey, 1975 b). When the distance r between the two wells is in the order of a few hundred feet, the corresponding time Δt to reach the radial flow regime can be as long as several weeks and many interference tests are stopped before data suitable for semi-log analysis is obtained. Only the log-log matching technique can be used to interpret such interference tests.

With interference test data, the log-log match is adjusted on a unique theoretical response and not a family of curves as for producing wells. The interpretation would appear to be very simple and to allow the reservoir parameters to be defined without difficulty but, in the next sections, it is shown that practical type-curve analysis of interference test is frequently difficult.

6.1.2 Typical interference responses

Figure 6.2 compares on a linear scale the response of a producing well to the response of an observation well some distance away. The test sequence is a 200 hours drawdown followed by a 300 hours build-up period. The two wells have the same wellbore storage and skin damage. The reservoir is homogeneous and infinite.

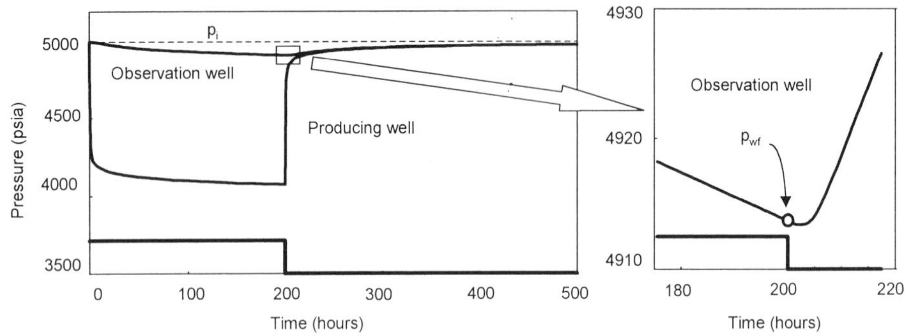

Figure 6.2. Response of a producing and an observation well. Linear scale, p versus t. On the second graph, the observation well pressure is presented on enlarged scale at time of shut-in.

At the start of the flow period, the pressure drop is instantaneous at the active well but, at the observation well, the drawdown response is only established very slowly. At the end of the 200 hrs of production, the pressure change is 1000 psi at the producer and less than 100 psi at the observation well. At shut-in time, the pressure at the active well increases immediately but not at the observation well. As shown on the expanded scale of Figure 6.2, the pressure continues to *fall* for several hours, until the influence of the shut-in has traveled the distance separating the two wells. Then it starts to turn upwards.

Figure 6.3 is the log-log pressure and derivative plot of the two build-up responses. For each well, the pressure difference $\Delta p = p_{ws} - p_{wf}$ is calculated with respect to the flowing pressure p_{wf} defined at the time of shut-in, Δt = 200hr. At the observation well, the pressure is still decreasing at the shut-in time (see Figure 6.2) and the pressure change Δp is *negative*. This first part of the response cannot be plotted on the log-log scales.

Figure 6.3 illustrates that the analysis of interference tests is more difficult than for producing wells:

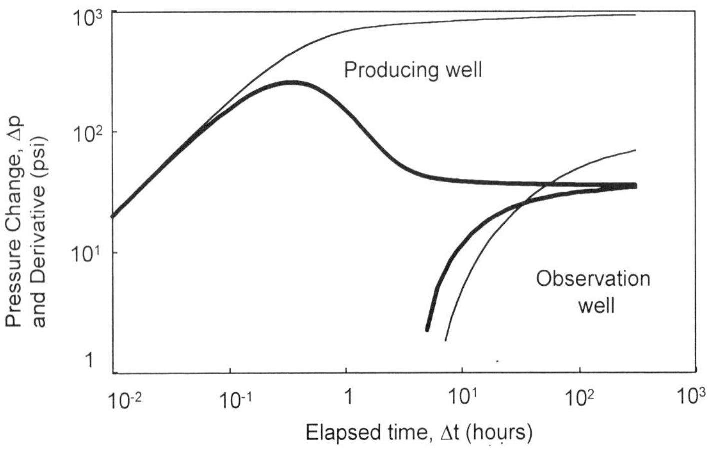

Figure 6.3. Build-up response of the producing and observation wells. Log-log scales, Δp versus Δt.

- at the observation well, the build-up response starts to be identified after 5 or 10 hours, the intersection of the pressure and derivative curves is seen at Δt =33 hours when Δp =25 psi, and the 300 hour long build-up period covers less than *two log-cycles* on the log-log plot. The semi-log approximation is reached only at the end of the 300 hours shut-in.
- In the producer, the build-up response curve extends over more than *four log-cycles*, and the semi-log approximation is reached after 5 hours of shut-in.

A 10 hour build-up response is interpretable for the producing well but, in the observation well, a period 10 times longer is needed in order to define the match uniquely. If the reservoir is affected by a general pressure trend or if the pressure gauge is drifting, the error of pressure and derivative log-log curves can frequently be neglected during the first 10 hours of shut-in but it becomes significant at later times, when the interference response is seen.

For the same reason, when the test period is long enough to reach the semi-log approximation, the straight line analysis is seldom very accurate. At large elapsed time Δt, the interference pressure is barely changing and the response must be strongly compressed on the logarithm scale in order to display the straight line of slope m (see discussion Figure 2.7 in Section 2.2.2). The noise in the pressure signal can become dominant compared to the transient reservoir response.

As an alternative, when the rate history of the producer is a simple drawdown and build-up sequence such as in the example of Figure 6.2, Ramey (1982) suggests matching the complete observation well response on the drawdown and build-up interference type-curve.

6.1.3 Influence of wellbore storage and skin effects at both wells

During the wellbore storage dominated regime, the sand-face flow rate at the active well is not fully established, and the reservoir pressure drop is smaller than for a well with no storage (Section 1.2.2). When the influence of the active well reaches the observation well, the early time pressure response may not follow the exponential integral curve of Figure 6.1.

As discussed in Section 3.1.3, the duration of the wellbore storage regime is a function of both the wellbore storage coefficient and the skin. For damaged wells, the effect of wellbore storage lasts longer. In interference responses, the influence of wellbore storage and skin at the active well was first investigated by Jargon in 1975. Later, Ogbe and Brigham (1984 b) considered the effect of wellbore storage and skin at the observation well. They concluded that the two well conditions have a similar influence on interference response and, when wellbore storage and skin are present in both wells, the early time deviation from the exponential integral solution can be doubled.

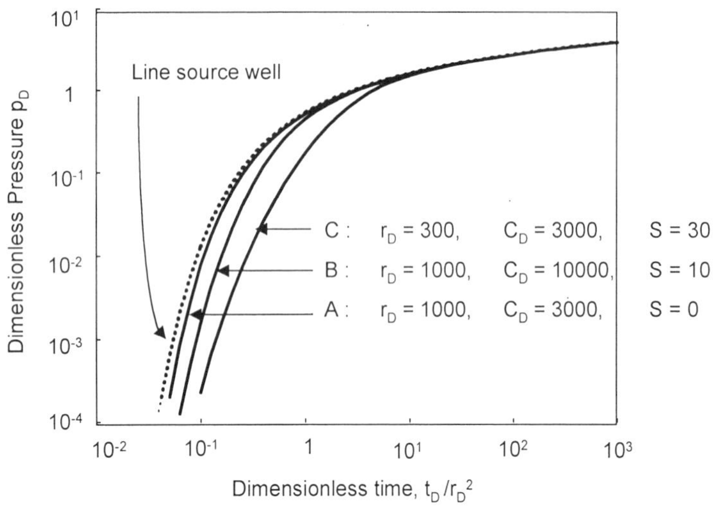

Figure 6.4. Influence of wellbore storage and skin effects.
Log-log scales, pressure responses versus t_D/r_D^2.

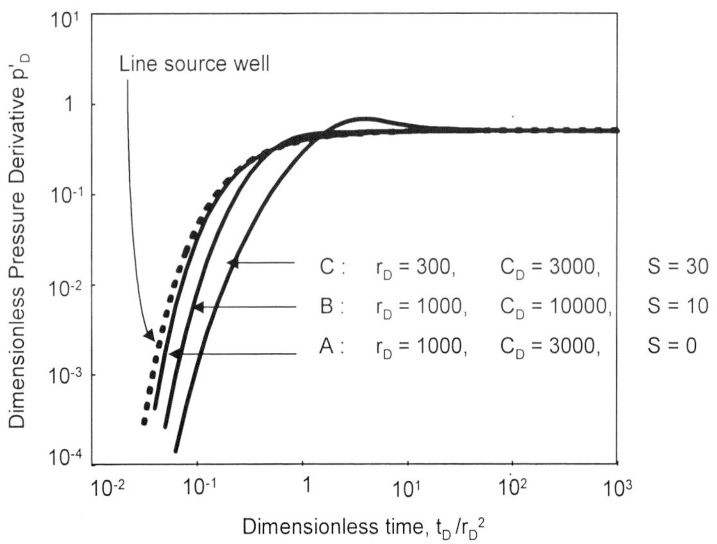

Figure 6.5. Influence of wellbore storage and skin effects.
Log-log scale, derivative responses of Figure 6.4 versus t_D/r_D^2.

In Figure 6.4, three interference pressure curves, generated with wellbore storage and skin at the active well, are compared to the original Theis solution of Figure 6.1 (shown by a dotted curve). At early time, the three responses are delayed until the influence of the wellbore effect becomes negligible and the constant sand face rate condition is established. The interference responses then follow the line-source type-curve. The responses **A** and **B** are generated with r_D =1000. The active well is defined with C_D=3000, S =0 for curve **A**, and C_D =10^4, S =10 for curve **B**. For larger inter-well distances r_D, the difference with the Theis solution would quickly become negligible.

Conversely, when the distance is very *small* such as on curve **C** (r_D =300) and, if the producer is strongly *damaged* (C_D =3000, S =30), the interference response shows a large deviation from the exponential integral curve. The same responses are presented with the derivative of pressure in Figure 6.5: with example **C**, the influence of wellbore storage starts to produce a *hump* above the 0.5 stabilization line.

In Figure 6.6, the two examples generated for r_D=1000 are presented with the pressure and derivative responses (derivative = thick curve, pressure = thin curve).

1. Example **A** (C_D =3000 and S =0), the influence of wellbore storage is finished at t_D/r_D^2=0.57, when the line source pressure and derivative curves intersect. This characteristic intersection point defines the match, independently of the deviation of the response earlier. In such a case, the line-source curve can be used to analyze the interference response accurately.
2. Example **B** (C_D=10000 and S=10), the wellbore effect is still acting at t_D/r_D^2=0.57 and the intersection between pressure and derivative is delayed. The match is not possible with the exponential integral solution, and an interference type-curve with wellbore storage and skin effect has to be used. The need for an interference model with wellbore storage is clear when the shapes of the two derivative curves are compared at early time: on curve **B** of Figure 6.5, the transition from the initial response to the 0.5 stabilization is short and a good match can not be obtained with the line source type-curve.

In 1972, Gringarten and Witherspoon envisaged the interference response of a vertically fractured active well, and Jargon (1976) considered in his study the effect of a negative skin at the producer, using a radial composite model. They showed that, with a stimulated well condition, the response reaches the observation well earlier than for a line source active well, and tends to produce a deviation above the exponential integral curve at early time. This conclusion is consistent with the discussion of interference tests in radial composite reservoirs presented in Section 6.3. The degree of deviation is related to the degree of stimulation of the well, but it decreases for a large distance r.

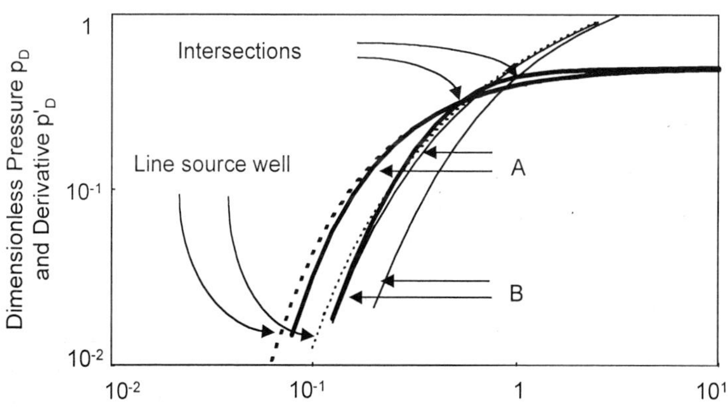

Figure 6.6. Influence of wellbore storage and skin effects. Log-log scale, pressure and derivative responses of Figure 6.4 examples **A** and **B** versus t_D/r_D^2.

In 1989, Meehan and Horne investigated the effect of the well orientation in interference tests between two wells intercepting a finite conductivity fracture. They concluded that the deviation above the exponential integral curve is larger when the path between the wells is parallel to the fracture orientation. Malekzadeh and Tiab considered interference tests between horizontal wells (1991). They reached a similar conclusion.

When the wellbore storage is also considered on the stimulated active well, the two well conditions have an opposite influence on the response. The result seems to be in most cases a delayed response (Jargon, 1976), similar to the examples of Figures 6.4 to 6.6.

Several correlating groups have been proposed to define the effect of wellbore storage and skin of one well. Jargon suggested to group the curves in terms of C_D/r_D^2 and Ogbe and Brigham (1984 b) preferred $\left(C_D e^{2S}\right)^{C_D/r_D^2}$. The different groups have been established from interference response data, they have not been theoretically demonstrated and no group has been proposed when both wells are affected by storage effect.

In conclusion, wellbore conditions can be *neglected* in most interference tests. Wellbore storage has to be considered only when the distance between the wells is relatively small and the wellbore effect large. The response then is delayed at early time compared to the original exponential integral response and the match becomes uncertain in particular if the radial flow regime is not reached at the end of the test period. When the wellbore storage effect is ignored and the line-source solution is used, the time match can be significantly underestimated and the resulting ($\phi\, c_t\, h$) product is then too large.

6.1.4 Semi-log analysis of interference responses

When t_D/r_D^2=5, the infinite acting radial flow regime is reached (Ramey, 1975 b) and the semi-log approximation is expressed, from Equation 1.19, as:

$$p_i - p_{wf} = \frac{162.6\, qB\mu}{kh}\left(\log\Delta t + \log\frac{k}{\phi\mu c_t r^2} - 3.2275\right) \tag{6.3}$$

The slope of the semi-log straight line provides the kh product and the intercept of the line with Δt =1hr is used to estimate the storativity product $\phi\, c_t\, h$.

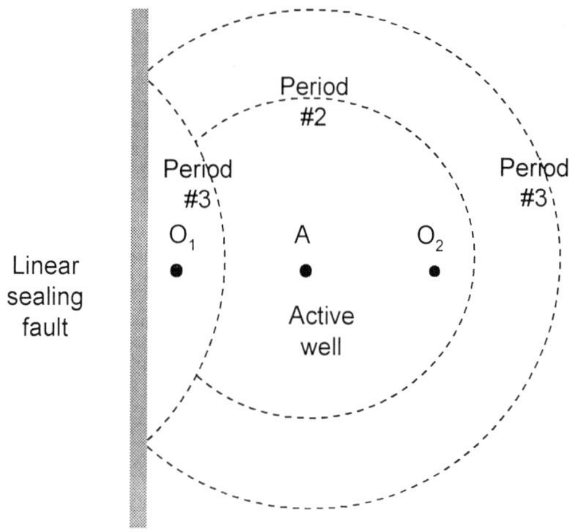

Figure 6.7. Interference in a reservoir with a sealing fault.
Location of the active well **A** and the two observation wells $\mathbf{O_1}$ and $\mathbf{O_2}$.

6.2 FACTORS COMPLICATING INTERFERENCE TESTS IN RESERVOIRS WITH HOMOGENEOUS BEHAVIOR

6.2.1 Influence of reservoir boundaries

As long as there is no flow barrier between the producing and the observation wells, the effect of outer reservoir boundaries has a similar influence on active and observation well responses. The different characteristic derivative behaviors presented in Chapter 5 can be observed on interference test data. In the case of one sealing fault, the derivative stabilizes at $p'_D=1$ at late time. When two parallel sealing faults are present, the pressure and derivative curves follow a straight line of slope 1/2. With a closed system, the pseudo steady state flow regime is described by a unit slope log-log straight line.

When an analytical model is used to generate interference responses with boundary effects, the relative location of the wells in the boundary geometry must be known. This is illustrated on the following interference test example:

In Figure 6.7, the reservoir is limited by one single sealing fault. The observation well "$\mathbf{O_1}$" is located between the sealing boundary and the active well "**A**", and the observation well "$\mathbf{O_2}$" is located in the opposite direction. It is assumed that the two observation wells are at the same distance r from the producer **A**. The boundary is observed at a different time on the three well responses. The sequence of regimes can bc summarized as follows:

1. The interference has not reached the observation wells. "**A**" shows the infinite acting reservoir behavior, "$\mathbf{O_1}$" and "$\mathbf{O_2}$" are at initial pressure.

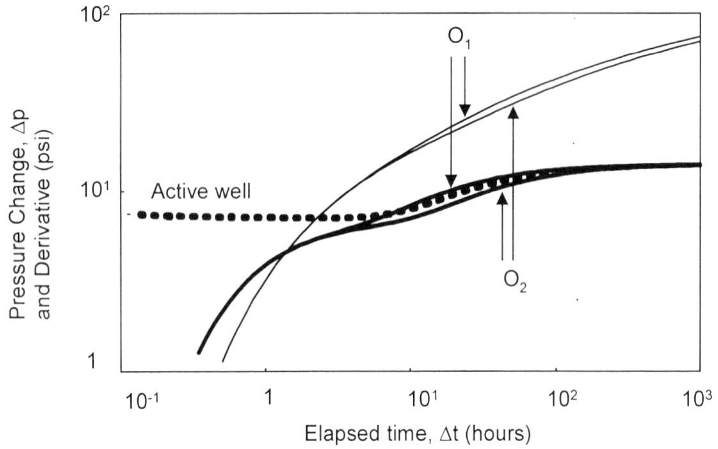

Figure 6.8. Interference in a reservoir with a sealing fault. Log-log scales, Δp versus Δt. Pressure and derivative curves of the two observation wells.

2. The interference has reached the two observation wells but not the sealing fault. The behavior of three wells "**A**", "$\mathbf{O_1}$" and "$\mathbf{O_2}$" show the infinite reservoir response (the inner circle in Figure 6.7 describes the radius of investigation reached at time of period #2).
3. The sealing fault influence is seen on "$\mathbf{O_1}$" but not on "**A**" and "$\mathbf{O_2}$", which are still in an infinite acting regime (period #3 corresponds to the outer circle).
4. The influence of the sealing fault reaches "**A**": "$\mathbf{O_1}$" and "**A**" change to the hemi-radial flow behavior but "$\mathbf{O_2}$" shows the infinite acting regime.
5. The influence of the sealing fault reaches "$\mathbf{O_2}$" and the three wells show the hemi-radial flow condition.

In the example of Figure 6.8, the sealing fault is located at $3r$ from A (the fault is therefore at $2r$ from $\mathbf{O_1}$ and $4r$ from $\mathbf{O_2}$). The derivative interference response of $\mathbf{O_1}$ shows the fault influence before that of $\mathbf{O_2}$. The derivative of the active well **A** (dotted line) describes the fault response at an intermediate time.

The analysis of interference test responses influenced by boundaries requires the use of a model generated for the specific reservoir geometry. The solution is frequently not unique, and the quality of the results depends upon the choice of the appropriate reservoir boundary system.

6.2.2 Interference tests in reservoirs with permeability anisotropy

On producing well responses, it is shown in Section 3.1.5 that the effect of permeability anisotropy can frequently be neglected. The resulting apparent negative skin S_{ani} is, for most practical cases, smaller than unity (Table 3.1). For interference tests, the time to reach the observation well is very dependent upon the directional permeability towards the active well, and results of interpretation are more sensitive to a small anisotropy of permeability.

It is assumed that the permeability anisotropy can be described by a tensor. According to Papadopulos (1965), the pressure distribution created by a line source well is changed from Equation 1.18 to Equation 6.4:

$$\Delta p(\Delta t, x, y) = 0.5 \frac{141.2\, qB\mu}{\sqrt{k_{max} k_{min}}\; h} \mathrm{Ei}\left[-\left(r_D^{\,2}/4t_D\right)_{x,y}\right] \tag{6.4}$$

As for producing well responses, the apparent permeability is defined as the geometric mean of the major and minor reservoir permeability (Equation 3.3), which is the *same* whatever the direction of the observation well. In Equation 6.4, the effect of the reservoir anisotropy is only present in the definition of the dimensionless time group $(t_D/r_D^{\,2})_{x,y}$. In other terms, the *apparent distance* $r_{D,x,y}$ is a function of the orientation of the observation well (Ma and Tiab, 1995).

With a coordinate system centered on the active well, the observation well location is defined as (x,y) shown in Figure 6.9. We call k_x, k_y and k_{xy} the components of the permeability tensor corresponding to the coordinate system. From Equation 3.3, the apparent permeability is defined as:

$$\bar{k} = \sqrt{k_{max}\, k_{min}} = \sqrt{k_x k_y - k_{xy}^2} \tag{6.5}$$

Ma and Tiab (1995) express the apparent distance $r_{D,x,y}$ as:

$$r_{D,x,y} = \frac{1}{r_w} \sqrt{\frac{k_x y^2 + k_y x^2 - 2k_{xy} xy}{\bar{k}}} \tag{6.6}$$

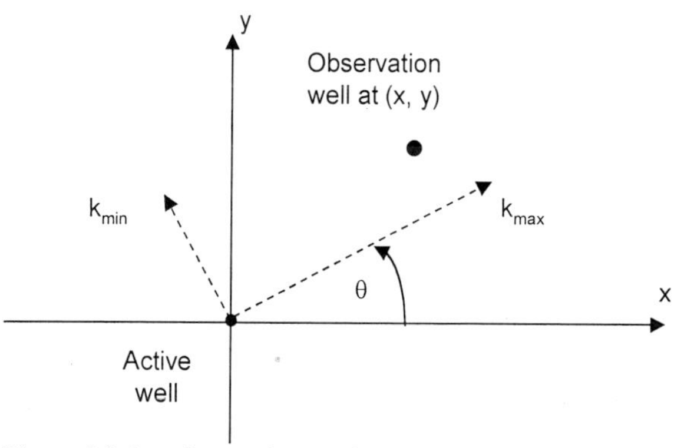

Figure 6.9. Interference in an anisotropic reservoir.
Location of the active well and the observation well.

and the dimensionless time corresponding to well (x,y) is:

$$\left(\frac{t_D}{r_D^{\,2}}\right)_{x,y} = \frac{0.000263\Delta t}{\phi\mu c_t}\left(\frac{k_{\max}k_{\min}}{k_x y^2 + k_y x^2 - 2k_{xy}xy}\right) \tag{6.7}$$

Ramey (1975 b) shows that observation wells are needed in three directions in order to estimate the reservoir anisotropy. When the three observation well responses are matched against the exponential integral type curve of Figure 6.1, the pressure match is the same for the three responses and only the time match changes. The product $k_{max}k_{min}$ is estimated from the pressure match.

Solving Equation 6.7 with the time match for three observation wells yields k_x, k_y and k_{xy} as a function of $\phi\mu\, c_t$. Equation 6.5 is then used with the pressure mach permeability $\sqrt{k_{\max}k_{\min}}$ to estimate $\phi\mu\, c_t$. Finally, the major and minor reservoir permeabilities k_{max} and k_{min} can be defined with:

$$k_{\max} = 0.5\left\{k_x + k_y + \left[\left(k_x - k_y\right)^2 + 4k_{xy}^2\right]^{1/2}\right\} \tag{6.8}$$

$$k_{\min} = 0.5\left\{k_x + k_y - \left[\left(k_x - k_y\right)^2 + 4k_{xy}^2\right]^{1/2}\right\} \tag{6.9}$$

The angle between the major permeability axis and the x axis of the coordinate system is expressed with (Ramey, 1975 b):

$$\theta = \arctan\left(\frac{k_{\max} - k_x}{k_{xy}}\right) \tag{6.10}$$

In practice, the solution of the system of the three Equations 6.7 is valid only when the two permeability terms k_{max} and k_{min}, and the $\phi\mu\, c_t$ product are constant over the reservoir region involved by the test. This condition, difficult to verify, is not always satisfied.

When the interference pressure is monitored in one observation well only, the reservoir anisotropy cannot be estimated. As the radial homogeneous reservoir solution is used for analysis, the average permeability $\sqrt{k_{\max}k_{\min}}$ is estimated from the pressure match but the $\phi\mu\, c_t$ product obtained from the time match with the radial Equation 6.2 can be wrong by a factor of two or even more.

6.3 INTERFERENCE TESTS IN COMPOSITE RESERVOIRS

The practical analysis of interference tests in composite reservoirs has not been fully discussed yet. In 1985, Satman presented an analytical solution for a radial composite reservoir with changing mobility and diffusivity from inner to outer zone. In 1985, Yaxley derived a linear composite solution with a variation of mobility between two reservoir regions separated by a partially communicating interface. Two years later, Ambastha et al. (1987) completed the linear composite reservoir description by introducing a change of diffusivity through the linear interface. In 1991, Chu and Grader presented an analytical three composite reservoir model with any location for the active and observation wells.

The interference pressure responses have been very briefly discussed with the different models, but the derivative behavior has not been presented in detail. Yet composite reservoir behavior must frequently be present during interference tests, for example when the active and the observation wells are in different fluid systems (change of viscosity, compressibility and temperature with water injection wells or wells in the aquifer). A more complete description of interference responses in radial composite reservoirs is given below.

The same definitions for mobility and storativity ratios M and F are used for interference models and for the active well solutions (see Section 4.3.1). By definition, the active well is in region 1, and the observer can be in region 1 or in region 2.

$$M = \frac{(k/\mu)_1}{(k/\mu)_2} \tag{4.78}$$

$$F = \frac{(\phi c_t)_1}{(\phi c_t)_2} \tag{4.79}$$

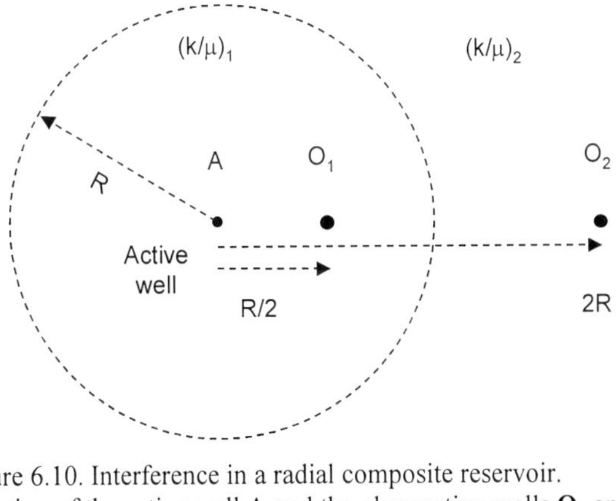

Figure 6.10. Interference in a radial composite reservoir.
Location of the active well **A** and the observation wells $\mathbf{O_1}$ and $\mathbf{O_2}$.

The distance from the active well to the radial and linear interface are given respectively by Equations 1.21 and 1.34.

6.3.1 Radial composite reservoirs

Influence of M

On the interference responses of Figure 6.11 and 6.12, the active well **A** is centered in a radial composite reservoir of radius R. As described in Figure 6.10, the observer $\mathbf{O_1}$ is at $R/2$ from **A** (in region 1), and the observer $\mathbf{O_2}$ at $2R$ (in region 2). It is assumed that the storativity ϕc_t is constant (F =1), but the mobility of the inner region is 4 times larger (Figure 6.11) or smaller (Figure 6.12). On the log-log plots, the interference responses of the two observation wells are shown with lines (derivative = thick, pressure = thin), the derivative of the active well **A** and the exponential integral derivative corresponding to $\mathbf{O_2}$ (homogeneous reservoir with region 2 parameters) are shown with dotted curves.

If the mobility of the inner zone is *larger* than the mobility in region 2 [$(k/\mu)_1>(k/\mu)_2$, i.e. $M>1$, Figure 6.11], the interference signal reaches the region 2 (observation well $\mathbf{O_2}$) *faster* than the equivalent homogeneous reservoir of mobility $(k/\mu)_2$ (the dotted derivative curve).

When the observation well is in the high permeability region 1 (well $\mathbf{O_1}$), the interference response tends to follow the line source response corresponding to region 1 parameters at early time, with possibly a first derivative stabilization defining $(k/\mu)_1$. Later, when the radial interface is reached, the derivative *increases* to follow the $(k/\mu)_2$ stabilization.

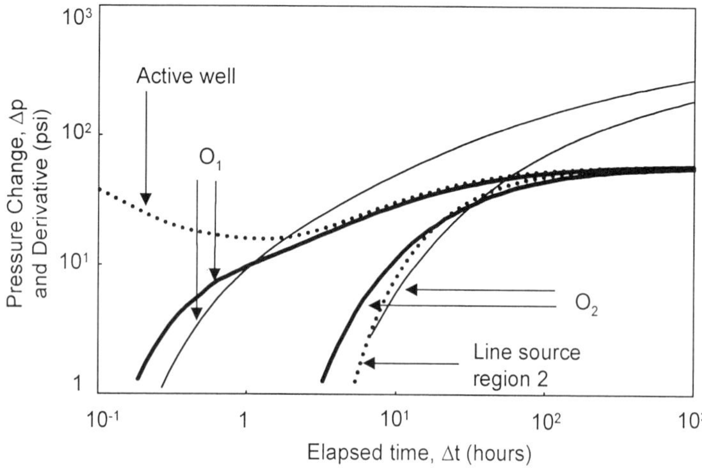

Figure 6.11. Interference responses in a radial composite reservoir. Log-log scales, Δp versus Δt. The mobility of the inner zone is 4 times larger (M=4).

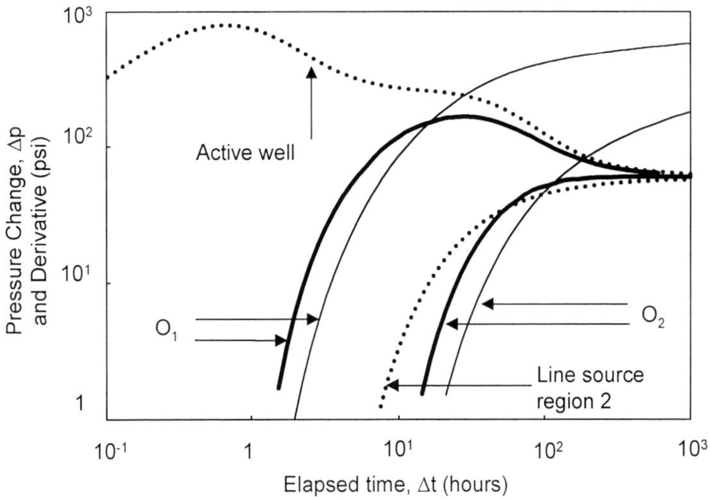

Figure 6.12. Interference responses in a radial composite reservoir. Log-log scales, Δp versus Δt. The mobility of the inner zone is 4 times smaller (M=1/4).

Conversely, when the mobility of the inner zone is *smaller* than the mobility of region 2 $[(k/\mu)_1<(k/\mu)_2$, i.e. $M<1$, Figure 6.12] the interference signal in the outer reservoir region is *delayed* at early time (well $\mathbf{O_2}$). In region 1, the derivative response of well $\mathbf{O_1}$ describes a hump, with a first stabilization corresponding to $(k/\mu)_1$ before the final $(k/\mu)_2$ stabilization.

In Figure 6.13, the pressure curves of the two observation wells are compared in the hypothesis $M>1$ and $M<1$:

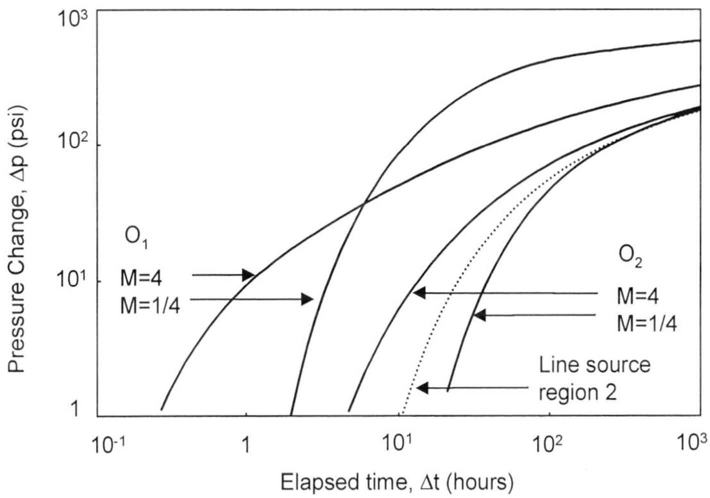

Figure 6.13. Interference responses in a radial composite reservoir. Log-log scales, Δp versus Δt. Pressure curves of examples Figures 6.11 and 6.12.

- In region 2, the pressure of well $\mathbf{O_2}$ is not affected after the early time deviation by the change of mobility near the active well. The curves *merge* at late time with the line source corresponding to region 2 parameters (dotted curve). When M>1, the early response deviates above the exponential integral response, as mentioned in Section 6.1.3 for a stimulated active well. (In 1976, Jargon evaluated the effect of a negative skin on the active well with a radial composite model).
- In region 1, the *amplitude of the interference response* describes the mobility hypothesis. When M<1, the observation well $\mathbf{O_1}$ shows a larger interference signal than when M>1. This reservoir configuration produces a geometrical positive skin on the active well **A** (Equation1.13).

Influence of *F*

It is now assumed that the mobility (k/μ) is constant in the reservoir (M=1), and only the storativity ϕc_t changes (in the inner zone, it is 4 times smaller in Figure 6.14 and 4 times larger in Figure 6.15). The derivative response of the observer $\mathbf{O_2}$ in region 2 (thick line) is compared to the derivative curves of well **A**, and to the exponential integral curve defined for region 2 parameters (shown with dotted curves). The pressure response of the observer $\mathbf{O_2}$ is shown with a thin line.

With *low storativity* in the inner zone $[(\phi\mu\ c_t)_1<(\phi\mu\ c_t)_2]$ the diffusivity is increased, and the interference signal reaches the observation well *faster* than for a homogeneous reservoir with region 2 properties (Figure 6.14). The response deviates above the exponential integral curve at early time.

If the inner zone has a *high storativity* $[(\phi\mu\ c_t)_1>(\phi\mu\ c_t)_2]$ and the reservoir mobility is constant, the early time response is *delayed* compared to the exponential integral homogeneous curve (Figure 6.15).

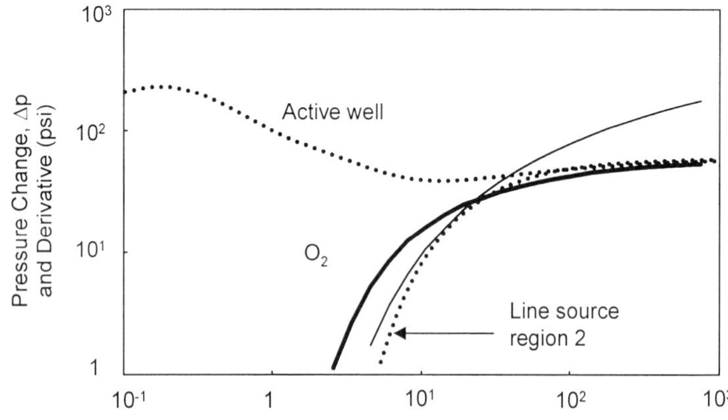

Figure 6.14. Interference responses in a radial composite reservoir. Well $\mathbf{O_2}$.
Log-log scales, Δp versus Δt. The storativity of the inner zone is 4 times smaller (M=1, F=1/4).

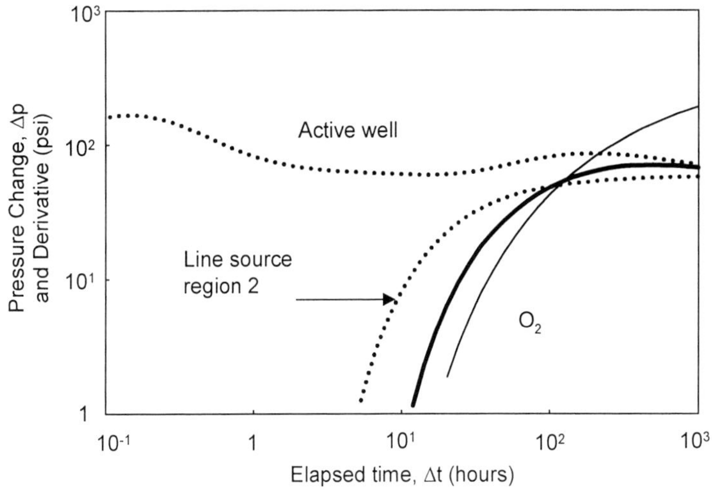

Figure 6.15. Interference responses in a radial composite reservoir. Well $\mathbf{O_2}$. Log-log scales, Δp versus Δt. The storativity of the inner zone is 4 times larger (M=1, F=4).

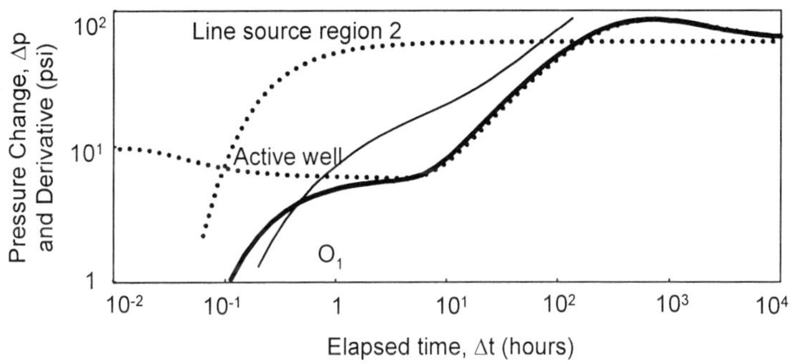

Figure 6.16. Interference responses in a radial composite reservoir. Well $\mathbf{O_1}$. Log-log scales, Δp versus Δt. The mobility and the storativity of the inner zone are 10 times larger (M=10, F=10).

Change of mobility and storativity

When both the mobility and the storativity change, it is difficult to summarize the different types of response illustrated by Satman (1985). It is noteworthy that when both the mobility ratio and the storativity ratio are high (M>10 and $F \geq 1$), the interference response tends to follow a unit slope straight line at early time. Satman describes this effect of the inner zone as similar to an equivalent wellbore storage behavior. As discussed in Chapter 4 for producing well responses (Figures 4.52 and 4.57), this 45° line could be also interpreted as a closed system response as illustrated in Figure 6.16.

In this example, the observation well is in region 1, $(k/\mu)_1=10(k/\mu)_2$ and $(\phi\, c_t)_1=10(\phi\, c_t)_2$ (M=10 and F=10). Since the diffusivity is constant, the radial composite interference response is seen at the same time as the exponential integral curve for region 2 parameters (when Δp<1 psi in Figure 6.16). Later, the derivative flattens and goes under the homogeneous response of the outer zone, then it increases along the unit slope line and merges with the derivative curve corresponding to the active well. Before the influence of the outer region is felt, the radial composite reservoir behaves like a closed circle of radius R and the response describes the pseudo steady state regime characterized by a unit slope pressure and derivative straight line.

In the case of a radial model such as in the examples of Figures 6.11 to 6.16, the derivative behavior of the observation wells tends to merge with that of the active well after the early time interference response. This demonstrates that, at any point in the drainage area, the rate of pressure change is the same.

6.3.2 Linear composite reservoirs

With the linear composite model, the interference model responses have not yet been addressed. The information available today can be summarized as follows (Yaxley, 1985; Ambastha et al., 1987):

- The late time equivalent homogeneous behavior provides an estimate of the average permeability thickness product, defined as $kh=0.5(k_1+k_2)h$ (see Section 4.3.3).
- The effect of changing storativity ($\phi\mu\ c_t$) between the two regions and the effect of the distance L_D between the interface and the producer have not been evaluated on interference responses.
- When the linear interface is partially sealing between the active and observation wells, the early time response observed before the equivalent homogeneous behavior is delayed.

6.4 INTERFERENCE TESTS IN DOUBLE POROSITY RESERVOIRS

In the following section, the pressure behavior of observation wells in double porosity reservoirs is discussed. It is shown that the presence of fissures affects the propagation of the interference signal around the producing well:

- During the fissure flow regime, the interference response travels very *fast* up to a certain distance to the producing well.
- When the distance between the active and the observation wells is large, the heterogeneous nature of the response *disappears* (Kazemi et al, 1969 b).

The two double porosity models introduced in Chapter 4, with restricted and transient interporosity flow, are used for interference test interpretation. The models are not described in the present Chapter, only their application to observation well responses is presented.

6.4.1 Pressure type-curve for restricted (pseudo-steady state) interporosity flow

In Figures 6.17, the pressure type-curve for an observation well in a double porosity reservoir with restricted interporosity flow is presented (Bourdet and Gringarten, 1980; Deruyck et al., 1982). The dimensionless time group t_{Df}/r_D^2 is defined with reference to the fissure system storativity $(\phi Vc_t)_f$ (the double porosity nomenclature has been presented in section 4.1.1):

$$\frac{t_{Df}}{r_D^2} = \frac{0.000264k}{(\phi Vc_t)_f \mu r^2}\Delta t \tag{6.11}$$

The type-curve is plotted with a set of exponential integral solutions labeled in terms of ω values for the two homogeneous flow regimes, and a family of transition curves, with typical values of λr_D^2 for the transition period.

The *first* exponential integral curve corresponds to $\omega = 1$ and it describes the fissure flow regime. When the transition starts, the response deviates from the fissure curve and follows a λr_D^2 *transition* curve. Later, the total system equivalent homogeneous regime is reached. A *second* exponential integral curve, displaced by the reservoir storativity ratio ω, is seen at late time.

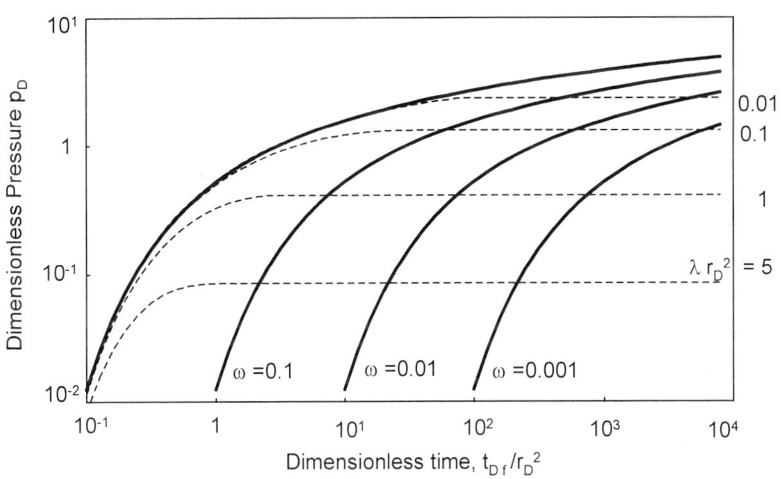

Figure 6.17. Pressure type-curve for double porosity reservoirs with restricted (pseudo-steady state) interporosity flow. Log-log scales, p_D versus t_{Df}/r_D^2.

6.4.2 Pressure behavior in double porosity reservoirs with restricted interporosity flow

The three typical regimes of the double porosity model are not always present on an observation well response. The level of the pressure change Δp during the transition between fissure and total system flow is defined by a λr_D^2 transition curve of Figure 6.17. For a large distance r_D the transition curve stabilization corresponds to a low Δp value on the pressure scale and, beyond a certain distance r_{iD}, Δp becomes less than the pressure gauge resolution. This distance r_{iD} represents the *radius of influence* of the fissures around the active well. It can be approximated from the minimum $\Delta p_{\min}$ measurable at test conditions (Deruyck et al., 1982):

$$\ln \frac{2}{1.78\sqrt{\lambda r_{iD}^2}} = \left(\Delta p_{\min}\right)_D \tag{6.12}$$

If the observation well is at a distance greater than r_{iD}, the interference signal is measurable only after the transition regime, when the response is in total system flow. In such a case, even though the active well response is clearly double porosity, the observation well shows the behavior of a homogeneous reservoir (Kazemi et al, 1969 b). The interpretation of the interference test provides the permeability and the total system storativity $(\phi V c_t)_{f+m}$ but the heterogeneous parameters ω and λ cannot be estimated.

If however the observation well is located inside the radius of influence r_{iD}, the fissure flow regime is observed before the equivalent homogeneous total system flow. On the log-log type-curve of Figure 6.17, the distance between the early time and the late time exponential integral curves defines the storativity ratio ω. For example with $\omega = 10^{-1}$, the fissure curve is seen on the logarithm time scale one cycle before the total system homogeneous response, and the time needed to observe the start of the interference signal is 10 times smaller than for the equivalent homogeneous reservoir. With $\omega = 10^{-2}$, the time ratio is 100.

Interference responses in fissured reservoirs can be relatively *fast*, and frequently they can be observed after only a few minutes, even when the observation well is several hundreds of feet away from the producer. This does not mean that the test duration can be reduced. In the next section, it is shown that the early time response does not define the match and a complete response of the total system radial flow regime is needed for interpretation. The time to reach the total system equivalent homogeneous behavior is as long as for interference tests in standard homogeneous reservoir.

6.4.3 Pressure and derivative analysis of interference tests in double porosity reservoirs with restricted interporosity flow

In Figures 6.18 and 6.19, the responses of two observation wells in a fissured reservoir are compared. Well **A** is at r_D =1000 from the producer and well **B** is farther away (r_D =5000). The interporosity flow parameter is λ=5x10^{-8} and the storativity ratio is ω=10^{-1}.

In Figure 6.18, the pressure and derivative curves are presented versus the dimensionless time t_{Df} and, in Figure 6.19, the usual t_{Df}/r_D^2 group of the type-curve Figure 6.17 is used (Equation 6.11). On t_{Df} time scale of Figure 6.18, the response of the well **B** at r_D =5000 is delayed by a factor of 25 compared to the response of the well **A** at r_D =1000. With the dimensionless time scale t_{Df}/r_D^2 of Figure 6.19, the two well responses are superposed at early time during the fissure flow regime.

As already mentioned in Section 6.3, the rate of pressure change in the reservoir is not space dependent, and it is the same at any point in the drainage area. In Figure 6.18, the derivative response at the active well is also presented. The *time of the transition* is the same for the three wells (the derivative curve of well **B** does not match exactly the derivative response of two other wells because, at this well, the radial flow regime has not been reached before the start of the transition).

When the interference is measured at r_D =1000 (well **A**), a long homogeneous fissure regime is seen at early time and the derivative stabilizes at 0.5 during the fissure radial flow. The point of intersection between the pressure and derivative curves can be used to define the match point. The permeability thickness product is estimated from the pressure match (Equation 2.9) and the fissure system storativity $(\phi\, Vc_t)_f$ from the time match with Equation 6.11.

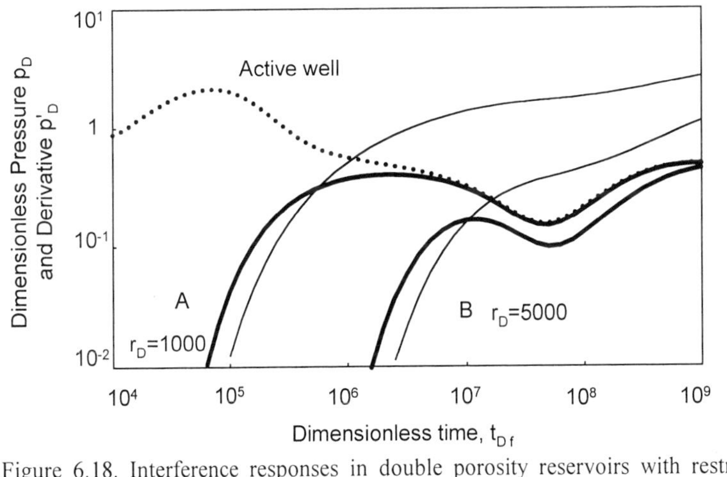

Figure 6.18. Interference responses in double porosity reservoirs with restricted interporosity flow. Log-log scales, p_D versus t_{Df}. ω = 0.1, λ = 5 x 10^{-8}, two distances.

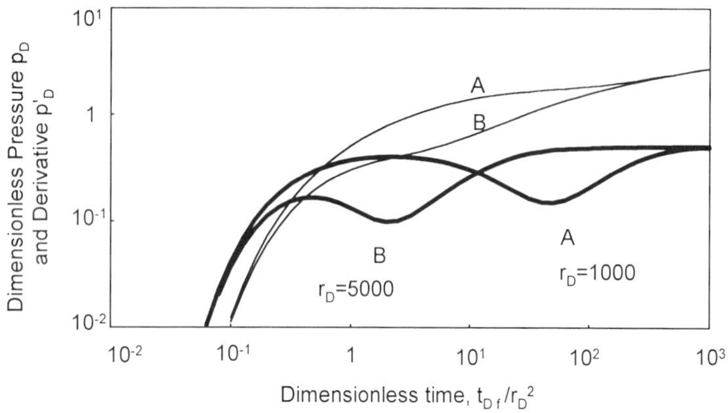

Figure 6.19. Interference responses in double porosity reservoirs with restricted interporosity flow. Log-log scales, p_D versus t_{Df}/r_D^2. Same as Figure 6.18.

After the fissure radial flow, the valley shape of the derivative curve during transition is characteristic of a double porosity response with restricted interporosity flow. The heterogeneity parameters ω and λ are adjusted in the same way as for a producing well (see Section 4.1.2).

More frequently, a response similar to the thick curves (well **B**, r_D =5000) is obtained. The fissure flow regime is short lived and the transition starts before the pressure and derivative curves intersect. The early time behavior does not describe a characteristic regime and the match has to be adjusted by trial and error on pressure and derivative theoretical responses.

6.4.4 Interference tests in double porosity reservoirs with restricted interporosity flow and permeability anisotropy

In double porosity reservoirs, the fractures are frequently more dominant in one direction than another. The permeability is then maximum in the direction of the fissures and minimum in the direction perpendicular. Interference tests are used to define the directional flow properties with the method presented in Section 6.2.2.

When several observation well interference responses are analyzed to estimate the permeability anisotropy, the same interporosity flow parameter λ should be used for all wells. Ma and Tiab (1995) propose using the effective distance $r_{D,x,y}$ of Equation 6.6 in the λr_D^2 group to define the match of the transition regime.

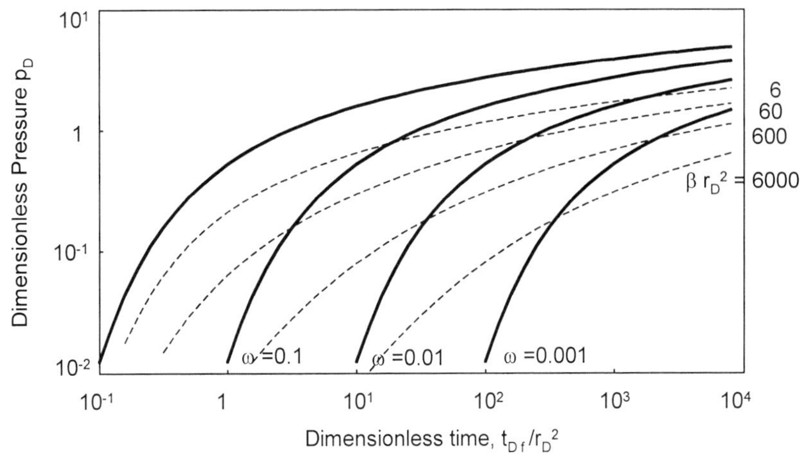

Figure 6.20. Pressure type-curve for double porosity reservoirs with unrestricted (transient) interporosity flow. Log-log scales, p_D versus t_{Df}/r_D^2.

6.4.5 Pressure type-curve for unrestricted (transient) interporosity flow

The type-curve of Figure 6.20 corresponds to a double porosity reservoir with unrestricted interporosity flow (Bourdet and Gringarten, 1980; Deruyck et al., 1982). The time scale is defined with the t_{fD}/r_D^2 group of Equation 6.11. A family of Theis solutions, labeled in ω values, is superimposed on transition curves labeled βr_D^2.

For slab matrix blocks, the β interporosity parameter is defined by:

$$\beta = \frac{3}{5}\frac{\lambda}{\omega} \tag{6.13}$$

and for sphere matrix blocks:

$$\beta = \frac{1}{3}\frac{\lambda}{\omega} \tag{6.14}$$

With unrestricted interporosity flow, the fissure flow is very short and seldom seen. The interference response starts on a *transition* βr_D^2 curve and reaches the exponential integral *total system* curve at late time. The shape of the βr_D^2 transition curve is smooth and not very characteristic compared to the final total system homogeneous response. When the derivative is not used, the match is sometimes difficult.

As for active well responses, a first straight line of slope $m/2$ can be seen on a semi-log scale, if the transition regime lasts long enough before the equivalent total system homogeneous regime.

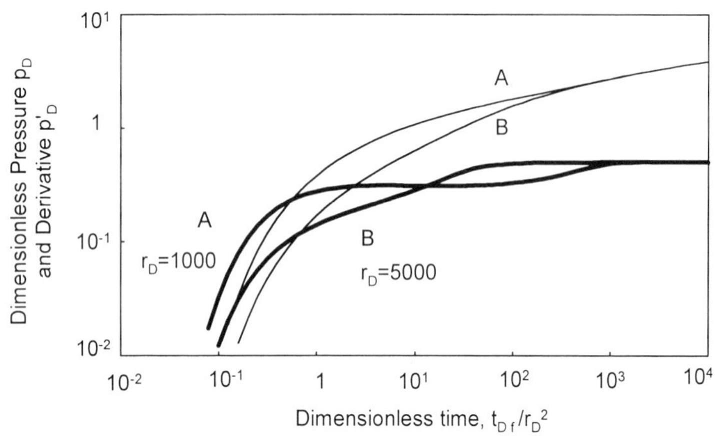

Figure 6.21. Interference responses in double porosity reservoirs with unrestricted interporosity flow. Log-log scales, p_D versus t_{Df}/r_D^2. Same parameters as on Fig. 6.19.

6.4.6 Pressure and derivative analysis of interference tests in double porosity reservoirs with unrestricted interporosity flow

The two observation well responses presented in Figure 6.21 are generated with the same parameters as in the examples of Figure 6.19 (the two distances are respectively r_D =1000 and r_D =5000 for wells **A** and **B** and the double porosity reservoir is defined with λ=5x10^{-8} and ω =10^{-1}). The reservoir cross flow is not restricted and the matrix blocks are assumed to be spherical. The time scale is t_{Df}/r_D^2.

For well **A**, the response shows a very short fissure flow regime, similar to the curve Figure 6.19 for a reservoir with restricted interporosity flow. When the pressure and derivative curves intersect, the response is already in transition behavior. In Figure 6.19, the intersection between the two curves **A** is defined at p_D =p'_D =0.32 as on a homogeneous response. In Figure 6.21, it is only p_D =p'_D =0.25. The point of intersection between the pressure and derivative curves cannot be used to define the match in reservoirs with unrestricted interporosity flow.

During the double porosity transition of example **A**, the derivative response tends to stabilize at approximately p'_D =0.3. Finally, when radial flow in the total system equivalent homogeneous behavior is seen, the derivative reaches the usual 0.5 plateau.

The response of well **B** is less characteristic. The fissure flow regime is not seen and the signal is delayed compared to curve **A**. The transition is short and no intermediate derivative stabilization is seen before the 0.5 plateau. If the test period is short and the final radial flow has not been reached, the analysis can be difficult to conclude.

6.5 INTERFERENCE TESTS IN LAYERED RESERVOIRS

With layered reservoirs, the interference test responses depend upon the two well conditions: if the same layers are perforated in the active and observation wells or not, and if there is cross flow in the observation well or not:

- When the perforated layers are different, the pressure response in the observation well is due to reservoir cross flow.
- If several layers are communicating with the observation well, the interference pressure drop can be different in the layers. Cross flow is then established though the observation well and, even though it is not producing at surface, the well is active and influences the reservoir pressure response (the flow lines are not radially symmetric around the active well any more).

Many combinations are possible depending upon the layered system and the completion of the wells. In 1981, Chu and Raghavan and, in 1989 Onur and Reynolds have discussed interference tests in non-communicating layers with crossflow in the observation well. They concluded than the effect of the skins at both wells can be greater than the effect of the contrast in the layers properties. Streltsova (1984 a) considered vertical crossflow in a two layers reservoir by assuming only one layer is producing to the well, and radial flow in the second layer is negligible.

In the following, a "two layer" configuration is considered with the double permeability model (Bourdet, 1985) described in Chapter 4. The theoretical responses of several interference test configurations are presented to illustrate some basic characteristic behaviors. The examples can be extrapolated to more complex configurations, and they show that a layered reservoir interference model is needed for analysis.

6.5.1 Line source well in a two layers with cross flow reservoir

On the double permeability responses presented in Figure 6.22 and 6.23, the contrast between the layers is small enough to produce an apparent homogeneous behavior in the active well (see Section 4.2): κ=0.7 and ω=0.4. The cross flow parameter is λ=10^{-6} and the distance of the observation well is r_D =1000.

The dimensionless pressure p_D and the dimensionless time group t_{D1+2}/r_D^2 are defined by the parameters of the total system (Equations 4.49 and 4.50). The active well is a line source communicating with the two layers and three different observation well configurations are considered: in Figure 6.22, only one of the two layers, respectively layer 1 or layer 2, is perforated (Bourdet, 1985), and in Figure 6.23, they are both perforated (Houzé and Viturat, 1985). The pressure and derivative of the Theis solution corresponding to the total equivalent homogeneous reservoir system are shown with the dotted curves.

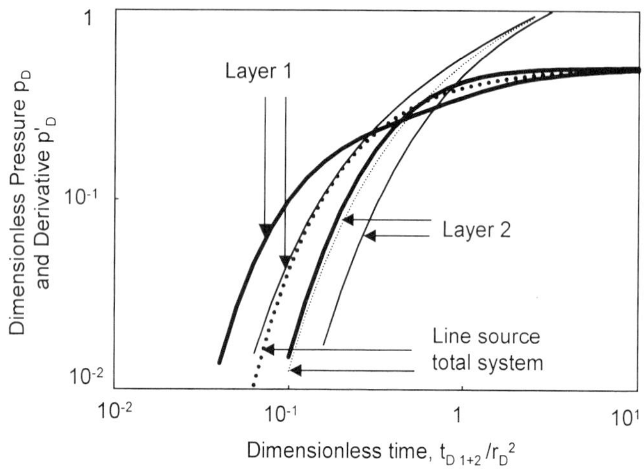

Figure 6.22. Interference responses in a double permeability reservoir, one layer is perforated in the observation well. Log-log scales, p_D versus t_{D1+2}/r_D^2. $\omega = 0.4$, κ=0.7, $\lambda = 10^{-6}$.

If the observation well is only perforated in the higher permeability layer (#1), the response is seen *before* the equivalent homogeneous solution for the total system (Figure 6.22). At t_{D1+2}/r_D^2=0.1 for example, the amplitude of the layer 1 pressure response is 10 times larger than the Theis solution. Later, about at t_{D1+2}/r_D^2=5, the layer 1 double permeability response reaches the equivalent homogeneous behavior.

In the opposite configuration, when only the low permeability layer (#2) is communicating with the observation well, the early time response is *delayed* compared to the Theis solution for the total system. The responses of layer 1 and that of layer 2 merge on the dotted pressure and derivative curves when the equivalent homogeneous behavior is reached.

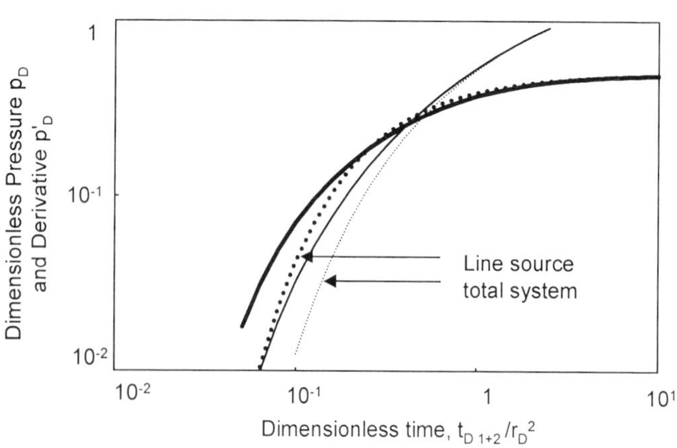

Figure 6.23. Interference responses in a double permeability reservoir, the two layers are perforated in the observation well. Log-log scales, p_D versus t_{D1+2}/r_D^2.Same parameters as on Fig. 6.22.

At the start of the response, the two layers show a different pressure. If they are both perforated, this pressure differential produces a cross flow in the observation well. The resulting response is shown in Figure 6.23: it is also ahead of the Theis solution for the total system, and it appears not very different from the layer 1 response. When several layers are perforated, the high permeability layer *dominates* the global observation well behavior. Chu and Raghavan (1981) reached a similar conclusion in their investigation of non communicating layered reservoirs. They concluded that if a homogeneous reservoir model is used for analysis, the total formation permeability thickness kh can be slightly overestimated but porosity compressibility product $\phi\ c_t$ can be underestimated by a factor of 2 or more.

The responses presented Figure 6.22 and 6.23 are generated with no skin effect on both wells. With this hypothesis, the time of start for the total system equivalent homogeneous behavior is defined, as for double porosity reservoirs, with the λr_D^2 group. If the skin is different between layer 1 and layer 2, the response follows a different behavior. It is difficult to differentiate the influence of the skin factors to that of the diffusivity ratio between the two layers (Chu and Raghavan, 1981), and the match is frequently not unique.

The interference examples of Figure 6.22 and 6.23 show that, even with a low contrast between the layers and an apparently homogeneous active well, the observation well response is clearly *heterogeneous*. This change of well behavior is inverse to that of double porosity reservoir responses, where the heterogeneity is visible in an active well but not always in the observation well.

6.6 PULSE TESTING

6.6.1 Advantages and limitations of the pulse testing method

With pulse testing, the active well is produced in a series of alternate flow and shut-in periods. The rate and the duration of each flow are the same. All shut-in periods also have the same duration, not necessarily equal to the flow time. With three or four pulses, the observation well response is easier to identify in a noisy pressure environment than a single pulse interference signal.

Pulse and interference tests may be used for the estimation of reservoir permeability and storativity, but the choice of the best procedure has not been completely discussed. Some of the parameters considered during the selection of the test procedure are summarized in the following, taking into account the recent advances in accurate pressure measurements and computing facilities.

Model identification

The analysis technique of pulse tests (Johnson et al., 1966; Brigham, 1970; Kamal and Brigham, 1975; Earlougher, 1977) is based on a completely different approach than that for interference tests. Only the amplitude and the time lag of the pressure pulses are considered and the type of reservoir behavior is not identified. A reservoir pressure trend or a possible drift of the gauge does not affect the analysis, but the type of *reservoir behavior* is assumed during the interpretation. Frequently the line source response for a homogeneous infinite reservoir is used but, with the modern analysis tools, the complete catalog of well and reservoir solutions is available.

The question of the choice of the proper interpretation model is not solved yet. When the reservoir behavior is not known from previous tests, only the interference procedure provides a diagnosis of the response by log-log pressure and derivative curve analysis.

Test history

It is frequently believed that the duration of pulse tests is shorter than the time usually needed for interference tests. In practice, the testing time is at least the same with both types of tests (Kamal, 1983) and only the shut-in time is less, which has some economic benefit. It is noteworthy that the period of the pulse test sequence should be different from 12 hours in order to prevent any confusion with a possible tidal effect in the observation well, and the measured time lag must be less than the duration of the flow or shut-in period (Brigham, 1970; El-Khatib, 1991).

Low diffusivity reservoirs

In the case of reservoirs with a low diffusivity ($k/\phi\mu\, c_t$) and a large distance r_D between the wells, the time needed to produce three pressure oscillations in the observation well can be so long that pulse testing is not possible (Bourdarot, 1998). Only the *interference procedure* can be envisaged in such conditions.

High diffusivity reservoirs

Conversely in high permeability reservoirs, the amplitudes of interference responses are low and they can be masked by the noise of the pressure signal. When the wells are close enough to produce a quick response at the observation point, the *pulse testing* approach is the recommended procedure.

6.6.2 Analysis of pulse tests

When the analysis is made by hand, the response amplitude and time lag are measured with the "tangent method" (Johnson, 1966) and the correlation curves presented by Kamal and Brigham (1975) are used to estimate the reservoir permeability and storativity. A homogeneous reservoir response is then considered, the active well is a line source. When the test or the reservoir response does not satisfy these conditions, results from hand analysis are not completely correct.

Several recent studies have discussed the effect of wellbore condition and reservoir heterogeneity on pulse test responses. No manual interpretation method has been proposed and the analysis is made by test history matching on simulated pressure responses.

Wellbore storage effect

The effect of wellbore storage at the active or observation well increases the time lag and reduces the amplitude of the pulse response (Prats and Scott, 1975; Ogbe and Brigham, 1984 a). When the magnitude of the wellbore storage is defined (from the analysis of the producing well response for example), it is possible to introduce this effect on the pulse test simulation used for analysis.

Wellbore storage is an important parameter to consider in the design of a pulse test: when a short pulse period appears feasible (if the inter-well distance is small for example), the flow and the shut-in periods must be longer than the effect of wellbore storage.

Heterogeneous reservoir behavior

The effect of reservoir anisotropy is treated in the same way as for interference tests (Kamal, 1983), the dimensionless time lag and time period are simply changed according to Equation 6.7. Little information is available on the effect of reservoir boundaries on pulse test responses (Vela, 1977) and, to our knowledge, pulse tests in double porosity reservoirs have not been considered yet. Prats (1986) used a numerical approach to investigate pulse test responses in double permeability reservoirs. He concluded that for large distance between wells and low reservoir heterogeneity, the analysis of pulse tests with the homogeneous solution yields a good estimate of the average reservoir properties. When the contrast between layers is large and the inter-well distance short, the results of homogeneous analysis provide a diffusivity ($k/\phi\mu\, c_t$) close to the better layer characteristics. A similar conclusion is presented in the discussion of interference responses in layered reservoirs (Figure 6.23 of Section 6.5).

6.7 CONCLUSION

When only mechanical pressure gauges were used, the main purpose of multiple well tests was to establish communication between wells. With the high accuracy pressure data available today, multiple well testing is recognized as a very powerful testing method, more sensitive to many types of reservoir heterogeneity than single well tests.

The operating conditions are very important in interference tests. In addition to the high accuracy pressure data requirement, the reservoir pressure trend must be defined accurately, and it must remain constant during the test time. The influence of other active wells in the reservoir must be considered, in particular when the production rates are not constant (see discussion of the influence of neighboring producers in Section 10.1.2).

Pulse testing is less affected by a noise in the reservoir pressure but, during the analysis, the model is not identified. Research is needed in pulse testing to develop practical testing procedures. A systematic analysis of pulse test responses generated for different well conditions and reservoir types would be useful to establish a general methodology for test design and interpretation. In case of fissured or layered reservoir for example, additional information is probably needed from cores, logs or producing well analysis, in order to conclude a pulse test interpretation successfully.

CHAPTER 7

APPLICATION TO GAS RESERVOIRS

Two different types of tests are used for gas wells. Historically, the first testing methods were only designed to define the *well deliverability* in order to predict the flow rate, as a function of the wellhead pressure. The results were used in the design of the surface production equipment, setting taxes and also for regulating production, particularly in North America. Backpressure tests and isochronal or modified isochronal tests are the usual deliverability testing methods. The theoretical rate at which the well would flow if the sand face was at atmospheric pressure is called the "*Absolute Open Flow Potential*," AOFP. The analysis of deliverability tests does not yield a description of the well nor of the reservoir.

More recently, *transient testing* has become current practice for gas wells. The analysis provides a description of the producing system, and therefore the well deliverability is also defined.

The transient response of gas wells is very different from that of liquid systems because the fluid properties vary with changes in pressure. The analytical models presented in earlier Chapters for liquid flow are not directly applicable, and the interpretation of transient tests in gas wells is more complex. The test data have to be modified before starting the analysis; in general, the pressure data is converted into "pseudo-pressures". Another difficulty of gas well test interpretation is the high velocity reached by the flow around the wellbore, which affects the well responses by inertial effects or turbulent flow conditions. The skin effect is then rate dependent and changes from one test period to the next.

In this Chapter, the behavior of natural gas is first described, the resulting transient analysis methods are presented and deliverability testing is discussed afterwards, together with the corresponding analysis plots. Only dry gas well responses are considered, test interpretation in multiphase reservoirs is the subject of Chapter 8.

7.1 DESCRIPTION OF GAS WELLS PRESSURE BEHAVIOR

The hypothesis of slightly compressible fluids, used in previous Chapters to describe liquid flow in a porous medium, is not valid for gas systems. In the following sections, the behavior of natural gas is presented, and the pressure responses of gas wells are compared to those of liquid wells.

7.1.1 Gas compressibility and viscosity

The compressibility of gas c_g is a function of the pressure. For a real gas, the equation of state is defined as:

$$pV = ZnRT \tag{7.1}$$

Where Z is the real gas deviation factor. For an ideal gas $Z = 1$, and the compressibility is $c_g = 1/p$. For a real gas, Z changes with the pressure, and the compressibility is expressed as:

$$c_g = \frac{1}{p} - \frac{1}{Z}\frac{\partial Z}{\partial p} \tag{7.2}$$

In gas systems, the viscosity μ is also a function of pressure. The resulting partial differential equation governing the pressure transient response of real gas is not linear and, as opposed to liquid flow, it cannot be solved by analytical methods.

7.1.2 Pseudo-pressure

As shown by Al-Hussainy et al. in 1966 (a and b), by changing the pressure variable to *pseudo-pressure* $m(p)$, the differential equation can be linearized approximately and it becomes very similar to the diffusivity equation for slightly compressible fluids.

The pseudo-pressure, also called "real gas potential", is defined as:

$$m(p) = 2\int_{p_0}^{p} \frac{p}{\mu(p)z(p)} dp \tag{7.3}$$

The reference pressure p_0 is an arbitrary constant, smaller than the lowest test pressure. All pressures are expressed in absolute units, the pseudo-pressure has the unit of $(\text{pressure})^2$ / viscosity, psia^2/cp with the usual system of units. For the calculation of $m(p)$, μ and Z must be known as a function of pressure. When the results of P.V.T. analysis are not available, gas property correlations are used.

For practical analysis of gas well tests, the complete pressure data is converted into pseudo-pressure and, by using $m(p)$ as the working variable, the solutions derived for liquid flow can be applied to the data. The pseudo-pressure change, expressed as $\Delta m(p) = m(p) - m[p(\Delta t = 0)]$, is independent of the reference pressure p_0 of Equation 7.3.

The pseudo-pressure corrects the effect of change of gas viscosity in the calculation of permeability. However, in the case of a large change of gas compressibility during the test, this transformation does not exactly reproduce an equivalent liquid behavior.

7.1.3 Pseudo-time

In 1979 (b), Agarwal introduced pseudo-time, t_{ps}, in order to linearize further the differential equation describing gas flow. The pseudo-time is defined as:

$$t_{ps} = \int_{t_0}^{t} \frac{dt}{\mu(p)c_t(p)} \tag{7.4}$$

μ and c_t are pressure dependent. The calculation of t_{ps} requires the pressure to be known during the *complete flow rate* sequence. Frequently, the pressure has not been recorded during part of the production history, and changing the test data into superposition of pseudo-time becomes cumbersome.

Except in very low pressure wells, when the gas compressibility changes significantly after a small variation of pressure, or during reservoir limit testing with a large depletion, using the pseudo-time does not really improve the quality of the analysis results. These situations are exceptional, and the pseudo time transformation is seldom necessary and usually ignored.

A typical example of low-pressure behavior can be observed in some damaged gas wells. When the last flowing pressure is only a few hundreds of psi, the build-up pressure and derivative curves immediately after shut-in do not follow the usual log-log unit slope wellbore storage straight line, but increase faster (see discussion of changing wellbore storage in Section 10.1.2). By using the pseudo-time, the first part of the response can be corrected to improve the aspect of the log-log match. As the match parameters are not changed, the results of log-log analysis are the same as when the pseudo-pressure is simply expressed versus the elapsed time.

In the case of significant depletion, Bourgeois et al. (1996 c) proposed correcting the analytical model instead of using the pseudo-time. Gas well material balance correction is discussed in Section 7.2.5.

7.2 PRACTICAL TRANSIENT ANALYSIS OF GAS WELL TESTS

7.2.1 Simplified pseudo-pressure for manual analysis

Before computers became generally available, the pseudo-pressure was calculated for each pressure data point by estimating the area of a graph of $2p/\mu Z$ versus p. Such a procedure is time consuming, and a simplified form of $m(p)$ was frequently preferred.

Figure 7.1 is a graph of μZ versus p for a typical natural gas at constant temperature:

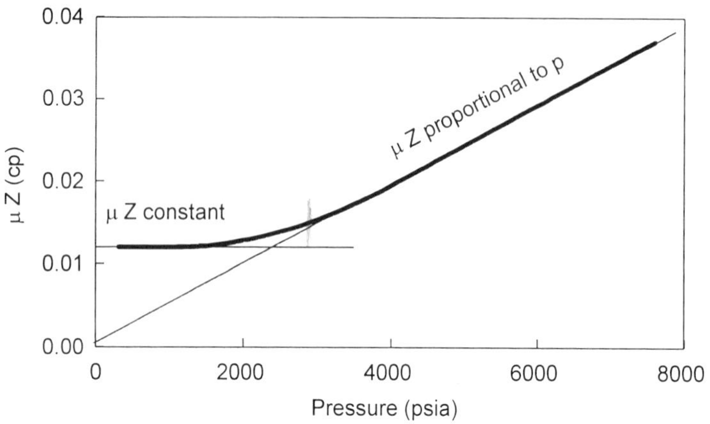

Figure 7.1. Isothermal variation of μZ with pressure. Linear scales.

- When the pressure is less than 2000 psia, the product μZ is almost *constant* and $m(p)$ simplifies into:

$$m(p) = \frac{2}{\mu z} \int_{p_0}^{p} p dp = \frac{p^2 - p_0^2}{\mu_i z_i} \tag{7.5}$$

On *low-pressure* gas wells, it is thus possible to analyze the test in terms of *pressure-squared* p^2.

- When the pressure is higher than 3000 psia, the product μZ tends to be *proportional* to p, $p/\mu Z$ can be considered as a constant and the pseudo-pressure $m(p)$ becomes:

$$m(p) = \frac{2p}{\mu z} \int_{p_0}^{p} dp = (p - p_0) \frac{p_i}{\mu_i z_i} \tag{7.6}$$

On *high-pressure* wells therefore, the gas behaves like a slightly compressible fluid, and the *pressure* data can be used directly for analysis.

- Between 2000 psi and 3000 psia, no simplification is available, and $m(p)$ must be used.

The two limits of validity of the simplified forms (<2000 psia and >3000 psia) are approximate, and depend upon the gas composition and temperature. When $m(p)$ can be estimated with a computer program, the pseudo-pressure is preferably used for the complete range of test pressure. However, the practical engineer sometimes prefers to see the analysis in real pressure or even in pressure squared, rather than $m(p)$ values of 10^{10} psia^2/cp, for which he or she has no "feel".

7.2.2 Definition of the dimensionless parameters

The *gas standard conditions* are used for the definition of the dimensionless terms. In the usual field units, the standard pressure is p_{sc} = 14.7 psia and the temperature is T_{sc} = 520°R (60°F, all temperatures are expressed in absolute units). The gas rate is expressed in standard condition as q_{sc} in 10^3 scft/D or Mscf/D.

The gas properties used for reference in the dimensionless terms depend upon the treatment applied to the raw data, namely the "pseudo-pressure", the "pressure-squared" or the "pressure". When the pseudo-pressure is considered, the dimensionless terms are defined with respect to the gas properties at *initial condition* (subscript *i*) whereas, with the pressure and pressure squared approaches, the properties are defined at the arithmetic *average pressure* of the test (symbol $\bar{\ }$).

In the following section, the dimensionless interpretation variables are given respectively for $m(p)$, p^2 and p.

Dimensionless pressure

The dimensionless pressure (Equation 2.3 for oil) is presented first in the general form, and a simplified expression is given for the usual standard conditions (p_{sc} = 14.7 psia and T_{sc} = 520°R).

$m(p)$:

$$p_D = \frac{kh}{50300 T q_{sc}} \frac{T_{sc}}{p_{sc}} \left[m(p_i) - m(p)\right] = \frac{kh}{1422 T q_{sc}} \left[m(p_i) - m(p)\right] \tag{7.7}$$

p^2:

$$p_D = \frac{kh}{50300 \overline{\mu Z} T q_{sc}} \frac{T_{sc}}{p_{sc}} \left(p_i^2 - p^2\right) = \frac{kh}{1422 \overline{\mu z} T q_{sc}} \left(p_i^2 - p^2\right) \tag{7.8}$$

p:

$$p_D = \frac{kh\bar{p}}{25150 \overline{\mu Z} T q_{sc}} \frac{T_{sc}}{p_{sc}} \left(p_i - p\right) = \frac{kh\bar{p}}{711 \overline{\mu Z} T q_{sc}} \left(p_i - p\right) \tag{7.9}$$

Dimensionless time

For gas wells, the dimensionless time of Equation 2.4 is expressed as:

$m(p)$:

$$t_D = \frac{0.000263k}{\phi \mu_i c_{ti} r_w^2} \Delta t \tag{7.10}$$

p^2 and p:

$$t_D = \frac{0.000263k}{\phi \overline{\mu c_t} r_w^2} \Delta t \tag{7.11}$$

Dimensionless wellbore storage

As for oil wells, the wellbore storage coefficient is expressed in Bbl/psi. In dimensionless terms, the Equation 2.5 for oil wells is changed into:

$m(p)$:

$$C_D = \frac{0.8936C}{\phi c_{ti} h r_w^2} \tag{7.12}$$

p^2 and p:

$$C_D = \frac{0.8936C}{\phi \overline{c_t} h r_w^2} \tag{7.13}$$

Dimensionless time group

The usual dimensionless time group t_D/C_D of Equation 2.6 becomes:

$m(p)$:

$$\frac{t_D}{C_D} = 0.000295 \frac{kh}{\mu_i} \frac{\Delta t}{C} \tag{7.14}$$

p^2 and p:

$$\frac{t_D}{C_D} = 0.000295 \frac{kh}{\overline{\mu}} \frac{\Delta t}{C} \tag{7.15}$$

7.2.3 Straight line parameters

The different characteristic flow regimes presented in previous Chapters (wellbore storage, linear flow, spherical flow, radial flow and, to some extent, pseudo steady state) are always described by straight lines on a plot of pressure versus a specialized time scale. For gas wells, the equation of the corresponding straight line can be obtained in terms of $m(p)$, p^2 or p, by introducing the appropriate definition (Equations 7.7 to 7.13) into the dimensionless Equations corresponding to these flow regimes (Equations 5.1 and 5.14 for example in the case of parallel sealing faults or a closed system).

In the following section, the straight-line equations for the wellbore storage regime and infinite acting radial flow are presented in the general form, and in terms of the usual standard conditions (p_{sc} = 14.7 psia and T_{sc} = 520°R).

Wellbore storage analysis

On a Cartesian plot, the wellbore storage straight-line slope m_{WB} of Equation 1.9 becomes, for gas:

$m(p)$:

$$m_{\text{WB}} = 14.85 \frac{T q_{sc}}{\mu_i C} \frac{p_{sc}}{T_{sc}} = 0.4197 \frac{T q_{sc}}{\mu_i C} \tag{7.16}$$

p^2:

$$m_{\text{WB}} = 14.85 \frac{\overline{Z} T q_{sc}}{C} \frac{p_{sc}}{T_{sc}} = 0.4197 \frac{\overline{Z} T q_{sc}}{C} \tag{7.17}$$

p:

$$m_{\text{WB}} = 7.425 \frac{\overline{Z} T q_{sc}}{\overline{p} C} \frac{p_{sc}}{T_{sc}} = 0.2098 \frac{\overline{Z} T q_{sc}}{\overline{p} C} \tag{7.18}$$

Semi-log analysis

The slope m of the semi-log straight line, expressed in Equation 1.15 for oil, is now defined as:

$m(p)$:

$$m = 5.791\text{x}10^4 \frac{Tq_{sc}}{kh}\frac{p_{sc}}{T_{sc}} = 1.637\text{x}10^3 \frac{Tq_{sc}}{kh} \tag{7.19}$$

p^2:

$$m = 5.791\text{x}10^4 \frac{\overline{\mu Z}Tq_{sc}}{kh}\frac{p_{sc}}{T_{sc}} = 1.637\text{x}10^3 \frac{\overline{\mu Z}Tq_{sc}}{kh} \tag{7.20}$$

p:

$$m = 2.896\text{x}10^4 \frac{\overline{\mu Z}Tq_{sc}}{\overline{p}kh}\frac{p_{sc}}{T_{sc}} = 8.183\text{x}10^2 \frac{\overline{\mu Z}Tq_{sc}}{\overline{p}kh} \tag{7.21}$$

and the skin equation 1.17 for drawdown is changed into:

$m(p)$:

$$S' = 1.151\left(\frac{\Delta m(p)_{1\,\text{hr}}}{m} - \log\frac{k}{\phi\mu_i c_{ti} r_w^2} + 3.23\right) \tag{7.22}$$

p^2 and p:

$$S' = 1.151\left(\frac{\Delta p_{1\,\text{hr}}^2 \text{ or } \Delta p_{1\,\text{hr}}}{m} - \log\frac{k}{\phi\overline{\mu c_t} r_w^2} + 3.23\right) \tag{7.23}$$

The pseudo-skin S' is the global skin, including the wellbore damage, possibly a geometrical skin component, and the inertial-turbulent flow effect discussed in the next section. It is frequently rate dependent, the higher is the flow rate, the higher is S'.

7.2.4 Non-Darcy flow

Inertial and turbulent flow effects

Due to the high velocity of the flow in the immediate surroundings of the well, *inertial effects* are frequently not negligible. In some cases, the flow is not even laminar but becomes *turbulent*. Darcy's law is then no longer applicable in the vicinity of the well, and the inertial and turbulent effects produce an additional pressure drop (Houpeurt, 1959; Wattenbarger, 1968; Mattar and Brar, 1975). The skin coefficient S', measured during well tests, is expressed with a rate dependent term as:

$$S' = S + Dq_{sc} \tag{7.24}$$

where D is called the non-Darcy flow coefficient. In order to separate the two components of the skin effect, S' has to be evaluated at several rates.

Estimation of S and D

During drawdown periods in a multiple rate test, the total skin effect of Equation 7.24 is, for period #n, $S'_n = S + D\, q_{n,sc}$, and during the previous period #n-1, $S'_{n-1} = S + D\, q_{n-1,sc}$. The corresponding pressure drops due to skin are $\Delta m(p)_{S'_n} = \left(1422T\, q_{n,sc}/kh\right)S'_n$ and $\Delta m(p)_{S'_{n-1}} = \left(1422\, T q_{n-1,sc}/kh\right)S'_{n-1}$. When the superposition method is used (see Chapter 2), the period #n is analyzed for a change of flow rate $(q_n - q_{n-1})_{sc}$, and the pressure at the start of the period is the last pressure of period #n-1. The skin pressure drop measured on period #n is expressed with the difference (Bourdarot, 1998):

$$\begin{aligned}\Delta m(p)_{skin} &= \left(1422T/kh\right)\left(q_{n,sc}S'_n - q_{n-1,sc}S'_{n-1}\right)\\ &= \left(1422T/kh\right)\left(q_n - q_{n-1}\right)_{sc} S + D\left(q_n^2 - q_{n-1}^2\right)_{sc}\\ &= \left(1422T/kh\right)\left(q_n - q_{n-1}\right)_{sc}\left[S + D\left(q_n + q_{n-1}\right)_{sc}\right]\end{aligned} \tag{7.25}$$

During *shut-in* periods ($q_n = 0$) and during the periods immediately *after shut-in* ($q_{n-1} = 0$), the measured skin coefficient corresponds to S' of equation 7.24 but, for all other flow periods it corresponds to $[S + D\,(q_n + q_{n-1})_{sc}]$.

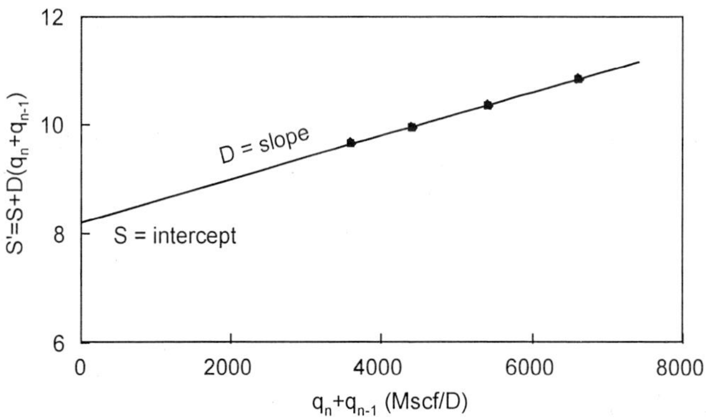

Figure 7.2. Variation of the pseudo skin with the rate (q_n+q_{n-1}). Linear scales.

In Figure 7.2, the pseudo-skin $[S + D\,(q_n + q_{n-1})_{sc}]$ is plotted on linear scales versus the rate term $(q_n + q_{n-1})_{sc}$. The straight-line intercept gives the true skin S and the slope gives the non-Darcy flow coefficient D. When only build-up periods are analyzed in a multiple rate test, the linear plot simplifies into S'_i versus q_{i-1}.

7.2.5 Material balance correction

During a reservoir limit test, the average pressure can be estimated with the usual material balance relationship:

$$\frac{\bar{p}}{\bar{Z}} = \frac{p_i}{Z_i}\left(1 - \frac{G_p}{G_i}\right) \tag{7.26}$$

where G_i is initial gas volume and G_p the cumulative gas production.

When the pseudo-steady state Equation 5.14 is used with the pseudo-pressure of Equation 7.7, the resulting average pressure $\bar{p}_{pp}$ is different (Equation 5.21 for oil wells) :

$$\begin{aligned} m(\bar{p}_{pp}) &= m(p_i) - 2.349\frac{T q_{sc}}{\phi \mu_i\, c_{ti}\, hA}\Delta t \\ &= m(p_i) - 2.349\frac{T}{\phi \mu_i\, c_{ti}\, hA}G_p \end{aligned} \tag{7.27}$$

As mentioned in Section 7.1.3, transforming the pressure variable into pseudo-pressure does not correct the changes in gas compressibility when there is significant depletion. Instead of using the pseudo-time, Bourgeois, and Wilson (1996 c) propose correcting the analytical model by $m(\bar{p}) - m(\bar{p}_{pp})$.

For each time step of the calculation, the cumulative gas production G_p is used to estimate $\bar{p}$ from the material balance relationship of Equation 7.26, and then $\bar{p}$ is changed into $m(\bar{p})$. The second term $m(\bar{p}_{pp})$ is estimated from Equation 7.27. With this correction, the model response follows the linear relationship of p/Z versus the cumulative production, and therefore the overall material balance is honored.

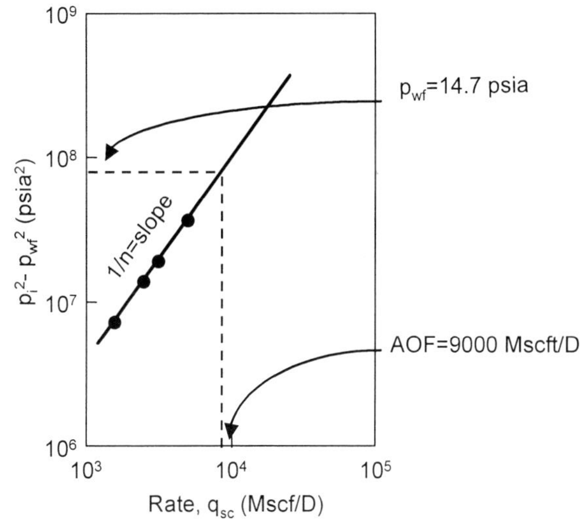

Figure 7.3. Deliverability plot for a backpressure test. Log-log scale, pressure-squared method.

7.3 DELIVERABILITY TESTS

Even though it is possible to predict gas flow rate against wellhead pressure from the results of transient test analysis, the deliverability testing procedure is still used on gas wells. By flowing the well at several different rates, deliverability tests are in fact well adapted to the analysis of the non-Darcy flow effect. The two analysis methods, transient and deliverability, can be applied to the data.

7.3.1 Deliverability equations

Empirical approach

In 1936, Rawlins and Schellhardt presented an empirical relationship between flow rates and the *stabilized* flowing pressures p_{wf}:

$$q_{sc} = C\left(p_i^2 - p_{wf}^2\right)^n \tag{7.28}$$

where the pressures p_i and p_{wf} are in absolute units, C and n are two constant terms. The coefficient n can vary from 1 in the case of laminar flow to 0.5 when the flow is fully turbulent.

As shown in Figure 7.3, a log-log plot of $(p_i^2 - p_{wf}^2)$ versus q_{sc} yields the "*stabilized deliverability straight line*", of slope $1/n$. The Absolute Open Flow (AOF) Potential is estimated by extrapolating the stabilized deliverability line to atmospheric pressure (p_{wf} = 14.7 psia). The deliverability analysis illustrated in Figure 7.3 is also called the "*C* & *n*" or Fetkovich (1973) method.

In infinite or closed systems, the pressure never stabilizes during drawdown: it first follows the semi-log approximation and, during pseudo-steady state regime, it is a linear function of time (see Chapter 5.4). A stabilized flowing pressure condition is seen only when the gas reservoir is in contact with a constant pressure support. In the empirical relationship of Equation 7.28, the flowing pressures p_{wf} is said to be stabilized when the pressure variation is no longer measurable. It is therefore related to the *sensitivity* of the pressure gauge.

Theoretical approach

A more rigorous treatment of deliverability tests in closed systems can be developed if the stabilized pressure is replaced by the pressure during the pseudo-steady state regime (see Chapter 5.4). In theory, the initial pressure p_i has to be replaced by the average reservoir pressure $\bar{p}$, but it is assumed that the depletion is negligible and $\bar{p}$ is constant. A new form of the deliverability equation, including a term q_{sc}^2 for turbulent flow is obtained.

In a closed system, the difference between the pseudo-steady state flowing pressure p_{wf} and the following shut-in average pressure $\bar{p}$ is constant (see Section 5.4.7). When the pseudo-pressure of Equation 7.7 is used, Equation 5.20 gives the laminar-inertial-turbulent (LIT) flow relationship:

$$m(\bar{p}) - m(p_{wf}) = 1637 \frac{T}{kh}\left(\log \frac{A/r_w^2}{C_A} + 0.351 + 0.87S\right) q_{sc} + 1422 \frac{T}{kh} D q_{sc}^2 \tag{7.29}$$

With a circular reservoir of radius r_e, C_A = 31.62 and the Equation 7.29 reduces to (Houpeurt, 1959; Mattar and Brar, 1975):

$$m(\bar{p}) - m(p_{wf}) = 1637 \frac{T}{kh}\left(2 \log \frac{0.472 r_e}{r_w} + 0.87S\right) q_{sc} + 1422 \frac{T}{kh} D q_{sc}^2 \tag{7.30}$$

Before the pseudo-steady state regime, the response follows the semi-log approximation and $\Delta m(p)$ is expressed as:

$$m(\bar{p}) - m(p_{wf}) = 1637 \frac{T}{kh}\left(\log \frac{k \Delta t}{\phi \mu_i c_{ti} r_w^2} + 3.23 + 0.87S\right) q_{sc} + 1422 \frac{T}{kh} D q_{sc}^2 \tag{7.31}$$

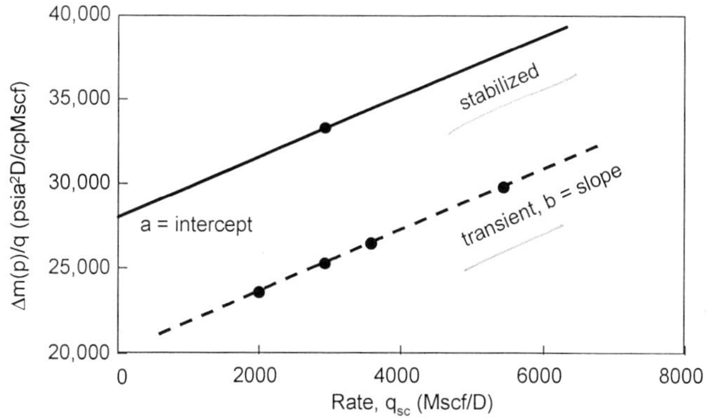

Figure 7.4. Deliverability plot for an isochronal or a modified isochronal test. Linear scale, pseudo-pressure method.

The two $\Delta m(p)$deliverability relationships can be expressed as $a(t)q_{sc} + bq_{sc}^2$. During the infinite acting regime, $a(t)$ of Equation 7.31 is an increasing function of the time whereas "a" is constant when pseudo-steady state is reached (Equation 7.30). The coefficient "b" is the same in the two equations.

In Figure 7.4, the ratio $\lfloor m(p_i) - m(p_{wf}) \rfloor / q_{sc}$ is expressed versus q_{sc} on linear coordinates (Houpeurt or LIT or "a & b" method). The two characteristic behaviors of Equations 7.29 and 7.31 are described by two parallel straight lines of slope "b". The lower line, of intercept $a(t_p)$, describes the *transient* deliverability response a*t* t_p, and the higher line is the "*stabilized deliverability straight line*", of intercept "a". The Absolute Open Flow Potential is estimated by solving the quadratic Equation 7.30 at p_{wf} = 14.7 psia:

$$q_{sc,\text{AOF}} = \frac{-a + \sqrt{a^2 + 4b\left(m(\bar{p}) - m(14.7)\right)}}{2b} \tag{7.32}$$

7.3.2 Back pressure tests

The well is produced to stabilized pressure at three or four increasing rates $q_{sc,j}$ (Figure 7.5), the different flow periods have the same duration. This testing sequence is also called a "Flow after flow test". In low permeability reservoirs, the total production time can be relatively long.

When the empirical approach is used, the stabilized pressures $p_{wf,j}$ and rates $q_{sc,j}$ are plotted on log-log scale with (p_i^2 - p_{wf}^2) versus q_{sc}, as shown Figure 7.3. If the pseudo-pressure is preferred, the deliverability plot is as shown in Figure 7.6, with $\lfloor m(p_i) - m(p_{wf}) \rfloor / q_{sc}$ versus q_{sc}. The intercept "a" and the slope "b" of the stabilized deliverability straight line are measured, and the AOF is estimated from Equation 7.32.

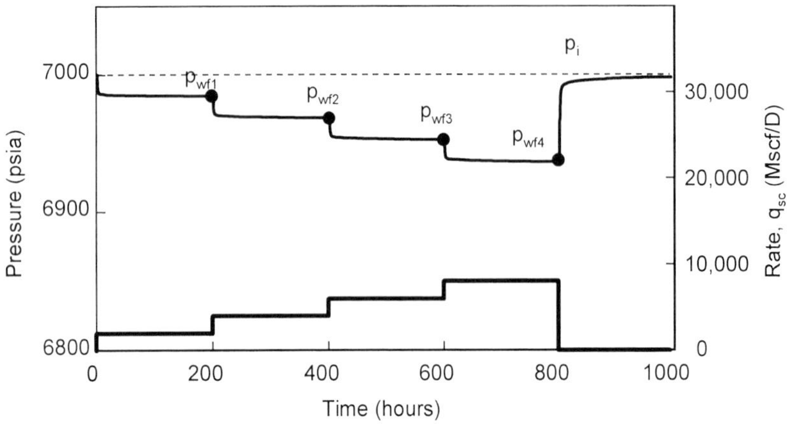

Figure 7.5. Pressure and rate history for a backpressure test.

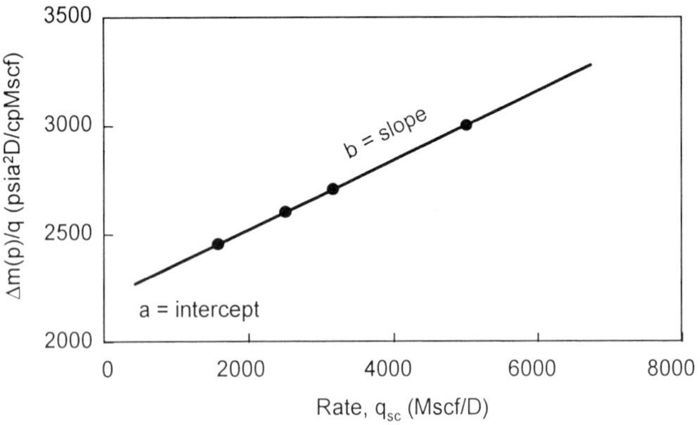

Figure 7.6. Deliverability plot for a backpressure test. Linear scale, pseudo-pressure method.

Backpressure tests were the first method to evaluate well deliverability. The testing procedure is time consuming, a large volume of gas is produced and there is only *one build-up* period. Since the procedure is not well adapted to transient analysis, the isochronal or modified isochronal tests are frequently preferred.

7.3.3 Isochronal tests

With isochronal tests, the well is again produced at three or four increasing rates but a shut-in period is introduced between each flow. The drawdown periods at $q_{sc,j}$ are stopped during the infinite acting regime after the same production time t_p, and the intermediate build-ups last until the pressure is back to *initial conditions* p_i. The final flow is extended, sometimes with a reduced flow rate, to reach the stabilized flowing pressure (Figure 7.6).

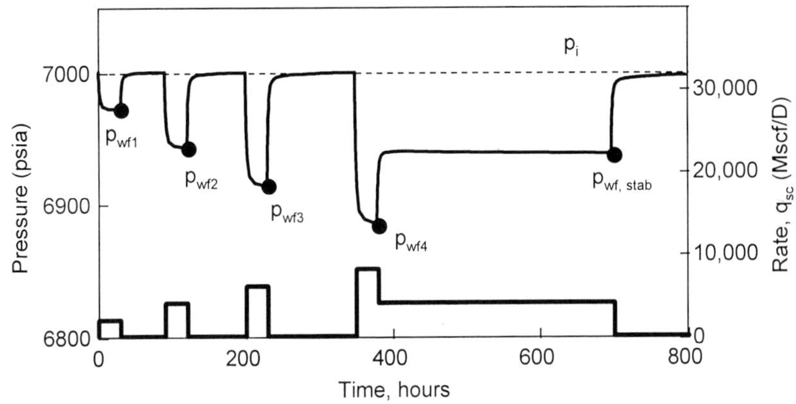

Figure 7.7. Pressure and rate history for an isochronal test.

The total produced gas is smaller during isochronal tests than with the back pressure procedure but, due to the shut-in periods, the test duration is relatively long. A complete transient analysis can be carried out on each of the build-up periods of the isochronal test procedure.

On the log-log plot (p_i^2 - p_{wf}^2) versus q_{sc} of Figure 7.7, the different pressure points $p_{wf,j}$ measured at t_p follow a transient deliverability straight line parallel to the stabilized deliverability line. The slope $1/n$ is then defined from the short flow periods, and the final deliverability line can be drawn with a slope of $1/n$ through the point corresponding to the stabilized pressure of the extended flow.

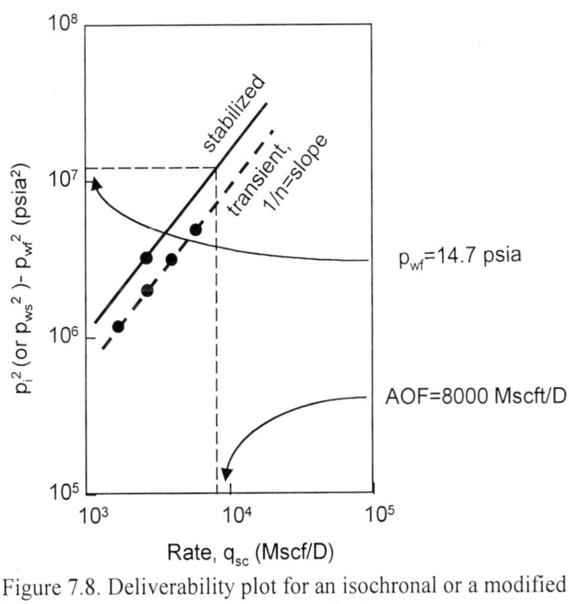

Figure 7.8. Deliverability plot for an isochronal or a modified isochronal test. Log-log scale, pressure-squared method.

With the pseudo-pressure method, the pressure points $p_{wf,j}$ follow a straight line on the linear plot of $\lfloor m(p_i) - m(p_{wf}) \rfloor / q_{sc}$ versus q_{sc} (Figure 7.4). The slope is "b" and intercept "$a(t_p)$". The stabilized deliverability straight line is then drawn parallel and passing through the point corresponding to the stabilized pressure of the extended flow period. The AOF is estimated from Equation 7.32.

7.3.4 Modified isochronal tests

With the modified isochronal sequence, the procedure is similar to isochronal tests except that the intermediate shut-in periods have the *same duration* as the drawdown periods and, as shown in Figure 7.9, only the last flow is extended until the stabilized flowing pressure is reached. The total test duration is relatively short, and several build-ups are available for transient analysis.

For the pressure points measured at t_p, the difference on the y axis of the deliverability plots are calculated with reference to the pressure at start of the period $p_{ws,j}$ (Katz et al., 1959), with $p_{ws,1} = p_i$. The point for the final stabilized flowing pressure is expressed with respect to the initial pressure p_i.

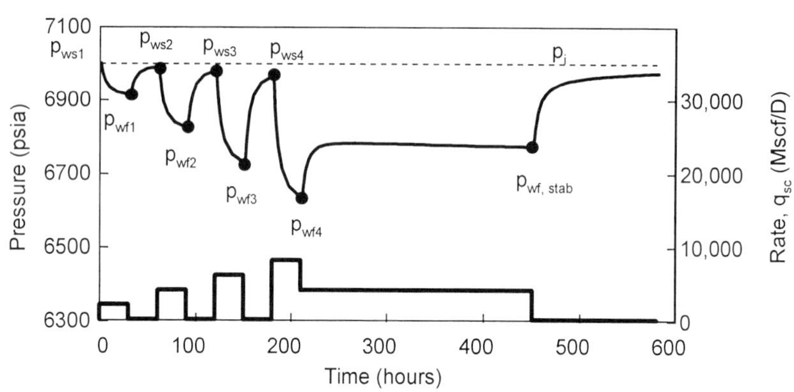

Figure 7.9. Pressure and rate history for a modified isochronal test.

With the empirical procedure, the log-log plot of $(p_{ws}^2 - p_{wf}^2)$ versus q_{sc} displays the transient deliverability straight line of slope $1/n$. The stabilized pressure of the extended flow period is used to define the stabilized deliverability straight line parallel to the first, and passing through the point $(p_i^2 - p_{wf}^2)$ (Figure 7.7). With the pseudo-pressure method (Figure 7.4), the plot of $\lfloor m(p_{ws}) - m(p_{wf}) \rfloor / q_{sc}$ versus q_{sc} defines the transient straight line of slope "b" and intercept "$a(t_p)$". The parallel stabilized deliverability straight line is drawn through the extended flow period point, $\lfloor m(p_i) - m(p_{wf}) \rfloor / q_{sc}$, and "$a$" is known. The AOF can be estimated from Equation 7.32.

7.4 FIELD EXAMPLE

Figure 7.10 presents an example of a four point modified isochronal test sequence. When the build-up periods are compared on a rate normalized log-log plot (Figure 7.11), the different derivative responses are correctly superposed during the radial flow stabilization. It can be concluded that the flow rates are accurately measured during the multiple rate test sequence.

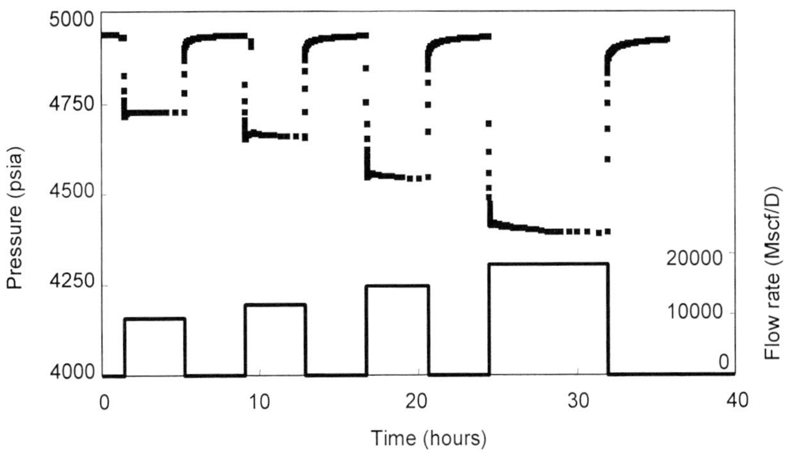

Figure 7.10. Modified isochronal test. Linear scale.

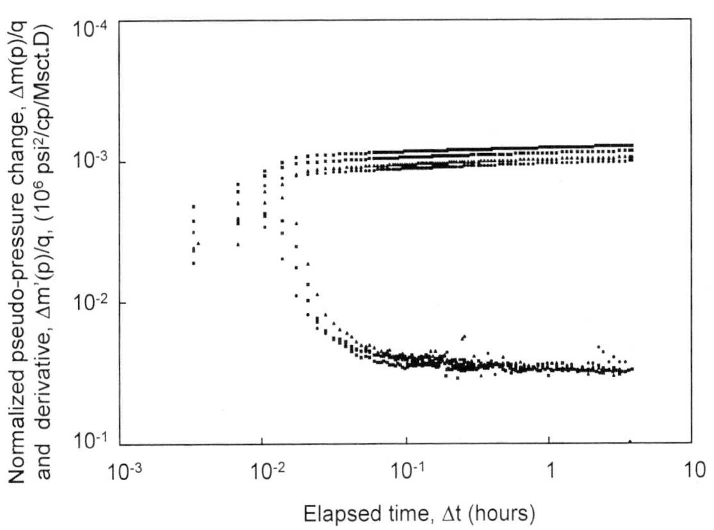

Figure 7.11. The four build-up periods.
Log-log scale with the rate normalized pseudo-pressure and derivative.

The following conclusions are established from the examination of the log-log plot of Figure 7.11:

1. The radial flow regime is reached after approximately 0.1 hours. The duration of the shut-in periods (and the three first flow periods) could have been reduced.
2. During the five initial minutes of shut-in, the wellbore storage effect is not constant. Due to the variable gas compressibility at early shut-in times, the shape of the derivative hump does not correspond to the usual wellbore storage behavior described on the type-curve of Figure 2.22 (see Section 10.1.2).
3. Even though the derivative responses are correctly superposed, the pressure curves are not. The skin is changing from one period to next; the well is affected by the non-Darcy flow condition common with gas wells.

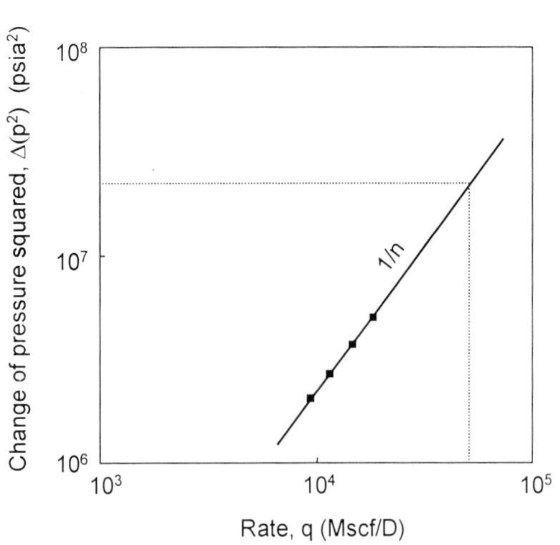

Figure 7.12. Deliverability plot of the modified isochronal test. Log-log scale of Δp^2 vs. q.

On the deliverability plot of Figure 7.12, the four flow periods define a single deliverability line: the flowing pressures are close to stabilization at times of shut-in. The straight-line slope indicates $n = 0.72$, confirming thus a moderate non-Darcy flow effect.

CHAPTER 8

APPLICATION TO MULTIPHASE RESERVOIRS

The well test interpretation methods described in previous Chapters are designed for wells producing a single-phase fluid, either liquid or gas. At surface, most wells produce oil and gas phases but the reservoir fluid can frequently be considered as a single phase fluid, and the usual interpretation methods for liquid or gas are applicable.

This Chapter describes test interpretation when the flow in the reservoir is multiphase. A two-phase flow condition is encountered in solution gas drive reservoirs, when the flowing pressure in the reservoir is below the bubble point, or in retrograde gas condensate reservoirs below dew-point pressure. When water is produced, the flowing fluid can be three phases. Multiphase flow is different from the wellbore phase segregation phenomenon discussed in Chapter 10, where the second phase appears in the wellbore.

When several phases are flowing together in a porous system, each reduces the ability of the other phases to flow and the effective permeability for each phase is *less* than the permeability for a single phase. The relative permeability of each phase is a function of its saturation, which is not constant, either in space or in time. For example, in reservoirs flowing below the bubble point, the effective permeability to oil is reduced by the high gas saturation near the well. In retrograde gas condensate reservoirs flowing below the dew point, the liquid saturation is highest around the well. The gas flow is more restricted in this area than in the outer reservoir region, and the presence of a liquid phase originated by retrograde condensation produces a *non-Darcy flow* condition (Kniazeff and Naville, 1965). As for gas wells (see Chapter 7), the skin effect S' can be *rate dependent* in multiphase systems and change from one test period to the next.

Several approaches have been proposed to analyze tests in multiphase reservoirs but they all make simplifications for modeling the flow behavior. The simplest and most widely used is the method presented by Perrine in 1956: an equivalent liquid flow rate is considered, and the analysis methods of oil wells are applied. Other methods have been presented, using multiphase pseudo-pressure (Fetkovich, 1973; Raghavan, 1976) or pressure squared (Al-Khalifah et al., 1987). These are more difficult to use, the quality of the results depends largely upon the validity of the saturation curves and the P.V.T. properties introduced in the models.

In this Chapter, only the Perrine method is described in detail, the pressure squared and the pseudo-pressure methods are simply introduced.

8.1 PERRINE'S METHOD

The method is a modified single-phase approach. An *equivalent liquid* of constant properties is defined as the sum of the three phases: oil, water and gas (Perrine, 1956; Martin, 1959). The analysis yields the effective mobility of this equivalent fluid, but it does not give the absolute reservoir permeability directly.

8.1.1 Hypothesis

1. The three phases are uniformly *distributed* in the reservoir
2. The *saturations* are constant and independent of the pressure
3. The *capillary pressures* are neglected; the pressure is the same in the different phases.

In practice, the saturation gradients are not always negligible and the equivalent single-phase fluid concept of Perrine can fail to describe multiphase flow (Weller, 1966; Raghavan, 1989). This is particularly the case in the low-pressure area around the wellbore, where the saturations can change significantly compared to the outer reservoir zone.

8.1.2 Definitions

The flow rate of the equivalent single-phase fluid is defined, at sand-face conditions, as:

$$\begin{aligned}(qB)_t &= q_o B_o + q_w B_w + 1{,}000 q_g B_g \\ &= q_o B_o + q_w B_w + \left(1{,}000 q_{sg} - q_o R_s\right) B_g\end{aligned} \tag{8.1}$$

where q_{sg} is the gas rate measured at surface (expressed in Mcf/D), and $q_o R_s$ the dissolved gas at bottom hole conditions.

Perrine uses an empirical definition for the total mobility $(k/\mu)_t$ of the equivalent single-phase fluid. It is expressed as the sum of the effective phase mobility:

$$(k/\mu)_t = k_o/\mu_o + k_w/\mu_w + k_g/\mu_g \tag{8.2}$$

The effective total compressibility c_t is expressed as (Martin, 1959):

$$c_t = c_f + S_o c_o + S_w c_w + S_g c_g + S_o \left(B_g/B_o\right)\frac{\partial R_s}{\partial p} + S_w \left(B_g/B_w\right)\frac{\partial R_{sw}}{\partial p} \tag{8.3}$$

The two last terms of Equation 8.3 correspond to free gas being liberated (or dissolved) in the oil and the water phases after a change of pressure. The component of the compressibility due to the change of gas-oil ratio at reservoir condition is frequently larger than the other terms of Equation 8.3.

8.1.3 Practical analysis

Using the equivalent single-phase liquid concept and the effective compressibility, all interpretation methods for oil wells can be applied directly to multiphase systems. The analysis does not provide the absolute permeability, but only the *total mobility* $(k/\mu)_t$.

Dimensionless terms

The multiphase dimensionless pressure (Equation 2.3 for oil) is:

$$p_D = \frac{(k/\mu)_t h}{141.2(qB)_t} \Delta p \tag{8.4}$$

The dimensionless time (Equation 2.4 for oil):

$$t_D = \frac{0.000264(k/\mu)_t}{\phi c_t r_w^2} \Delta t \tag{8.5}$$

The dimensionless wellbore storage coefficient (Equation 2.5 for oil) is not changed :

$$C_D = \frac{0.8936C}{\phi c_t h r_w^2} \tag{2.5}$$

The dimensionless time group t_D/C_D of Equation 2.6 is:

$$\frac{t_D}{C_D} = 0.000295 \frac{(k/\mu)_t h}{C} \Delta t \tag{8.6}$$

Semi-log analysis

The slope m of the semi-log straight line, expressed in Equation 1.15 for oil, is:

$$m = 162.6 \frac{(qB)_t}{(k/\mu)_t h} \tag{8.7}$$

and, for drawdown periods, the skin equation 1.17 is expressed:

$$S = 1.151\left(\frac{\Delta p_{1\,\mathrm{hr}}}{m} - \log\frac{(k/\mu)_t}{\phi c_t r_w^2} + 3.23\right) \tag{8.8}$$

8.1.4 Discussion of the Perrine's method

The results of Perrine's method become less reliable with increasing gas saturation (Weller, 1966). Furthermore, by ignoring the changes of saturation around the wellbore, the skin can be *over-estimated* and the effective permeability is frequently underestimated (Al-Khalifah, 1987; Raghavan, 1989).

When the relative permeability $k_{r"o,w,g"}$ of the different phases is known, the absolute permeability can be estimated from the effective mobility $(k/\mu)_t$:

$$(k/\mu)_t = k\left(k_{ro}/\mu_o + k_{rw}/\mu_w + k_{rg}/\mu_g\right) \tag{8.9}$$

The individual phase mobility $(k/\mu)_{"o,w,g"}$ is sometimes estimated by introducing the corresponding phase flow rate in the permeability Equations 8.4 and 8.7. This procedure appears questionable when the saturations are not uniform or the producing fluid ratio is not representative of the investigated zone.

8.2 PSEUDO-PRESSURE METHOD

Multiphase pseudo pressure functions have been proposed for tests in solution gas drive reservoirs and gas condensate reservoirs. As opposed to the real gas pseudo-pressure of Al-Hussainy et al. (1966 a) presented in Chapter 7, the multiphase pseudo-pressure uses *relative permeability* data.

The sensitivity of the pseudo-pressure methods to the quality of the relative permeability curves deserves further examination. To our knowledge, the multiphase analysis methods presented below have not been used intensively in the industry. The methods need to be validated with documented examples.

8.2.1 Solution gas drive reservoirs

Definitions

In 1973, Fetkovich defined the multiphase pseudo-pressure as:

$$m(p) = \int_0^p \frac{k_{ro}(S_o)}{\mu_o B_o} dp \qquad (8.10)$$

where the relative oil permeability k_{ro} is a function of the oil saturation S_o.

The relationship between pressure and saturation must be known to calculate the integral $m(p)$. During drawdown periods in solution gas drive reservoirs, the pressure gradients and the saturation profiles around the well depend upon the rate sequence. Thus, the multiphase pseudo-pressure function is not only dependent on pressure, but also on the test history. During shut-in periods, the saturation-pressure relationship used for the $m(p)$ calculation is different. The pressure build-up reflects the initial conditions (Raghavan, 1976; Boe et al., 1981), namely the pressure distribution and saturation profile at time of shut-in (Δt=0). Contrarily to the gas pseudo-pressure of Al-Hussainy et al., the multiphase pseudo-pressure is *not the same* for drawdown and for build-up periods.

Interestingly, $m(p)$ can be generated with wellbore data only as the saturation at sand face S_o is used in Equation 8.10 (Raghavan, 1976; Boe et al., 1981). Raghavan (1989) presented a method in which the permeability ratio k_g/k_o is estimated from the gas-oil ratio Equation 8.11. For drawdown, the instantaneous producing gas-oil ratio R versus the pressure is used. For build-up, it is assumed that R is constant during the shut-in period, and it is equal to the producing gas-oil ratio at time of shut-in (Δt =0).

$$R = R_s + \frac{k_g \mu_o B_o}{k_o \mu_g B_g} \qquad (8.11)$$

The relative permeability curves are used to convert k_g/k_o into the oil saturation at sand face S_o, and finally into k_{ro} versus the wellbore pressure.

Discussion

The multiphase dimensionless pressure (Equation 2.3 for oil) is:

$$p_D = \frac{kh}{141.2 q_o} \Delta m(p) \qquad (8.12)$$

and the dimensionless time (Equation 2.4 for oil) is defined with the initial system properties :

$$t_D = \frac{0.000264k}{\phi \mu_i c_{ti} r_w^2} \Delta t \qquad (8.13)$$

The validity of the multiphase pseudo-pressure function has been checked against numerically simulated pressure data (Raghavan, 1989). During the transient period, the method gives the absolute formation permeability and corrects for the skin effect due to free gas saturation near the wellbore. During pseudo steady state flow the results are less accurate.

The main drawback of the pseudo-pressure is that, as the calculation of the pseudo pressure is not the same during flow and shut-in periods, drawdown and build-up are not exactly reversible. This implies that the concept of superposition cannot be used, and therefore the analysis of transient tests becomes questionable.

8.2.2 Gas condensate reservoirs

Definitions

For gas condensate reservoirs, Jones et al. (1988, 1989) developed a pseudo pressure function with the same concept as the solution gas drive Equation 8.10, but it is expressed with the two phases:

$$m(p) = \int_{p0}^{p} \left(\rho_o \frac{k_{ro}}{\mu_o} + \rho_g \frac{k_{rg}}{\mu_g} \right) dp \tag{8.14}$$

where $\rho_{o,g}$ is the molar density of the oil and gas phases. It is assumed that the pressure drops below dew-point pressure around the well, but the outer reservoir region is still above dew point, in single-phase gas.

In order to express the saturation with the pressure, Jones and Raghavan (1988) suggest using a steady-state relationship between the relative permeability for oil (k_{ro}) and for gas (k_{rg}):

$$\frac{k_{ro}}{k_{rg}} = \frac{\rho_g \mu_o L}{\rho_o \mu_g V} \tag{8.15}$$

Here, L and V are the mole fractions of liquid and vapor, for each step of equilibrium of a Constant-Composition-Expansion test.

In the calculation of the integral Equation 8.14 with respect to the wellbore pressure, an equation of state is used to define molar density and viscosity, and relative permeability curves are needed.

Discussion

In transient analysis of gas condensate reservoirs, the estimated skin factor appears smaller when the multiphase pseudo pressure function is used instead of the usual gas pseudo pressure of Al-Hussainy et al (1966). The multiphase analysis takes into account the saturation profile when the pressure is below dew point, and therefore it corrects for the "damage" caused by condensate drop-out near the well. On the other hand, the steady-state approximation used in Equation 8.15 can introduce errors in the total mobility.

When the single phase gas pseudo pressure is used for analysis, the radial changes of mobility due to variable saturations are not accounted for, and test pressure responses usually describe a *composite* reservoir behavior. As for falloff tests in injection wells (see Section 4.3.5), Gringarten et al. (2000) defines several zones around a well flowing below dew point. Between the outer reservoir region, at initial liquid saturation, and the near wellbore region where the two phases are flowing and the apparent gas mobility is reduced, a transition zone with increasing condensate saturation develops. Possibly, a fourth region with a low liquid saturation and an increased gas mobility can be present in the immediate vicinity of the wellbore.

8.3 PRESSURE SQUARED METHOD

Because of the limitations of the pseudo-pressure approach discussed earlier, and when relative permeability curves are not available, the pressure squared method of Al-Khalifah et al. (1987) offers an interesting alternative for the analysis of multiphase flow tests.

Definitions

Al-Khalifah et al. showed that, by approximating the group ($k_o/\mu_o B_o$) to a linear function of the pressure, the multi-phase flow equation can be simplified to a diffusivity equation expressed in terms of pressure squared. Therefore, multi-phase pressure data can be analyzed with p^2 instead of p, using the usual interpretation solutions.

Calling "a" the constant proportionality,

$$\frac{k_o}{\mu_o B_o} = a\,p \tag{8.16}$$

the multiphase dimensionless pressure (Equation 2.3 for oil) is defined with respect to the oil parameters a and q_o:

$$p_D = \frac{ah}{282.4 q_o} \Delta\left(p^2\right) \tag{8.17}$$

The effective total compressibility and the dimensionless time are the same as with Perrine's method (respectively Equation 8.3 and, using the effective mobility $(k/\mu)_t$, Equation 8.5).

For semi-log analysis with p^2 versus log Δt, the straight-line slope m is expressed as:

$$m = 325.2 \frac{q_o}{ah} \tag{8.18}$$

and, for drawdown periods, the skin equation is:

$$S = 1.151\left(\frac{p_{1\,\mathrm{hr}}^2 - p_i^2}{m} - \log \frac{(k/\mu)_t}{\phi c_t r_w^2} + 3.23 \right) \tag{8.19}$$

Several choices of reference pressure p_{ref} are proposed for evaluating the empirical slope "a" in Equation 8.16. In the case of high volatile oil, and in low volatile oil when the drawdown is low, the authors suggest estimating this constant at initial pressure p_i for drawdown periods and, for build-ups, they recommend using the average reservoir pressure $\bar{p}$. When the oil volatility is low and the drawdown high, $p_{ref}=p_{wf}(\Delta t=0.1\mathrm{hr})$ for drawdown, and $p_{ref}=p_{ws}(\Delta t=10\mathrm{hr})$ for shut-in.

The effective oil permeability is obtained from the pressure match of Equation 8.17 (or the semi-log straight line Equation 8.18)

$$k_o = \frac{282.4 q_o\, p_{ref} \left(\mu_o B_o\right)_{p_{ref}}}{h} \mathrm{PM} \tag{8.20}$$

the gas and water effective permeability are estimated from the following:

$$k_g = \left(GOR - R_s\right) \frac{\mu_g B_g}{\mu_o B_o} k_o \tag{8.21}$$

$$k_w = WOR \frac{\mu_w B_w}{\mu_o B_o} k_o \tag{8.22}$$

where the viscosity $\mu_{g,w}$ and the formation volume factor $B_{g,w}$ are evaluated at p_i for drawdown, and $\bar{p}$ for build-up periods. The total mobility is then expressed with Equation 8.2.

Discussion

As with Perrine's equivalent fluid, the producing fluid ratio must be representative of the investigated zone. Only the effective permeability can be estimated with the pressure-squared method, not the absolute permeability. The results depend directly on the choice of the constant of proportionality for the $(k_o/\mu_o B_o)$ group. Al-Khalifah et al. (1987) report that the oil effective permeability and the skin are more accurately estimated than with the Perrine's approach, especially for build-up tests.

CHAPTER 9

SPECIAL TESTS

In the previous Chapters, well pressure responses to constant flow rates were considered for analysis. Other testing procedures are used in some cases, and the interpretation methods must be adapted accordingly. In the first part of this Chapter, different types of well tests with varying rates are described.

In Sections 9.1 and 9.2, the pressure response to a *step pressure change* is considered, such as in tests with down hole shut-in devices (drill stem tests) and impulse tests, when the well is opened for only a few minutes. The DST procedure is presented and the analysis methods are reviewed with the hypothesis of the well flowing to surface or not. Specific techniques are required for the interpretation of DST's. The impulse test procedure is then described as an alternative.

Section 9.3 covers tests of wells flowing at *constant pressure*. The transient flow rate analysis methods are presented for oil and gas reservoirs, and the practical limitations of constant pressure tests are discussed.

In a second part, Section 9.4, *vertical interference* tests are briefly reviewed. It is shown that this single-well testing technique can be used for accurate determination of the vertical permeability.

9.1 DST

9.1.1 Test description

With the drillstem testing technique, the well is controlled by a down hole shut-in valve. For safety reason, the drill string is not usually used for the test, and production tubing is preferred.

Before the test, the well is partially filled with a liquid cushion designed to apply a hydrostatic pressure p_0 above the valve smaller than the formation pressure p_i. When the tester valve is opened, an *instantaneous* drop of pressure is transmitted to the sand face, and the formation fluids start to flow into the well.

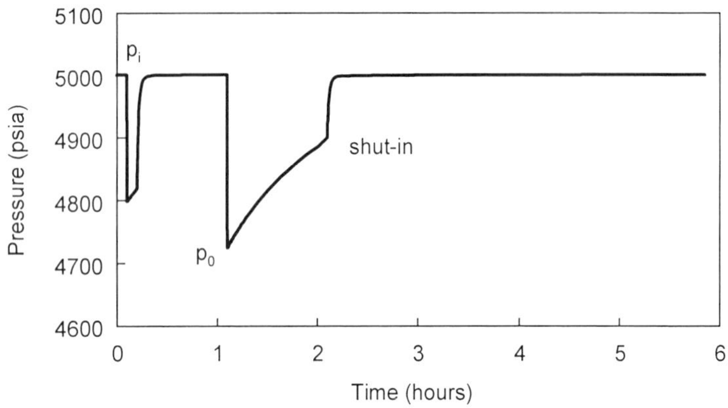

Figure 9.1. Example of DST pressure response. The sequence is initial flow, initial shut-in, flow period and final shut-in. The rate is less than critical.

In case of liquid flow, the *level rises* in the production string and the backpressure due to the liquid column increases. As long as the liquid level has not reached the surface (and provided the flow rate is less than critical, Ramey et al., 1975 a), the rate decreases. This is called a "*slug test*", which requires specific analysis techniques.

The well is then shut-in for a pressure build-up. When no flow to surface is desired, the down hole valve is closed before the liquid level has reached the surface. As illustrated in Figure 9.1, the usual drill stem test procedure consists of a first short initial flow followed by the initial shut-in to reach p_i. The well is then opened for the slug test and, due to the backpressure of the rising liquid column, the bottom hole pressure increases. Finally, the well is shut-in for a build-up period.

If surface production is possible, the flow time is extended until the well produces at surface and the rate tends to stabilize. The DST procedure then becomes similar to that of a standard production test.

In low-pressure wells, the flowing pressure can reach the initial reservoir pressure before the down hole valve is closed. In these cases, the well kills itself and the pressure build-up cannot be monitored after the liquid flow has stopped. Only a slug test analysis can be attempted.

When the flowing condition is *critical*, the rate is not controlled by the downstream pressure but by the completion or perforations configuration. The rate is *constant* and the pressure increases linearly with time during the flow. The flowing bottom hole pressure is not suitable for interpretation and only the shut-in period can be used for analysis of such tests.

9.1.2 Slug test analysis

Slug test type-curves

In 1975 (a), Ramey et al. presented a set of pressure type curves for the analysis of slug tests on log-log and semi-log scales (the log-log type curve is presented in Figure 9.2). The pressure, expressed by the dimensionless ratio p_{DR}, is presented versus the dimensionless time t_D/C_D. The different curves describe the well condition with the usual correlating $C_D e^{2S}$ group.

The dimensionless pressure ratio p_{DR} is defined by the drop of pressure $(p_i - p_{wf})$, normalized by the instantaneous drop of pressure $(p_i - p_0)$ applied when opening the valve:

$$p_{DR} = \frac{p_i - p_{wf}(t)}{p_i - p_0} \tag{9.1}$$

When the well is opened, the pressure ratio p_{DR} is 1. As the liquid level rises in the well, $(p_i - p_{wf})$ decreases and the ratio becomes less than one.

With the slug test type curves of Figure 9.2, the *same* pressure ratio is used for the data and the dimensionless curves. Only the time match and the curve match have to be adjusted.

The permeability thickness product is accessed from the *time match* (Equation 2.6):

$$kh = \frac{\mu C}{0.000295}\left(\frac{t_D/C_D}{\Delta t}\right)_{\text{MATCH}} \tag{9.2}$$

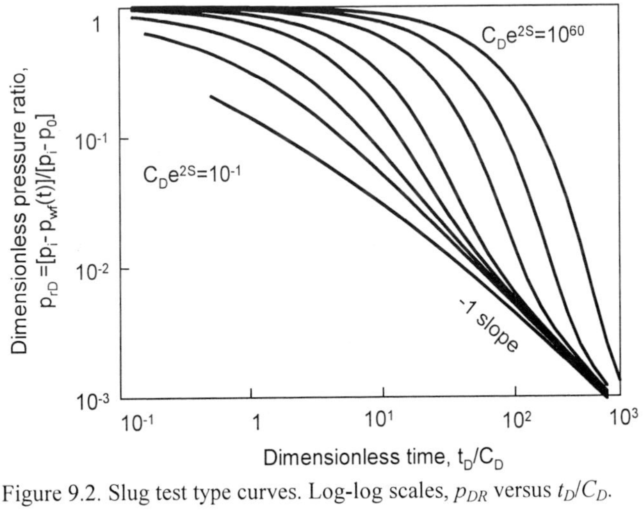

Figure 9.2. Slug test type curves. Log-log scales, p_{DR} versus t_D/C_D.

The wellbore storage coefficient must be known. With a changing liquid level, C is expressed, in oil field units:

$$C = 144 \frac{V_u}{\rho(g/g_c)} \tag{1.8}$$

where V_u is the wellbore volume per unit length in barrels per foot, ρ the density of the liquid in pounds per cubic foot and g/g_c the gravitational acceleration in lb_f/lb_m. The skin is estimated from Equation 2.11:

$$S = 0.5 \ln \frac{\left(C_D e^{2S}\right)_{\text{MATCH}}}{0.8936C/\phi c_t h r_w^2} \tag{9.3}$$

Analysis of slug test pressure with derivative type-curves

The pressure behavior during a slug test can be expressed with the usual drawdown pressure response Δp of Equation 1.1 by a simple relationship: the response to an instantaneous change of pressure is proportional to the *time derivative* of the pressure response to a constant flow rate (Ramey et al., 1975 a; Cinco-Ley, et al., 1986; de Franca Correa and Ramey, 1987; Ayoub et al. 1988).

Using the dimensionless pressure p_D of Equation 2.3, Peres et al. (1993) express the pressure drop $(p_i - p_{wf})$ during a slug test as:

$$\frac{0.000295kh}{C\mu(p_i - p_0)} \Delta t\left(p_i - p_{wf}(t)\right) = \frac{dp_D}{d\ln t_D} \tag{9.4}$$

From Equation 9.4 it can be seen that, the product of the slug test pressure change $(p_i - p_{wf})$ and the elapsed time Δt can be matched directly against a *derivative type-curve*, without having to differentiate the data. Any of the derivative type-curves presented in previous Chapters can be used to analyze slug test responses, in homogeneous and heterogeneous reservoirs.

The permeability thickness product is estimated either from the time match with Equation 9.2, or from the pressure match of Equation 9.4:

$$kh = \frac{\mu C\left(p_i - p_0\right)}{0.000295} \left(\frac{dp_D / d\ln t_D}{\Delta t\left(p_i - p_{wf}(t)\right)} \right)_{\text{MATCH}} \tag{9.5}$$

As with the slug test type-curves of Ramey et al., the wellbore storage and the skin coefficients are calculated from Equations 1.8 and 9.3.

Limitations of slug test analysis

Several other analysis methods have been presented for slug test interpretation. Peres et al. (1993) proposed integrating the slug test pressure response and using the conventional pressure type curves as a complement to the derivative curves. Conversely, Ostrowki and Kloska (1989) suggested using the time derivative of Ramey et al. slug test type curves to match the derivative of slug pressure.

Whatever method is used, the analysis of slug tests is not as accurate as the analysis of a standard drawdown response with a constant surface rate. The following two factors are identified as the main limitations for slug tests interpretation:

1. Definition of the *initial pressure*: All slug test analysis methods use the dimensionless pressure ratio of Equation 9.1. With this group, the initial reservoir pressure p_i must be known accurately and a small error in p_i can introduce a large distortion on the late time data curve. This is especially true in low-pressure wells, when the flowing pressure becomes close to the initial pressure and the difference (p_i - p_{wf}) relatively small.
2. *Wellbore storage* effect: Due to changing liquid level, the wellbore storage coefficient C is large and, as the duration of the flowing period is in general short, the response is often influenced by the wellbore storage effect for the entire flowing time. The radial flow regime is not seen, and the match is difficult to conclude uniquely.

9.1.3 Build-up analysis

At the end of the drill stem test procedure, the well is shut-in down-hole for a pressure build-up. Two possibilities have to be considered, depending on whether the well is shut in before the liquid level has reached the surface or not:

1. If the surface flow rate is well established before shut-in, the build-up is analyzed in the same way as the shut-in period of a producing well (Section 2.2.2).
2. When, as illustrated in Figure 9.1, the well is shut-in while the liquid level is still rising, the decreasing sand face rate has to be known as a function of time in order to analyze the subsequent build-up. Due to the high cost of down-hole flow metering, and for safety reasons, rate measurements are in general not considered for non-surface flowing DST's and flow rates must be *estimated.*

Rate estimation

In the following section, an example of flow rate estimation for a DST with no flow at surface is presented. As shown in Figure 9.3 with the simulated DST example of Figure 9.1, the slug test response is described by several pressure steps. Knowing the liquid gravity and neglecting inertial and frictional effects in the production string (which is reasonable since fluid velocity is low), the pressure difference between each step is converted into the corresponding height of fluid. From the capacity of the production pipe, the height is converted into volume. With this hydrostatic calculation, the final build-up period can be analyzed with the multiple rate methods described in Section 2.2.2.

In this example, the one-hour flow period is divided into 6 time intervals. During each interval j, the constant pressure $\bar{p}_j$ is defined as the average pressure between the start and the end of the time interval $\bar{p}_j = (p_{j-1} + p_j)/2$. Seven pressure points are used: p_0 is the pressure in the string immediately before opening, and p_6 is the last flowing pressure p (Δt =0). The rise of liquid level is estimated from pressure difference between two steps $(\bar{p}_j - \bar{p}_{j-1})$. Assuming a specific gravity of 0.75, the oil gradient is estimated at 0.325 ft/psi. The capacity of the drill string is 0.007 bbl/ft.

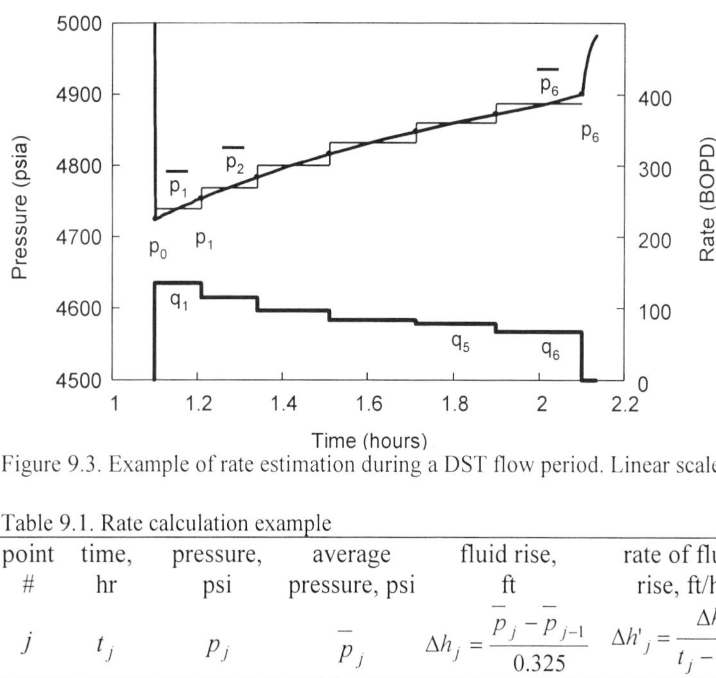

Figure 9.3. Example of rate estimation during a DST flow period. Linear scales.

Table 9.1. Rate calculation example

point #	time, hr	pressure, psi	average pressure, psi	fluid rise, ft	rate of fluid rise, ft/hr	flow rate, BOPD
j	t_j	p_j	$\bar{p}_j$	$\Delta h_j = \frac{\bar{p}_j - \bar{p}_{j-1}}{0.325}$	$\Delta h'_j = \frac{\Delta h_j}{t_j - t_{j-1}}$	$q_j = 0.007 \times 24 \times \Delta h'$
0	1.1000	4724.51				
1	1.2095	4753.13	4738.82	44.05 (*)	805.06	135.25
2	1.3414	4783.17	4768.15	90.25	684.00	114.91
3	1.5110	4816.47	4799.82	97.45	576.04	96.77
4	1.7120	4848.19	4832.33	100.04	497.42	83.57
5	1.8983	4873.02	4860.61	87.00	466.09	78.30
6	2.1000	4900.73	4886.88	80.83	399.73	67.15

(*) the first rate q_1 is estimated with $\bar{p}_0 = p_0$ and the time rate of fluid rise $\Delta h'$ is doubled.

As shown in Table 9.1, the flow rate decreases from 135.25 to 67.15 BOPD, the average rate during the one-hour flow period being 91.05 BOPD.

Inertial and frictional effects can be ignored for the usual DST conditions. In high rate producing wells, they could become significant (Saldana-C and Ramey, 1986).

Practical analysis of DST's

As for any build-up test, the final shut in period is analyzed using log-log and semi-log methods. Due to the down hole shut-in device, the effect of the wellbore storage is *short lived* during a DST build-up period. Frequently, the response reaches the radial flow regime before the first recorded point and, on log-log scale, the pressure and derivative responses tend to be flat. As a result, both the time match and the curve match $C_D e^{2S}$ can be difficult to fix on the p_D versus t_D/C_D dimensionless type curve. The match of the build-up data is not uniquely defined, and the analysis does not provide the wellbore storage coefficient C but only the kh product and the skin coefficient S.

Computer analysis allows a simulation of the complete test to be made as a checking procedure but, with most interpretation programs, the wellbore storage coefficient C used in the simulation is constant during the complete test sequence. With DST's, when the down-hole valve is closed for build-up, C is two or three orders of magnitude smaller than the wellbore storage of a flowing well. As a result, the DST simulation, generated from the build-up analysis results, does not match the flow period accurately.

de Franca Correa et al. (1987) view the flow / shut-in sequence of non-surface flowing DST's as a single slug test period, with a sharp change of wellbore storage at time of shut-in. After the initial instantaneous pressure drop, the wellbore storage first corresponds to the changing liquid level of Equation 1.8. When the down-hole valve is closed, the storage becomes the compressibility term of Equation 1.7. With this description of the DST sequence, the complete test can be simulated accurately.

9.2 IMPULSE TEST

9.2.1 Test description

In impulse tests, the well is produced from or injected into the reservoir for only a few minutes and then closed. The impulse technique is a variation of the DST analysis method:

- For a DST, a step pressure drop is applied to the formation and, during the resulting slug test period, the well is considered to be flowing. A build-up test is made after the flow (second flow and shut-in of Figure 9.1).

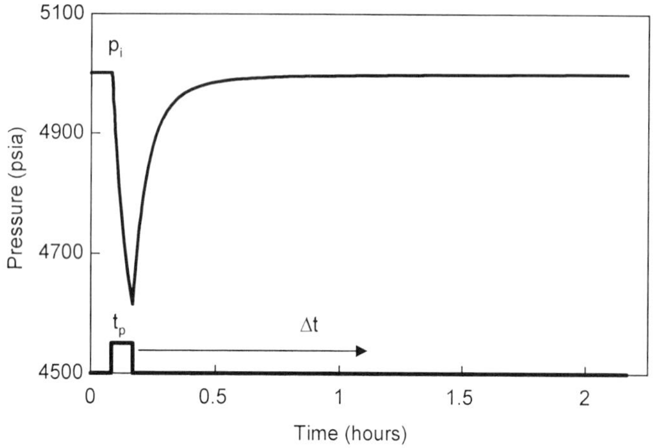

Figure 9.4. Impulse pressure response.

- With impulse tests, the pressure drop is generated by a short production, and the following pressure increase is analyzed as a shut-in period (first flow and shut-in of Figure 9.1, or Figure 9.4).

9.2.2 Impulse analysis

With the impulse analysis method of Ayoub et al. (1988), a *single* analysis plot of the complete well pressure response is used to obtain all well and reservoir parameters. During the short flow, the impulse response is expressed as $(p_i - p_{wf})t_p$ and, during the shut-in, as $(p_i - p_{ws})(t_p + \Delta t)$.

As with the Perez et al. method, the *derivative* type curves are used to analyze the pressure response. The impulse match of the homogeneous reservoir example of Figure 9.4 (t_p=5 min., shut-in = 2 hr.) is shown in Figure 9.5. During the flowing time, the impulse response is matched on a *pressure* type curve and, during the shut-in period, the response deviates from the usual pressure response to reach the derivative curve with the same $C_D e^{2S}$.

The pressure match is adjusted to the 0.5 line of the derivative type-curve. Since the flow rate is not measured during the short flow period and only the amount of fluid Q_t produced or injected is known, the pressure match is expressed by replacing $C(p_i - p_0)$ in Equation 9.4 by Q_t (Ayoub et al., 1988):

$$\frac{0.000295kh}{Q_t \mu}\left(t_p + \Delta t\right)\left(p_i - p\right) = \frac{dp_D}{d \ln t_D} \tag{9.6}$$

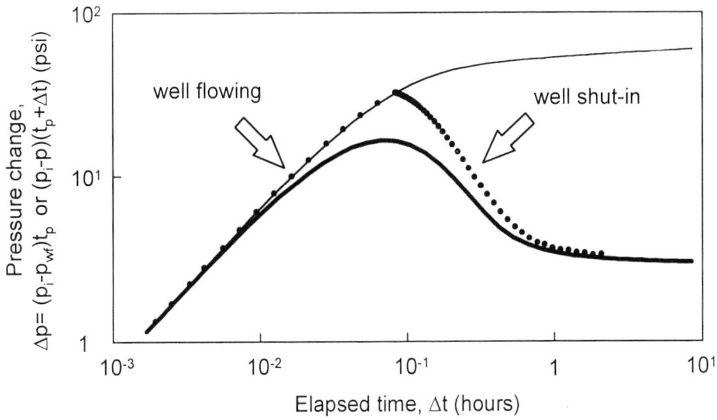

Figure 9.5. Impulse match. Log-log scales, pressure-time group versus Δt.

The time match is obtained by matching the early time data with the pressure type-curve, giving the wellbore storage coefficient C with Equation 2.10. The skin is then estimated with Equation 2.11.

9.2.3 Discussion of impulse analysis

Frequently, the rate is not constant during the short flow period and the early time match is not accurate. In these cases, the response is properly defined only during the radial flow regime at shut-in time, with a stabilization of the pressure response on the 0.5 derivative line. Both the wellbore storage coefficient C and the type curve parameter $C_D e^{2S}$ are difficult to define, and the calculation of skin is approximate.

For accurate analysis of impulse tests, the initial pressure must be known. When p_i is not known, Cinco et al. (1986) propose starting by analyzing the derivative of the impulse, on a second derivative type-curve.

The impulse method has also been used (rather than the Horner method) to analyze build-up tests after relatively long production periods (Soliman, 1982; Cinco et al., 1986), when $\Delta t >> t_p$. The pressure change $(p_i - p_{ws})$ is analyzed versus $(t_p + \Delta t)$, thus avoiding the compression effect of the Horner time on the analysis plot.

The conventional pressure derivative (Chapter 2) does not compress the time scale either. When the derivative response can be accurately defined, it offers the best alternative for the analysis of build-up tests, whatever the duration of the flowing period t_p, and the shut-in time Δt. The derivative analysis is not affected by a possible error in initial pressure, and the pressure curve can be used to estimate the skin accurately. This is illustrated in Figure 9.6, where the conventional build-up analysis of the shut-in period after 5 minutes flow of the example Figure 9.4 is presented.

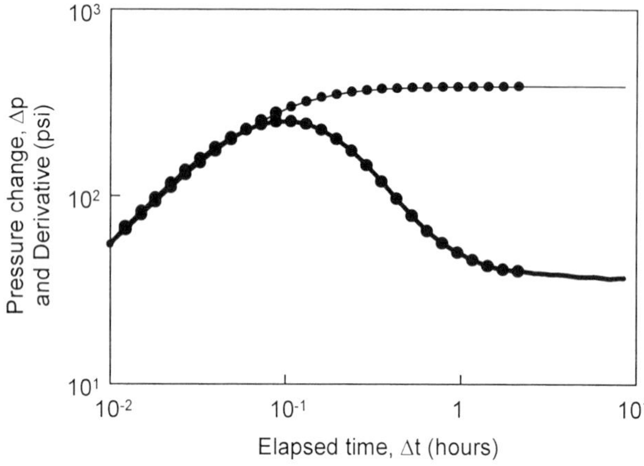

Figure 9.6. Pressure and derivative analysis of the impulse shut-in period. Log-log scales, Δp and $\Delta p'$ versus Δt.

9.2.4 Well responses after an instantaneous source: summary

Slug and impulse test theories are based on the response to an *instantaneous source*. During non-surface flowing DST's, slug test responses are affected by a large wellbore storage effect and the analysis is frequently not unique. With the impulse technique, the effect of wellbore storage is shorter, and the start of the characteristic radial flow regime is seen earlier. The main limitation of the impulse technique during early time is the change of behavior, between the short flow and the shut-in period. The best analysis data is obtained during build-up tests with down-hole shut-in.

The response to an instantaneous source can be used not only for specifically designed tests, but also for a quick and reliable analysis of the well pressure behavior after underbalanced perforation, after backsurging operations, or for a repeated formation test (Ayoub et al., 1988; Cinco-Ley et al., 1986; Waller and Krase, 1986).

9.3 CONSTANT PRESSURE TEST, AND RATE DECLINE ANALYSIS

Wells are sometimes producing at constant wellbore pressure. This condition is obtained for example when the well produces into a separator or a pipeline at constant pressure. In such cases, the rate is not constant but *declines* with time.

As shown by van Everdingen and Hurst (1949), there is a simple relationship between the constant rate solution and the constant pressure solution. Any of the constant rate p_D functions presented in previous chapters can be changed into a constant pressure q_D function. Transient rates can be analyzed using techniques analogous to that of transient pressures for constant rate flow.

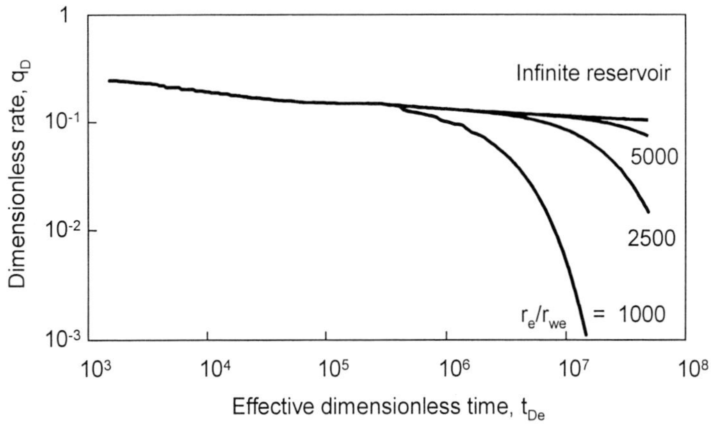

Figure 9.7. Decline curves on log-log scale. q_D versus t_{De}.

9.3.1 Radial homogeneous oil reservoir response

Several analysis methods of rate versus time have been proposed, using log-log type curves or using semi-log straight line analysis. In 1952, Jacob and Lohman presented the type curve shown in Figure 9.7, as the infinite reservoir response.

For log-log analysis, the dimensionless flow rate is expressed as:

$$q_D = \frac{141.2 B\mu}{kh\left(p_i - p_{wf}\right)} q(t) \tag{9.7}$$

and the effective dimensionless time t_{De}, based on the equivalent wellbore radius of Equation 1.14, is used (Uraiet and Raghavan, 1980; Ehlig-Economides and Ramey Jr., 1981 a). The kh product and the skin factor S are estimated from the rate and time matches.

Except at very early time, the rate does not change significantly with time, and the shape of the q versus Δt log-log curve is not very characteristic. Frequently, the uniqueness of the match is difficult to establish.

For semi-log analysis, the *reciprocal* of the rate $1/q$ is plotted vs. log Δt.

$$\frac{1}{q} = 162.6 \frac{B\mu}{kh(p_i - p_{wf})} \left[\log \Delta t + \log \frac{k}{\phi \mu c_t r_w^2} - 3.23 + 0.87S \right] \tag{9.8}$$

The relationship expressed in Equation 9.8 is similar, except for the slope m_q on the right hand side, to the usual semi-log pressure response of Equation 1.15 (homogeneous reservoir and constant rate production). However, the time to reach the semi-log straight line is *longer* with constant pressure tests than with the constant rate pressure responses (Uraiet and Raghavan, 1980; Ehlig-Economides and Ramey Jr., 1981 a).

The permeability is estimated from the slope m_q and the skin is estimated from the intercept of the $1/q$ straight line at 1 hour:

$$kh = 162.6 \frac{B\mu}{m_q\,(p_i - p_{wf})} \tag{9.9}$$

$$S = 1.151\left[\frac{1/q(1\text{hr})}{m_q} - \log\frac{k}{\phi\mu c_t r_w^2} + 3.23\right] \tag{9.10}$$

Fetkovich (1980) presented a set of flow rate type curves for closed circular reservoir (Figure 9.7). Dimensionless flow rate and time are the same as on the Jacob and Lohman type curve (1952), the curves are labeled in terms of dimensionless reservoir radius r_e/r_{we}. Before boundary effects, the rate response is similar to the Jacob and Lohman type curve. During pseudo steady state flow, the decline of flow rate is exponential when a constant pressure is imposed on the well.

The practical aspects of long-term constant pressure tests make the identification of boundary effects by decline curves analysis difficult. A late time deviation from the infinite reservoir curve can be caused by a small change in the flowing pressure, or a variation in the wellbore skin damage during the production. Fetkovich decline curves have been designed not only for the analysis of constant pressure tests, but also for production forecasting.

9.3.2 Other well and reservoir configurations

Double porosity reservoirs

Ozkan et al. (1987) discuss deliverability responses for double porosity reservoirs with transient interporosity flow. The rate decline curves follow the same pattern as double porosity constant rate pressure responses. On a semi-log plot of $1/q$ vs. Δt, the double porosity response exhibits two parallel straight lines of slope m_q. During transition, the semi-log straight line has a slope of $m_q/2$.

Finite conductivity fracture

Decline type curves for a well intercepting a finite conductivity fracture were first generated with a numerical simulator by Agarwal et al. (1979 a) and, in 1981, Guppy et al. used a semi-analytical solution. On the log-log type curves, the reciprocal of the dimensionless rate $1/q_D$ is presented as a function of the dimensionless t_{Df} of Equation 3.8, and different fracture conductivities $k_{fD}w_{fD}$ are considered. The dimensionless reciprocal rate $1/q_D$ has the same form as the usual dimensionless pressure of Equation 2.3, but Δp is a constant (p_i - p_{wf}), and q varies with time:

$$\frac{1}{q_D}=\frac{kh\left(p_i-p_{wf}\right)}{141.2B\mu}\frac{1}{q(t)} \tag{9.11}$$

Interestingly, the $1/q_D$ deliverability curves look similar in shape to the p_D constant rate pressure type curve of Figure 3.9, and therefore are easy to understand. The values of the $1/q_D$ curves are greater than the p_D curves, but on the log-log scale, they show the same characteristic 1/4 and 1/2 straight line slopes during the bi-linear and linear flow regimes.

Massive hydraulic fracturing is a common practice in low permeability gas reservoirs. Such wells normally produce at constant well pressure. The use of the $1/q_D$ finite conductivity fracture log-log type curves on gas well responses is discussed in the next Section "Gas reservoirs".

Multiple well production

Other well and reservoir configurations can be envisaged for decline rate analysis. Any well test interpretation model, designed for pressure analysis, can be used to generate decline curves. In addition, Bourgeois and Couillens (1994) proposed a general superposition method to predict pressure or flow rate responses in the case of multiple well production. The van Everdingen and Hurst (1949) relationship, between the constant rate p_D function and the constant pressure q_D function of a single well, is generalized to the case of "n" interfering wells producing with mixed constant pressure or flow rate constraints. A computer program is required to generate the responses, but the simulations predict the rate behavior accurately for complex producing system and any well test model.

Gas reservoirs

In 1989, Berumen et al. investigated gas rate decline analysis. When radial flow is established, and assuming the skin is *not rate dependent*, $1/q_{sc}$ versus log Δt follows a semi-log straight line. The slope m_q is expressed with the gas pseudo pressure $m(p)$ (as in Equation 7.19 for constant gas rate pressure response, Agarwal et al., 1979 a):

$$m_q=1.63\text{x}10^3\frac{T}{kh\left[m(p_i)-m(p_{wf})\right]} \tag{9.12}$$

The skin is calculated with Equation 9.10, where μ and c_t are estimated at initial pressure p_i.

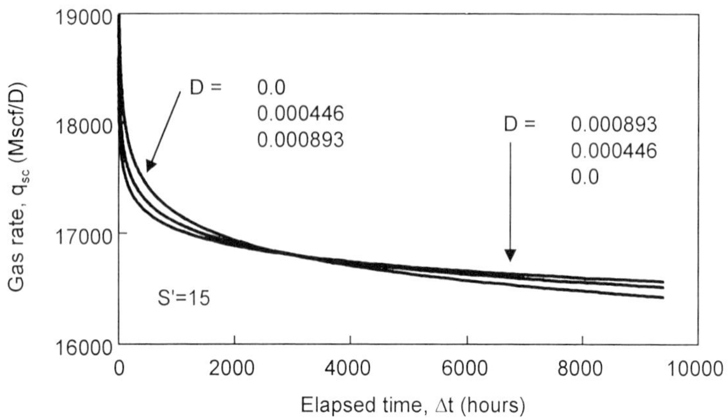

Figure 9.8. Transient gas rate with rate dependent skin factors. Linear scale.

By incorporating a rate dependent skin ($S' = S + Dq$) in the analytical solution used to generate decline rate curves, Bourgeois and Wilson (1996 c) showed that the *shape* of the rate transient is changed significantly compared to the constant skin response.

This is illustrated on the simulated example of Figure 9.8, with a well flowing at constant pressure in a homogeneous infinite reservoir. First, a constant skin is assumed (D=0, $S' = S$ =15) and the rate at 3000 hours is estimated at 17,180Mscf/D. Two additional simulations are generated with the same global skin (S'=15) at 3000 hours. In one case, S =7.5 and D=0.000446 and, in the other case, S =0 and D=0.000893. On the two simulations with non-Darcy flow effect, the flow rate starts to be smaller than in the case of constant skin, at 3000 hours the curves cross each other, and it becomes larger. If the reservoir was closed, the rate curves would intersect again during the pseudo steady state regime, and the constant skin rate would end-up the larger (Bourgeois and Wilson, 1996 c).

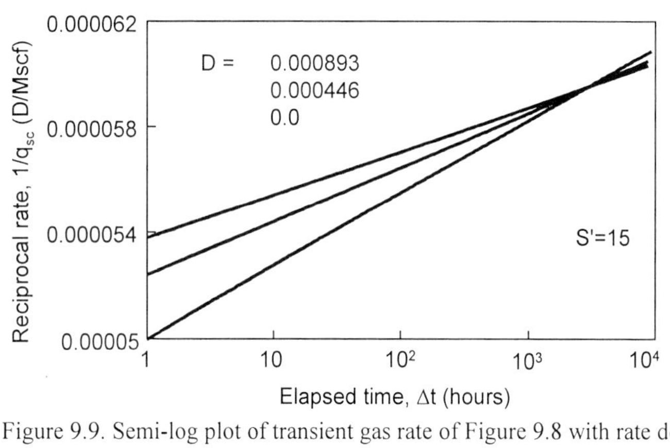

Figure 9.9. Semi-log plot of transient gas rate of Figure 9.8 with rate dependent skin factors.

In Figure 9.9, the data of Figure 9.8 is presented on a semi-log scales in terms of the reciprocal of the rate $1/q_{sc}$. The three simulations follow an apparent semi-log straight line but, when $D \neq 0$, the slope is smaller than m_q defined in Equation 9.12 for $D=0$ (by 25% and 40% respectively in this example). When the skin is rate dependent, the permeability is over estimated by semi-log straight-line analysis.

In the case of finite conductivity fractured gas wells, when the well response is not affected by non-Darcy flow effects, the deliverability type curves of Agarwal et al. (1979 a) or Guppy et al. (1981) can be used to estimate the well and reservoir parameters (namely kh, x_f and $k_f w_f$).

Using the gas pseudo pressure function $m(p)$, the dimensionless reciprocal rate $1/q_D$ is expressed as (Equation 7.7 for the dimensionless pressure) :

$$\frac{1}{q_D} = \frac{kh\left[m(p_i) - m(p_{wf})\right]}{1422\,T} \frac{1}{q_{sc}(t)} \qquad (9.13)$$

In the dimensionless time t_{Df}, μ and c_t are estimated at initial pressure p_i.

During the bilinear flow regime, a Cartesian plot of $1/q_{sc}$ versus $(\Delta t)^{1/4}$ displays a straight line of slope m_{BLF} (similar to Equation 1.27 for the constant oil rate pressure response, Guppy et al., 1981):

$$m_{\mathrm{BLF}} = \frac{493.8T}{\left[m(p_i) - m(p_{wf})\right] h \sqrt{k_f w_f} \sqrt[4]{(\phi \mu c_t)_i k}} \qquad (9.14)$$

In practice, when the fracture is long and k is small, radial flow is not reached before several years, and the transient rate data does not exhibit the three characteristic flow regimes. A complete analysis with the deliverability curve frequently requires a previous knowledge of the kh product from a pre-fracturing build-up test (Agarwal et al., 1979 a). Alternatively, when the three parameters kh, x_f and $k_f w_f$ are known, the curves can be used for performance prediction of fractured gas wells.

Guppy et al. (1981) report that when the well response is affected by non-Darcy flow effects, the 1/4 slope straight line of the of $1/q_{sc}$ log-log plot is distorted, resulting in an apparent varying fracture conductivity.

Build-up analysis after a constant pressure flow

When a build-up test is conducted after a period of constant pressure flow, the time superposition should be used in order to take into account the declining rate prior to shut-in. As shown Section 9.1.3 for DST analysis, the superposition function is generated by approximating the rate curve with several constant flow rate periods.

When the equivalent Horner time is used for simplification, the production time is defined as the cumulative production divided by the last flow rate. Ehlig-Economides and Ramey (1981 b) report that the analysis results are sufficiently accurate in the majority of cases.

9.3.3 Discussion

Even although tests at constant pressure yield in theory the same information as constant rate pressure data, they appear to have limited applications due to practical problems:

- Maintaining a constant flowing pressure can be difficult, especially at early time, when the analysis of near wellbore effects is desired. In addition, a constant wellhead pressure does not imply a constant sand face pressure. The pressure drop from flowing friction in the wellbore varies with the transient rate, and a correction is needed for interpretation of constant wellhead pressure tests (Ehlig-Economides and Ramey, 1981 a).
- The wellbore skin must be *constant* for rate decline analysis. This point must be checked before analysis of long-term production tests. For gas wells, the occurrence of non-Darcy flow distorts the transient rate response, thus significantly reducing the accuracy of constant pressure test analysis.
- The technology for measuring flow rates does not provide data as accurate as the pressure gauges now available.

Caution must be exercised when planning or interpreting transient rate tests.

9.4 VERTICAL INTERFERENCE TEST

The knowledge of vertical permeability is of prime importance for field development strategy. Some usual examples of the application of this knowledge are:

- Completion decisions, such as the evaluation of possible productivity increase by horizontal drilling,
- Production strategy, in the event of gas or water coning,
- Enhanced recovery projects in the case of layered systems, when a tight zone is separating two permeable intervals.

Feasibility studies related to underground storage in aquifers also require an accurate evaluation of the vertical flow properties.

Vertical interference testing has been proposed by many authors for homogenous or stratified formations, with different approximations of the vertical flow properties (Bremer et al., 1983; Kamal, 1984; Ehlig-Economides and Ayoub, 1984). The vertical interference test model described in the following text is an extension of the Kuchuk et al. multiple layer horizontal well analytical solution (1991 b). Both homogenous and layered systems are envisaged, with a spherical flow geometry.

9.4.1 Test description

For a single-well vertical interference test, the well is perforated at two different depths, but only one interval is active (flowing or injecting). The pressure is monitored on the two perforated segments, isolated in the wellbore (see first diagram of Figure 9.10). Several types of well completion can be used during such tests.

With the double-stage testing approach, which has been developed for gas storage studies in low permeability formations, two tests are performed: one test on a discrete short interval, and the other on a longer interval that includes the short interval. A coupled interpretation of both tests allows independent calculation of horizontal and vertical permeability.

9.4.2 Vertical interference test responses

On the examples of vertical interference responses of Figure 9.11, a homogenous layer is assumed. The active segment is centered in the interval, the corresponding dotted pressure and derivative curves are the same as the partial penetration example of Figure 3.17 for z_w/h=0.5. Several vertical distances are considered for the observation segment. If the vertical distance is small ($z_{w\text{-}obs}/h$=0.6), the interference response describes the spherical flow regime and the final radial flow over the entire reservoir thickness. When the vertical distance is increased ($z_{w\text{-}obs}/h$=0.8), the response is delayed and only the final radial flow regime can be observed. As already discussed in the multiple well interference test, Chapter 6, the three interference derivative curves of Figure 9.11 show that, within the drainage volume, the rate of pressure change is the same at any point.

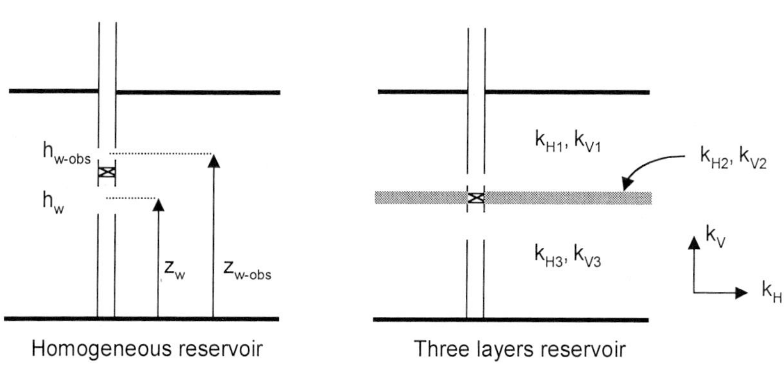

Figure 9.10. Well and reservoir configurations.

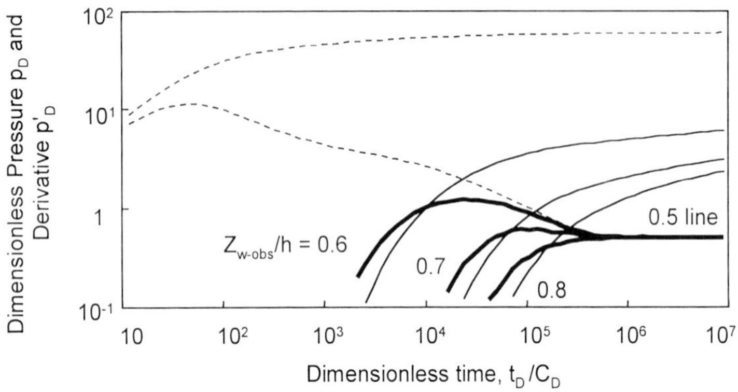

Figure 9.11. Vertical interference responses from a well in partial penetration with wellbore storage. Log-log scale, p_D versus t_D/C_D. $C_D = 6$, S_w=0, $k_V/k_H = 0.005$.
Several distances. Producing segment: h_w/h = 1/10, z_w/h = 0.5; observation segment: $h_{w\text{-}obs}/h$ = 1/100, $z_{w\text{-}obs}/h$ = 0.6, 0.7, 0.8.

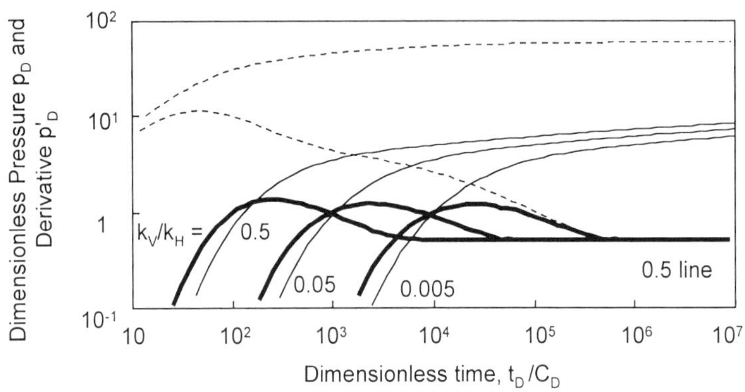

Figure 9.12. Vertical interference responses from a well in partial penetration with wellbore storage. Log-log scale, p_D versus t_D/C_D. $C_D = 6$, S_w=0. Producing segment: h_w/h = 1/10, z_w/h = 0.5; observation segment: $h_{w\text{-}obs}/h$ = 1/100, $z_{w\text{-}obs}/h$ = 0.6.
Several vertical permeability: k_V/k_H = 0.5, 0.05, 0.005.

With the same well configuration, the influence of the vertical permeability k_V is demonstrated in Figure 9.12 (the dotted pressure and derivative curves correspond to the same example of Figure 3.17, for k_V/k_H = 0.005). When k_V is increased, the time of the start of the vertical interference effect is reduced, and the magnitude of the pressure response is larger.

Figure 9.13 presents three examples of vertical interference response across a low permeability interval. The reservoir configuration, depicted on the second diagram of Figure 9.10, is the same as for the horizontal well example of Figure 3.42 where a low permeability interval divides the pay zone in two main layers. The dotted pressure and derivative curves describes the response of a partial penetration well perforated in the bottom layer. The three interference responses, shown by the solid lines, are monitored in the top layer, above the low permeability interbed.

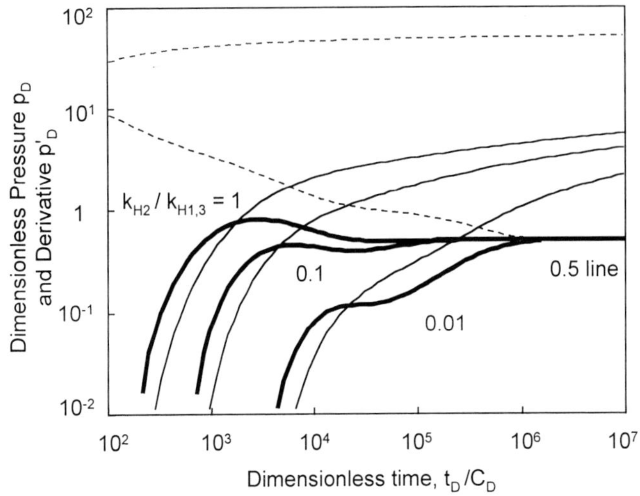

Figure 9.13. Vertical interference responses from a well in partial penetration with wellbore storage in a reservoir with a low permeability interbed. Log-log scale, p_D versus t_D/C_D. $C_D = 6$, $S_w=0$, $h_1/h=0.45$, $h_2/h=0.05$, $h_1/h=0.5$. Producing segment: $h_w/h = 1/10$, $z_w/h = 0.45$; observation segment: $h_{w\text{-}obs}/h = 1/100$, $z_{w\text{-}obs}/h = 0.6$.
Several permeability: $k_{H1}=k_{H3}=k_{H2}$ or $=10k_{H2}$, or $=100\ k_{H2}$, $(k_V/k_H)_i=0.1$.

When the permeability reduction in the central interval is large ($k_{H2} = 10^{-2}\ k_{H1,3}$), the magnitude of the vertical interference response is significantly smaller than when there is no flow restriction ($k_{H2} = k_{H1,3}$). The tight zone delays the start of the interference response and, on the derivative curve $k_{H2} = 10^{-2}\ k_{H1,3}$, no spherical flow regime is evident. Only a long transitional behavior is seen before the final horizontal radial flow regime.

CHAPTER 10

PRACTICAL ASPECTS OF WELL TEST INTERPRETATION

In this final Chapter, the practical application of the theory presented earlier is discussed. First, different types of problems with the test data are envisaged. Preparation and validation of the raw data used for analysis are thoroughly discussed. Then, usual distortions in the pressure response are described, and guidelines are established for proper analysis of such test responses.

The second part deals with the interpretation methodology. After a synthesis of the diagnosis methodology, the different model responses are summarized and two examples of interpretation consistency checks are presented.

Finally, the question "how representative are the interpretation results" is addressed. The implication of using analytical models for interpretation is reviewed, the meaning and the accuracy of the interpretation results are evaluated and, for illustration, a discussion of the radius of investigation is presented.

10.1 FACTORS COMPLICATING WELL TEST ANALYSIS

Results of interpretation are directly dependent upon the quality of the pressure and rate data used for analysis. Data preparation is crucial in well test interpretation, and frequently it takes more time than the analysis of well pressure responses. In the following section, the usual problems encountered when preparing the data for analysis are discussed in detail, and data checking and validation are presented.

In the second part of this section, it is shown that the well or reservoir conditions can affect, in some cases, the pressure recorded down hole. The identification of wellbore problems is discussed for test responses that do not follow the usual wellbore storage behavior described in Chapters 1 to 3, and the recommended analysis approach is presented. In developed fields, the effect on well test responses of interference from neighboring producing wells is addressed.

10.1.1 Data preparation and validation

The final build-up of the test sequence presented in Figure 10.1 is used in this Section to illustrate several possible errors during the data processing. In this example, the well has been produced for 100 hours, shut-in for 50 hours and re-opened at the same flow rate for a 20 hour flow test before the final build-up. During the test sequence, the reservoir behavior corresponds to the infinite acting radial flow regime.

Rate history definition

Two different difficulties can be encountered when the rate history is prepared for a well test analysis:
1. The well production history is not complete, or accurate. The rate must be *estimated* during some flow periods of the test sequence.
2. Too many rate changes occurred, sometimes for a very long period before the test period of interest. The history has to be *simplified.*

First, the influence of an inaccurate production history is illustrated with the simple test example of Figure 10.1. Then guidelines are presented for the practical definition of the well production history used for interpretation.

Rate simplification example

Two approaches are currently used in order to simplify the rate history of a test:
1. An equivalent production time is defined as the ratio of the cumulative production divided by the last rate (called *equivalent Horner time*). On the example in Figure 10.1, the final build-up period is analyzed with a previous rate history simplified into a single drawdown of t_p=120 hr.
2. When there has been a shut-in period in the rate history and if the bottom hole pressure has almost reached the initial pressure p_i, it is wrongly assumed that the rate history prior to this shut-in has no effect on the final build-up response and it is *ignored.* On the test example, t_p =20 hr with this method.

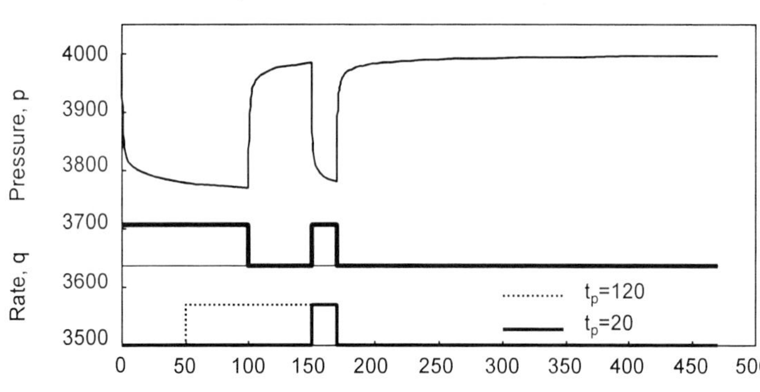

Figure 10.1. Example of test sequence with two drawdown periods. Linear scale.

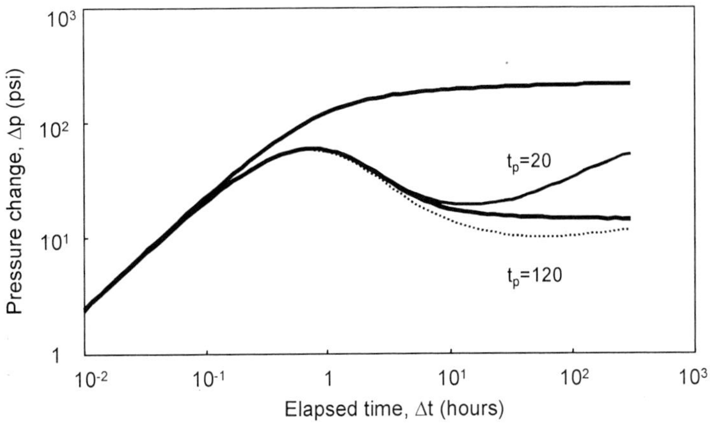

Figure 10.2. Log-log plot of the final build-up. The derivative is generated with three different rate histories.

On the log-log plot of Figure 10.2, the correct multiple rate derivative response is compared to the curves generated with the two simplified rate sequences. When t_p =20 hr is used, the correction introduced by the superposition time is too large (see discussion of Figure 2.7), and the derivative deviates *above* the theoretical stabilization corresponding to radial flow. Conversely, with t_p=120 hr, the intermediate shut-in from 100 hr to 150 hr is ignored, the time superposition function does not completely correct the influence of the previous rate sequence, and the derivative curve drops *below* the stabilization at intermediate time. As this later approximation honors the cumulative production, the derivative response is correct on late time data.

Definition of the rate history

In practice, it is possible to simplify the rate history when the production changes occurred a long time before the analyzed period, but not if the rate variations happened immediately before the test period. The closer to the time of the start of the test period, the more accurate must be the flow rate profile. Bourdarot (1998) proposes the following rule of thumb: if the duration of the analyzed period is Δt, any rate changes that occurred at more than 2 Δt before the start of the period can be simplified. The equivalent Horner time is then used to reduce the number of rate changes, keeping only the most significant rate variations and long shut-in periods.

When the total production time prior to shut-in is long compared to the duration of a build-up test, the interpretation of the model extrapolated pressure is difficult. In Chapter 5, it is shown with the discussion of Figure 5.25, that the slope of the Horner semi-log straight line is not affected by initial flow periods of a long production history, but the extrapolated pressure p^* is.

Introducing years of accurate production data in the well flow history does not always improve the quality of the interpretation results. With the time superposition method presented in Section 2.2.2, when the first flow periods are extrapolated into the time of the final shut-in, the total production time for the initial flow rates becomes very large compared to the final shut-in time, and the corresponding radius of investigation also becomes large. Many changes of reservoir characteristics and boundaries may affect the extrapolated flow periods, when they are not visible on the short build-up test. In such a case, extrapolation at very long times of the reservoir model defined from the short build-up analysis may not be valid.

This point is further discussed in Section 10.2.3 with two examples. It is shown that, when a long production history is used in a multiple rate sequence, the *reservoir* and *boundary* model must be applicable to the longest extrapolated period, and include all changes of reservoir properties and limits within the large investigated area of the first extrapolated period. If not, the model extrapolated pressure (and the initial pressure p_i) estimated from the build-up analysis is not correct. The radius of investigation for build-up periods is further discussed in Section 10.3.3, at end of this chapter.

Frequently, some rate data are missing, such as during a clean-up, or any flow period where no separator measurements are made. When the missing rates have to be introduced in the production history, they must be estimated. Usually, well head pressure and choke size are used but, if pressure measurements are available during these flow periods, it is possible to validate the estimated flow rate. As discussed in Section 3.1.3, the comparison of different test periods on a *normalized* log-log plot ($\Delta p/q$ and $\Delta p'/q$ versus Δt) is very efficient for checking the flow rates.

The *test simulation* on a linear scale is another good quality control plot. When the interpretation model, defined on a single period, does not match the complete test sequence, three hypotheses can be considered:
1. Either the model is not applicable for long periods (with for example a difference in the initial pressure, see Chapter 5 or the next Section 10.2.3),
2. Or the well condition has changed during the production (well cleaning / being damaged, or rate dependent skin as discussed in Chapter 7),
3. Or the rate history is not correct.

Error of start of the period

Once rate history is defined, the pressure data measured down hole is spit into individual test periods, and the different quantities Δp, $\Delta p'$ and Δt are estimated for log-log analysis. If a well test interpretation software is used, all test periods are usually extracted automatically from the rate changes defined in the rate history. During this process, several errors can be introduced on the period response curves:

1. When the pressure and the rate data are not perfectly *synchronized*, the time of start of the test period can be earlier or later than the true change of rate.
2. In some cases, the pressure is noisy or *oscillating* at the time of shut-in. The program uses the pressure point at the time of the rate change for the start of the new period $p(\Delta t\!=\!0)$. This point can be higher or lower the true stabilized pressure at the end of the previous period, and the resulting calculated pressure change is wrong.

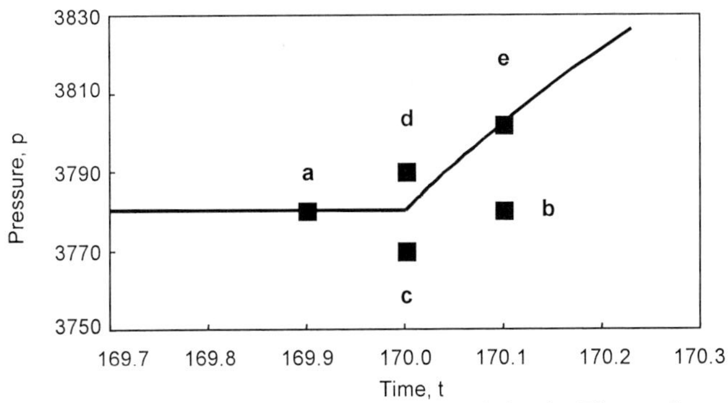

Figure 10.3. Example of Figure 10.1 at time of shut-in. Time and pressure errors.
Shut-in time error: curve **a** = 0.1 hr before and curve **b** = 0.1 hr after the actual shut-in time.
Shut-in pressure error: curve **c** = 10 psi below and curve **d** = 10 psi above the last flowing pressure. Error in time and pressure: curve **e**.

Figure 10.3 is a magnified plot of Figure 10.1 at the time of shut-in of the final build-up. Five possible errors are considered. Cases **a** and **b** describe a 0.1 hr time error before and after the shut-in time, cases **c** and **d** a 10 psi pressure error below and above the last flowing pressure and, with case **e** corresponding to a time and pressure error, a build-up point is used for the start of the period.

In the case of a time error, the calculated elapsed time Δt is either too large (case **a**), or too small (case **b**). In the first case, the pressure curve is displaced towards the right and, at early time, the pressure curve increases with a slope higher than unity (Figure 10.4). When the shut-in time used to extract the test period is too late, the pressure curve is displaced towards the left, and at early time, it is distorted as shown in Figure 10.5. If the quality of the pressure data is poor, this error can suggest the presence of a linear flow regime at early time. Interestingly, the derivative curves are *not distorted* as severely as the pressure responses, thus allowing the diagnosis of an error at the start of the period.

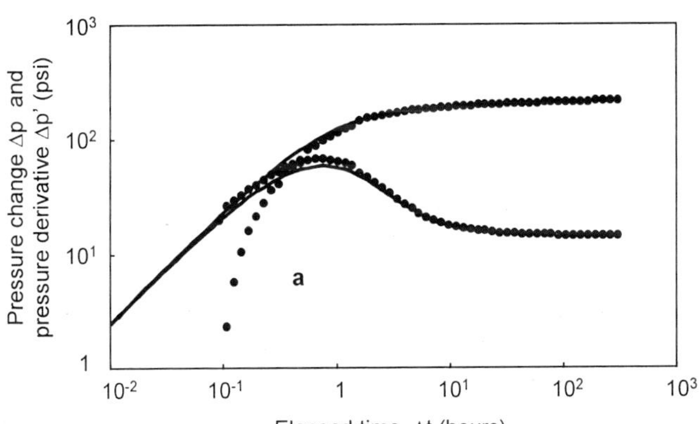

Figure 10.4. Case **a**: shut-in time too early.

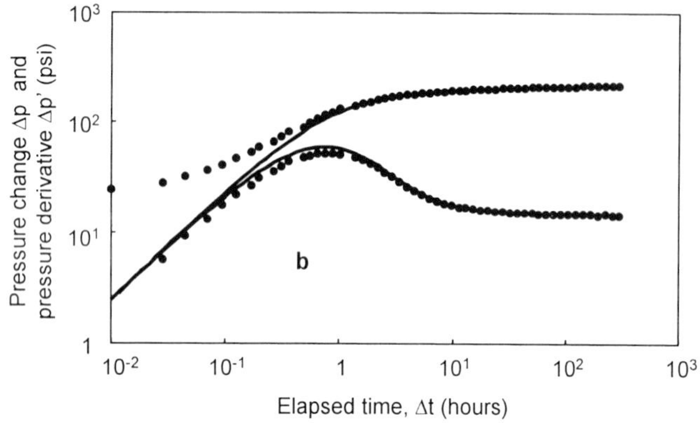

Figure 10.5. Case **b**: shut-in time too late.

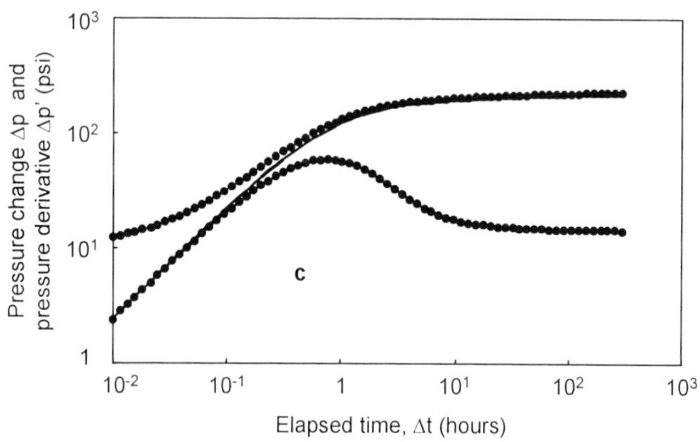

Figure 10.6. Case **c**: last flowing pressure too low.

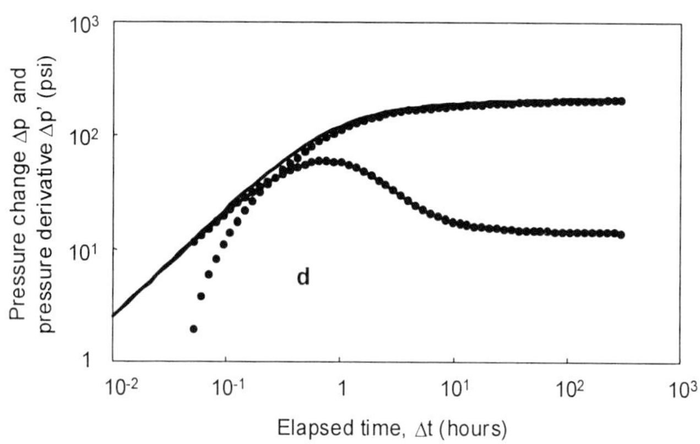

Figure 10.7. Case **d**: last flowing pressure too high.

With pressure errors, a similar distortion is observed on the pressure curves. When Δp is over estimated (case **c**, Figure 10.6), the pressure curve is displaced upwards and, at early time, the distortion is very similar to the case **b** of Figure 10.5. With case **d**, Δp is under estimated (Figure 10.7) and the response looks the same as case **a** of Figure 10.4.

When a build-up point is used for the start of the period, the error can be difficult to identify on a log-log scale. With case **e**, a point during the pure wellbore storage regime has been selected for the calculation of Δp, $\Delta p'$ and Δt. The resulting pressure and derivative curves follow a unit slope straight line at early time, the response appears correct. A good match can be obtained on such a test period but, as Δp is too small, the resulting skin is under estimated. When the build-up point used for the start of the period is taken after the pure wellbore storage, the distortion of the response is easier to identify.

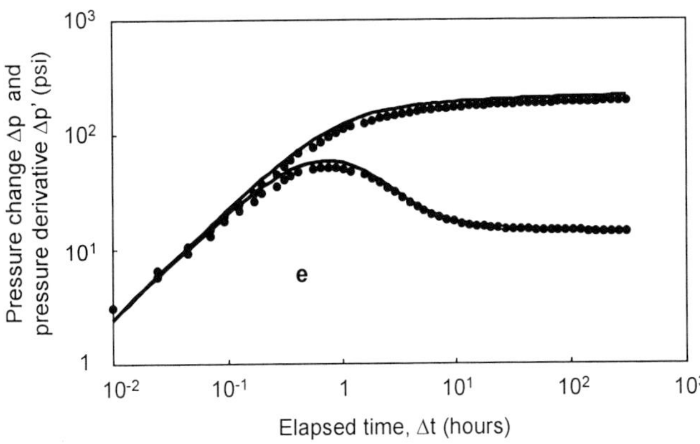

Figure 10.8. Case **e**: shut-in time too late, last flowing pressure is taken in the build-up data, during the wellbore storage regime.

When a log-log plot suggests a time or pressure error such as in Figures 10.4 to 10.7, several correction methods are available. If the response is affected by a wellbore storage effect, a linear scale plot of Δp versus Δt such as in Figure 1.4 can be used. The wellbore storage straight line of slope m_{WBS} must intercept at the *origin*. If not, a pressure or time correction must be applied but the linear scale plot does not indicate which parameter has to be changed. An examination of the *test history plot* on expanded scale, such as Figure 10.3, more accurately defines the correction to apply. With time and pressure error such as case **e**, a good log-log match can be obtained but, on the test simulation match, the underestimated skin coefficient shows clearly with a reduced amplitude on the simulated curve.

Pressure gauge drift

In order to minimize the risk of a gauge drift, several pressure sensors are usually run down hole during testing. Before pressure transient analysis, the gauge responses are compared by estimating the difference δp between the pressure signals. When δp is not constant, either one gauge is effected by a drift, or the weight of the fluid column between the two sensors is not constant (see next Section 10.1.2).

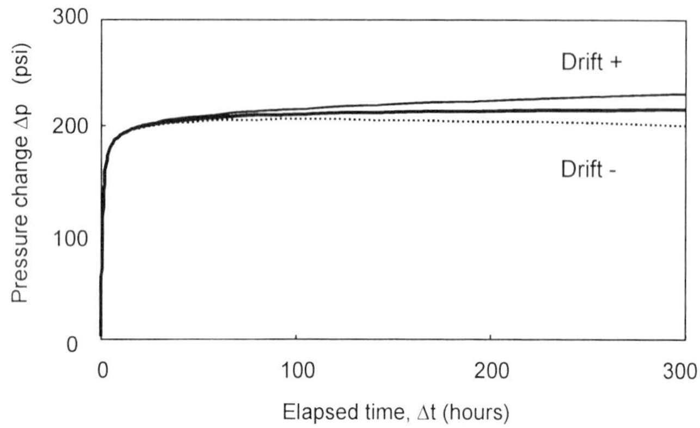

Figure 10.9. Final build-up of Figure 10.1. Drift of ± 0.05 psi/hr. Linear scale.

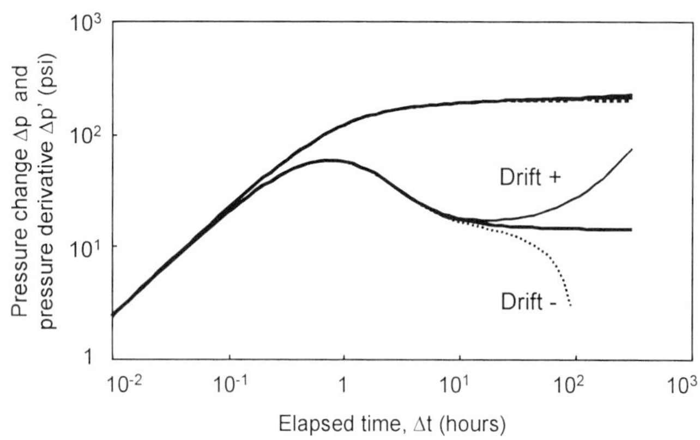

Figure 10.10. Log-log plot of the build-up example. Drift of ± 0.05 psi/hr.

Drift can be positive, when the recorded pressure increases, or negative when it drops. In Figure 10.9, a constant drift of ± 0.05 psi/hr is introduced on the build-up example of Figure 10.1. On the resulting log-log plot of Figure 10.10, the derivative curves suggest the presence of an apparent boundary effect, sealing in the case of positive drift, and constant pressure when it is negative.

The effect of a constant drift is *inverse* during flow and shut-in periods. For example, an increase of derivative on build-up responses is transformed into a pressure stabilization during drawdown, therefore a falling derivative curve. This fact can help identify a problem of constant drift.

When only one pressure gauge is available for analysis, pressure gauge drift can be identified by comparing the flow and shut-in periods on a normalized log-log plot ($\Delta p/q$ and $\Delta p'/q$ versus Δt). When the responses are not symmetrical, a gauge error can be envisaged. Examination of the test simulation on linear scale also gives a clear indication of a pressure drift. If for example a build-up response shows a pressure stabilization corresponding to a declining derivative response, such as on the negative drift curve of Figure 10.10, two boundary conditions can be possibly envisaged: closed system or constant pressure:

1. With a negative drift, the magnitude of drawdown responses is amplified, and the flowing pressure does not stabilize as expected with the constant pressure hypothesis.
2. When a closed reservoir model is used to match the build-up, the model response can possibly reproduce both the drawdown and build-up data but, in such a case, the size of the reservoir is usually very small and not compatible with geological reservoir data.

Pressure gauge noise

In most cases, pressure gauge noise can be reduced by using the smoothing algorithm described in Section 2.3.5. This technique is efficient with a random noise but, when the raw data has been processed manually before generating the derivative plots, a regular noise may have been introduced. This may be the case for example when the pressure points are grouped by pairs, with a very short time difference (a few seconds), each pair being separated by a much longer time interval, such as a few minutes.

In the following case, a regular noise is introduced in the final build-up example of Figure 10.1 by adding 1 psi every 2 pressure points (Figure 10.11).

On the resulting log-log plot Figure 10.12, the derivative response is very scattered, and it starts to oscillate after one hour with an increasing amplitude. As a result, the derivative curves seem to split into two smooth branches. In some cases, the time of departure of the two apparent branches is much earlier than on the example of Figure 10.12 (generated with a low density of pressure points), and the lower branch is out of scale. Then, only 50% of the data is displayed on the log-log derivative curve, but the general aspect is smooth. This configuration can be misleading because, apparently no smoothing is needed, and the increasing trend of the upper branch can be interpreted as a reservoir response when it is only a truncated response. In such a case, the log-log pressure response does not confirm the derivative signature, and no consistent match can be obtained.

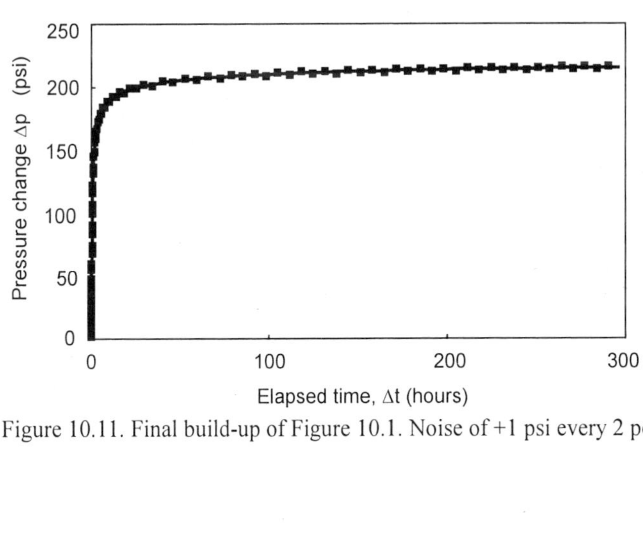

Figure 10.11. Final build-up of Figure 10.1. Noise of +1 psi every 2 points. Linear scale.

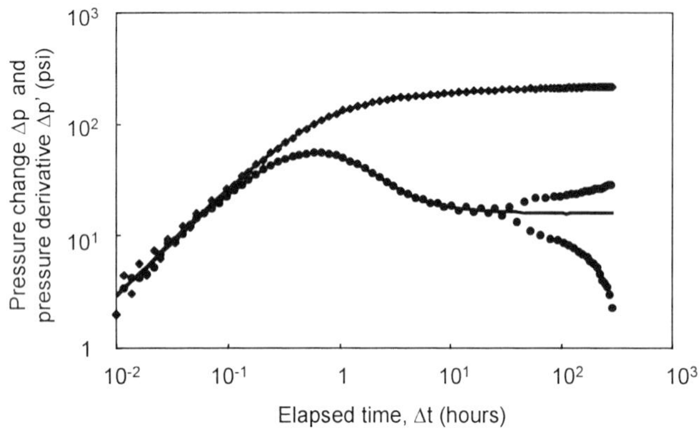

Figure 10.12. Log-log plot of the build-up example. Noise of +1 psi every 2 points. Three points derivative algorithm. No smoothing.

In order to prevent a diagnosis on a truncated derivative curve, it is convenient to indicate, on the log-log plot, the derivative points that are not plotted because they are negative or simply out of scale. Frequently, this is achieved by showing the missing derivative data points on the bottom x axis of the graph, with a different color.

10.1.2 Effect of the well and reservoir condition on pressure responses

Changing wellbore storage

Changing wellbore storage happens when the compressibility of the fluid in the wellbore is not constant. It is observed for example when, in a damaged oil well, free gas is liberated in the production string: the reservoir is flowing above bubble point but, after Δp_{skin}, the fluid becomes two phases.

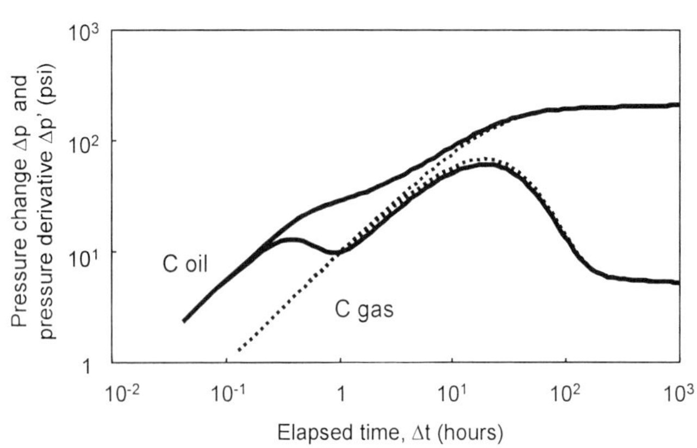

Figure 10.13. Log-log plot of a drawdown example of changing wellbore storage.

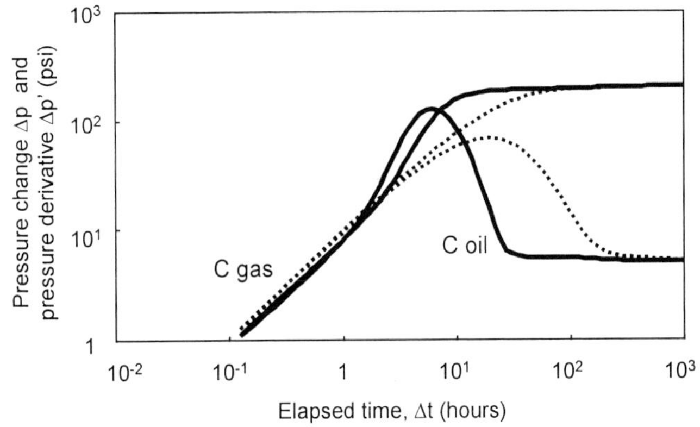

Figure 10.14. Log-log plot of a build-up example of changing wellbore storage

During drawdown, the pressure is high at early time and no free gas is liberated in the wellbore. First the response describes the compressibility of the oil. Later, when the wellbore pressure drops below bubble point, the gas compressibility dominates and the wellbore storage coefficient of Equation 1.7 is *increased* by the change from c_o to c_g. On a log-log scale, an increase of the wellbore storage coefficient C is shown by a second unit slope straight line at later times, as shown in Figure 10.13. During the transition between the oil compressibility wellbore storage C_o, and that for the gas C_g, the pressure tends to stabilize and the derivative can show a short declining trend.

During build-up periods, the response corresponds to the gas wellbore storage coefficient immediately after shut-in, and changes to the lower oil wellbore storage later. This produces a steep increase of derivative and, in some cases, the derivative follows a slope *greater than unity* at the end of the gas dominated early time response as illustrated in Figure 10.14.

When a large drawdown is applied on gas wells, changing wellbore storage can also be observed due to the variable gas compressibility. The compressibility variation being smoother than for oil wells below bubble point, the distortion is less characteristic on the pressure and derivative curves. With build-up data, the wellbore storage derivative hump is shorter than on the theoretical models with constant wellbore storage. In such a case, when the early time unit slope straight line is correctly matched, a constant wellbore storage interpretation model suggests a long derivative transition hump, and it reaches the derivative stabilization later than the data. It is then preferable to ignore the early time unit slope straight line, and to adjust the wellbore storage coefficient on later time data, in order to correctly describe the *start of radial flow* at the beginning of the derivative stabilization. The match is not good at early time, but the reservoir response is correctly described.

High temperature gas wells can also show changing wellbore storage effects because of variable temperature in the wellbore during shut-in.

Two phases liquid level

For wells producing different fluid phases (oil + water, or gas + condensate), a phase redistribution happens in the wellbore during shut-in, producing a characteristic "humping" effect. This is illustrated in the following with an example of well producing oil and water.

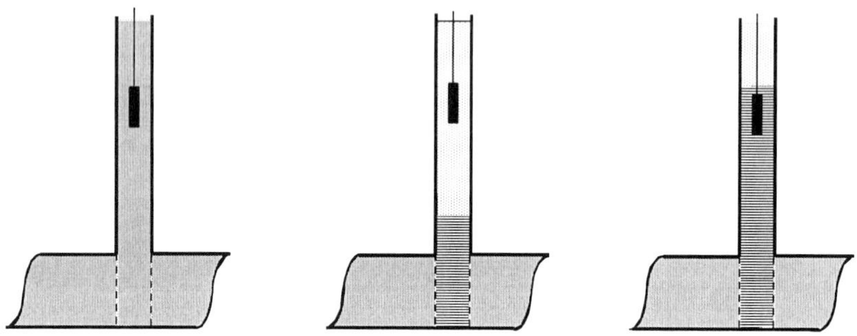

Figure 10.15. Changing liquid level after phase segregation.

In the example Figure 10.15, the depth of the gauge is above formation. When, after shut-in, the water droplets fall to the bottom of the well, the weight of the fluid column between the pressure gauge and the formation is not constant, but increases as long as the water level rises. Initially the hydrostatic weight corresponds to a low percentage of water, to ultimately reach 100% of water if the interface reaches the gauge depth. In some cases, the build-up pressure can show a temporary decreasing trend after some shut-in time as illustrated Figure 10.16.

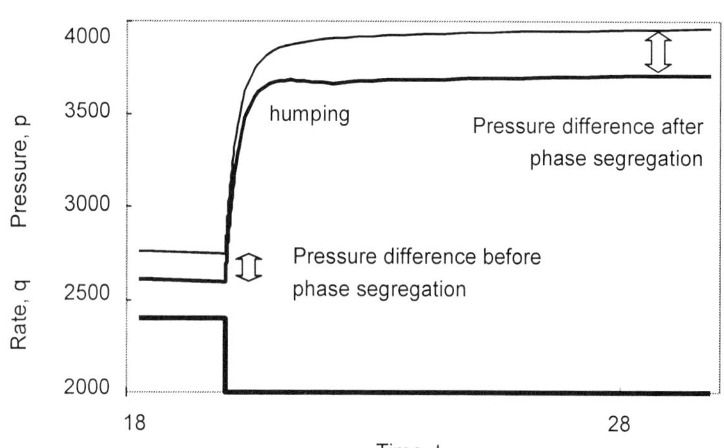

Figure 10.16. Example of build-up response distorted by phase segregation. Humping effect.

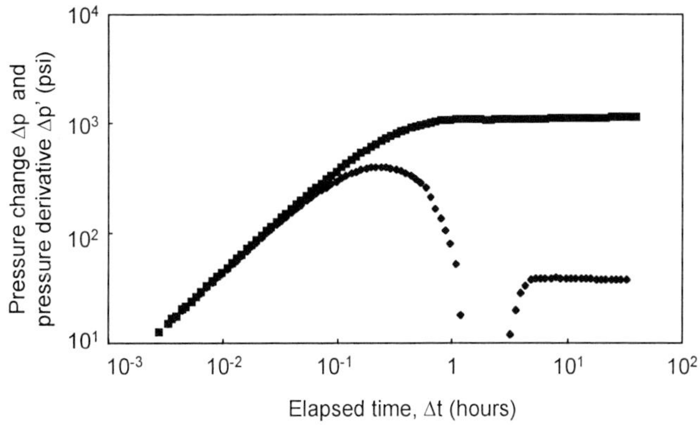

Figure 10.17. Log-log plot of the build-up example of phase segregation.

When the interface between the two phases stabilizes or reaches the depth of the pressure gauge, the pressure difference between gauge and formation returns to a constant, and the remaining build-up data can be properly analyzed. During the hump when the build-up pressure is declining, the derivative becomes *negative* (Figure 10.17).

In some cases, the water cushion created during the first hours of shut-in is slowly re-injected back into the reservoir at later times. Changing liquid level effects can then dominate the entire build-up response, and only drawdown periods are suitable for analysis (Gringarten, 2000).

As a general rule, the pressure gauge should always be positioned as close as practically possible to the perforations or producing interval. When phase redistribution is expected in a well producing several phases, the duration of the humping effect is shortened by reducing the distance between the pressure gauge and the reservoir.

Interference effects from neighboring wells

When testing wells in producing fields, interference effects from neighboring producers can affect the analyzed pressure data. Ideally, a multiple well simulation model should be used for analysis. Using the proper rate history for each producer, and accurate reservoir geometry, the combined effect of neighboring wells is added to the response of the tested well. This procedure is cumbersome, and frequently many approximations have to be made. For example, the different wells may not produce from exactly the same layers, or the well spacing and the geometry of the reservoir boundaries are difficult to describe with an analytical model. In many cases, tests are analyzed with a single well model. It is then recommended to minimize as far as possible the pressure disturbance generated by other wells.

As most well responses follow a logarithmic time relationship, the transient effect is clearly reduced as the time increases. When a well test is planned in a multiple well reservoir environment, it is preferable to maintain *unchanged* the flowing condition of all other wells before the test. If a neighboring well is opened or closed just before or during the well test, its possible interference effect is larger than if no change is made in its flow rate.

10.2 INTERPRETATION PROCEDURE

In this Section, a summary of the interpretation methodology, introduced in Section 3.1 and described through the other Chapters, is initially presented. Typical responses of the usual interpretation models are then reviewed, and two examples of interpretation consistency checks with the test history plot are discussed.

10.2.1 Methodology

Well test analysis is a three step process:

1. *Identification* of the interpretation model. The derivative plot is the primary identification tool.
2. *Calculation* of the interpretation model. The log-log pressure and derivative plot is used to make the first estimates.
3. *Verification* of the interpretation model. The simulation is adjusted on the three usual plots: log-log, semi-log superposition and test history on linear scale.

Log-log and superposition scale plots focus on a single test period, as opposed to the test history plot that applies the interpretation model to a larger time interval, the complete test sequence.

- The main purpose of the semi-log superposition match is to refine the initial log-log results. On log-log scales, the pressure Δp curve is not very sensitive to small variations in the response (see discussion Section 2.1) and, on the derivative curve, the constant skin factor is only present on early time data. Furthermore, the derivative response can be affected by noise. With the linear y-axis of the semi-log superposition scale, the definition of the pressure response is improved, without being affected by data processing such as smoothing.
- The test history plot can indicate discrepancies in the data such as in the rate history, or in the start of the analyzed period (see discussion Section 10.1.1). Alternatively, it is also a good verification plot for the interpretation model, as illustrated next in Section 10.2.4.

The consistency of the interpretation model is finally checked against non-testing information.

10.2.2 The diagnosis: typical pressure and derivative shapes

In the log-log diagnosis process, not only the different characteristic flow regimes are identified in the derivative signature, but also changes of flow properties during a given regime are evident. The flow regime identification, and the effect of a change of mobility or storativity during a regime are discussed in the following.

Flow regime identification

In the description of the different interpretation models presented in previous Chapters, it is shown that the number of typical flow regimes identifiable on pressure responses is limited. Only six different *flow geometries* produce a characteristic response on the derivative and, in some cases, also on log-log pressure curves. These regimes may happen at different times in a test response, depending on the interpretation model. In the Table 10.1 below, the characteristic shape (pressure = gray and derivative = black) for the six flow regimes are summarized, together with the most usual well or reservoir configurations, tentatively classified into early, intermediate or late time response.

Table 10.1. Flow regimes summary

Geometry	**Log-log scale**		**Time range**		
	Shape	***Slope***	***Early***	***Intermediate***	***Late***
Radial :	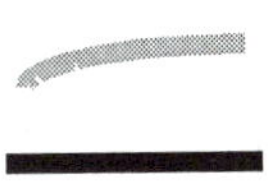	No 0	Double porosity restricted	Homogeneous behavior	Semi-infinite reservoir
Linear :		1/2 1/2	Infinite conductivity fracture	Horizontal well	Two parallel sealing boundaries
Bi-linear :		1/4 1/4	Finite conductivity fracture	Finite conductivity fault	Double porosity unrestricted with linear flow
Spherical :		No -1/2		Well in partial penetration	
Pseudo Steady State :		1 1	Wellbore storage	Layered no crossflow with boundaries	Closed reservoir (drawdown)
Steady State :		0 -1 (-∞)		Conductive fault	Constant pressure boundary

Changes of properties during radial flow

During a given flow regime, some reservoir properties may apparently change in the course of the response. In the following, the effect of a change of mobility k/μ, and storativity ϕc_t, during the radial flow regime is presented. The influence is described both in terms of derivative, with a deviation from the radial flow stabilization, and on semi-log scales. For each case, some examples of corresponding interpretation models are given.

Mobility decreases: Sealing boundaries, composite reservoirs, horizontal well with a long drain hole.

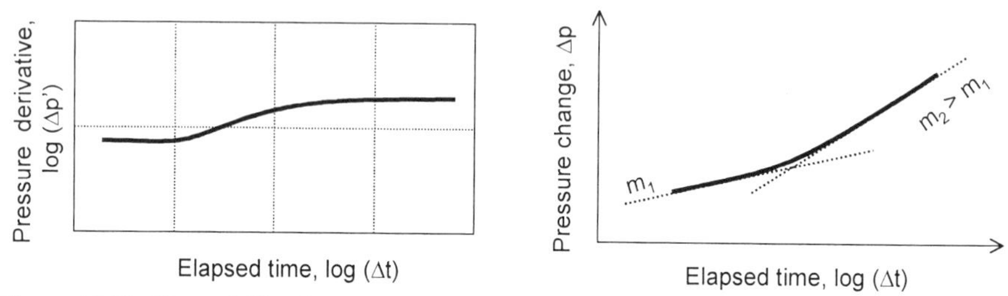

Figure 10.18. The mobility decreases ($k/\mu \downarrow$). Log-log and semi-log scales.

Mobility increases: Composite reservoirs, constant pressure boundaries, layered systems, wells in partial penetration.

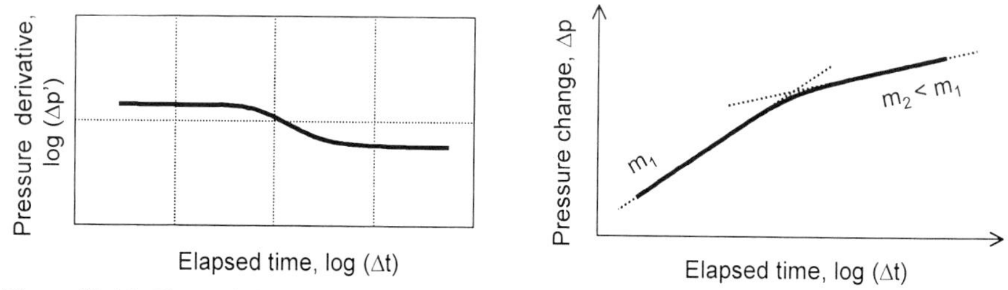

Figure 10.19. The mobility increases ($k/\mu \uparrow$). Log-log and semi-log scales.

During radial flow, the change of mobility is described by a *vertical displacement* of the derivative stabilization. When the mobility is decreased, the second derivative plateau is higher than the first and, conversely, when the mobility is increased it is lower. On semi-log scales, the slope of the second straight line is respectively higher or lower than the first slope.

Storativity increases: Double porosity reservoirs, layered and composite reservoirs.

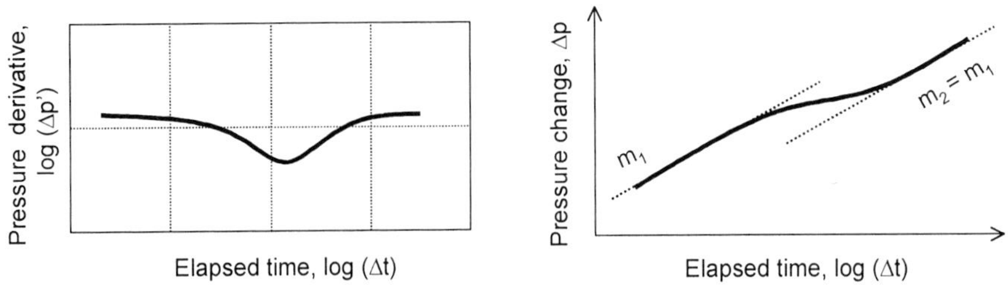

Figure 10.20. The storativity increases ($\phi\, c_t \uparrow$). Log-log and semi-log scales.

Storativity decreases: Composite systems.

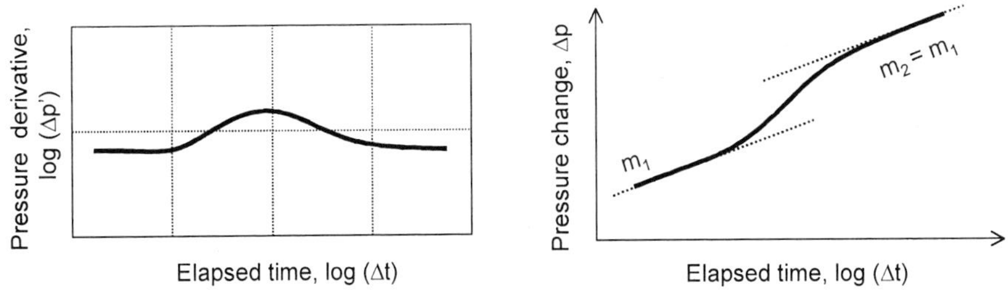

Figure 10.21. The storativity decreases ($\phi\, c_t \downarrow$). Log-log and semi-log scales.

When only the storativity changes, the two derivative stabilizations are at the same level but a *valley* shaped transition is observed when the storativity increases and a *hump* above the radial flow stabilization in the case of a storativity decrease. On semi-log scales, the derivative valley is transformed into a pressure stabilization at transition time between the two semi-log straight lines, and the hump into a sharp pressure rise. Change of storativity affects the time of occurrence of the characteristic regime: the second semi-log straight line is displaced towards late time in Figure 10.20 and towards early time in Figure 10.21.

The above discussion can be extended to any of the five other characteristic regimes. For example, the changing wellbore storage examples of Figures 10.13 and 10.14 can be described as a change of storativity during a pseudo steady state regime. An example of a decrease of the apparent mobility has been presented for the spherical flow regime in Figure 3.17, and during linear flow in Figure 5.20. It is easy to predict the effect of an increase of storativity, due for example to a double porosity system with restricted interporosity flow, during the same flow regimes (see discussion Sections 4.1.4 and 5.7).

Summary of usual log-log responses

In the Appendix 1, a summary of the basic interpretation models is presented. They are classified in well models, reservoir models and boundary models. For the first and the last group, only a homogeneous reservoir is considered and, for the second group, the well is affected by wellbore storage and skin, the reservoir is infinite acting. Combined interpretation models are not considered in this summary.

For each model, the chronology of the characteristic flow regimes is given, together with the controlling parameter(s) or group of parameter(s). Except for the first model, the wellbore storage regime is not listed for reasons of conciseness.

10.2.3 Consistency check with the test history simulation

In this Section, two examples of well test interpretation adjustment with the test history plot are presented. For both examples, the initial pressure is 5000 psi. The first choice of interpretation model, defined from log-log analysis of the short shut-in period, is inconsistent when applied to the complete rate history. A second model has to be used for a consistent description of the well behavior.

Increase of derivative response after the last build-up point (second sealing boundary)

The log-log derivative plot Figure 10.22 suggests the presence of a sealing fault but, when this model is applied with the extended production history (Figure 10.23), the initial pressure used to correctly describe the build-up test is lower than the original initial pressure.

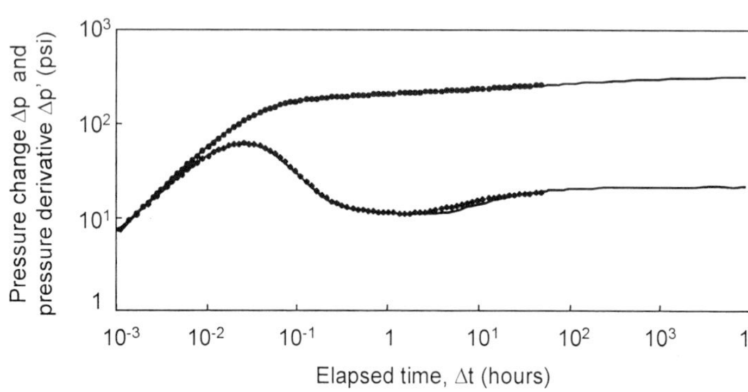

Figure 10.22. Log-log plot of the final build-up. Homogeneous reservoir with a sealing fault.

The single sealing fault model does not generate enough pressure drop during the extended flow period and, in a second estimation, a second fault, parallel to the first is introduced (Figures 10.24 and 10.25). This second boundary is not seen during the short build-up test, only the test simulation match suggests the first model is not appropriate.

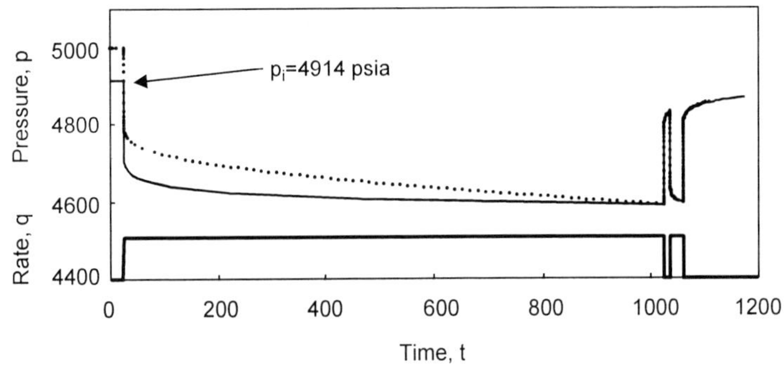

Figure 10.23. Test history simulation. Linear scale. Homogeneous reservoir with a sealing fault.

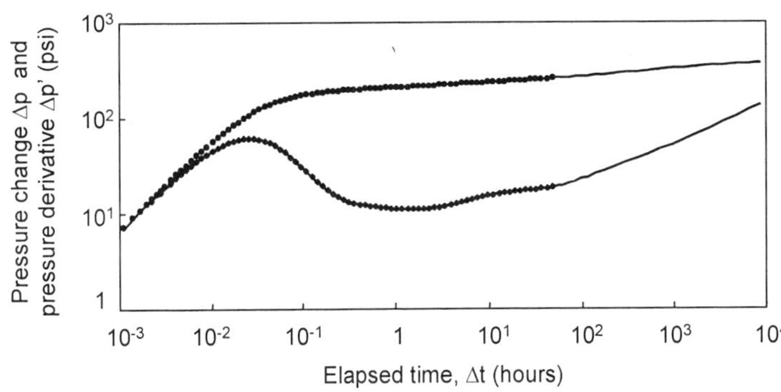

Figure 10.24. Log-log plot of the final build-up.
Homogeneous reservoir with two parallel sealing faults.

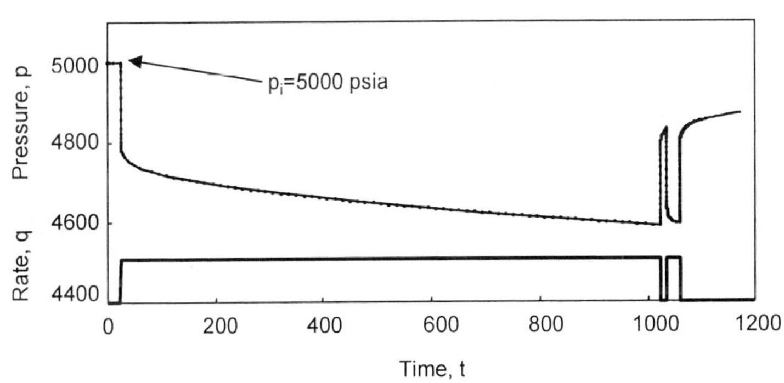

Figure 10.25. Test history simulation. Linear scale.
Homogeneous reservoir with two parallel sealing faults.

Decrease of derivative response after the last build-up point (Layered semi infinite reservoir)

On this example, the opposite diagnosis is made: the log-log derivative plot Figure 10.26 suggests the presence of two parallel sealing faults.

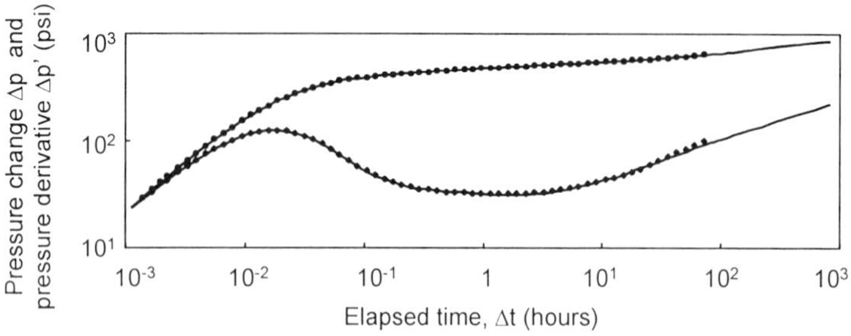

Figure 10.26. Log-log plot of the final build-up.
Homogeneous reservoir with two parallel sealing faults.

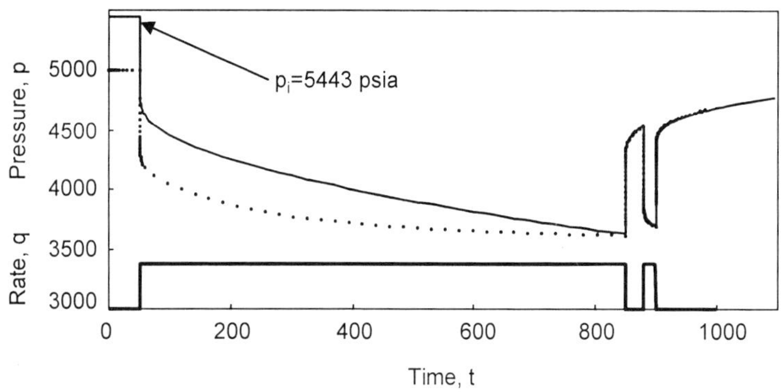

Figure 10.27. Test history simulation. Linear scale.
Homogeneous reservoir with two parallel sealing faults.

On the test history simulation of Figure 10.27 with the parallel sealing faults model, the initial pressure before the production history is too high, showing this boundary model is not applicable. The reservoir is layered and a two layer no crossflow model is used, one layer is closed. At late time, the derivative stabilizes to describe the radial flow regime in the infinite layer. The hump at intermediate time corresponds to the storage of the limited zone, and not to the linear flow regime as originally believed.

The two examples illustrate the importance of an accurate initial pressure measurement.

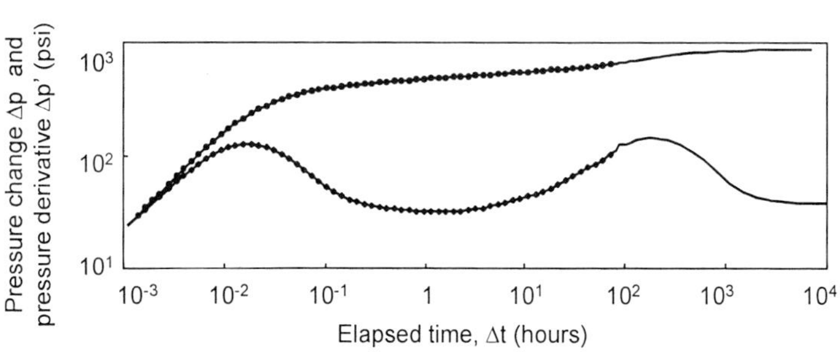

Figure 10.28. Log-log plot of the final build-up.
Two layers reservoir, one infinite and one closed layer.

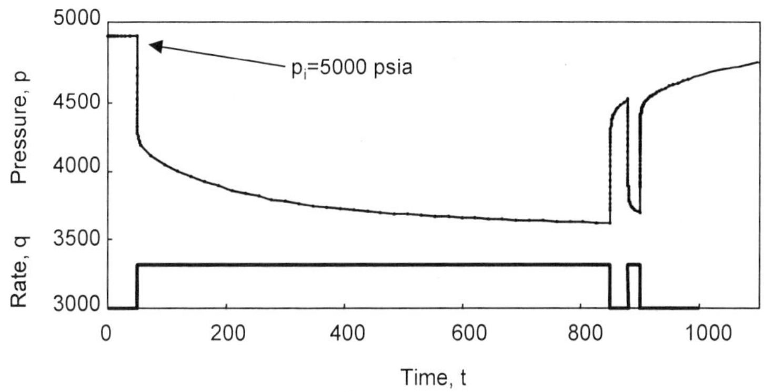

Figure 10.29. Test history simulation. Linear scale.
Two layers reservoir, one infinite and one closed layer.

10.3 WELL AND RESERVOIR CHARACTERISATION- INTERPRETATION RESULTS

In this Section, the significance and the accuracy of the interpretation results are addressed. As illustrated with the final discussion of the radius of investigation, well test interpretation makes use of simple concepts, to describe a relatively complex system. When well test analysis results are transformed into reservoir properties, an appropriate understanding of the actual meaning and limitations of the interpretation concepts is required.

10.3.1 Interpretation model

Well test analysis consists of defining the interpretation model(s) that best describe the available pressure data, recorded during a given flow rate history. As discussed in Section 1.1.2, the interpretation model is simply a transfer function, it describes the well and reservoir *behavior*, not the real nature of the producing system. For example, the homogeneous reservoir model is the most frequently used in well test interpretation, when almost no reservoir can be considered strictly homogenous. Oil bearing formations are in general made up of several strata with different characteristics, the permeability is not uniform over the pay zone thickness and, sometimes, it is not isotropic but it varies with the direction. Still, the well and reservoir behavior can be described with the homogeneous reservoir model.

The relationship between the parameters estimated from well test interpretation (the model parameters), and static measurements such as those obtained from core or log data is not easy to establish. The various discussions of the vertical permeability measurements are an illustration of the difference between interpretation results, and reservoir properties (see Sections 3.4.4 for partially penetrating wells, 3.6.11 for horizontal wells, 4.2.6 for layered systems and 9.4.2 for vertical interference testing).

Well test analysis results provide an *average* of the reservoir properties. With the permeability for example, the apparent vertical permeability k_V is a *harmonic* average (Equation 3.56) whereas, for the horizontal permeability k_H, well testing gives the *arithmetic* average of each layer permeability (Equation 4.42 for example). Furthermore, in the case of permeability anisotropy, each layer's horizontal permeability is defined as the *geometric* average of Equation 3.3.

The skin discussion of Section 3.7.1 illustrates another example of potentially unclear interpretation results, as many components can be introduced into this basic well test interpretation concept. In Section 10.3.3, it is shown that the radius of investigation is only a practical concept without a clear theoretical justification, and it should be considered as approximate, especially with build-up periods.

In Section 1.1.2, the interpretation of well test data is described as an inverse problem, whose solution is in general not unique, and several models can be found applicable to describe the pressure response. The lack of uniqueness in well test interpretation has been frequently illustrated in previous Chapters, where different models are shown to produce similar responses. The model verification and validation is made by integrating data from other sources, such as geological, geophysical, petrophysical or fluid descriptions.

10.3.2 Errors in static parameters

The input data required for analysis has been summarized in Section 1.1.2. Possible errors in well test data are discussed in Section 10.1.1. In the following, the accuracy of the well and reservoir parameters is envisaged.

Errors in the static parameters directly influence the calculated interpretation results, but in most cases they do not affect the choice of the interpretation model. When static parameters are not known precisely, it is always possible to make a first analysis with approximate values, and to refine the results with adjusted values later, without significantly changing the interpretation model.

The net thickness h for example is frequently not accurately defined. From open-hole log interpretation results, the range of uncertainty for h can be as high as 30%. Other usual configurations, such as when a well is found to be behaving partially perforated due to the guns not going off over the complete formation thickness, or when the complete formation is not flowing into the wellbore can possibly lead to an error in the thickness. Similarly, when the oil viscosity μ used for analysis is estimated from correlations, the reliability of the fluid property correlation can be relatively low. Since well test interpretation provides the kh/μ group, any error on h or μ directly influences the permeability estimate k.

With other interpretation results, an error in h or μ does not have the same influence. For example, with a fixed m (and kh/μ group) in Equation 1.17, the estimated skin factor S is only slightly dependent upon h (with a logarithm relationship), but not upon the viscosity μ.. In the same way, the radius of investigation, and the distance to a possible boundary, are dependent upon h (with the square root relationship of Equation 1.23 or 1.24), but independent of μ .

The influence, on the interpretation results, of an error in one static parameter can be easily evaluated from the corresponding equations. Two parameters are frequently a subject of discussion:

1. The wellbore radius r_w is usually defined either as the drilled hole diameter, or as the casing ID. Strictly speaking, the radial flow hypothesis is not valid near the perforations (see for example the discussion of Equation 4.41), and the real wellbore radius is not clearly defined. This parameter only influences the calculated skin factor (in a logarithm, as described in equation 1.17). The influence of any error in r_w is not significant but, when the purpose of well testing is the variation of skin between tests, the same reference diameter should be used for consistency: the skin of a well should be defined relative to a choice of r_w. With horizontal wells for example, several skin definitions have been proposed for different reference wellbore radius (see discussion of Equations 3.38 and 3.41).
2. The total compressibility c_t is a function of the saturations for each phase (Equation 1.3). This parameter is difficult to estimate, in particular in a reservoir near the bubble point pressure where the gas saturation changes (see Chapter 8). In infinite acting homogeneous reservoirs, c_t only influences the skin factor S (in the logarithm term of Equation 1.17) and the distances (with a square root relationship of Equations 1.23, 1.24). An error in c_t by a factor of 10 changes the radius of investigation r_i by a factor of 3.16, and the skin by only 1.151.

Another example of a difficult parameter has been discussed in the Section 3.6 for horizontal wells. It is shown that the horizontal drain length, which can be both an input value or a derived value, directly influences the derived value of vertical permeability.

10.3.3 Discussion of Pressure Profile and Radius of Investigation

The radius of investigation r_i, which is used as an input value for proven hydrocarbons, is frequently viewed as a minimum radial distance to any event that would not be observed during the test period. In practice, for an initial flow period, when a linear boundary is introduced with an interpretation software to estimate the radius of investigation, the resulting distance is similar to that estimated from Equation 1.23 or 1.24. This apparently contradicts the sequence of flow geometry near a sealing fault described in Figure 1.22, where the pressure transient actually reaches the fault 4 times earlier than the boundary can be observed on the producing well pressure (i.e. when the transient radius - different to the radius investigation shown in Figure 10.30 below - is *double* that of the fault distance). As discussed in Section 1.2.3, the mathematical concept of radius of investigation is approximate. By examining Figure 10.30 below, clearly, the transient radius is larger than the radius of investigation estimated from well testing with the above methods. Due to the averaging effects of the interpretation results discussed previously, well pressure responses do not describe the complete reservoir area affected by the well production but, practically, a smaller circular area around the wellbore.

This is confirmed by Oliver's study of radial changes of permeability around the well (1990): he established that the pressure response is *primarily* affected by the permeability at a distance approximately half the radius of investigation estimated from Equation 1.23 or 1.24 (see Equation 4.95, Section 4.3.4). It can be concluded that the maximum influence is at a distance 1/4 of the sequence of pressure transient radii illustrated in Figure 1.22, whereas the *start* of a fault influence is seen when its distance is 1/2 of the transient radius.

The definition of the time to the start of a boundary effect is also subject to errors. With a linear boundary, the change of reservoir characteristic is not radial but in one direction only. On the pressure response, the effect of the change of property is more diluted than in the case of a radial symmetrical configuration, and the influence of the boundary is first seen by a long transition before the hemi-radial flow regime. In the discussion of the sealing fault Equation 1.33 (Section 1.2.7 and 5.1.2), the time Δt_x used for estimating the fault distance is defined at the *midpoint* of the transition from radial (0.5 on log-log plot) to hemi-radial flow (1.0 on log-log plot), and not at the start of this transition. Comparing Equations 1.24 and 1.33 shows that the radius of investigation at Δt_x is more than twice the distance of the fault (Δt_x corresponds approximately to time t_4 in Figure 1.22, with a transient radius four time larger than the fault distance, or to the elapsed time 22 hr on the interference responses of Figure 6.8, when the observation well $\mathbf{O_2}$ is already affected by the fault influence).

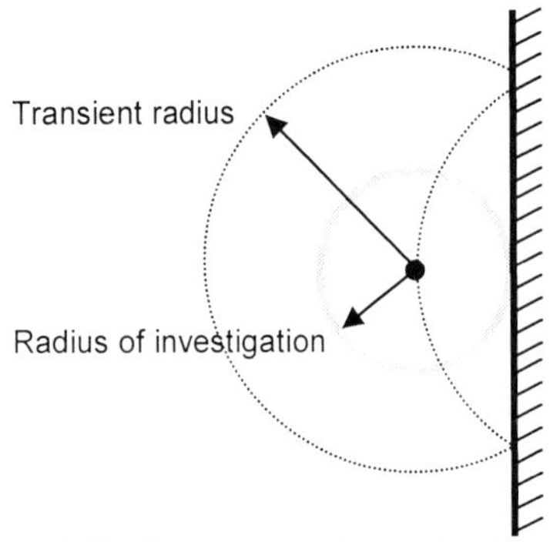

Figure 10.30. Transient radius and radius of investigation at time the sealing fault starts to be seen.

When the criteria for detecting a possible reservoir limit is only a 10% deviation of the derivative above the infinite acting radial flow stabilization, Daungkaew (2000) reports that the resulting fault distance is still 20% smaller than the radius estimated from Equation 1.24. A distance similar to the calculated radius of investigation is obtained when the boundary distance is adjusted to show the start of the deviation on the *last point* of the test period. (In his horizontal well study, Kuchuk (1991 a) uses a similar criteria during vertical radial flow, see the discussion of Equation 3.44).

For shut-in periods, the discussion of Figure 2.20 shows that the concept of radius of investigation is not clear. On this example with a sealing fault, the boundary is reached first with the extrapolated drawdown period, and later during the injection period starting at shut-in time. More generally, at a given shut-in time Δt after a multiple rate sequence, a radius of investigation can be defined for each extrapolated flow period (see definition of the rate history Section 10.1.1). At shut-in time Δt, the Equation 1.23 or 1.24 only describes the radius of investigation corresponding to the injection period; the influence of a possible reservoir limit on the extrapolated flow periods is ignored.

In practice, the radius of investigation during shut-in periods is approximated, as for drawdown, by the distance to a boundary effect introduced at the end of the build-up period. If a constant pressure boundary is be used, the distortion of the derivative response is sharper than with the sealing fault model, and the distance to the change of reservoir property is easier to define.

APPENDIX 1

SUMMARY OF USUAL LOG-LOG RESPONSES

Table A. 1. Well models

Name and characteristic regimes	Log-log pressure and derivative curves
Wellbore storage and Skin (3.1) 1 Wellbore storage, C 2 Radial, kh and S	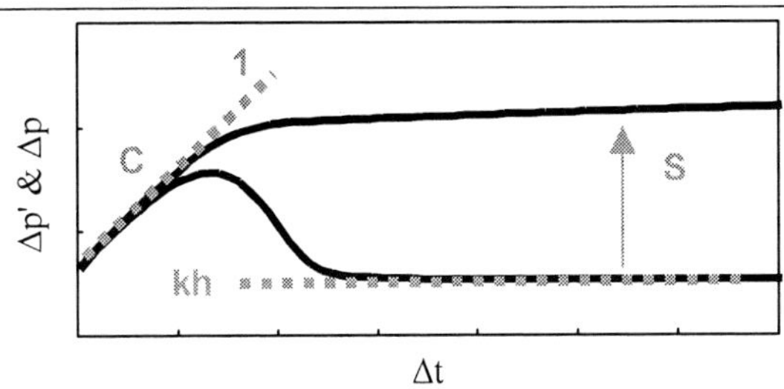
Infinite conductivity fracture (3.2) 1 Linear, x_f 2 Radial, kh and S_{HKF}	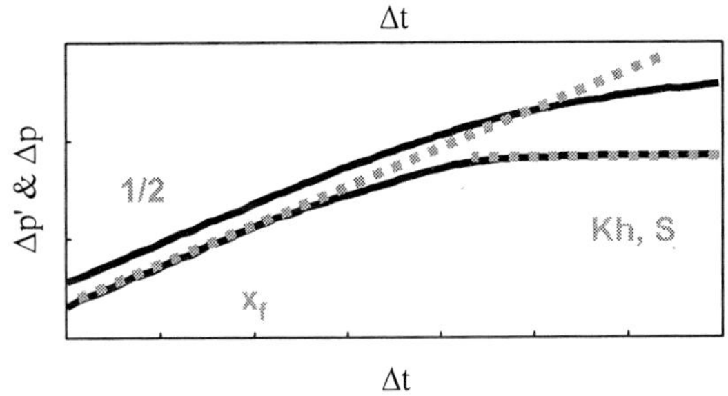
Finite conductivity fracture (3.3) 1 Bi-linear, $k_f w_f$ 2 Linear, x_f 3 Radial, kh and S_{LKF}	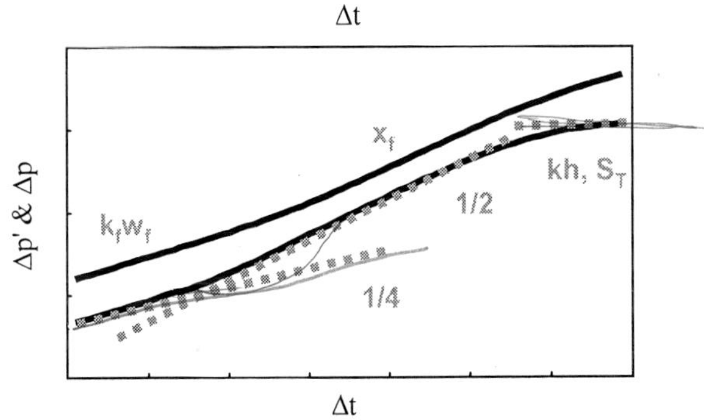
Partial penetration (3.4) 1 Radial, h_w and S_w 2 Spherical (mobility ↑), k_V 3 Radial, kh and S_T	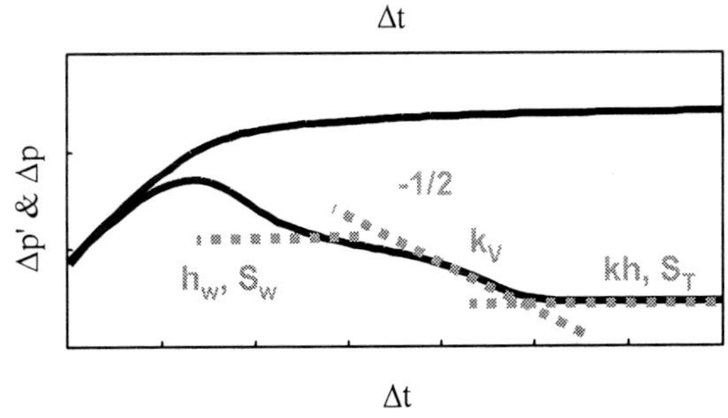

Horizontal well (3.6)

1 Radial vertical, k_V and S_w
2 Linear (mobility ↓)
3 Radial, kh and S_T

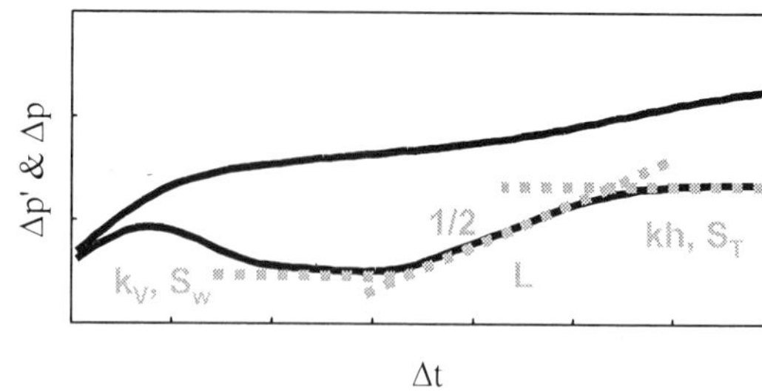

Table A. 2. Reservoir models

Name and characteristic regimes	Log-log pressure and derivative curves

Double porosity, restricted interporosity flow (4.1.2)

1 Radial fissures, kh
2 Transition (storativity ↑), ω and λ
3 Radial fissures + matrix, kh and S

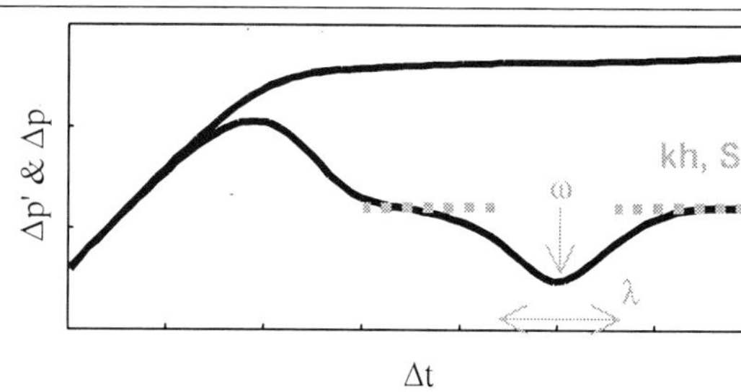

Double porosity, unrestricted interporosity flow (4.1.3)

1 Transition, λ
2 Radial fissures + matrix, kh and S

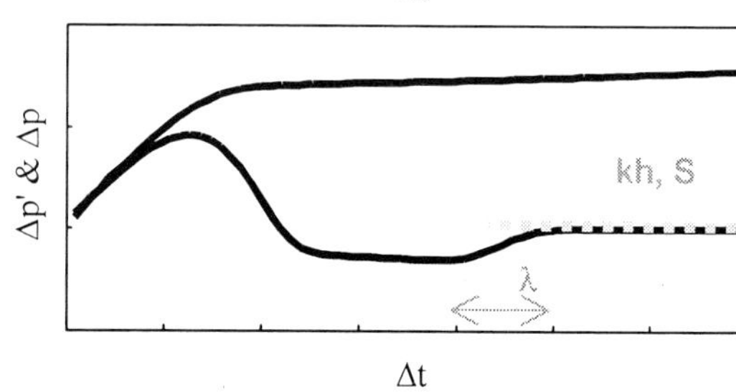

Double permeability, same skin S_1=S_2 (4.2.2)

1 No crossflow
2 Transition (storativity ↑), ω, κ and λ (k_V)
3 Radial, kh_1+kh_2 and S_T

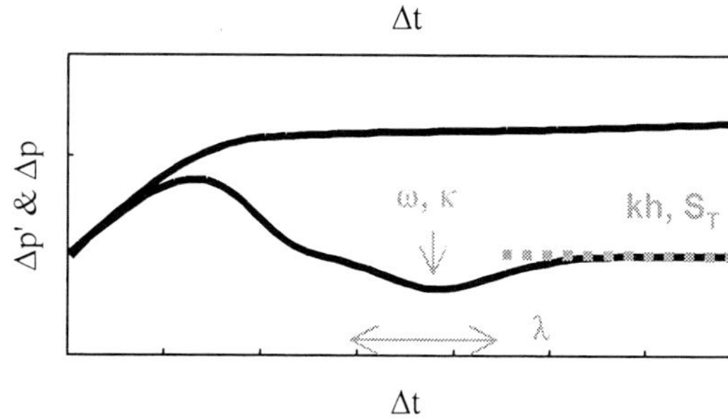

Double permeability, partial penetration $S_1 = \infty$ (4.2.3)

1 Radial, k_2h_2 and S_2
2 Transition (mobility ↑), λ (k_V)
3 Radial, kh_1+kh_2 and S_T

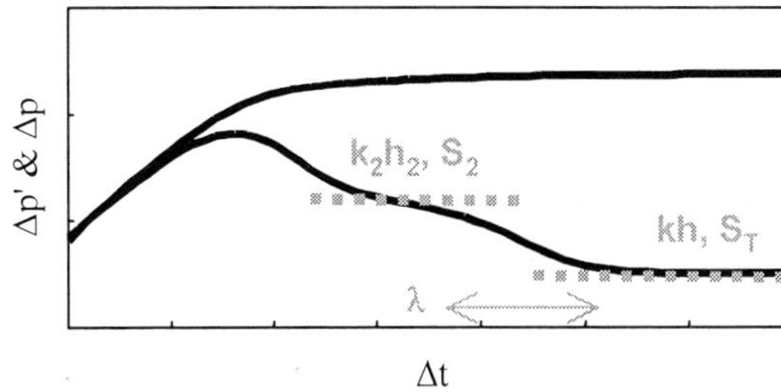

Radial composite (4.3.1)

1 Radial inner, k_1h and S_w
2 Transition (mobility ↑, ↓), r
3 Radial outer, k_2h and S_{RC}

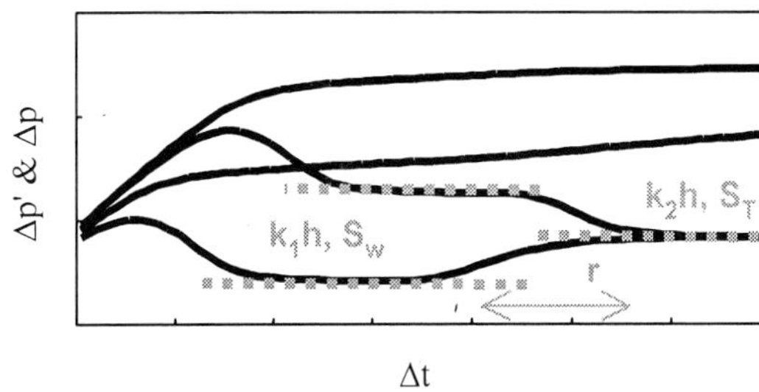

Linear composite (4.3.3)

1 Radial inner, k_1h and S_w
2 Transition (mobility ↑, ↓), L
3 Radial total, $(k_1h + k_2h)/2$ and S_T

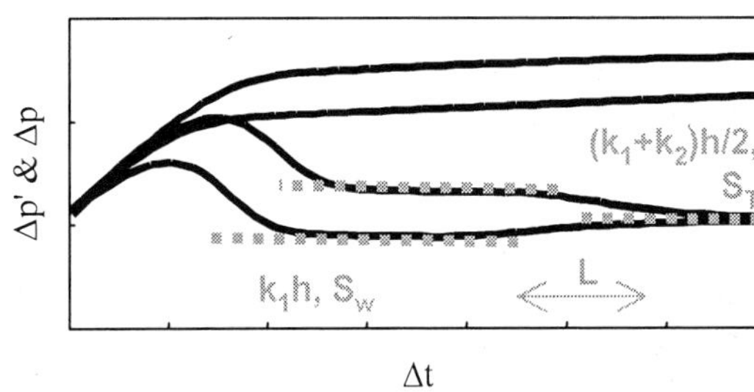

Table A. 3. Boundary models

Name and characteristic regimes	Log-log pressure and derivative curves
Sealing fault (5.1) 1 Radial, kh and S 2 Transition (mobility ↓), L 3 Hemi-radial	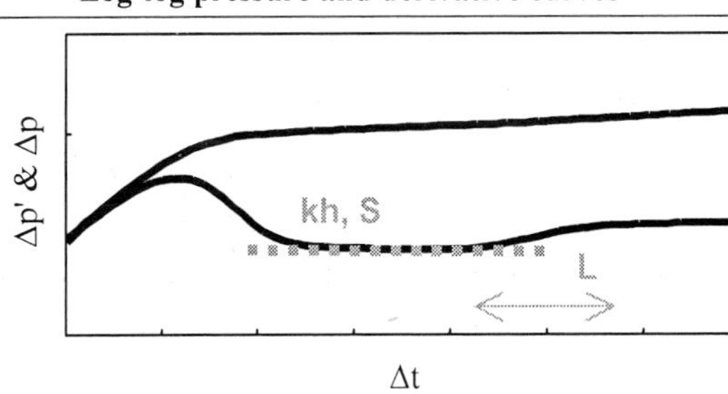

Channel (5.2)
Centered
1 Radial, *kh* and *S*
2 Linear, L_1+L_2
Off-centered
1 Radial, *kh* and *S*
2 Hemi-radial, L_1
3 Linear, L_1+L_2

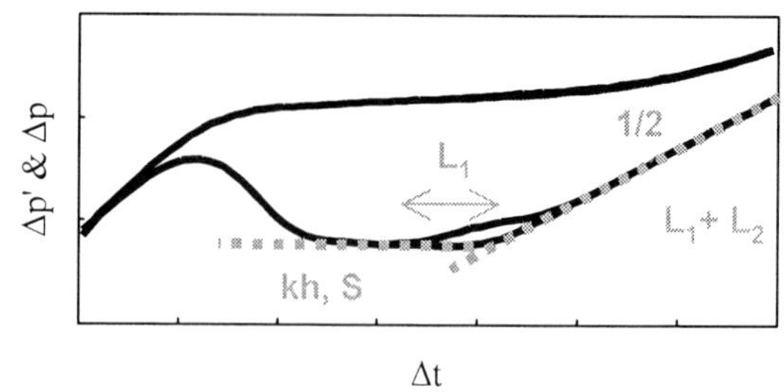

Channel closed at one end (5.4)
Centered
1 Radial, *kh* and *S*
2 Linear, L_1+L_2
3 Transition (mobility ↓), L_3
4 Hemi- linear

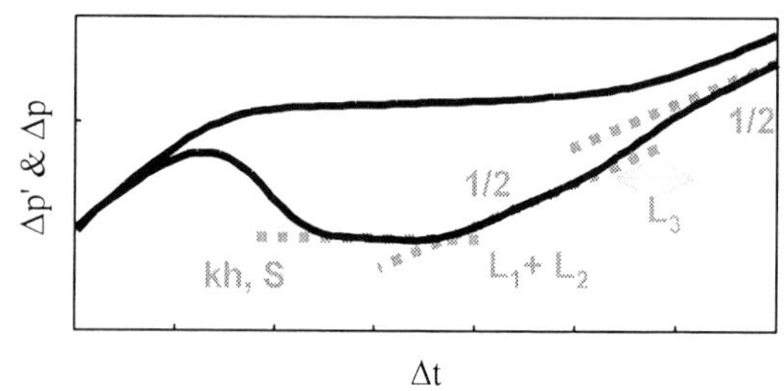

Intersecting faults (5.3)
Centered
1 Radial, *kh* and *S*
2 Linear, L_1+L_2
3 Fraction of radial, θ
Off-centered
1 Radial, *kh* and *S*
2 Hemi-radial, L_1
3 Linear, L_1+L_2
4 Fraction of radial, θ

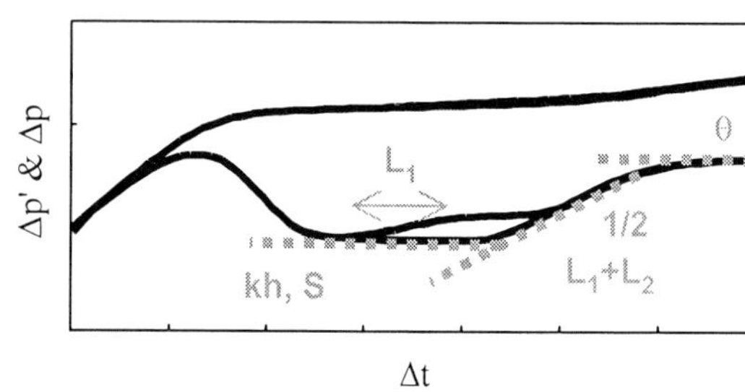

Closed system centered (5.4)
Drawdown
1 Radial, *kh* and *S*
2 Pseudo steady state, *A*
Build-up
1 Radial, *kh* and *S*
2 Average pressure, $\bar{p}$ and *A*

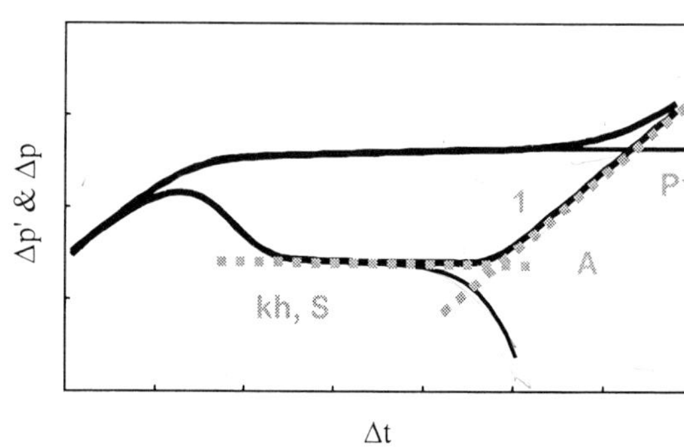

Closed channel (5.4)
Drawdown
1 Radial, *kh* and *S*
2 Linear, L_1+L_2
3 Pseudo steady state, *A*
Build-up
1 Radial, *kh* and *S*
2 Linear, L_1+L_2
3 Average pressure, $\bar{p}$ and *A*

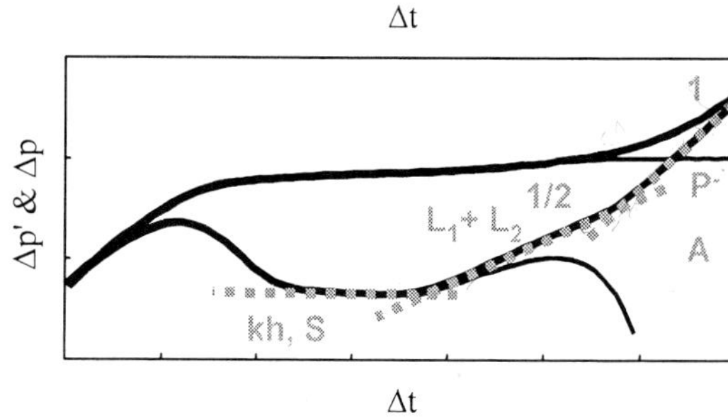

Closed and intersecting faults (5.4)

Drawdown

1 Radial, *kh* and *S*
2 Linear, L_1+L_2
3 Fraction of radial, θ
4 Pseudo steady state, *A*

Build-up

1 Radial, *kh* and *S*
2 Linear, L_1+L_2
3 Fraction of radial, θ
4 Average pressure, $\bar{p}$ and *A*

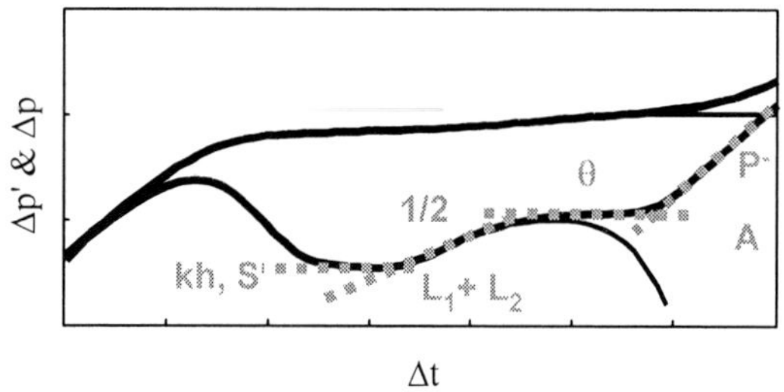

Constant pressure boundaries (5.5)

1 Radial, *kh* and *S*
2 Transition (mobility ↓), *L*

One boundary or multiple boundaries

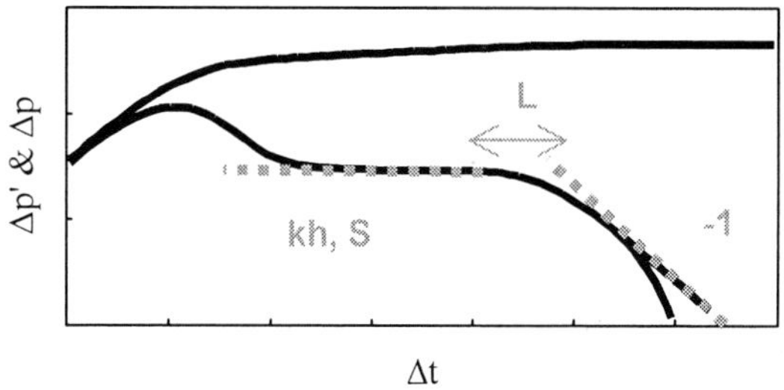

PRACTICAL METRIC SYSTEM OF UNITS

CONVERSION FACTORS

Table A. 4. Conversion factors from oilfield to metric system, and inverse

Quantity	Oilfield units	Practical metric units	Multiply by	Inverse
Gas rate	Mscf/D	m^3/D	$2.831\,685\text{x}10^{-2}$	$3.531\,466\text{x}10^{1}$
Length	ft	m	$3.048\,000\text{x}10^{-1}$	3.280 840
Liquid rate	BOPD	m^3/D	$1.589\,873\text{x}10^{-1}$	6.289 811
Permeability	md	md	1.000 000	1.000 000
Pressure	psi	Bar	$6.894\,757\text{x}10^{-2}$	$1.450\,377\text{x}10^{1}$
Temperature	°R	°K	$5.555\,555\text{x}10^{-1}$	1.800 000
Time	hr	hr	1.000 000	1.000 000
Viscosity	cp	cp	1.000 000	1.000 000
Volume	cuft	m^3	$2.831\,685\text{x}10^{-2}$	$3.531\,466\text{x}10^{1}$

EQUATIONS

Chapter 1

$$C = 10197 \frac{V_u}{\rho(g/g_c)} \tag{1.8}$$

$$S = \frac{kh}{18.66 q B \mu} \Delta p_{\text{Skin}} \tag{1.11}$$

$$p_{w,S} - p_{w,S=0} = \frac{18.66 q B \mu}{k_S h} \ln \frac{r_S}{r_w} - \frac{18.66 q B \mu}{kh} \ln \frac{r_S}{r_w} \tag{1.12}$$

$$\Delta p = 21.5 \frac{q B \mu}{kh} \left[\log \Delta t + \log \frac{k}{\phi \mu c_t r_w^2} - 3.10 + 0.87 S \right] \tag{1.15}$$

$$kh = 21.5\frac{qB\mu}{m} \tag{1.16}$$

$$S = 1.151\left(\frac{\Delta p_{1\,\mathrm{hr}}}{m} - \log\frac{k}{\phi\mu c_t r_w^2} + 3.10\right) \tag{1.17}$$

$$\Delta p(\Delta t, r) = -0.5\frac{18.66qB\mu}{kh}\mathrm{Ei}\left(-\frac{\phi\mu c_t r^2}{0.0001423k\Delta t}\right) \tag{1.18}$$

$$\Delta p(\Delta t, r) = \frac{21.5qB\mu}{kh}\left[\log\frac{0.000356k\Delta t}{\phi\mu c_t r^2} + 0.809\right] \tag{1.19}$$

$$\frac{0.000356k\Delta t}{\phi\mu c_t r_i^2} = \frac{1}{4} \text{ or } \frac{1}{\gamma^2} \tag{1.20}$$

$$r_i = 0.034\sqrt{k\Delta t/\phi\mu c_t} \tag{1.23}$$

$$r_i = 0.037\sqrt{k\Delta t/\phi\mu c_t} \tag{1.24}$$

$$\Delta p = 0.623\frac{qB}{hx_f}\sqrt{\frac{\mu}{\phi c_t k}}\sqrt{\Delta t} \tag{1.25}$$

$$x_f = 0.623\sqrt{\frac{\mu}{\phi c_t k}}\frac{qB}{hm_{\mathrm{LF}}} \tag{1.26}$$

$$\Delta p = 6.28\frac{qB\mu}{h\sqrt{k_f w_f}\sqrt[4]{\phi\mu c_t k}}\sqrt[4]{\Delta t} \tag{1.27}$$

$$k_f w_f = 39.46\sqrt{\frac{1}{\phi\mu c_t k}}\left(\frac{qB\mu}{hm_{\mathrm{BLF}}}\right)^2 \tag{1.28}$$

$$\Delta p = 9.33\frac{qB\mu}{k_S r_S} - 279.3\frac{qB\mu\sqrt{\phi\mu c_t}}{k_S^{3/2}\sqrt{\Delta t}} \tag{1.29}$$

$$k_S = \left(279.3qB\mu\frac{\sqrt{\phi\mu c_t}}{m_{\mathrm{SPH}}}\right)^{2/3} \tag{1.31}$$

$$L = 0.0141\sqrt{\frac{k\Delta t_x}{\phi\mu c_t}} \tag{1.33}$$

$$\Delta p = 0.0417\frac{qB}{\phi c_t hA}\Delta t + 21.5\frac{qB\mu}{kh}\left[\log\frac{A}{r_w^2} - \log(C_A) + 0.351 + 0.87S\right] \tag{1.35}$$

$$\phi hA = 0.0417\frac{qB}{c_t m^*} \tag{1.36}$$

$$\mathrm{PI} = \frac{kh}{21.5B\mu\left(\log\Delta t + \log\frac{k}{\phi\mu c_t r_w^2} - 3.10 + 0.87S\right)} \tag{1.39}$$

$$\mathrm{PI} = \frac{kh}{21.5B\mu\left(\log\frac{A}{r_w^2} - \log(C_A) + 0.351 + 0.87S\right)} \tag{1.40}$$

Chapter 2

$$p_D = \frac{kh}{18.66qB\mu}\Delta p \tag{2.3}$$

$$t_D = \frac{0.000356k}{\phi\mu c_t r_w^2}\Delta t \tag{2.4}$$

$$C_D = \frac{0.1592C}{\phi c_t h r_w^2} \tag{2.5}$$

$$\frac{t_D}{C_D} = 0.00223\frac{kh}{\mu}\frac{\Delta t}{C} \tag{2.6}$$

$$C_D e^{2S} = \frac{0.1592C}{\phi c_t h r_w^2}e^{2S} \tag{2.7}$$

$$kh = 18.66qB\mu(\mathrm{PM}) \tag{2.9}$$

$$C = 0.00223\frac{kh}{\mu}\left(\frac{1}{\mathrm{TM}}\right) \tag{2.10}$$

$$[\Delta p(\Delta t)]_{BU} = 21.5\frac{qB\mu}{kh}\left[\log\frac{t_p\Delta t}{t_p+\Delta t}+\log\frac{k}{\phi\mu c_t r_w^2}-3.10+0.87S\right] \tag{2.13}$$

$$p_{ws} = p_i - 21.5\frac{qB\mu}{kh}\log\frac{t_p+\Delta t}{\Delta t} \tag{2.14}$$

$$S = 1.151\left(\frac{\Delta p_{1\,hr}}{m}-\log\frac{k}{\phi\mu c_t r_w^2}+\log\frac{t_p+1}{t_p}+3.10\right) \tag{2.15}$$

$$p_{ws}(\Delta t) = p_i - 21.5\frac{B\mu}{kh}\sum_{i=1}^{n-1}(q_i-q_{i-1})\log(t_n+\Delta t-t_i)+(q_n-q_{n-1})\log(\Delta t) \tag{2.18}$$

$$\Delta p' = 9.33\frac{qB\mu}{kh} \tag{2.24}$$

$$\Delta p' = 0.311\frac{qB}{hx_f}\sqrt{\frac{\mu}{\phi c_t k}}\sqrt{\Delta t} \tag{2.29}$$

$$\Delta p' = 1.571\frac{qB\mu}{h\sqrt{k_f w_f}\sqrt[4]{\phi\mu c_t k}}\sqrt[4]{\Delta t} \tag{2.30}$$

$$\Delta p' = 139.6\frac{qB\mu\sqrt{\phi\mu c_t}}{k_S^{3/2}\sqrt{\Delta t}} \tag{2.31}$$

$$\Delta p' = 0.0417\frac{qB}{\phi c_t hA}\Delta t \tag{2.32}$$

Chapter 3

$$C_S = C + \pi\left(r_S^2 - r_w^2\right)\phi c_t h \tag{3.1}$$

$$t_{Df} = \frac{0.000356k}{\phi\mu c_t x_f^2}\Delta t \tag{3.8}$$

$$x_f = \sqrt{\frac{0.000264k}{\phi\mu c_t}\frac{1}{TM}} \tag{3.9}$$

$$C_{Df} = \frac{0.1592C}{\phi c_t h x_f^2} \tag{3.10}$$

$$t_D = \frac{0.000356k}{\phi \mu c_t L^2} \Delta t \tag{3.29}$$

$$\Delta p = \frac{21.5\, qB\mu}{2\sqrt{k_V k_H}\, L}\left[\log \frac{\sqrt{k_V k_H}\, \Delta t}{\phi \mu c_t r_w^{é}} - 3.10 + 0.87 S_w - 2\log \frac{1}{2}\left(\sqrt[4]{\frac{k_V}{k_H}} + \sqrt[4]{\frac{k_H}{k_V}}\right)\right] \tag{3.31}$$

$$\Delta p = \frac{1.246\, qB}{2Lh}\sqrt{\frac{\mu \Delta t}{\phi c_t k_H}} + \frac{18.66\, qB\mu}{2\sqrt{k_V k_H}\, L} S_w + \frac{18.66\, qB\mu}{k_H h} S_z \tag{3.33}$$

$$\Delta p = 21.5 \frac{qB\mu}{k_H h}\left[\log \frac{k_H \Delta t}{\phi \mu c_t L^2} - 2.40\right] + \frac{18.66\, qB\mu}{2\sqrt{k_V k_H}\, L} S_w + \frac{18.66\, qB\mu}{k_H h} S_{zT} \tag{3.35}$$

$$\Delta p = 21.5 \frac{qB\mu}{k_H h}\left[\log \frac{k_H \Delta t}{\phi \mu c_t r_w^2} - 3.10 + 0.87 S_{TH}\right] \tag{3.37}$$

$$\sqrt{k_V k_H}\, L = \frac{10.74\, qB\mu}{m_{VRF}} \tag{3.42}$$

$$S_w = 1.151\left[\frac{p(1\text{hr}) - p(\Delta t = 0)}{m_{VRF}} - \log \frac{\sqrt{k_V k_H}}{\phi \mu c_t r_w^2} + 2\log \frac{1}{2}\left(\sqrt[4]{\frac{k_V}{k_H}} + \sqrt[4]{\frac{k_H}{k_V}}\right) + 3.10\right] \tag{3.43}$$

$$k_V = \frac{\phi \mu c_t}{0.000356 \pi \Delta t_{end}} \min\left\{z_w^2, (h - z_w)^2\right\} \tag{3.44}$$

$$k_H L^2 = 0.389\left(\frac{qB}{m_{LF} h}\right)^2 \frac{\mu}{\phi c_t} \tag{3.45}$$

$$S_w = \frac{2\sqrt{k_V k_H}\, L}{18.66\, qB\mu}\left[p(0hr) - p(\Delta t = 0)\right] + 2.303 \log\left[\frac{\pi r_w}{h}\left(1 + \sqrt{\frac{k_V}{k_H}}\right)\sin\left(\frac{\pi z_w}{h}\right)\right] \tag{3.46}$$

$$k_H h = \frac{21.5 qB\mu}{m_{\text{HRF}}} \tag{3.47}$$

$$S_{TH} = 1.151\left[\frac{p(1\text{hr}) - p(\Delta t = 0)}{m_{\text{HRF}}} - \log \frac{k_H}{\phi \mu c_t r_w^2} + 3.10\right] \tag{3.48}$$

$$m_{\text{RLF}} = 10.75 \frac{qB\mu}{k_f w_f} \tag{3.50}$$

$$m_{\mathrm{BLF}} = 6.28 \frac{qB\mu}{x_f \sqrt{k_f w_f} \sqrt[4]{\phi \mu c_t k_H}} \quad (3.51)$$

$$m_{\mathrm{LF}} = 0.793 \frac{qB}{hr_f} \sqrt{\frac{\mu}{\phi c_t k_H}} \quad (3.52)$$

$$m_{\mathrm{LF}} = 0.623 \frac{qB}{hx_f} \sqrt{\frac{\mu}{\phi c_t k_H}} \quad (3.53)$$

Chapter 4

$$t_{Df} = \frac{0.000356k}{(\phi V c_t)_f \mu r_w^2} \Delta t \quad (4.11)$$

$$t_{Df+m} = \frac{0.000356k}{(\phi V c_t)_{f+m} \mu r_w^2} \Delta t \quad (4.12)$$

$$C_{Df} = \frac{0.1592C}{(\phi V c_t)_f h r_w^2} \quad (4.13)$$

$$C_{Df+m} = \frac{0.1592C}{(\phi V c_t)_{f+m} h r_w^2} \quad (4.14)$$

$$\frac{t_D}{C_D} = \frac{t_{Df}}{C_{Df}} = \frac{t_{Df+m}}{C_{Df+m}} = 0.00223 \frac{kh}{\mu} \frac{\Delta t}{C} \quad (4.17)$$

$$\left(C_D e^{2S}\right)_f = \frac{0.1592Ce^{2S}}{(\phi V c_t)_f h r_w^2} \quad (4.18)$$

$$\left(C_D e^{2S}\right)_{f+m} = \frac{0.1592Ce^{2S}}{(\phi V c_t)_{f+m} h r_w^2} \quad (4.19)$$

$$\Delta p = 21.5 \frac{qB\mu}{kh} \left[\log \Delta t + \log \frac{k}{(\phi V c_t)_f \mu r_w^2} - 3.10 + 0.87S \right] \quad (4.27)$$

$$\Delta p = 21.5\frac{qB\mu}{kh}\left[\log\Delta t + \log\frac{k}{(\phi V c_t)_{f+m}\mu r_w^2} - 3.10 + 0.87S\right] \quad (4.28)$$

$$p_D = \frac{k_1 h_1 + k_2 h_2}{18.66 qB\mu}\Delta p \quad (4.49)$$

$$t_D = 0.000356\frac{k_1 h_1 + k_2 h_2}{\left[(\phi c_t h)_1 + (\phi c_t h)_2\right]\mu r_w^2}\Delta t \quad (4.50)$$

$$C_D = \frac{0.1592C}{\left[(\phi c_t h)_1 + (\phi c_t h)_2\right]r_w^2} \quad (4.51)$$

$$\frac{t_D}{C_D} = 0.00223\frac{k_1 h_1 + k_2 h_2}{\mu}\frac{\Delta t}{C} \quad (4.52)$$

$$k_1 h_1 + k_2 h_2 = 18.66 qB\mu(\text{PM}) \quad (4.53)$$

$$C = 0.00223\frac{k_1 h_1 + k_2 h_2}{\mu}\left(\frac{1}{\text{TM}}\right) \quad (4.54)$$

$$p_D = \frac{k_i h_i}{18.66 qB\mu}\Delta p \quad (4.58)$$

$$t_D = \frac{0.000356 k_i h_i}{(\phi c_t h)_i \mu r_w^2}\Delta t \quad (4.59)$$

$$C_D = \frac{0.1592C}{(\phi c_t h)_i r_w^2} \quad (4.60)$$

$$\frac{t_D}{C_D} = 0.00223\frac{k_i h_i}{\mu}\frac{\Delta t}{C} \quad (4.61)$$

$$k_i h_i = 18.66 qB\mu(\text{PM}) \quad (4.62)$$

$$C = 0.00223\frac{k_i h_i}{\mu}\left(\frac{1}{\text{TM}}\right) \quad (4.63)$$

$$\Delta p = 21.5\frac{qB\mu}{k_i h_i}\left[\log\Delta t + \log\frac{k_i}{(\phi c_t)_i \mu r_w^2} - 3.10 + 0.87S_i\right] \quad (4.64)$$

$$\Delta p = 21.5\frac{qB\mu}{kh_{\text{TOTAL}}}\left[\log\Delta t + \log\frac{kh_{\text{TOTAL}}}{(\phi c_t h)_{\text{TOTAL}}\,\mu r_w^2} - 3.10 + 0.87S\right] \tag{4.65}$$

$$p_D = \frac{k_1 h}{18.66qB\mu_1}\Delta p \tag{4.80}$$

$$t_D = \frac{0.000356k_1}{(\phi\mu c_t)_1 r_w^2}\Delta t \tag{4.81}$$

$$C_D = \frac{0.1592C}{(\phi c_t)_1 h r_w^2} \tag{4.82}$$

$$\frac{t_D}{C_D} = 0.00223\frac{k_1 h}{\mu_1}\frac{\Delta t}{C} \tag{4.83}$$

$$S = \frac{k_1 h}{15.66qB\mu_1}\Delta p_{\text{skin}} \tag{4.84}$$

$$k_1 h = 18.66\,qB\mu_1(\text{PM}) \tag{4.85}$$

$$C = 0.00223\frac{k_1 h}{\mu_1}\left(\frac{1}{\text{TM}}\right) \tag{4.86}$$

$$\Delta p = 21.54\frac{qB\mu_1}{k_1 h}\left(\log\Delta t + \log\frac{k_1}{(\phi\mu c_t)_1 r_w^2} - 3.10 + 0.87S\right) \tag{4.88}$$

$$\Delta p = 21.5\frac{qB\mu_2}{k_2 h}\left(\log\Delta t + \log\frac{k_2}{(\phi\mu c_t)_2 r_w^2} - 3.10 + 0.87S_{\text{RC}}\right) \tag{4.89}$$

$$S_{RC} = \frac{k_2 h}{18.66qB\mu_2}\Delta p_{\text{skin RC}} \tag{4.90}$$

$$r_i = R_n\left(1-\sqrt{\frac{k_{n+1}}{k_n}}\right) + \sqrt{k_{n+1}}\left[\frac{R_{n-1}}{\sqrt{k_n}}\left(1-\sqrt{\frac{k_n}{k_{n-1}}}\right) + \ldots + \frac{R_i}{\sqrt{k_{j+1}}}\left(1-\sqrt{\frac{k_{j+1}}{k_j}}\right) + \ldots \right.$$
$$\left. \ldots + \frac{R_1}{\sqrt{k_2}}\left(1-\sqrt{\frac{k_2}{k_1}}\right)\right] + 0.034\sqrt{\frac{k_{n+1}\Delta t}{\phi\mu c_t}} \tag{4.94}$$

$$\lambda_t = k\left(\frac{k_{ro}}{\mu_o} + \frac{k_{rw}}{\mu_w}\right)_r = 0.5\frac{18.66qB}{h\Delta p'} \tag{4.96}$$

$$r = 0.028\sqrt{\frac{\lambda_t}{\phi c_t}\Delta t} = 0.028\sqrt{\frac{18.66qB}{\phi c_t h}\frac{\Delta t}{2\Delta p'}} \tag{4.97}$$

Chapter 5

$$t_{DL} = \frac{0.000356k}{\phi\mu c_t\left(L_1 + L_2\right)^2}\Delta t \tag{5.2}$$

$$\Delta p = 1.246\frac{qB}{h(L_1 + L_2)}\sqrt{\frac{\mu\Delta t}{k\phi c_t}} + 18.66\frac{qB\mu}{kh}\left(S_{\text{ch}} + S\right) \tag{5.4}$$

$$m_{\text{ch}} = 1.246\frac{qB}{h(L_1 + L_2)}\sqrt{\frac{\mu}{k\phi c_t}} \tag{5.5}$$

$$L_1 + L_2 = 1.246\frac{qB}{hm_{\text{ch}}}\sqrt{\frac{\mu}{k\phi c_t}} \tag{5.6}$$

$$S_{\text{ch}} = \frac{kh}{18.66qB\mu}\Delta p_{\text{ch int}} - S \tag{5.7}$$

$$t_{DA} = \frac{0.000356k}{\phi\mu c_t A}\Delta t \tag{5.15}$$

$$m_{\text{hch}} = 2.494\frac{qB}{h(L_1 + L_2)}\sqrt{\frac{\mu}{k\phi c_t}} \tag{5.16}$$

$$p_i - \bar{p} = 0.0417\frac{qB}{\phi c_t hA}\Delta t \tag{5.21}$$

$$L = 0.0616\sqrt{\frac{m_{\text{cp}}h}{qB\phi c_t}} \tag{5.26}$$

$$\Delta p = C + 6.28\frac{qB\mu}{h\sqrt[4]{\phi\mu c_t k}\sqrt{2k_f w_f\frac{\sqrt{M}+1}{\sqrt{M}}}}\sqrt[4]{\Delta t} \tag{5.33}$$

Chapter 6

$$\frac{t_D}{r_D^2} = \frac{0.000356k}{\phi \mu c_t r^2} \Delta t \tag{6.2}$$

$$p_i - p_{wf} = \frac{21.5\, qB\mu}{kh}\left(\log \Delta t + \log \frac{k}{\phi \mu c_t r^2} - 3.10 \right) \tag{6.3}$$

$$\Delta p(\Delta t, x, y) = 0.5 \frac{18.66\, qB\mu}{\sqrt{k_{\max} k_{\min}}\, h} \mathrm{Ei}\left[-(r_D{}^2/4t_D)_{x,y} \right] \tag{6.4}$$

$$\left(\frac{t_D}{r_D{}^2} \right)_{x,y} = \frac{0.000356\Delta t}{\phi \mu c_t} \left(\frac{k_{\max} k_{\min}}{k_x y^2 + k_y x^2 - 2k_{xy} xy} \right) \tag{6.7}$$

$$\frac{t_{Df}}{r_D^2} = \frac{0.000356k}{(\phi V c_t)_f \mu r^2} \Delta t \tag{6.11}$$

Chapter 7

Standard conditions: p_{sc} = 1 Bar and T_{sc} = 288.15°K.

$$\begin{aligned} p_D &= \frac{kh}{37.33T q_{sc}} \frac{T_{sc}}{p_{sc}} \left[m(p_i) - m(p) \right] \\ &= \frac{kh}{0.1296T q_{sc}} \left[m(p_i) - m(p) \right] \end{aligned} \tag{7.7}$$

$$\begin{aligned} p_D &= \frac{kh}{37.33\overline{\mu Z} T q_{sc}} \frac{T_{sc}}{p_{sc}} \left(p_i^2 - p^2 \right) \\ &= \frac{kh}{0.1296\overline{\mu z} T q_{sc}} \left(p_i^2 - p^2 \right) \end{aligned} \tag{7.8}$$

$$\begin{aligned} p_D &= \frac{kh\overline{p}}{18.66\overline{\mu Z} T q_{sc}} \frac{T_{sc}}{p_{sc}} (p_i - p) \\ &= \frac{kh\overline{p}}{0.0648\overline{\mu Z} T q_{sc}} (p_i - p) \end{aligned} \tag{7.9}$$

$$t_D = \frac{0.000356k}{\phi \mu_i c_{ti} r_w{}^2} \Delta t \tag{7.10}$$

$$t_D = \frac{0.000356k}{\phi \overline{\mu} \overline{c}_t r_w^2} \Delta t \tag{7.11}$$

$$C_D = \frac{0.1592C}{\phi c_{ti} h r_w^2} \tag{7.12}$$

$$C_D = \frac{0.1592C}{\phi \overline{c}_t h r_w^2} \tag{7.13}$$

$$\frac{t_D}{C_D} = 0.00223 \frac{kh}{\mu_i} \frac{\Delta t}{C} \tag{7.14}$$

$$\frac{t_D}{C_D} = 0.00223 \frac{kh}{\overline{\mu}} \frac{\Delta t}{C} \tag{7.15}$$

$$\begin{aligned} m_{\mathrm{WB}} &= 0.0834 \frac{T q_{sc}}{\mu_i C} \frac{p_{sc}}{T_{sc}} \\ &= 0.000289 \frac{T q_{sc}}{\mu_i C} \end{aligned} \tag{7.16}$$

$$\begin{aligned} m_{\mathrm{WB}} &= 0.0834 \frac{\overline{Z} T q_{sc}}{C} \frac{p_{sc}}{T_{sc}} \\ &= 0.000289 \frac{\overline{Z} T q_{sc}}{C} \end{aligned} \tag{7.17}$$

$$\begin{aligned} m_{\mathrm{WB}} &= 0.0417 \frac{\overline{Z} T q_{sc}}{\overline{p} C} \frac{p_{sc}}{T_{sc}} \\ &= 0.000145 \frac{\overline{Z} T q_{sc}}{\overline{p} C} \end{aligned} \tag{7.18}$$

$$\begin{aligned} m &= 42.98 \frac{T q_{sc}}{kh} \frac{p_{sc}}{T_{sc}} \\ &= 0.1492 \frac{T q_{sc}}{kh} \end{aligned} \tag{7.19}$$

$$\begin{aligned} m &= 42.98 \frac{\overline{\mu} \overline{Z} T q_{sc}}{kh} \frac{p_{sc}}{T_{sc}} \\ &= 0.1492 \frac{\overline{\mu} \overline{Z} T q_{sc}}{kh} \end{aligned} \tag{7.20}$$

$$m = 21.49 \frac{\overline{\mu}\overline{Z}T q_{sc}}{\overline{p}kh} \frac{p_{sc}}{T_{sc}} = 0.0746 \frac{\overline{\mu}\overline{Z}T q_{sc}}{\overline{p}kh} \tag{7.21}$$

$$S' = 1.151\left(\frac{\Delta m(p)_{1\,\text{hr}}}{m} - \log\frac{k}{\phi\mu_i c_{ti} r_w^2} + 3.10\right) \tag{7.22}$$

$$S' = 1.151\left(\frac{\Delta p_{1\,\text{hr}}^2 \text{ or } \Delta p_{1\,\text{hr}}}{m} - \log\frac{k}{\phi\overline{\mu}\overline{c}_t r_w^2} + 3.10\right) \tag{7.23}$$

$$\begin{aligned}\Delta m(p)_{skin} &= (0.1296T/kh)(q_{n,sc}S'_n - q_{n-1,sc}S'_{n-1}) \\ &= (0.1296T/kh)(q_n - q_{n-1})_{sc}S + D(q_n^2 - q_{n-1}^2)_{sc} \\ &= (0.1296T/kh)(q_n - q_{n-1})_{sc}\left[S + D(q_n + q_{n-1})_{sc}\right]\end{aligned} \tag{7.25}$$

$$\begin{aligned}m(\overline{p}_{pp}) &= m(p_i) - 0.000289\frac{Tq_{sc}}{\phi\mu_i c_{ti} hA}\Delta t \\ &= m(p_i) - 0.000289\frac{T}{\phi\mu_i c_{ti} hA}G_p\end{aligned} \tag{7.27}$$

$$m(\overline{p}) - m(p_{wf}) = 0.1491\frac{T}{kh}\left(\log\frac{A/r_w^2}{C_A} + 0.351 + 0.87S\right)q_{sc} + 0.1296\frac{T}{kh}Dq_{sc}^2 \tag{7.29}$$

$$m(\overline{p}) - m(p_{wf}) = 0.1491\frac{T}{kh}\left(2\log\frac{0.472r_e}{r_w} + 0.87S\right)q_{sc} + 0.1296\frac{T}{kh}Dq_{sc}^2 \tag{7.30}$$

$$m(\overline{p}) - m(p_{wf}) = 0.1491\frac{T}{kh}\left(\log\frac{k\Delta t}{\phi\mu_i c_{ti} r_w^2} + 3.10 + 0.87S\right)q_{sc} + 0.1296\frac{T}{kh}Dq_{sc}^2 \tag{7.31}$$

$$q_{sc,\text{AOF}} = \frac{-a + \sqrt{a^2 + 4b\left(m(\overline{p}) - m(14.7)\right)}}{2b} \tag{7.32}$$

Chapter 8

$$\begin{aligned}(qB)_t &= q_o B_o + q_w B_w + q_g B_g \\ &= q_o B_o + q_w B_w + (q_{sg} - q_o R_s)B_g\end{aligned} \tag{8.1}$$

$$p_D = \frac{(k/\mu)_t h}{18.66 (qB)_t} \Delta p \tag{8.4}$$

$$t_D = \frac{0.000356 (k/\mu)_t}{\phi c_t r_w^2} \Delta t \tag{8.5}$$

$$\frac{t_D}{C_D} = 0.000223 \frac{(k/\mu)_t h}{C} \Delta t \tag{8.6}$$

$$m = 21.5 \frac{(qB)_t}{(k/\mu)_t h} \tag{8.7}$$

$$S = 1.151 \left(\frac{\Delta p_{1\,\mathrm{hr}}}{m} - \log \frac{(k/\mu)_t}{\phi c_t r_w^2} + 3.10 \right) \tag{8.8}$$

$$p_D = \frac{kh}{18.66 q_o} \Delta m(p) \tag{8.12}$$

$$t_D = \frac{0.000356 k}{\phi \mu_i c_{ti} r_w^2} \Delta t \tag{8.13}$$

$$\frac{k_o}{\mu_o B_o} = a p \tag{8.16}$$

$$p_D = \frac{ah}{37.33 q_o} \Delta (p^2) \tag{8.17}$$

$$m = 42.97 \frac{q_o}{ah} \tag{8.18}$$

$$S = 1.151 \left(\frac{p_{1\,\mathrm{hr}}^2 - p_{\mathrm{i}}^2}{m} - \log \frac{(k/\mu)_t}{\phi c_t r_w^2} + 3.10 \right) \tag{8.19}$$

$$k_o = \frac{37.33 q_o p_{ref} (\mu_o B_o)_{p_{ref}}}{h} \mathrm{PM} \tag{8.20}$$

Chapter 9

$$kh = \frac{\mu C}{0.00223} \left(\frac{t_D / C_D}{\Delta t} \right)_{\mathrm{MATCH}} \tag{9.2}$$

$$S = 0.5 \ln \frac{\left(C_D e^{2S}\right)_{\text{MATCH}}}{0.1592C/\phi c_t h r_w^2} \tag{9.3}$$

$$\frac{0.00223kh}{C\mu(p_i - p_0)} \Delta t\left(p_i - p_{wf}(t)\right) = \frac{dp_D}{d \ln t_D} \tag{9.4}$$

$$kh = \frac{\mu C\left(p_i - p_0\right)}{0.00223} \left(\frac{dp_D / d \ln t_D}{\Delta t\left(p_i - p_{wf}(t)\right)}\right)_{\text{MATCH}} \tag{9.5}$$

$$\frac{0.00223kh}{Q_t \mu}\left(t_p + \Delta t\right)\left(p_i - p\right) = \frac{dp_D}{d \ln t_D} \tag{9.6}$$

$$q_D = \frac{18.66B\mu}{kh\left(p_i - p_{wf}\right)} q(t) \tag{9.7}$$

$$\frac{1}{q} = 21.5 \frac{B\mu}{kh(p_i - p_{wf})}\left[\log \Delta t + \log \frac{k}{\phi \mu c_t r_w^2} - 3.10 + 0.87S\right] \tag{9.8}$$

$$kh = 21.5 \frac{B\mu}{m_q\,(p_i - p_{wf})} \tag{9.9}$$

$$S = 1.151\left[\frac{1/q\left(1\text{hr}\right)}{m_q} - \log \frac{k}{\phi \mu c_t r_w^2} + 3.10\right] \tag{9.10}$$

$$\frac{1}{q_D} = \frac{kh\left(p_i - p_{wf}\right)}{18.66B\mu} \frac{1}{q(t)} \tag{9.11}$$

$$m_q = 0.1492 \frac{T}{kh\ \left[m(p_i) - m(p_{wf})\right]} \tag{9.12}$$

$$\frac{1}{q_D} = \frac{kh\left[m(p_i) - m(p_{wf})\right]}{0.1296T} \frac{1}{q_{sc}(t)} \tag{9.13}$$

$$m_{\text{BLF}} = \frac{0.0449T}{\left[m(p_i) - m(p_{wf})\right] h\sqrt{k_f w_f}\ \sqrt[4]{(\phi \mu c_t)_i k}} \tag{9.14}$$

a = intercept of the stabilized deliverability straight line, Fig. 7.4, 7.8, psia^{2}D/cp/Mscf, or constant in Eq. 8.16
A = area, Eq. 5.13, sq ft
b = slope of the stabilized deliverability straight line, Fig. 7.4, 7.8, psia^{2}D^2/cp/(Mscf^2)
B = formation volume factor, RB/STB
B_g = gas formation volume factor, Eq. 8.1, RB/scf
B_o = oil formation volume factor, Eq. 8.1, RB/STB
B_w = water formation volume factor, Eq. 8.1, RB/STB
C = wellbore storage coefficient, Eq. 1. 7, Bbl/psi, or constant in Eq. 7.28
C_A = shape factor
C_D = dimensionless wellbore storage coefficient, Eq. 2.5, 7.12, 7.13
C_{De} = dimensionless wellbore storage coefficient based on equivalent wellbore radius, Eq. 3.13
C_{Df} = dimensionless wellbore storage coefficient based on half fracture length, Eq. 3.10, or on fissure parameters, Eq. 4.13
C_{Df+m} = dimensionless wellbore storage coefficient based on total system parameters, Eq. 4.14
c_f = formation compressibility, psi^{-1}
c_g = gas compressibility, Eq. 7.2, psi^{-1}
c_o = oil compressibility, psi^{-1}
C_S = apparent wellbore storage coefficient for a stimulated well, Eq. 3.1, Bbl/psi
c_t = total compressibility, Eq. 1.3 and 8.3, psi^{-1}
$\overline{c_t}$ = total compressibility at average test pressure, section 7.2.2, psi^{-1}
c_{ti} = total compressibility at initial pressure, section 7.2.2, psi^{-1}
$c_{t1,2,j}$ = total compressibility in layered, Fig. 4.41, or composite reservoir, Fig. 4.51, psi^{-1}
c_w = water compressibility, psi^{-1}
D = turbulent, Eq. 7.24, D/Mscf, or friction flow coefficient, Eq. 3.49, D/Bbl
d_{iD} = dimensionless distance of linear composite interface i, Eq. 1.34
e = exponential (2.7182 . . .)
Ei = exponential integral
F = storativity ratio (inner zone / outer zone), Eq. 4.79
F_{cD} = dimensionless conductivity of finite conductivity fault, Eq. 5.28
F_i = storativity ratio in radial composite layer i, Eq. 4.79
g = gravitational acceleration, ft/sec^2
G = dimensionless pressure drop in a low conductivity fracture, Eq. 3.15
g_c = gravitational acceleration conversion factor
G_i = initial gas volume, Eq. 7.26, Mscft
G_p = cumulative volume of gas produced, Eq. 7.26, Mscft
h = formation thickness, ft
h_a = apparent formation thickness, Eq. 3.27, ft
h_d = matrix skin thickness, Fig. 4.27, ft
h_D = dimensionless reservoir thickness-anisotropy group, for a well in partial penetration Eq. 3.19, for a horizontal well Eq. 3.30
h_f = fissures thickness, Eq. 4.1, ft
h_w = perforated interval length, Fig. 3.15, ft
$h_{w\text{-}obs}$ = observation interval length, Fig. 9.10, ft
$h_{1,2,j}$ = layer thickness, Fig. 4.41, ft
h' = semi-permeable wall thickness, Fig. 4.41, ft
Δh = change of liquid level depth, section 1.2.2, ft
Δh_j = change of liquid level during interval j, Table. 9.1, ft
$\Delta h'_j$ = time rate of change of liquid level during interval j, Table. 9.1, ft
k = permeability, md
$\bar{k}$ = average horizontal permeability in anisotropic system, Eq. 3.3 and 6.5, md
k_a = altered permeability near a finite conductivity fault, Eq.5.29, md
k_d = matrix skin permeability, Fig. 4.27, md

k_f = permeability in fracture, Fig. 1.14, or fissures, Eq. 4.1, or semi-permeable, Fig. 5.32, or finite conductivity fault, Fig. 5.37, md

$k_{fD}w_{fD}$ = dimensionless fracture conductivity, Eq. 3.14

k_g = effective permeability to gas, Eq. 8.2, md

k_H = horizontal permeability, md

$\bar{k}_H$ = average horizontal permeability in layered system, Eq. 3.55 , md

k_m = matrix permeability, Fig. 4.27, md

$k_{\max}$ = maximum directional permeability, Eq. 3.3, Fig. 6.9, md

$k_{\min}$ = minimum directional permeability, Eq. 3.3, Fig. 6.9, md

k_o = effective permeability to oil, Eq. 8.2, md

$k_{rg,o,w}$ = relative permeability to gas, oil and water, Eq. 8.9, fraction

k_S = permeability in the skin zone, Fig. 1.12, or spherical permeability, Eq. 1.30, md

k_V = vertical permeability, Fig. 3.15 & 3.24, md

$\bar{k}_V$ = average vertical permeability in layered system, Eq. 3.56 , md

$k_{x,y,z}$ = directional permeability, Fig. 3.40, or components of permeability tensor, Eq. 6.5, md

k_w = effective permeability to water, Eq. 8.2, md

$k_{z1,2,j}$ = layer vertical permeability, Fig. 4.41, md

$k_{1,2,j}$ = permeability in layered, Fig. 4.41, or composite reservoir, Fig. 4.51, md

k'_z = semi-permeable wall vertical permeability, Fig. 4.41, md

L = distance (Fig. 1.21, 4.51), or half length of an horizontal well, Fig. 3.24, ft, or smoothing coefficient, Fig. 2.21, or liquid mole fraction, Eq. 8.15

L_a = apparent effective half length of an horizontal well in anisotropic system, Eq. 3.54

L_{cD} = dimensionless critical distance of a sealing fault, Eq. 5.34

L_{cp} = apparent distance of constant pressure boundary in a finite conductivity fault response, Eq. 5.32

L_D = dimensionless distance, Eq. 1.34

L_{eff} = effective half length of an horizontal well producing in several segments

L_{jD} = dimensionless distance of boundary j, Eq. 1.34

m = straight line slope during radial flow, Eq. 1.15, psi/cycle, Eq. 7.19, psia^2/cp/cycle, Eq. 7.20, psia^2/ cycle, Eq. 7.21, psi/ cycle

M = mobility ratio (inner zone / outer zone), Eq. 4.78

m_{BLF} = straight line slope during bilinear flow, Eq. 1.27 or 3.51, psi/(hr$^{1/4}$), or Eq. 9.14, D/(Mscf.hr$^{1/4}$)

m_{ch} = straight line slope during linear flow in a channel, Eq. 5.5, psi/(hr$^{1/2}$)

m_{cp} = derivative straight line slope in case of a constant pressure boundary, Eq. 5.26, psi.hr

m_{hch} = straight line slope during semi-linear flow in a channel, Eq. 5.16, psi/(hr$^{1/2}$)

m_{HRF} = straight line slope during horizontal radial flow, Eq. 3.35, psi/cycle

M_i = mobility ratio in radial composite layer i, Eq. 4.78

m_{LF} = straight line slope during linear flow to a fracture, Eq. 1.25, psi/(hr$^{1/2}$)

m_q = $1/q$ straight line slope during radial flow, Eq. 9.8, 9.12, 1/(BOPD. cycle)

m_{RLF} = straight line slope during radial linear flow, Eq. 3.50, psi/cycle

m_{SPH} = straight line slope during spherical flow, Eq. 1.29, psi.hr$^{1/2}$

m_{VRF} = straight line slope during vertical radial flow, Eq. 3.31, psi/cycle

m_{WBS} = straight line slope during wellbore storage effect, Eq. 1.9, psi/hr, Eq. 7.16, psia^2/cp/hr, Eq. 7.17, psia^2/hr, Eq. 7.18, psi/hr

m_{wedge} = straight line slope during fraction of radial flow in a wedge, Eq. 5.12, psi/cycle

m^* = straight line slope during pseudo steady state, Eq. 1.35, psi/hr

$m(p)$ = pseudo-pressure (or real gas potential), Eq. 7.3, psia^2/cp, Eq. 8.10 and 8.14, psia/cp

n = number of fissure plane directions, Fig. 4.27, or laminar - turbulent coefficient in Eq. 7.28

p = pressure, psia

$\overline{p}$ = average pressure, Eq. 5.21, in layered system, Eq. 4.77, or average test pressure, section 7.2.2, psia
p_{BUD} = dimensionless build-up pressure, Eq. 2.12
p_D = dimensionless pressure, Eq. 2.3, 7.7, 7.8, 7.9
$\overline{p_D}$ = dimensionless average pressure, Fig. 5.25, Eq. 5.22 & 5.24
$p_{D,MBH}$ = M.B.H. dimensionless pressure, Eq. 5.22
p_{DR} = dimensionless pressure ratio, Eq. 9.1
p_f = pressure in the fissures, psia
p_i = initial pressure, psia
$\overline{p_i}$ = average initial pressure in layered system, Eq. 4.76, psia
PI = productivity index, Eq. 1.37, (STB/D)/psi
p_j = pressure at end of interval j, Table. 9.1, psia
$\overline{p_j}$ = average pressure during interval j, Table. 9.1, psia
p_m = pressure in the matrix, psia
PM = pressure match, Eq. 2.22, psi^{-1}
p_{MRD} = dimensionless multirate pressure, Eq. 2.17
$\overline{p}_{pp}$ = average reservoir pressure of Eq. 7.27, psia
p_{ref} = reference pressure for Eq. 8.16, psia
p_{sc} = standard pressure condition (14.7 psia)
p_w = bottom-hole pressure, psia
p_{wf} = flowing bottom-hole pressure, psia
p_{wfj} = flowing bottom-hole pressure at end of period j, section 7.3, psia
p_{ws} = shut-in bottom-hole pressure, psia
p_{wsj} = shut-in bottom-hole pressure at end of period j, Fig. 7.9, psia
$p_{wf,stab}$ = stabilized flowing bottom-hole pressure, section 7.3, psia
p_0 = reference pressure used for pseudo-pressure, Eq. 7.3, or pressure above a DST valve, Fig. 9.1, psia
p'_D = dimensionless pressure derivative
p^* = extrapolated Horner pressure, Fig. 2.8, psia
p_D^* = dimensionless extrapolated Horner pressure, Fig. 5.25 & 5.26
δp = vertical distance between two semi-log straight lines, Eq. 4.29, psi

Δp = pressure change, Eq. 1.1 & 1.2, psi
Δp_{BU} = build-up pressure change, Eq. 1.2, psi
Δp_{skin} = pressure change due to skin, Eq. 1.11, psi
$\Delta p'$ = pressure derivative, Eq. 2.23, psi
Δp^*_{int} = time zero intercept of the pseudo-steady state straight line m^*, Eq. 5.17 & 5.18, psi
$\Delta p_{1\,hr}$ = pressure change at 1 hour on the semi-log straight line, Fig. 1.9, psi
q = flow rate, STB/D or Mscf/D (= 10^3scft/D)
q_D = dimensionless flow rate, Eq. 9.7
q_{fD} = dimensionless flux per unit of fracture length, Eq. 3.16
q_g = sand face gas rate at standard conditions, Eq. 8.1, Mscf/D
q_{iD} = fractional rate of layer i, Eq. 4.70, 4.71, 4.72
q_j = rate during the period j in a multiple rate sequence, Fig. 2.9 or Table. 9.1, STB/D
q_{iD}^{LT} = late time stabilized fractional rate of layer i, Eq. 4.66
$q_{j,sc}$ = gas rate at standard conditions during period j, Eq. 7.25, Mscf/D
q_o = oil rate, Eq. 8.1, STB/D
q_{sc} = gas rate at standard conditions, Mscf/D
q_{sg} = surface gas rate at standard conditions, Eq. 8.1, Mscf/D
Q_t = cumulative volume produced, Eq. 9.6, STB
q_w = water rate, Eq. 8.1, STB/D
r = radial distance to the well, ft
R = distance of radial composite interface, Fig. 4.51, ft, producing gas - oil ratio, Eq. 8.11
r_D = dimensionless radius, Eq. 1.21
R_D = dimensionless distance of radial composite interface, Eq. 1.21
$r_{D,x,y}$ = dimensionless apparent distance of observation well, Eq. 6.6
r_e = reservoir radius in Eq. 7.30, ft
r_f = fracture radius, Eq. 3.52, ft
r_i = radius of investigation, Fig. 1.5, Eq. 1.23 & 1.24, ft
r_{iD} = dimensionless radius of investigation, Eq. 1.22, or of influence of the fissures, Eq. 6.12

R_{jD} = dimensionless interface distance in multiple radial composite, or in radial composite layer j

r_m = matrix blocks size, Fig. 4.27, Eq. 4.7, ft

r_S = radius of the skin zone, Fig. 1.8, or equivalent spherical radius, Eq. 1.29, ft

R_s = dissolved GOR, Eq. 8.1, scf/STB

R_{sw} = dissolved gas - water ratio, Eq. 8.3, scf/STB

r_w = wellbore radius, Fig. 1.5, ft

r_{we} = equivalent wellbore radius, Eq. 1.14 or 3.6, ft

S = skin coefficient, Eq. 1.11

$\overline{S}$ = average skin in layered system, Eq. 4.75

S_{ani} = anisotropy skin, Eq. 3.7

S_{ch} = geometrical skin during linear flow in a channel, Eq. 5.3

S_f = false skin calculated from total system parameters, Eq. 4.30 or skin factor across a finite conductivity fault, Eq. 5.29

S_g = gas saturation, Eq. 8.3, fraction

S_G = geometrical skin Table 3.7, for horizontal well, Eq.3.39 and Fig. 3.31, 3.32

S_i = skin of layer i, Fig. 4.41

S_{HKF} = geometrical skin of an infinite conductivity fracture, Eq. 3.12

S_L = pseudo-skin in layered system, Eq. 4.74

S_{LKF} = geometrical skin of a low conductivity fracture, Eq. 3.15

S_m = matrix skin, Eq. 4.36

S_o = oil saturation, Eq. 8.3, fraction

S_{or} = oil residual saturation, fraction

S_{PP} = geometrical skin of a well in partial penetration, Eq. 3.18

S_{RC} = radial composite skin, Eq. 4.91

S_T = total skin of a well in partial penetration, Eq. 3.17, or slanted Eq. 3.26

S_{TH} = total skin during horizontal radial flow, Eq. 3.38

S_{TV} = total skin during vertical radial flow, Eq. 3.32

S_{UFF} = geometrical skin of a uniform flux fracture, Eq. 3.11

S_w = water saturation, fraction, or skin factor in front of the perforated interval

S_{wi} = initial water saturation, fraction

S_z = partial penetration skin during linear flow to a horizontal well, Eq. 3.34

S_{zT} = partial penetration skin during horizontal radial flow to a horizontal well, Eq. 3.36

$S_{2\phi}$ = double porosity skin, Eq. 4.41

S_θ = geometrical skin of a slanted well, Eq. 3.23

S' = rate dependent skin coefficient, Eq. 7.22, 7.23, 7.24

S'_j = rate dependent skin coefficient during period j, Eq. 7.25

S'_w = apparent skin during vertical radial flow based on vertical well model, Eq. 3.41

t = time, hr

T = absolute temperature, °R

t_D = dimensionless time, Eq. 2.4, 7.10, 7.11

t_{DA} = dimensionless time based on drainage area, Eq. 5.15

t_{Df} = dimensionless time based on half fracture length, Eq. 3.8, or on fissure parameters, Eq. 4.11

t_{Df+m} = dimensionless time based on total system parameters, Eq. 4.12

t_{DL} = dimensionless time based on half well length, Eq. 3.29, or on channel width, Eq. 5.2

t_i = time at start of the period i in a multiple rate sequence, Eq. 2.17, hr

TM = time match, hr^{-1}

t_p = production time, Fig. 2.4, hr

t_{pD} = dimensionless production time

t_{pDA} = dimensionless production time based on drainage area

t_{ps} = pseudo-time, Eq. 7.4, hr

t_{pss} = time of start of pseudo-steady state regime, hr

T_{sc} = standard temperature condition, (520°R=60 °F)

Δt = elapsed time or build-up time, Fig. 2.4, hr

Δt_x = intersection time of two semi-log straight lines or midpoint of derivative transition, Eq. 1.33, hr

V = volume, bbl, or vapor mole fraction, Eq. 8.15

V_f = fissures volume per unit reservoir volume, Eq. 4.2, fraction

V_i = pore volume of closed layer i, Eq. 4.77, cu ft

V_m = matrix volume per unit reservoir volume, Eq. 4.2, fraction
V_u = wellbore volume per unit length, Eq. 1.8, bbl/ft
V_w = wellbore volume, Eq. 1.7, bbl
ΔV = change in volume, bbl
x = coordinate, Fig. 6.9, ft
x_e = half length of a rectangular reservoir, Fig. 3.4, ft
x_f = half fracture length, Fig. 1.11, ft
y = coordinate, Fig. 6.9, ft
y_e = half width of a rectangular reservoir, Fig. 3.4, ft
w_a = thickness of the altered permeability region near a finite conductivity fault, Eq. 5.29
w_f = fracture, or semi-permeable, or finite conductivity fault width, Fig. 1.14 or 5.32 or 5.37, ft
Z = real gas deviation factor, Eq. 7.1
$\overline{Z}$ = real gas deviation factor at average test pressure, section 7.2.2
Z_i = real gas deviation factor at initial pressure, Eq. 7.5 and section 7.2.2
z_w = distance from the center of the perforated interval, Fig. 3.15, or from a horizontal well, Fig. 3.24, to the lower reservoir boundary, ft
z_{wa} = apparent distance from a horizontal well to the lower reservoir boundary, Eq. 3.28, ft
$z_{w\text{-}obs}$ = distance from the center of the observation interval to the lower reservoir boundary, Fig. 9.10, ft
α = geometric coefficient in λ, Eq. 4.6, or transmissibility ratio of a semi-permeable fault, Eq. 5.27
β = parameter in transition curve for an observation well with double porosity transient interporosity flow, Fig. 6.20, Eq. 6.13 and 6.14
β' = transition curve for a well with wellbore storage and skin, double porosity transient interporosity flow, Fig. 4.16, Eq. 4.31
γ = Euler's constant (1.78 . . .)
Δ = difference
δ_j = storage of one group of matrix blocks per unit of matrix storage, Eq. 4.40, fraction
δ' = constant of in β', Eq. 4.32 & 4.33
θ = angle between a slanted well and the vertical, or between two intersecting faults, Fig. 5.14
θ_w = well location in a wedge, Fig. 5.14
θ_w' = transformed angle for a slanted well in anisotropic system, Eq. 3.24
κ = mobility ratio, Eq. 4.44
κ_j = mobility ratio of layer j, Eq. 4.67
λ = interporosity flow coefficient, Eq. 4.5, or layer, Eq. 4.46
λ_{eff} = effective interporosity flow coefficient, Eq. 4.38 & 4.39
λ_j = interlayer flow coefficient of layer j, Eq. 4.69, or in radial composite layered system, Eq. 4.46, or interporosity flow coefficient in fissured layer, or in fissured composite system, Eq. 4.5
λ_t = total mobility, Eq. 4.96
μ = viscosity, cp
μ_g = gas viscosity, Eq. 8.2, md
μ_i = viscosity in composite reservoir, Fig. 4.51, or at initial pressure, Eq. 7.5 and section 7.2.2, cp
μ_o = oil viscosity, Eq. 8.2, md
μ_w = water viscosity, Eq. 8.2, md
$\overline{\mu}$ = viscosity at average test pressure, section 7.2.2, cp
ρ = density, Eq. 1.8, lb/cu ft
$\rho_{o,g}$ = molar density of oil or gas, Eq. 8.14
ϕ = porosity, fraction
ϕ_f = fissures porosity, Eq. 4.2, fraction
ϕ_m = matrix porosity, Eq. 4.2, fraction
$\phi_{1,2,j}$ = porosity in layered, Fig. 4.41, or composite reservoir, Fig. 4.51
$\Psi(p)$ = $m(p)$
ω = storativity ratio in fissured, Eq. 4.4, or layered system, Eq. 4.45
ω_i = storativity ratio of layer i, Eq. 4.68, or in radial composite layered system, Eq. 4.45, or fissured layered or fissured radial composite system, Eq. 4.4

REFERENCES

Abbaszadeh, M. and Hegeman, P., 1988. Pressure Transient Behavior of Slanted Limited Entry Well in a Reservoir with a Gas Cap and/or Bottomwater Drive. *SPE paper* 19045, *unsolicited.*

Abbaszadeh, M. and Kamal, M.M., 1989. Pressure-Transient Testing of Water-Injection Wells. *SPE-RE, Feb.*, 115-124.

Abbaszadeh, M.D., Tomita, N. and Tanaka, T., 1993. Transient Testing of Limited-Entry Wells in Communicating Layered Reservoirs. *SPE paper* 25664, *Middle East Oil Technical Conference, Bahrain.*

Abbaszadeh, M.D. and Cinco-Ley, H., 1995. Pressure Transient Behavior in a Reservoir With a Finite-Conductivity Fault. *SPEFE, March,* 26-32.

Abdassah, D. and Ershaghi, I., 1986. Triple-Porosity Systems for Representing Naturally Fractured Reservoirs. *SPEFE, April,* 113-127.

Acosta, L.G. and Ambastha, A.K., 1994. Thermal Well Test Analysis Using an Analytical Muli-Regions Composite Reservoir Model. *S.P.E. paper* 28422, *69th Annual Fall Meeting, New Orleans, La.*

Agarwal, B., Chen, H-Y. and Raghavan, R., 1992. Buildup Behaviors in Commingled Reservoirs Systems With Unequal Initial Pressure Distributions: Interpretation. *SPE paper* 24680, *67th Annual Fall Meeting, Washington, DC.*

Agarwal, R.G., Al-Hussainy, R. and Ramey, H. J., Jr., 1970. An Investigation of Wellbore Storage and Skin Effect in Unsteady Liquid Flow. I: Analytical Treatment. *SPEJ, Sept,* 279.

Agarwal, R.G., Carter, R.D. and Pollock, C.B., 1979 a. Evaluation and Performance Prediction of Low-Permeability Gas Wells Stimulated by Massive Hydraulic Fracturing," *J. Pet. Tech., March,* 362-372 *and May,* 651-654.

Agarwal, R.G., 1979 b. Real Gas Pseudo-Time - A New Function for Pressure Build-up Analysis of MHF Gas Wells. *S.P.E. paper* 8279, *54th Annual Fall Meeting, Las Vegas, Nev.*

Agarwal, R. G. , 1980. A New Method to Account for Production Time Effects When Drawdown Type Curves Are Used to Analyze Buildup and Other Test Data. *SPE paper* 9289, *55th Annual Fall Meeting, Dallas, Tex.*

Aguilera, R. and Ng, M. C., 1991. Transient Pressure Analysis of Horizontal Wells in Anisotropic Naturally Fractured Reservoirs. *SPEFE, March,* 95-100.

Alagoa, A., Bourdet, D. and Ayoub, J.A., 1985. How to Simplify The Analysis of Fractured Well Tests. *World Oil, Oct. Gulf Publishing Co., Houston.*

Al-Ghamdi, A. and Ershaghi, I., 1994. Pressure Transient Analysis of Dually Fractured Reservoirs. SPE paper 26959, *3rd Latin American Conference, Buenos Aires, Argentina.*

Al-Hussainy, R., Ramey, H.J. Jr. and Crawford. P. B., 1966 a. The Flow of Real Gases Through Porous Media. *J. Pet. Tech., May,* 624-636; *Trans. AIME,* 237.

Al-Hussainy, R. and Ramey, H.J. Jr., 1966 b. Application of Real Gas Flow Theory to Well Testing and Deliverability Forecasting. *J. Pet. Tech., May,* 637-642; *Trans. AIME,* 237.

Al-Khalifah, A.A., Aziz, K. and Horne, R.N., 1987. A New Approach to Multiphase Well Test Analysis. *S.P.E. paper* 16473, *62nd Annual Fall Meeting, Dallas, Tex.*

Aly, A., Chen, H.Y. and Lee, W.J., 1994. A New Technique for Analysis of Wellbore Pressure From Multi-Layered Reservoirs With Unequal Initial Pressures To Determine Individual Layer Properties. *SPE paper* 29176, *Eastern Regional Conference, Charleston, WV.*

Ambastha, A.K., McLeroy, P.G. and Sageev, A., 1987. Effects of a Partially Communicating Fault in a Composite Reservoir on Transient Pressure Testing. *S.P.E. paper* 16764, *62nd Annual Fall Meeting, Dallas, Tex.*

Anbarci, K., Grader A.S. and Ertekin, T., 1989. Determination of Front Locations in a Multilayer Composite Reservoir. *SPE paper* 19799, *64th Annual Fall Meeting, San Antonio, Tex.*

Anisur Rahman, N.M. and Ambastha, A.K., 1997. A Generalized 3D Analytical Model for Transient Flow in Compartmentalized Reservoirs. *SPE paper* 38675, *70th Annual Fall Meeting, San Antonio, Tex.*

Ayoub, J.A., Bourdet, D.P. and Chauvel, Y.L., 1988. Impulse Testing. *SPEFE., Sept.,* 534-46; *Trans. AIME,* 285.

Barenblatt, G. E., Zheltov, I.P. and Kochina, I.N., 1960. Basic Concepts in the Theory of Homogeneous Liquids in Fissured Rocks. *J. Appl.. Math. Mech. (USSR)* 24 (5) 1286-1303.

Barua, J. and Horne, R.N., 1987. Computerized Analysis of Thermal Recovery Well Test Data. *SPEFE, Dec.*, 560-566.

Belani, A.K. and Yazdi, Y.J., 1988. Estimation of Matrix Block Size Distribution in Naturally Fractured Reservoirs. *SPE paper* 18171, *63rd Annual Fall Meeting, Houston, Tex.*

Bennet, C.O., Rosato, N.D., Reynolds, A.C. and Raghavan, R., 1983. Influence of Fracture Heterogeneity and Wing Length on the Response of Vertically Fractured Wells. *SPEJ, Apr.* 219-230.

Berumen-C., S. Samaniego, V. and Cinco-Ley, H., 1989. Transient Pressure Analysis and Performance of Gas Wells Producing Under Constant Pressure Conditions. *S.P.E. paper* 19098, *Gas Technology Symposium, Dallas, Tex.*

Bidaux, P., Whittle, T.M., Coveney, P.J. and Gringarten, A.C., 1992. Analysis of Pressure and Rate Transient Data From Wells in Multilayered Reservoirs: Theory and Application. *S.P.E. paper* 24679, *67th Annual Fall Meeting, Washington, DC.*

Bixel, H.C., Larkin, B.K. and Van Poollen, H.K., 1963. Effect of Linear Discontinuities on Pressure Build-Up and Drawdown Behavior. *J. Pet. Tech., Aug.,* 885-895; *Trans., AIME,* 228.

Blasingame, T.A., Johnston, J.L. and Lee, W.J., 1989. Type-Curves Analysis Using the Pressure Integral Method. *SPE paper* 18799, *California Regional Meeting, Bakersfield.*

Blasingame, T.A., Johnston, J.L. Rushing, J.A., Thrasher, T.S. Lee, W.J. and Raghavan, R., 1990. Pressure Integral Type-Curves Analysis-II: Applications and Field Cases. *SPE paper* 20535, *Annual Fall Meeting, New Orleans, La.*

Boe, A., Skjaeveland, S.M. and Whitson C.H., 1981. Two-Phase Pressure Test Analysis. *S.P.E. paper* 10224, *56th Annual Fall Meeting, San Antonio, Tex.*

Bourdarot, G., 1998. Well Testing: Interpretation Methods. *Editions Technip, Institut Français du Pétrole.*

Bourdet, D. and Gringarten, A. C., 1980. Determination of Fissure Volume and Block Size in Fractured Reservoirs by Type-Curve Analysis. *S.P.E. paper* 9293, *55th Annual Fall Meeting, Dallas, Tex.*

Bourdet, D. P., Whittle, T. M., Douglas, A. A. and Pirard, Y. M., 1983 a. A New Set of Type Curves Simplifies Well Test Analysis. *World Oil, May,* 95-106. *Gulf Publishing Co., Houston.*

Bourdet, D., Ayoub, J. A, Whittle, T. M., Pirard, Y. M. and Kniazeff V., 1983 b. Interpreting Well Test in Fractured Reservoirs. *World Oil, Oct.*, 77-87. *Gulf Publishing Co., Houston.*

Bourdet, D., Alagoa A., Ayoub J. A. and, Pirard, Y. M., 1984. New Type Curves Aid Analysis of Fissured Zone Well Tests. *World Oil, April,* 111-124. *Gulf Publishing Co., Houston.*

Bourdet, D., 1985. Pressure Behavior of Layered Reservoirs with Crossflow. *S.P.E. paper* 13628, *California Regional Meeting, Bakersfield, Ca.*

Bourdet, D. Ayoub, J. A. and Pirard, Y. M., 1989. Use of Pressure Derivative in Well-Test Interpretation. *SPEFE, June,* 293-302.

Bourgeois, M. and Horne, R.N., 1991. Well Test Recognition Using Laplace Space Type Curves. *S.P.E. paper* 22682.

Bourgeois, M. and Couillens, P., 1994. Use of Well Test Analytical Solutions for Production Prediction. *S.P.E. paper* 28899, *Europec, London, UK.*

Bourgeois, M.J., Boutaud de la Combe, J-L. and Deboaisne, R.M., 1996 a. Field Cases of Horizontal Well Pressure Transient Analysis in West Africa. *S.P.E. paper* 36526, *Annual Fall Meeting, Denvers, Co.*

Bourgeois, M.J., Daviau, F.H. and Boutaud de la Combe, J-L., 1996 b. Pressure Behavior in Finite Channel-Levee Complexes. *SPEFE, Sept.* 177-183.

Bourgeois, M.J. and Wilson, MR., 1996 c. Additional Use of Well Test Analytical Solutions for Production Prediction. *S.P.E. paper* 36820, *Europec, Milan, Italy.*

Boutaud de la Combe, J.-L., Deboaisne, R.M. and Thibeau, S., 1996. Heterogeneous Formation: Assessment of Vertical Permeability Through Pressure Transient Analysis - Field Example. *SPE paper* 36530, *Annual Fall Meeting, Denvers, Co.*

Bratvold, R.B. and Horne, R.N., 1990. Analysis of Pressure-Falloff Tests Following Cold-Water Injection. *SPEFE, Sept.,* 293-302.

Bremer, R.E., Winston, H. and Vela, S., 1983. Analytical Model for Vertical Interference Tests Across Low-Permeability Zones. *SPE paper* 11965, *58th Annual Fall Meeting, San Francisco, Ca.*

Brigham, W.E., 1970. Planning and Analysis of Pulse-Tests. *J. Pet. Tech., May,* 618-624; *Trans. AIME,* 249.

Brigham, W. E., 1990. Discussion of Productivity of a Horizontal Well. *SPERE, May,* 254-255.

Brons, F. and Marting, V. E., 1961 a. The Effect of Restricted Fluid Entry on Well Productivity. *J. Pet. Tech., Feb.,* 172-174; *Trans. AIME,* 222.

Brons, F. and Miller, W. C., 1961 b. A Simple Method for Correcting Spot Pressure Readings. *J. Pet. Tech., Aug.,* 803-805; *Trans. AIME,* 222.

Camacho-V., R.G. and Raghavan, R., 1994. Well-Test Analysis in Solution-Gas-Drive Systems for Fractured Reservoirs. *SPEFE, March,* 15-22.

Carter, R.D., 1966. Pressure Behavior of a Limited Circular Composite Reservoir. *SPEJ, Dec.,* 328-334; *Trans. AIME,* 237.

Carvalho, R. S. and Rosa, A.J., 1988. Transient Pressure Behavior of Horizontal Wells in Naturally Fractured Reservoir. *S.P.E paper* 18302, *63rd Annual Fall Meeting, Houston, Tex.*

Chen, C-C. and Raghavan, R., 1995. Modeling a Fractured Well in a Composite Reservoir. *SPEFE, Dec.,* 241-246.

Chen C-C., Chu W-C; and Saighi S., 1996. Pressure-Transient Testing of Gas Reservoirs With Edge-Waterdrive. *SPEFE, Dec.,* 251-256.

Chen, H-Y., Poston, S.W. and Raghavan, R., 1990. The Well Response in a Naturally Fractured Reservoir: Arbitrary Fracture Connectivity and Unsteady Fluid Transfer. *S.P.E. paper* 20566, *65th Annual Fall Meeting, New Orleans, La.*

Chen, H-Y., Raghavan, R. and Poston, S.W., 1993. Average Reservoir Pressure Estimation of a Layered Commingled Reservoir. *S.P.E. paper* 26460, *68th Annual Fall Meeting, Houston, Tex.*

Chu, L. and Grader, A.S., 1991. Transient Pressure Analysis of Three Wells in a Three-Composite Reservoir. *S.P.E. paper* 22716, *66th Annual Fall Meeting, Dallas, Tex.*

Chu, W.C. and Raghavan, R., 1981. The Effect of Noncommunicating Layers on Interference Test Data. *J. Pet. Tech., Feb.,* 370-382.

Chu, W-C. and Shank, G.D., 1993. A New Model for a Fractured Well in a Radial, Composite Reservoir. *SPEFE, Sept.,* 225-232.

Cinco-Ley, H., Miller, F.G. and Ramey, H.J., Jr., 1975. Unsteady-State Pressure Distribution Created by a Directionally Drilled Well. *J.Pet.Tech., Nov.,* 1392-1400; *Trans. AIME,* 259.

Cinco, L.H., Samaniego-V., F. and Dominguez, A.N., 1976. Unsteady-State Flow Behavior for a Well Near a Natural Fracture. *S.P.E. paper* 6019, *51st Annual Fall Meeting, New Orleans, La.*

Cinco-Ley, H., Samaniego-V., F. and Dominguez, N., 1978 a. Transient Pressure Behavior for a Well with a Finite Conductivity Vertical Fracture. *SPEJ, Aug.,* 253-264.

Cinco-Ley, H. and Samaniego-V., F., 1978 b. Transient Pressure Analysis for Fractured Wells. *S.P.E. paper* 7490, *53rd Annual Fall Meeting, Houston, Tex.*

Cinco-Ley, H. and Samaniego-V., F., 1981. Transient Pressure Analysis: Finite Conductivity Fracture Case Versus Damaged Fracture Case. *S.P.E. paper* 10179, *56th Annual Fall Meeting, San Antonio, Tex.*

Cinco, H., 1982. Evaluation of Hydraulic Fracturing By Transient Pressure Analysis Methods. *S.P.E. paper* 10043, *Bejing Meeting, China.*

Cinco-Ley, H., Samaniego-V., F. and Kuchuk, F., 1985. The Pressure Transient Behavior for Naturally Fractured Reservoirs With Multiple Block Size. *S.P.E. paper* 14168, *60th Annual Fall Meeting, Las Vegas, Nev.*

Cinco-Ley, H, Kuchuk, F., Ayoub, J., Samaniego-V., F. and Ayestaran, L., 1986. Analysis of Pressure Tests Through the Use of Instantaneous Source Response Concepts. *S.P.E. paper* 15476, *61st Annual Fall Meeting, New Orleans, La.*

Cinco-Ley, H. and Menh, H.-Z., 1988. Pressure Transient Analysis of Wells With Finite Conductivity Fractures in Double Porosity Reservoirs. *S.P.E. paper* 18172, *63rd Annual Fall Meeting, Houston, Tex.*

Clark, D. G. and Van Golf-Racht, T. D., 1985. Pressure Derivative Approach to Transient Test Analysis: A High-Permeability North Sea Reservoir Example. *J. Pet. Tech., Nov.,* 2023-2039.

Clark, K. K., 1968. Transient Pressure Testing of Fractured Water Injection Wells. *J. Pet. Tech., June,* 1639-643; *Trans. AIME,* 243.

Clonts, M. D. and Ramey, H. J. Jr., 1986. Pressure Transient Analysis for Wells with Horizontal Drainholes. *S.P.E. paper* 15116, *56th California Regional Meeting, Oakland, Ca.*

Daungkaew, S., Hollaender, F. and Gringarten, A. C., 2000. Frequently Asked Questions in Well Test Analysis. *S.P.E. paper* 63077, *Annual Fall Meeting, Dallas, Tex.*

Daviau, F., Mouronval, G., Bourdarot, G and Curutchet P., 1985. Pressure Analysis for Horizontal Wells. *S.P.E. paper* 14251, *60th Annual Fall Meeting, Las Vegas, Nev.*

Daviau, F., 1986. Interprétation des Essais de Puits – Les Méthodes Nouvelles. *Editions Technip, Institut Français du Pétrole.*

de Franca Correa A.C. and Ramey, H.J. Jr., 1987. A Method for Pressure Buildup Analysis of Drillstem Tests. *S.P.E. paper* 16808, *62nd Annual Fall Meeting, Dallas, Tex.*

Deruyck, B.G., Bourdet, D.P., DaPrat G. and Ramey, H.J. Jr., 1982. Interpretation of Interference Tests in Reservoirs with Double Porosity Behavior - Theory and Field Examples. *S.P.E. paper* 11025, *57th Annual Fall Meeting, New Orleans, La.*

de Swaan, O. A., 1976. Analytic Solutions for Determining Naturally Fractured Reservoir Properties by Well Testing. *SPEJ, June,* 117-122; *Trans. AIME,* 261.

Dietz D.N., 1965. Determination of Average Reservoir Pressure From Build-Up Surveys. *J. Pet. Tech., Aug.,* 955-959.

Dikken, B.J., 1990. Pressure Drop in Horizontal Wells and Its Effect on Production Performance. *J. Pet. Tech., Nov.* 1426-1433.

Ding, W. and Reynolds, A. C. Jr., 1994. Computation of the Pseudoskin Factor for a Restricted-Entry Well. *SPEFE, March,* 9-14.

Du, F-K. and Stewart, G., 1992. Transient Pressure Response of Horizontal Wells in Layered and Naturally Fractured Reservoirs with Dual-Porosity Behavior. *S.P.E. paper* 24862, *67th Annual Fall Meeting, Washington, DC.*

Duong, A.N., 1987. A New Set of Type Curves for Well Test Interpretation Using the Pressure Derivative Ratio. *S.P.E. paper* 16812, *62nd Annual Fall Meeting, Dallas, Tex.*

Earlougher, R.C. Jr., 1971. Estimating Drainage Shapes From Reservoir Limit Tests. *J. Pet. Tech., Oct.* 1266-1268; *Trans. AIME,* 251.

Earlougher, R. C., Jr. and Kersh, K. M., 1974. Analysis of Short-Time Transient Test Data by Type-Curve Matching. *J. Pet. Tech., Jul.,* 793-800; *Trans. AIME,* 257.

Earlougher, R. C., Jr., 1977. Advances in Well Test Analysis. *Monograph Series No 5, Society of Petroleum Engineers of AIME, Dallas.*

Ehlig-Economides, C.A. and Ramey, H.J. Jr., 1981 a. Transient Rate Decline Analysis for Wells Produced at Constant Pressure. *SPEJ, Feb.,* 98-104.

Ehlig-Economides, C.A. and Ramey, H.J. Jr., 1981 b. Pressure Buildup for Wells Produced at Constant Pressure. *SPEJ, Feb.,* 105-114.

Ehlig-Economides, C.A. and Ayoub, J.A., 1984. Vertical Interference Testing Across a Low-Permeability Zone. *S.P.E. paper* 13251, *59th Annual Technical Conference, Houston, Tex.*

Ehlig-Economides, C.A. and Joseph, J.A., 1985. A New Test for Determination of Individual Layer Properties in a Multilayered Reservoir. *S.P.E. paper* 14167, *60th Annual Fall Meeting, Las Vegas, Nev.*

El-Khatib, N.A.F., 1991. New Correlations for Time Lags and Pressure Response Amplitude in Pulse-test Analysis. *S.P.E. paper* 21404, *Middle East Oil Show, Bahrain.*

Feitosa, G.S., Chu, L., Thompson, L.G. and Reynolds, A.C., 1993. Determination of Reservoir Permeability Distributions From Pressure Buildup Data. *S.P.E. paper* 26457, *68th Annual Fall Meeting, Houston, Tex.*

Fetkovich, M.J., 1973. The Isochronal Testing of Oil Wells. *S.P.E. paper* 4529, *48th Annual Fall Meeting, Las Vegas, Nev.*

Fetkovich, M.J., 1980. Decline Curve Analysis Using Type Curves. *J. Pet. Tech., June,* 1065-1077.

Fleming, C.H., Perchke, D.R., Gilman, J.R. and Kazemi, H., 1994. Pressure Transient Testing of Horizontal Wells With an Overlying Gas Cap. *S.P.E. paper* 28391, *69th Annual Fall Meeting, New Orleans, La.*

Gao, C-T., 1984. Single-Phase Fluid Flow in a Stratified Porous Medium With Crossflow. *SPEJ, Feb.,* 97-106.

Gao, C., Jones, J.R., Raghavan, R. and Lee, W.J., 1994. Responses of Commingled Systems With Mixed Inner and Outer Boundary Conditions Using Derivatives. *SPEFE, Dec.,* 264-271.

Gomes, E. and Ambastha, A.K., 1993. An Analytical Pressure-Transient Model for Multilayered, Composite Reservoirs With Pseudosteady-State Formation Crossflow. *S.P.E. paper* 26049, *Western Regional Meeting, Anchorage, Alaska.*

Goode, P. A. and Thambynayagam, R. K. M., 1987. Pressure Drawdown and Buildup Analysis of Horizontal Wells in Anisotropic Media. *SPEFE, Dec.,* 683-697.

Goode, P.A. and Wilkinson, D.J., 1991 .Inflow Performance of Partially Open Horizontal Wells. *J. Pet. Tech., Aug.,* 983-987.

Gray, K.E., 1965. Approximating Well-to-Fault Distance From Pressure Build-up Tests. *J. Pet. Tech., July,* 761-767.

Gringarten, A.C. and Witherspoon, P.A., 1972. A Method of Analyzing Pump Test Data From Fractured Aquifers. *Proc., Symposium on Percolation Through Fissured Rocks, International Society for Rocks Mechanics, Stuttgart.*

Gringarten, A. C., Ramey, H. J., Jr. and Raghavan, R., 1974 a. Unsteady-State Pressure Distribution Created by a Well with a Single Infinite Conductivity Fracture. *SPEJ, Aug.,* 347-360.

Gringarten, A. C. and Ramey, H. J. Jr., 1974 b. Unsteady Pressure Distribution Created by a Well With a Single Horizontal Fracture, Partial Penetration, or Restricted Entry. *Soc.Pet.Eng. J., Aug.* 413-426.

Gringarten, A.C., Ramey, H.J. Jr. and Raghavan, R., 1975 a. Applied Pressure Analysis for Fractured Wells. *J. Pet. Tech., July,* 887-892.

Gringarten, A. C. and Ramey, H. J. Jr., 1975 b. An Approximate Infinite Conductivity Solution for a Partially Penetrating Line-Source Well. *Soc.Pet.Eng. J., Apr.* 347-360.

Gringarten, A. C., 1978. Reservoir Limit Testing for Fractured Wells. *S.P.E. paper* 7452, *53rd Annual Fall Meeting, Houston, Tex.*

Gringarten, A. C., Bourdet D. P., Landel, P. A. and Kniazeff, V. J., 1979. A Comparison between Different Skin and Wellbore Storage Type-Curves for Early-Time Transient Analysis. paper *S.P.E. paper* 8205, 54th *Annual Fall Meeting, Las Vegas, Nev.*

Gringarten, A. C., 1984. Interpretation of Tests in Fissured and Multilayered Reservoirs with Double-Porosity Behavior: Theory and Practice. *J. Pet. Tech., April,* 549-564.

Gringarten, A. C., Al-Lamki, S., Daungkaew, S., Mott, R. and Whittle, T.M., 2000. Well Test Analysis in Gas-Condensate Reservoirs. *S.P.E. paper* 62920, *Annual Fall Meeting, Dallas, Tex.*

Guppy, K.H., Cinco, H. and Ramey, H.J. Jr., 1981. Transient Flow Behavior of a Fractured Well Producing at Constant Pressure. *S.P.E. paper* 9963, *unsolicited.*

Hatzignatiou, D.G., Ogbe, D.O., Dehghani, K. and Economides, M.J., 1987. Interference Pressure Behavior in Multilayered Composite Reservoirs. *S.P.E. paper* 16766, *62nd Annual Fall Meeting, Dallas, Tex.*

Horner, D. R., 1951. Pressure Build-ups in Wells. *Proc., Third World Pet. Cong., E. J. Brill, Leiden* II, 503-521. *Also, Reprint Series, No.* 9 *Pressure Analysis Methods, Society of Petroleum Engineers of AIME, Dallas* (1967) 25-43.

Houpeurt A., 1959. On the Flow of Gas in Porous Medias. *Revue de l'Institut Français du Pétrole,* XIV (11), 1468-1684.

Houzé, O. and Viturat, D., 1985. *Flopetrol-Schlumberger Internal Report.*

Houzé, O.P., Horne, R. and Ramey, H. J. Jr., 1998. Infinite Conductivity Vertical Fracture in a Reservoir With Double Porosity Behavior. *SPEFE, Sept.,* 510-518.

Hurst, W., 1960. Interference Between Oil Fields. *Trans. AIME,* 219, 175-192.

Jacob, C.E. and Lohman, S.W., 1952. Nonsteady Flow to a Well of Constant Drawdown in an Extensive Aquifer. *Trans., AGU, Aug,* 559-569.

Jacquard, P., 1960. Etude Mathématique du Drainage d'un Reservoir Hétérogène. *Revue de l'Institut Français du Pétrole,* XV (10).

Jargon, J.R., 1976. Effect of Wellbore storage and Wellbore Damage at the Active Well on Interference Test Analysis. *J. Pet. Tech., Aug.,* 851-858.

Johnson, C.R., Greenkorn, R.A. and Woods, E.G., 1966. Pulse-Testing: A New Method for Describing Reservoir Flow Properties Between Wells. *J. Pet. Tech., Dec.,* 1599-1604; *Trans. AIME,* 237.

Jones, J.R. and Raghavan, R., 1988. Interpretation of Flowing Well Responses in Gas-Condensate Wells. *SPEFE, Sep.,* 578-594.

Jones, J.R., Vo, D.T. and Raghavan, R., 1989. Interpretation of Pressure Build-up Responses in Gas-Condensate Wells. *SPEFE, March,* 93-104.

Jones, P., 1956. Reservoir Limit Tests. *Oil and Gas J., June* 18, 54, *No* 59, 184.

Joseph, J., Bocock, A., Nai-Fu, F. and Gui, L.T., 1986. A Study of Pressure Transient Behavior in Bounded Two-Layered Reservoirs: Shengli Field, China. *S.P.E. paper* 15418, *61st Annual Fall Meeting, New Orleans, La.*

Kamal, M. and Brigham, W.E., 1975. Pulse-Testing Response for Unequal Pulse and Shut-In Periods. *SPEJ, Oct.,* 399-410; *Trans. AIME,* 259.

Kamal, M., 1983. Interference and Pulse Testing - A Review. *J. Pet. Tech., Dec.,* 2257-70.

Kamal, M.M., 1984. The Effects of Wellbore Storage and Skin on Vertical Permeability Testing. *S.P.E. paper* 13250 *59th Annual Technical Conference, Houston, Tex.*

Kamal, M.M., Braden, J.C. and Park, H., 1992. Use of Pressure Transient Testing To Identify Reservoir Damage Problems. *S.P.E. paper* 24666, *67th Annual Fall Meeting, Washington, DC.*

Kamal, M.M., Buhidma, I.M., Smith, S.A. and Jones, W.R., 1993. Pressure-Transient Analysis for a Well With Multiple Horizontal Sections. *S.P.E. paper* 26444, *68th Annual Fall Meeting, Houston, Tex.*

Katz, D.L., Cornell, D., Kobayashi, R., Poettmann, F.H., Vary, J.A., Elenbaas, J.R. and Weinaug, C.F., 1959. Handbook of Natural Gas Engineering. *McGraw-Hill Book Co.,Inc., New York.*

Kazemi, H., 1969 a. Pressure Transient Analysis of Naturally Fractured Reservoirs with Uniform Fracture Distribution. *SPEJ, Dec.,* 451-462; *Trans. AIME,* 246.

Kazemi, H., Seth, M.S. and Thomas, G.W., 1969 b. The Interpretation of Interference Tests in Naturally Fractured Reservoirs with Uniform Fracture Distribution. *SPEJ, Dec.,* 463-472.

Khachatoorian, R., Ershaghi, I. and Shikari, Y., 1995. Complexities in the Analysis of Pressure-Transient Response in Faulted Naturally Fractured Reservoirs. *SPEFE, Sept.,* 167-172.

Kikani, J. and Walkup, G.W., 1991. Analysis of Pressure-Transient Tests for Composite Naturally Fractured Reservoirs. *SPEFE, June,* 176-182.

Kniazeff, V.J. and Naville, S.A., 1965. A Contribution to the Problem of Two Phase Flow of Volatile Hydrocarbons. *SPEJ, March,* 37.

Kuchuk, F.J. and Kirwan, P.A., 1987. New Skin and Wellbore Storage Type Curves for Partially Penetrated Wells. *SPEFE, Dec.,* 546-554.

Kuchuk, F.J., Goode, P.A., Brice, B.W., Sherrard, D.W. and Thambynayagam, R.K.M., 1990. Pressure-Transient Analysis for Horizontal Wells. *J. Pet. Tech, Aug.,* 974-1031.

Kuchuk, F. J., Goode, P.A., Wilkinson, D.J. and Thambynayagam, R. K. M., 1991 a. Pressure-Transient Behavior of Horizontal Wells With and Without Gas Cap or Aquifer. *SPEFE, March,* 86-94.

Kuchuk, F.J., 1991 b. Pressure Behavior of Horizontal Wells in Multilayer Reservoirs with Crossflow. *S.P.E. paper* 22731, *66th Annual Fall Meeting, Dallas, Tex.*

Kuchuk, F.J., Habashy, T.M. and Torres-Verdín, C., 1993. Nonlinear Approximation for the Pressure Behavior of Heterogeneous Reservoirs. *S.P.E. paper* 26456, *68th Annual Fall Meeting, Houston, Tex.*

Kuchuk, F., 1995. Well Testing and Interpretation for Horizontal Wells. *J. Pet. Tech., Jan.,* 36-41.

Kuchuk, F.J. and Habashy, T., 1996. Pressure Behavior of Horizontal Wells in Multilayer Reservoirs With Crossflow. *SPEFE, March,* 55-64.

Kuchuk, F.J. and Habashy, T.M., 1997. Pressure Behavior of Laterally Composite Reservoir. *SPEFE, March,* 547-564.

Larsen, L., 1981. Wells Producing Commingled Zones with Unequal Initial Pressures and Reservoir Properties. *S.P.E. paper* 10325, *56th Annual Fall Meeting, San Antonio, Tex.*

Larsen, L., 1983. Limitations on the Use of Single- and Multiple-Rate Horner, Miller-Dyes-Hutchinson, and Matthews-Brons-Hazebroek Analysis. *S.P.E. paper* 12135, *58th Annual Fall Meeting, San Francisco, Ca.*

Larsen, L., and Hovdan, M., 1987. Analysis of Well Test Data from Linear Reservoirs by Conventional Methods", *S.P.E. paper* 16777, *62d Annual Fall Meeting, Dallas, Tex.*

Larsen, L., 1988. Similarities and Differences in Methods Currently Used to Analyze Pressure-Transient Data From Layered Reservoirs. *S.P.E. paper* 18122, *63rd Annual Fall Meeting, Houston, Tex.*

Larsen, L., 1989. Boundary Effects in Pressure-Transient Data From Layered Reservoirs. *S.P.E. paper* 19797, *64th Annual Fall Meeting, San Antonio, Tex.*

Larsen, L. and Hegre, T.M., 1991. Pressure-Transient Behavior of Horizontal Wells With Finite-Conductivity Vertical Fractures. *S.P.E. paper* 22076, *International Arctic Technology Conference, Anchorage, Alaska.*

Larsen, L., 1993. Pressure-Transient Behavior of Reservoirs Forming a Pattern of Linear Segments. *S.P.E. paper* 26459, *68th Annual Fall Meeting, Houston, Tex.*

Larsen, L. and Hegre, T.M., 1994 a. Pressure Transient Analysis of Multifractured Horizontal Wells. *S.P.E. paper* 28389, *69th Annual Fall Meeting, New Orleans, La.*

Larsen, L., 1994 b. Experiences With Combined Analyses of PLT and Pressure-Transient Data From Layered Reservoirs. *S.P.E. paper* 27973, *University of Tulsa Centennial Symposium, Tulsa, Ok.*

Larsen, L., 1996 a. Productivity Computations for Multilateral, Branched and Other Generalized and Extended Well Concepts. *S.P.E. paper* 36754, *Annual Fall Meeting, Denvers, Co.*

Larsen, L., 1996 b. Pressure-Transient Behavior of Wells Producing From Complex Patterns of Interconnected Linear Segments. *S.P.E. paper* 36549, *Annual Fall Meeting, Denvers, Co.*

Lee, J., 1982. Well Testing. *Textbook Series, Vol.* 1, *Society of Petroleum Engineers of AIME, Dallas.*

Lefkovits, H. C., Hazebroek, P., Allen, E. E. and Matthews, C.S., 1961. A Study of the Behavior of Bounded Reservoirs Composed of Stratified Layers. *SPEJ, March,* 43-58; *Trans. AIME,* 222.

Levitan, M.M. and Crawford, G.E., 1995. General Heterogeneous Radial and Linear Models for Well Test Analysis. *S.P.E. paper* 30554, *70th Annual Fall Meeting, Dallas, Tex.*

Liu, C-q. and Wang, X-D., 1993. Transient 2D Flow in Layered Reservoirs With Crossflow. *SPEFE, Dec.,* 287-291.

Liu, X., Chen, Z and Jiang, L., 1987. Exact Solution of Double-Porosity, Double-Permeability Systems Including Wellbore Storage and Skin Effect. *S.P.E. paper* 16849, *62nd Annual Fall Meeting, Dallas, Tex.*

Loucks, T.L. and Guerrero, E.T., 1961. Pressure Drop in a Composite Reservoir. *SPEJ, Sept.,* 170-176; *Trans. AIME,* 222.

Ma, Q. and Tiab, D., 1995. Interference Test Analysis in Naturally Fractured Reservoirs. *S.P.E. paper* 29514, *Production Operations Symposium, Oklahoma City, Ok.*

Malekzadeh, D. and Tiab, D., 1991. Interference Testing of Horizontal Wells. *S.P.E. paper* 22733, *66th Annual Fall Meeting, Dallas, Tex.*

Martin, J.C., 1959. Simplified Equations of Flow in Gas Drive Reservoirs and the Theoretical Foundation of Multiphase Pressure Buildup Analyses. *Trans. AIME,* 216, 309-311.

Mattar, L. and Brar, G.S., 1975. Theory and Practice of the Testing of Gas Wells. *Energy Resources Conservation Board, Calgary, Alta., Canada.*

Matthews, C.S., Brons, F. and Hazebroek, P., 1954. A Method for Determination of Average Pressure in a Bounded Reservoir. *Trans. AIME,* 201, 182-191. *Also Reprint Series N°9 - Pressure Analysis Methods, SPE,* 51-60.

Matthews, C. S. and Russell, D.G.,1967. Pressure Build-up and Flow Tests in Wells. *Monograph Series No 1, Society of Petroleum Engineers of AIME.*

Mavor, M. J. and Cinco, H., 1979. Transient Pressure Behavior of Naturally Fractured Reservoirs. *S.P.E. paper* 7977, *California Regional Meeting, Ventura, Ca.*

McKinley, R. M., 1971. Wellbore Transmissibility from Afterflow Dominated Pressure Build-up Data. *J. Pet. Tech., July,* 863.

Meehan, D. N., Horne, R.N. and Ramey, H.J. Jr., 1989. Interference Testing of Finite Conductivity Hydraulically Fractured Wells. *S.P.E. paper* 19784, *64th Annual Fall Meeting, San Antonio, Tex.*

Miller, C. C., Dyes, A. B., and Hutchinson, C. A., 1950. Estimation of Permeability and Reservoir Pressure from Bottom-Hole Pressure Build-up Characteristics. *Trans. AIME,* 189, 91-104.

Miller, F. G., 1962. Theory of Unsteady-State Influx of Water in Linear Reservoirs. *J. Inst. Pet., Vol.* 48, *No* 467.

Moench, A. F., 1984. Double-Porosity Models for a Fissured Groundwater Reservoir With Fracture Skin. *Water Resources Res., Vol.* 20, *No* 7, *July*, 831-846.

Moran, J. H. and Finklea, E. E., 1962. Theoretical Analysis of Pressure Phenomena Associated with the Wireline Formation Tester. *J. Pet. Tech., Aug.,* 899-908. *Trans. AIME,* 225.

Muskat, M., 1934. The Flow of Compressible Fluids Through Porous Media and some Problems in Heat Conduction. *Physics 5, March.*

Najurieta, H.L., 1980. A Theory for Pressure Transient Analysis in Naturally Fractured Reservoirs. *J. Pet. Tech., July,* 1241.

Nutakki, R. and Mattar, L., 1982. Pressure Transient Analysis of Wells in very Long Narrow Reservoirs. *S.P.E. paper* 11221, *57th Annual Fall Meeting, New Orleans, La.*

Odeh, A.S., 1965. Unsteady-State Behavior of Naturally Fractured Reservoirs. *SPEJ, March,* 60-64; *Trans. AIME,* 234.

Odeh, A.S., 1980. An Equation for Calculating Skin Factor Due to Restricted Entry. *J. Pet. Tech., June,* 964-65.

Ogbe, D.O. and Brigham, W.E., 1984 a. Pulse Testing with Wellbore Storage and Skin Effects. *S.P.E. paper* 12780, *California Regional Meeting, Long Beach, Ca.*

Ogbe, D.O. and Brigham, W.E., 1984 b. A Model for Interference Testing with Wellbore Storage and Skin Effects at Both Wells. *S.P.E. paper* 13253, *59th Annual Fall Meeting, Houston, Tex.*

Olarewaju, J.S. and Lee, W.J., 1989. A Comprehensive Application of a Composite Reservoir Model to Pressure-Transient Analysis. *SPERE, Aug.,* 325-231.

Oliver, D.S., 1990. The Averaging Process in Permeability Estimation From Well-Test Data. *SPEFE, Sept.,* 319-324.

Oliver, D.S., 1992. Estimation of Radial Permeability Distribution From Well-Test Data. *SPEFE, Dec.,* 290-296.

Onur, M. and Reynolds, A.C., 1988. A New Approach for Constructing Derivative Type Curves for Well Test Analysis. *SPEFE, March,* 197-206.

Onur, M. and Reynolds, A.C., 1989. Interference Testing of a Two-Layers Commingled Reservoir. *SPEFE., Dec.,* 595-603.

Ostrowski, L.P. and Kloska, M.B., 1989. Use of Pressure Derivative in Analysis of Slug Test or DST Flow Period Data. *S.P.E. paper* 18595, *Production Operations Symposium, Oklahoma City, Ok.*

Ozkan, E., Ohaeri, U. and Raghavan, R., 1987. Unsteady Flow to a Well Produced at a Constant Pressure in a Fractured Reservoir. *SPEFE., June,* 186-200.

Ozkan, E., Raghavan, R. and Joshi, S.D., 1989. Horizontal Well Pressure Analysis. *SPEFE, Dec.,* 567-575.

Ozkan, E., Sarica, C., Haciislamoglu, M. and Raghavan, R., 1995. Effect of Conductivity on Horizontal Well Pressure Behavior. *SPE Advanced Technology Series, Vol.* 3, *March,* 85-94.

Ozkan , E. and Raghavan, R., 1997. Estimation of Formation Damage in Horizontal Wells. *S.P.E. paper* 37511, *Production Operations Symposium, Oklahoma City, Ok.*

Papadopulos, I.S., 1965. Nonsteady Flow to a Well in an Infinite Anisotropic Aquifer. *Proc. Dubrovnik Symposium on Hydrology of Fractured Rocks.*

Papatzacos, P., 1987. Approximate Partial-Penetration Pseudoskin for Infinite-Conductivity Wells. *SPER.E., May,* 227-234.

Park, H. and Horne, R.N., 1989. Well Test Analysis of a Multilayered Reservoir With Crossflow. *S.P.E. paper* 19800, *64th Annual Fall Meeting, San Antonio, Tex.*

Peres, A.M.M., Onur, M. and Reynolds, A.C., 1993. A New General Pressure-Analysis Procedure for Slug Tests. *SPEFE., Dec.,* 292-98.

Perrine, R.L., 1956. Analysis of Pressure Build-up Curves. *Drill. and Prod. Prac., API,* 482-509.

Polubarinova-Kocina, P. Y., 1962. Theory of groundwater Movement. *Princeton Univ. Press., Princeton, N. J.*

Poon, D.C.C., 1984. Pressure Transient Analysis of a Composite Reservoir With Uniform Fracture Distribution. *S.P.E. paper* 13384 *available at SPE, Richardson, Tex.*

Prassad, R.K., 1975. Pressure Transient Analysis in the Presence of Two Intersecting Boundaries. *J. Pet. Tech., Jan.,* 89-96.

Prats, M. and Scott, J.B., 1975. Effect of Wellbore Storage on Pulse Test Pressure Response. *J. Pet. Tech., June,* 707-09.

Prats, M., 1986. Interpretation of Pulse Tests in Reservoirs With Crossflow Between Contiguous Layers. *SPEFE., Oct.,* 511-20.

Prijambodo, R., Raghavan, R. and Reynolds, A.C., 1985. Well Test Analysis for Wells Producing Layered Reservoirs With Crossflow. *SPEJ, June,* 380-396.

Pucknell, J.K. and Clifford, PJ., 1991. Calculation of Total Skin Factors. *S.P.E. paper* 23100 *Offshore Europe Conference, Aberdeen, UK.*

Raghavan, R., 1976. Well Test Analysis: Wells Producing by Solution Gas Drive Wells. *SPEJ, Aug.,* 196-208; *Trans. AIME,* 261.

Raghavan, R., 1989. Well Test Analysis for Multiphase Flow. *SPEFE, Dec.,* 585-594.

Raghavan, R.S., Chen, C.C. and Agarwal, B., 1997. An Analysis of Horizontal Wells Intercepted by Multiple Fractures. *SPEJ, Vol.* 2, Sept., 235-245.

Rahon, D., Edoa, P.F. and Masmoudi, M., 1997. Inversion of Geological Shapes in Reservoir Engineering Using Well-Test and History Matching of Production Data. *S.P.E. paper* 38656 *70th Annual Fall Meeting, San Antonio, Tex.*

Ramey, H. J., Jr., 1970. Short-Time Well Test Data Interpretation in The Presence of Skin Effect and Wellbore Storage. *J. Pet. Tech., Jan.,* 97.

Ramey, H.J. Jr., Agarwal, R.G. and Martin, I., 1975 a. Analysis of 'Slug Test' or DST Flow Period Data. *J. Cdn. Pet. Tech., July-Sept.,* 14, 37.

Ramey, H.J. Jr., 1975 b. Interference Analysis for Anisotropic Formations - A Case History. *J. Pet. Tech., Oct.,* 1290-98; *Trans. AIME,* 259.

Ramey, H. J., Jr., 1982. Pressure Transient Testing. *J. Pet. Tech., July,* 1407-1413.

Rawlins, E.L. and Schellardt, M.A., 1936. Back-Pressure Data on Natural-Gas Wells and Their Application to Production Practices. *Monograph 7, USBM.*

Rosa, A.J. and Carvalho, R.S., 1989. A Mathematical Model for Pressure Evaluation in an Infinite-Conductivity Horizontal Well. *SPEFE, Dec.* 559-566.

Russell, D.G. and Prats, M., 1962. The Practical Aspects of Interlayer Crossflow. *J. Pet. Tech., June,* 53-67; *Trans. AIME,* 225.

Russell, D. G. and Truitt, N. E., 1964. Transient Pressure Behavior in Vertically Fractured Reservoirs. *J. Pet. Tech., Oct.,* 1159-1170.

Salas, R.J., Clifford, P.J. and Jenkins, D.P., 1998. Multilateral Well Performance Prediction. *S.P.E. paper* 35711, *Western Regional Meeting, Anchorage, Alaska.*

Saldana-C, MA and Ramey, H.J. Jr., 1986. Slug test and Drillstem Test Flow Phenomena Including Wellbore Inertial and Frictional Effects. *S.P.E. paper* 15118, *56th California Regional Meeting, Oakland, Ca.*

Satman, A., 1980. An Analytical Study of Transient Flow in Systems With Radial Discontinuities. *S.P.E. paper* 9399, *55th Annual Fall Meeting, Dallas, Tex.*

Satman, A., 1985. An Analytical Study of Interference in Composite Reservoirs. *SPEJ, Apr.,* 281-290.

Satman, A., 1991. Pressure-Transient Analysis of a Composite Naturally Fractured Reservoir. *SPEFE, June,* 169-175.

Schildberg, Y., Bandiziol, D., Deboaisne, R., Laffont, F. and Vittori, J., 1997. Integration of Geostatistics and Well Test to Validate a Priori Geological Models for the Dynamic Simulation : Case Study", *S.P.E. paper* 38752, *Annual Fall Meeting, San Antonio, Tex.*

Shah, P.C. and Thambynayagam, R.K.M., 1992. Transient Pressure Response of a Well With Partial Completion in a Two-Layer Crossflowing Reservoir. *S.P.E. paper* 24681, *67th Annual Fall Meeting, Washington, DC.*

Slevinsky, B.A., 1997. Well Test Imaging - A New Method for Determination of Boundaries From Well Test Data. *AAPG/DOE 4th International Conference on Reservoir Characterization. Also, AAPG monograph 71, Advances in Reservoir Characterization.*

Soliman, M.Y., 1982. Analysis of Buildup Tests With Short Producing Time. *S.P.E. paper* 11083, *57th Annual Fall Meeting, New Orleans, La.*

Soliman, M.Y., Hunt, J.L. and El Rabas, A.M., 1990. Fracturing Aspects of Horizontal Wells. *J. Pet. Tech., Aug.* 966-973.

Stehfest, H, 1970. Algorithm 368, Numerical Inversion of Laplace Transforms. *Communication of the ACM , Jan.,* 47, *No* 1, 13.

Stewart, G. and Ascharsobbi, F., 1988. Well Test Interpretation for Naturally Fractured Reservoirs. *S.P.E. paper* 18173, *63rd Annual Fall Meeting, Houston, Tex.*

Stewart, G., 1997. Future Developments In Well Test Analysis: Introduction of Geology. *Hart's Petroleum Engineer International, Sept.,* 73-76.

Streltsova, T.D., 1979. Pressure Drawdown in a Well With Limited Flow Entry. *J.Pet.Tech., Nov.,* 1469-1476.

Streltsova-Adams, T.D., 1981. Pressure Transient Analysis For Afterflow-Dominated Wells Producing From a Reservoir With a Gas Cap. *J. Pet. Tech., April,* 743-754.

Streltsova, T.D., 1983. Well Pressure Behavior of a Naturally Fractured Reservoir. *SPEJ, Oct.,* 769.

Streltsova, T.D., 1984 a. Buildup Analysis for Interference Tests in Stratified Formations. *J. Pet. Tech., Feb.* 301-310.

Streltsova, T.D. and McKinley, R.M., 1984 b. Effect of Flow Time Duration on Build-Up Pattern for Reservoirs with Heterogeneous Properties. *SPEJ, June,* 294-306.

Suzuki, K. and Nanba, T., 1991. Horizontal Well Pressure Transient Behavior in Stratified Reservoirs. *S.P.E. paper* 22732, *66th Annual Fall Meeting, Dallas, Tex.*

Tariq, S. M. and Ramey, H. J., Jr., 1978. Drawdown Behavior of a Well with Storage and Skin Effect Communicating with Layers of Different Radii and Other Characteristics. *S.P.E. paper* 7453, *53rd Annual Fall Meeting, Houston, Tex.*

Tauzin, E. and Horne, R.N., 1994. Influence Functions for the Analysis of Well Test Data From Heterogeneous Permeability Distributions. *S.P.E. paper* 28433, *69th Annual Fall Meeting, New Orleans, La.*

Theis, C.V., 1935. The Relation Between the Lowering of the Piezometric Surface and the Rate and Duration of Discharge of a Well Using Ground-Water Storage. *Trans. AGU,* 519-524.

Thompson, L.G. and Reynolds, A.C., 1997. Well Testing for Radially Heterogeneous Reservoirs Under Single and Multiphase Flow Conditions. *SPEFE, March,* 57-64.

Tiab, D. and Kumar, A., 1980 a. Application of p_D' Function to Interference Analysis. *J. Pet. Tech., Aug.,* 1465-70.

Tiab, D. and Kumar, A., 1980 b. Detection and Location of Two Parallel Sealing Faults Around a Well. *J. Pet. Tech., Oct.,* 1701-1708.

Tiab, D. and Puthigai, S. K., 1988. Pressure-Derivative Type Curves for Vertically Fractured Wells. *SPEFE, March,* 156-158.

Uraiet, A.A. and Raghavan, R., 1980. Unsteady Flow to a Well Producing at a Constant Pressure. *J. Pet. Tech., Oct.* 1803-1812.

van Everdingen, A. F. and Hurst, W., 1949. The Application of the Laplace Transformation to Flow Problems in Reservoirs. *Trans. AIME,* 186, 305-324.

van Everdingen, A. F., 1953. The Skin Effect and its Influence on the Productive Capacity of a Well. *Trans. AIME,* 198, 171-176.

van Poollen, H. K., 1965. Drawdown Curves give Angle between Intersecting Faults. *The Oil and Gas J., Dec.,* 71-75.

Vela, S., 1977. Effect of Linear Boundary on Interference and Pulse Tests - The Elliptical Influence Area. *J. Pet. Tech., Aug.,* 947-50.

Vrbik, J., 1991. A Simple Approximation to the Pseudoskin Factor Resulting From Restricted Entry. *SPEFE, Dec.,* 444-446.

Waller H.N. Jr. and Krase, L.D., 1986. A North Sea Application of Slug Testing for Quick Reservoir Analysis. *S.P.E. paper* 15480, *61st Annual Fall Meeting, New Orleans, La.*

Warren , J. E. and Root, P. J., 1963. Behavior of Naturally Fractured Reservoirs. *SPEJ, Sept.,* 245-255; *Trans. AIME,* 228.

Wattenbarger, R.A. and Ramey, H.J. Jr., 1968. Gas Well Testing with Turbulence, Damage and Wellbore Storage. *J. Pet. Tech., Aug.,* 877-887.

Weller, W.T., 1966. Reservoir Performance During Two-Phase Flow. *J. Pet. Tech., Feb.,* 240-246; *Trans. AIME,* Vol. 240.

Wijesinghe, A.M. and Culham, W.E., 1984. Single-Well Pressure Testing Solutions for Naturally Fractured Reservoirs With Arbitrary Fracture Connectivity. *S.P.E. paper* 13055, *59th Annual Fall Meeting, Houston, Tex.*

Wong, D.W., Harrington, A.G. and Cinco-Ley, H., 1985. Application of the Pressure-Derivative Function in the Pressure-Transient Testing of Fractured Wells. *SPEFE, Oct.,* 470-480.

Wong, D.W., Mothersele, C.D., Harrington, A.G. and Cinco-Ley, H., 1986. Pressure Transient Analysis in Finite Linear Reservoirs Using Derivative and Conventional Techniques: Field Examples. *S.P.E. paper* 15421, *61st Annual Fall Meeting, New Orleans, La.*

Yaxley, L.M., 1985. The Effect of a Partially Communicating Fault on Transient Pressure Behavior. *S.P.E. paper* 14311, *60th Annual Fall Meeting, Las Vegas, Nv.*

Yeh, N-S. and Agarwal, R.G., 1989. Pressure Transient Analysis of Injection Wells in Reservoirs With Multiple Fluid Banks. *S.P.E. paper* 19775, *64th Annual Fall Meeting, San Antonio, Tex.*

Yildiz, T. and Ozkan, E., 1994. Transient Pressure Behavior of Selectively Completed Horizontal Wells. *S.P.E. paper* 28388, *69th Annual Fall Meeting, New Orleans, La.*

AUTHOR INDEX

T

U

V

W

Y

A

B

C

D

E

F

G

H

I

L

M

N

O

P

Q

R

S

T

U

V

W

Z